Sustainable Development Goals Series

World leaders adopted Sustainable Development Goals (SDGs) as part of the 2030 Agenda for Sustainable Development. Providing in-depth knowledge, this series fosters comprehensive research on these global targets to end poverty, fight inequality and injustice, and tackle climate change.

The sustainability of our planet is currently a major concern for the global community and has been a central theme for a number of major global initiatives in recent years. Perceiving a dire need for concrete benchmarks toward sustainable development, the United Nations and world leaders formulated the targets that make up the seventeen goals. The SDGs call for action by all countries to promote prosperity while protecting Earth and its life support systems. This series on the Sustainable Development Goals aims to provide a comprehensive platform for scientific, teaching and research communities working on various global issues in the field of geography, earth sciences, environmental science, social sciences, engineering, policy, planning, and human geosciences in order to contribute knowledge towards achieving the current 17 Sustainable Development Goals.

This Series is organized into eighteen subseries: one based around each of the seventeen Sustainable Development Goals, and an eighteenth subseries, "Connecting the Goals," which serves as a home for volumes addressing multiple goals or studying the SDGs as a whole. Each subseries is guided by an expert Subseries Advisor.

Contributions are welcome from scientists, policy makers and researchers working in fields related to any of the SDGs. If you are interested in contributing to the series, please contact the Publisher: Zachary Romano [Zachary.Romano@springer.com].

More information about this series at http://www.springer.com/series/15486

Joel C. Gill · Martin Smith
Editors

Geosciences and the Sustainable Development Goals

Editors
Joel C. Gill
Environmental Science Center
British Geological Survey
Keyworth, Nottinghamshire, UK

Geology for Global Development
Loughborough, UK

Martin Smith
The Lyell Centre
British Geological Survey (BGS Global Geoscience)
Edinburgh, UK

ISSN 2523-3084   ISSN 2523-3092   (electronic)
Sustainable Development Goals Series
ISBN 978-3-030-38814-0   ISBN 978-3-030-38815-7   (eBook)
https://doi.org/10.1007/978-3-030-38815-7

This Springer imprint is published by the registered company Springer Nature Switzerland AG
The registered company address is: Gewerbestrasse 11, 6330 Cham, Switzerland

For Chloe, Leo, Aaron, and Eilidh

'Future generations will judge us not by what we say, but what we do.'

—Ellen Johnson Sirleaf (President of Liberia, 2006 to 2018)

Preface

In writing this book, we hope to catalyse greater engagement of the geological science (or geoscience) community in implementing the Sustainable Development Goals (SDGs), as set out in a Resolution adopted by the United Nations General Assembly on 25 September 2015: *Transforming our world: the 2030 Agenda for Sustainable Development*.[1] We set out to constructively engage with this agenda, and to illustrate how geoscientists can facilitate the ambitions of the SDGs, monitor progress, and ensure the ongoing translation and integration of geoscience to support sustainable growth, well-being, and environmental protection in the decades following 2030. Our desire is that this book will enhance teaching on the societal relevance of geoscience. Sustainability concepts are notably lacking from the traditional education of many geoscientists, and in their research communities, limiting their ability to engage in the SDGs and other global development frameworks. Each chapter includes educational resources to help those with teaching responsibilities to support students to contextualise and apply the substance of this book.

While seeking to focus on the role of geoscientists in delivering the SDGs, we are acutely aware that complex, multifaceted development problems require interdisciplinary solutions, inclusive engagement, and participation by diverse groups from across different sectors and disciplines. Setting out how geoscientists can support these efforts requires an understanding of the political, economic, social, cultural, technological, and environmental contexts in which we seek to engage. Balancing the tension between delving into aspects of geology and the economic and social drivers underlying the SDGs has not been easy. We have not attempted to capture every aspect of social, economic, and environmental science relevant to addressing any given SDG in this one volume. We hope that our approach helps readers to understand how geoscience sits within the bigger picture of sustainable development, and that the suggested further reading in each chapter enables them to continue exploring relevant themes and build new partnerships. We also hope that this book enhances understanding outside the geoscience community of how geoscientists can support sustainable growth and decent jobs, resilient cities and infrastructure, access to basic services, food and water security, and effective environmental management.

---

[1]www.un.org/en/development/desa/population/migration/generalassembly/docs/globalcompact/A_RES_70_1_E.pdf.

Our philosophy in editing this volume has been that ensuring lasting and positive change not only depends on *what* we as geoscientists do, but also *how* we do this work and engage in sustainable development. For example, geoscientists' actions can advance the inclusion of vulnerable and marginalised groups, or could exacerbate existing inequalities; geoscientists can recognise and build on existing expertise when working internationally, or undermine local leadership and science institutions. This book is, therefore, about both science and the professional practice of science. We cover themes linked to ethics, equity, conduct, and partnerships, as well as water, minerals, engineering geology, and geological hazards. Where possible we have used examples and images from the Global South to illustrate the themes in this book, but we recognise that actions towards the SDGs require engagement from all countries and regions.

What This Book Includes

Following an introduction, this book explores each of the 17 SDGs in 17 corresponding chapters (i.e., **SDG 1** is explored in Chap. 1; **SDG 2** is explored in Chap. 2, etc.). We bring together learning, emerging themes, and recommendations in the conclusions (Chap. 18).

Through each of Chaps. 1–17, we refer to links with other chapters in order to demonstrate the SDG interlinkages and how progress in one goal can drive progress in another. We use the SDG number (e.g., **SDG 6**, **SDG 10**) rather than stating Chap. 6 or Chap. 10 to make things easier for the reader.

In the chapters relating to **SDGs 1–17**, we include a visual abstract that sums up the key content of the chapter and illustrates how geoscience can help deliver its ambitions. In addition to the main text, we also include (i) key learning concepts, a series of bullet points summarising the chapter, (ii) educational resources, to support contextualisation of the information in this book in the classroom (aimed at undergraduates/taught postgraduates), (iii) further reading, directing you to resources that complement the chapter theme, and (iv) a full reference list at the end of each chapter.

Forty-two authors have contributed to this book, collectively coming from every inhabited continent of the world. We started this project desiring that the final book would have a 'global voice'. While we recognise that we can always do more to improve representation, we are delighted to present a book with authors from diverse countries and sectors. We have diverse gender representation, and include early career scientists, experienced professionals, and voices from diverse sectors.

Introduction to Supporting Organisations

The **British Geological Survey** (BGS), part of UK Research and Innovation (UKRI) and a research centre under the Natural Environment Research Council (NERC), is the UK's principal supplier of objective, impartial, and up-to-date geological expertise and information for decision-making for governmental, commercial, and individual users. The BGS maintains and develops the nation's understanding of its geology to improve policymaking, enhance national wealth and reduce risk. It also collaborates with the national and international scientific community in carrying out research in strategic areas, including decarbonisation and resource management; environmental change, adaptation, and resilience; and multi-hazards and resilience. You can read more about the BGS at www.bgs.ac.uk.

Geology for Global Development (GfGD) is a registered charity, based in the UK, existing to champion the role of geology in sustainable development, mobilising and reshaping the geology community to help deliver the SDGs. GfGD organise conferences and training, support international projects working to achieve the SDGs, and advocate for the importance of Earth science at local, national, and international forums. GfGD is an affiliated organisation of the International Union of Geological Sciences and a contributing organisation to the UNESCO/IUGS International Geoscience Programme Project 685 (Geoscience for Sustainable Futures). You can read more about GfGD at www.gfgd.org.

Keyworth, UK   Joel C. Gill
Edinburgh, UK   Martin Smith

Acknowledgements

We are grateful to Springer Nature for the invitation to prepare this volume, and for their support throughout the drafting and publishing process. The resulting chapters, characterising the role of geoscience in **SDGs 1–17**, are a product of many hours of research and writing by a team of authors. We thank them all for contributing to this book, sharing their experiences and ideas, and responding rapidly and constructively to the requests of Editors. Past and ongoing projects and partnerships have shaped many of the reflections in this book, and we thank all those who have generously shared their time and helped to enrich what we present.

We have had great support from Henry Holbrook, Ian Longhurst, and Craig Woodward (all at or formerly at BGS) in the preparation of figures for this book. The visual abstracts at the start of each chapter—coordinated by Ian and Henry—provide an excellent way to explore how geoscience relates to each SDG. Support was also provided to the Editors by Bryony Chambers-Towers (BGS Intellectual Property Rights), and a review completed by John Rees (BGS Chief Scientist, Multi-Hazards and Resilience).

Photographs and graphics have kindly been provided by Sarah Boulton (University of Plymouth/Girls into Geoscience); Stafford McKnight (Federation University Australia); Solmaz Mohadjer (Parsquake); The Villuercas Ibores Jara UNESCO Global Geopark; the American Geosciences Institute (AGI); Chris Rochelle (BGS); the Mixteca Alta, Oaxaca UNESCO Global Geopark; the Qeshm Island UNESCO Global Geopark; the Observatory of Rural Change, OCARU, Ecuador; and Andrew Bloodworth (BGS). We also acknowledge our gratitude to the Our World in Data resource (https://ourworldindata.org/) for generating useful content and making this freely available to use. The analysis and images on this site have informed many of the chapters in this book.

We are grateful to all those who have provided information and ideas to enrich this book. Keely Mills (BGS) and Laura Hunt (University of Nottingham, BGS) provided information and images for a case study in SDG 15. Tom Bide and Teresa Brown (both of the BGS) helped inform the Hanoi Material Flow Analysis in SDG 12. Bob Macintosh and Brighid Ó Dochartaigh (both at BGS) kindly shared their experiences and insights to inform SDG 17. Laura Hunt (University of Nottingham/BGS) also skilfully assisted with some final editorial tasks.

The writing and editing of this book (and contributions by BGS staff members to Chaps. 1, 8–12, 14–17) were supported by the British Geological Survey NC-ODA grant NE/R000069/1: *Geoscience for Sustainable Futures*. All authors based at the British Geological Survey publish with the permission of the Executive Director, British Geological Survey (UKRI).

Preparing a book of this size and scope unsurprisingly requires work to spill into many evenings, weekends, and holidays. Our deep thanks go to Stephanie and Jan for their patience, understanding, and constant support as we completed this work.

Introduction: Geoscience for Sustainable Futures

Science and the 2030 Agenda for Sustainable Development

In September 2015, UN member states formally adopted the 2030 Agenda for Sustainable Development, also known as the Sustainable Development Goals (SDGs). This set of 17 goals (Fig. 1) and 169 targets aim to eradicate global poverty, end unsustainable consumption patterns, and facilitate sustained and inclusive economic growth, social development, and environmental protection by 2030 (United Nations, 2015). The SDGs are complemented by a suite of associated development strategies relating to disaster risk reduction (Sendai Framework for Disaster Risk Reduction), climate change (COP21 Paris Climate Change Agreement), and sustainable urban development (New Urban Agenda). Achieving the SDGs by 2030 will require a concerted and sustained effort from many communities and sectors across the globe.

The British Geological Survey supports the Sustainable Development Goals

Fig. 1 The 17 Sustainable Development Goals. Each goal has an associated set of targets, means of implementation, and indicators (United Nations, 2015)

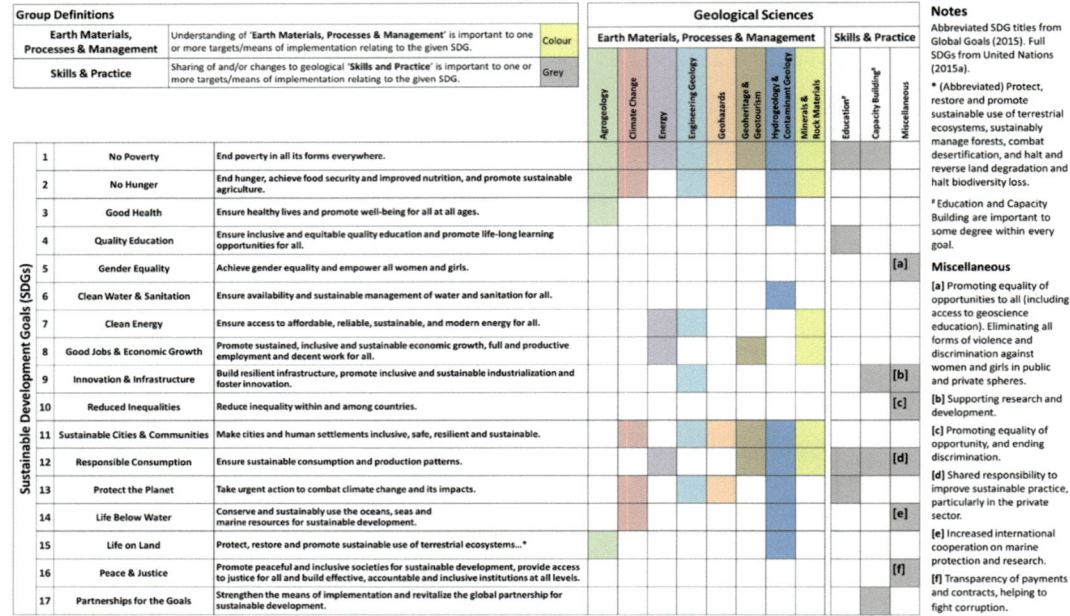

Fig. 2 Geology and the Sustainable Development Goals. From Gill (2017), used with permission

At the time of publication, the COVID-19 pandemic has had a devastating impact on families and communities around the world. This includes major loss of live, but also threats to livelihoods, education, and efforts to ensure gender equality in all contexts. Alongside conflict and other humanitarian disasters, COVID-19 is a serious threat to the development gains made in recent years and our ability to deliver the SDGs by 2030. This pandemic also highlights the need for delivery of the SDGs if we are to reduce the impact of future global health emergencies. Tackling poverty (**SDG 1**) and inequalities (**SDG 10**), improving health and wellbeing (**SDG 3**), increasing access to clean water (**SDG 6**), building safer communities (**SDG 11**), and protecting and restoring natural capital (**SDG 15**) all contribute to risk reduction and more resilient societies.

The SDGs are science intensive, emphasising the need for research, innovation, capacity building, and technology transfer. Meeting the SDG targets requires contributions by those scientists focused on understanding, monitoring, protecting, managing, and restoring the natural environment, including geoscientists. Geoscience is the study of the Earth's structure, processes and resources, and how life (including humans), interacts with Earth (American Geosciences Institute, 2019). Humans are extending their three-dimensional footprint on Earth (for example, through agriculture, infrastructure development, and urban expansion), inducing environmental change, and consuming greater volumes of natural resources.

In its broadest definition, demonstrated by the range of scientific divisions of organisation such as the European Geoscience Union, geoscience includes the study of the oceans, atmosphere, rivers and lakes, ice sheets and glaciers, soils, complex and dynamic surface, rocky interior, and metallic core

(American Geosciences Institute, 2019). Geological processes, including plate tectonics, basin development, and surface geomorphology, control the formation and distribution of resources, the generation of geological hazards and the flow of sediment across our landscapes through rivers and erosion, 'feeding' our oceans and supporting diverse ecosystems. Geoscience is, therefore, an essential part of the integrated research needed for development, and delivery of the SDGs as illustrated in Fig. 2 and expanded on through this book.

Geoscience engagement in the SDGs will be needed across academia, industry, government, and civil society, working in close partnership with other disciplines (e.g., engineering, ecology, social sciences, anthropology, psychology, health), and ensuring effective translation of knowledge into tools to inform policy and practice. A challenge for geoscientists is to demonstrate and communicate the relevance of our studies to policy and decision-makers now and into the future. For example, this includes ensuring the subsurface is considered in development discourses on urbanisation (see **SDG 11**), considering the availability of critical metal resources when developing energy, climate, and decarbonisation policies (see **SDGs 7, 12**, and **13**), and improving public health by understanding links to the natural environment (see **SDG 3**). Embedding *public relations* as a theme in geoscience education, has long been advocated for (Stow and Laming, 1991) to strengthen connections between geoscientists and policymakers, but it is still largely missing in the core training provided to geoscientists around the world.

Case Study 1: Sustainable Development in Eastern Africa

One region that exemplifies the challenges of and multiple impacts from climate change, human activities, and development faced by the Global South is eastern Africa. Kenya, Tanzania, and Ethiopia are all striving for economic stability and growth, vying to be the regional hub for business and research and development, each with an ambitious development strategy (i.e., Kenya Vision 2030[2], Tanzania Vision 2025[3], Ethiopia Growth and Transformation Plan[4]).

Development in eastern Africa is envisaged to occur along geographical corridors, where infrastructure is developed that facilitates the movement of goods between sites of production (e.g., a copper mine, a gas field), processing zones, and national and international economic hubs (Enns, 2018). In northern Kenya (Fig. 3), a combination of recent discoveries of hydrocarbons in buried rift structures (Tullow Oil, 2019) and the construction of new wind farms (Dahir, 2019), together with existing knowledge of major aquifers and geothermal power, is driving infrastructure development and

---

[2]http://vision2030.go.ke/.

[3]http://www.mof.go.tz/mofdocs/overarch/vision2025.htm.

[4]https://www.greengrowthknowledge.org/national-documents/ethiopia-growth-and-transformation-plan-ii-gtp-ii.

Fig. 3 The Gilgel Gibe III Dam on the Omo River in Ethiopia *Credit* Mimi Abebayehu (CC-BY-SA 4.0, https://creativecommons.org/licenses/by-sa/4.0/)

presents significant potential for economic growth. This development corridor will extend northwards into Ethiopia, and connect Uganda to the Indian Ocean with new roads, railways, and projected pipelines to carry oil and gas to Lamu, on the Kenyan coast.

Much of this region is a semi-desert environment, inhabited by nomadic pastoralists, with Lake Turkana to the north, the world's largest permanent desert lake. Lake Turkana provides a source of much needed protein, and increased income from tourism. It is a UNESCO World Heritage Site, and important anthropological and archaeological sites, with the discovery of Hominin fossils of some of the earliest human ancestors (e.g., Feibel et al., 1991; Wood and Leakey, 2011). Lake Turkana is a closed basin and its sole water supply comes from the Omo River in Ethiopia. Plans for hydropower schemes and increased use of water from the Omo River for irrigation will affect the long-term water supply to Lake Turkana. The Gilgel Gibe III Dam in Ethiopia (Fig. 3), for example, is predicted to have a significant impact upon the sustainability of the lake (Avery 2012; Ojwang et al., 2017) and increase trans-boundary tensions.

The collective and diverse impacts of corridor development will bring significant change to this region of Kenya. It can be regarded as a microcosm, one of many around the world, exemplifying the challenges of sustainable development. In seeking to implement the SDGs, it is fundamental to understand their impact on each other at the local level (i.e., the ways actions to support one goal could catalyse or hinder progress in another goal), and both planned and unintended consequences on people, wildlife, and the wider natural environment (Fig. 4).

Geoscience research into the evolution of the East African Rift, an active continental rift zone where tectonic plates are gradually diverging, can support a wide array of development ambitions. It can inform our understanding and the development of groundwater resources (**SDG 6**), with the cascading impacts of improved health through reductions in diarrhoeal diseases (**SDG 3**), improved agriculture through greater means of

Fig. 4 Braided river in the Suguta Valley, Northern Kenya Rift, draining into Lake Logipi. *Credit* Martin Trauth (distributed via imaggeo.egu.eu), CC-BY 3.0 (https://creativecommons.org/licenses/by/3.0/)

irrigation (**SDG 2**), and improved economic growth through reducing the time spent collecting water (**SDG 8**). The geology of the East African Rift also determines the availability of energy resource in the region, including both geothermal and hydrocarbon discoveries (**SDG 7**). The management of both water and energy resource, together with the metals and minerals required for construction, manufacturing, and infrastructure development requires careful planning to ensure responsible consumption and production (**SDG 12**), action on climate change (**SDG 13**), and strengthened diplomatic relations between neighbouring countries with trans-boundary resources (**SDG 16**). The hazards associated with the East African Rift include volcanic eruptions, earthquakes, and landslides on steep topographical features. Characterising this multi-hazard landscape, integrating seismology, volcanology, and engineering geology, can inform the actions required to reduce risk, helping to develop resilient infrastructure (**SDG 9**), sustainable communities (**SDG 11**), and reduce poverty (**SDG 1**). Geoscience communities of eastern Africa, spanning all countries and specialisms, should therefore be integrated into the groups and processes shaping development planning and implementation, but also equipped to contribute to supporting and facilitating sustainability in a full and effective way.

Resourcing Geoscientists to Support Sustainable Futures

This book is not the first publication to make claim that geoscientists should be a major partner in the endeavour to transition to a sustainable way of inhabiting Earth. Since the birth of geoscience as a scientific discipline, sustainable development has been part of its DNA, with James Hutton noting in the 1788 volume 'Theory of the Earth' that *'this globe of the earth is a habitable world, and on its fitness for this purpose, our sense of wisdom in its formation must depend'* (Stewart and Gill, 2017). Geoscientists possess skills and understanding that make us well-suited to support development initiatives, with geology being fundamentally important to improving lives and supporting sustainability (Stow and Laming, 1991; Cordani, 2000; Mora, 2013).

After UN member states agreed to the SDGs in 2015, Gill (2017) completed an initial mapping of their dependence on geoscience, Gill and Bullough (2017) provided a broader discussion of *how* geoscientists can engage in the SDGs and other global development frameworks, and Schrodt et al. (2019) have mapped the SDGs to eight *essential geodiversity variables*. The UN Development Programme, World Economic Forum, and Columbia Center on Sustainable Investment have set out the links between mining and the SDGs (Sonesson et al., 2016). IPIECA, the global oil and gas industry association for advancing environmental and social performance, the International Finance Corporation and UN Development Programme have done the same for the oil and gas industry (2017). The International Association of Hydrogeologists (IAH) have published a note showing how groundwater links to the SDGs (IAH, 2017).

These analyses all show significant linkages between the targets of the SDGs and geoscience. In this book, we have collated perspectives from the authors who live and have worked around the world to expand on these works and set out why achieving all of the SDGs requires the study and practice of geoscience, and what steps the geoscience sector can take to accelerate progress towards these goals. While structured around the 17 interdependent SDGs, we recognise that their ambitions not only require concerted action in the months and years to 2030, but an ongoing commitment to pursue knowledge and adhere to frameworks that enable humankind to live sustainably well beyond 2030. We, therefore, aim to

1. *Raise awareness among both geoscientists and the development community of the role of geoscience in realising sustainable development, framed in the context of the 17 SDG priorities.* We do this by describing direct contributions geoscientists can make to the SDGs (e.g., in **SDG 6** we describe how the characterisation of groundwater resources helps to ensure universal access to safe and reliable water supplies), and links between development challenges and the wider natural environment, which geoscience helps to characterise (e.g., in **SDG 10** we outline how environmental degradation can exacerbate inequalities).

2. *Explore how the geoscience community needs to reform to help deliver the SDGs.* We recognise that issues of quality education (**SDG 4**), gender equality (**SDG 5**), equitable access to knowledge (**SDG 10**), safe and secure work environments (**SDG 8**), and effective science partnerships (**SDG 17**) require individual disciplines and sectors to take responsibility, identify weaknesses, and put into place the measures required to deliver these aspects of sustainable development. While government policies (local, regional, or national) are necessary to drive these agendas forward, disciplines and sectors (through professional bodies, scientific unions, and individual organisational policies) also have an ability to influence and contribute to their delivery.

3. *Set out critical aspects of socio-economic context that help broaden geoscientists' understanding of development challenges, the actions needed to address these, and how geoscience sits in that bigger picture.* We do not set out every aspect of economics or social reform relevant to each SDG, but we do introduce concepts that help to contextualise the input of geoscientists. For example, **SDG 1** (end poverty) describes the causes and catalysts of poverty relating to conflict, governance, economics, history, and the environment. The latter is set out in much more detail (covering spatial poverty traps, natural resources, environmental change, pollution, and natural hazards), but we believe it aids the reader to see how these sit alongside other themes.

In helping to deliver on these three ambitions (awareness, reform, context), we hope to accelerate engagement of geoscientists in implementing Agenda 2030, and encourage the embedding of geoscientists into sustainable development initiatives. Throughout this book, we highlight three key themes (equity, knowledge exchange, and interdisciplinarity), which the Agenda 2030 and SDGs also emphasise.

Equity. Leaving no one behind is emphasised throughout the SDGs, acknowledging the importance of supporting the least developed and low-income countries, landlocked developing countries, and small island developing states. We have integrated perspectives from scientists in many of these settings into this book, and selected case studies that demonstrate challenges and opportunities associated with sustainable development. For example, **SDG 14** (life below water) has a focus on small-island developing states in the Pacific Ocean, **SDG 9** (industry, innovation and infrastructure) includes an example from Nepal, a landlocked developing country, and **SDG 17** (partnerships) includes a science-for-development programme in Afghanistan, one of the world's least developed countries. Equity is also needed *within* countries. There are individuals, groups, and communities that do not currently have equitable access to services, infrastructure, or resources. Across many chapters, we highlight initiatives that are widening access to geoscience. **SDG5** (gender equality) includes details of inspiring engagement and mentorship activities such as the African Association of Women Geoscientists and Girls into Geoscience (Fig. 5), **SDG 8** (decent work and economic growth) outlines how 'geoparks' are increasing public understanding of geoscience and creating livelihood opportunities for marginalised groups. **SDG 16** (peace, justice, and strong institutions)

Fig. 5 Girls into Geoscience Fieldtrip to Dartmoor, UK. © Sarah Boulton (University of Plymouth/Girls into Geoscience), used with permission

describes the role of scientific unions and professional societies in tackling harassment and discrimination.

Knowledge Exchange. The creation and exchange of knowledge, skills, and technologies can accelerate progress towards the SDGs. We have previously highlighted the emphasis on research, capacity building, and technology transfer within the 2030 Agenda. This book includes examples of knowledge exchange across countries. **SDG 4** (quality education) profiles projects to strengthen understanding of seismic hazards in Central Asia, and **SDG 9** (industry, innovation and infrastructure) describes how geoscientists in the United States collaborates with scientists around the world to improve understanding of and response to volcanic hazards. **SDG 17** (partnerships) includes examples of how geoscientists can engage in the UN Technology Facilitation Mechanism, with the specific objective of increasing access to and understanding of science, technology and innovation.

Interdisciplinary and Multisectoral Partnerships. While this book demonstrates why geoscience matters when addressing sustainable development challenges, it also recognises that we will increasingly be working in partnership with other disciplines and across sectors. Many geoscientists already work with engineers, ecologists, and chemists, but we will increasingly need to collaborate with economists, human geographers, anthropologists, psychologists, and public affairs professionals. These partnerships take time to develop, but are necessary to develop responses to the complex challenges that communities around the world are facing. We highlight in this book how networks and organisations fostering collaborations between geoscientists and other disciplines can help deliver improved health and well-being (**SDG 3**), restoration of biodiversity (**SDG 15**), and strengthened ocean management (**SDG 14**).

Stewart (2016) notes that *geologists possess a valuable synoptic and temporal conceptual framework for evaluating Earth's sustained viability for life.* This, together with thematic knowledge of Earth systems, natural resources, Earth hazards, and environmental management places geoscientists in a strong position to be key partners in sustainable development and champions of change. To leverage this opportunity, geoscientists should evaluate our contribution, our systems, and our role. As you read the following 17 chapters, one for each of the SDGs, we invite you to reflect on your own contribution to sustainable development, and how you can influence other geoscientists to fulfil our shared responsibility to support society in achieving a sustainable future.

Joel C. Gill
Martin Smith

References

American Geosciences Institute (2019) What is geoscience? Available at: www.americangeosciences.org/critical-issues/faq/what-is-geoscience. Accessed on 1 Oct 2019

Avery ST (2012) Lake Turkana and the Lower Omo: hydrological impacts of Gibe III and lower Omo irrigation development, vols. I and II. African Studies Centre/University of Oxford. http://www.africanstudies.ox.ac.uk/what-future-lake-turkana

Cordani UG (2000) The role of the earth sciences in a sustainable world: Episodes, 23 (3):155–162

Dahir AL (2019) Africa's largest wind power project is now open in Kenya. Available at: https://qz.com/africa/1671484/kenya-opens-africas-largest-wind-power-project-in-turkana/. Accessed on 30 Oct 2019

Enns C (2018) Mobilizing research on Africa's development corridors. Geoforum 88:105–108

Feibel CS, Harris JM, Brown FH (1991) Neogene paleoenvironments of the Turkana Basin, In: Harris JM, (eds), Koobi Fora research project, Volume 3. Stratigraphy, artiodactyls and paleoenvironments: Oxford, UK, Clarendon Press, p 321–370

Gill JC (2017) Geology and the sustainable development goals. Episodes 40(1):70–76

Gill JC, Bullough F (2017) Geoscience engagement in global development frameworks. Annals of geophysics, 60

IPIECA, IFC, UNDP (2017) Mapping the oil and gas industry to the Sustainable Development Goals: An Atlas. Available at: https://www.undp.org/content/undp/en/home/librarypage/poverty-reduction/mapping-the-oil-and-gas-industry-to-the-sdgs–an-atlas.html. Accessed on 1 Oct 2019

Mora G (2013) The need for geologists in sustainable development: GSA Today, 23 (12):36–37

Ojwang WO, Obiero KO, Donde OO, Gownaris N, Pikitch EK, Omondi R, … Avery ST (2017) Lake Turkana: World's Largest Permanent Desert Lake (Kenya). In: Finlayson C, Milton G, Prentice R, Davidson N (eds) The Wetland Book. Springer, Dordrecht

Schrodt F, Bailey JJ, Kissling WD, Rijsdijk KF, Seijmonsbergen AC, Van Ree D, Hjort J, Lawley RS, Williams CN, Anderson MG, Beier P, Van Beukering P, Boyd DS, Brilha J, Carcavilla L, Dahlin KM, Gill JC, Gordon JE, Gray M, Grundy M, Hunter ML, Lawler JJ, Mongeganuzas M, Royse KR, Stewart I, Record S, Turner W, Zarnetske PL, Field R (2019) To Advance Sustainable Stewardship, We Must Document Not Only Biodiversity But Geodiversity. Proceedings Of The National Academy Of Sciences Of The United States Of America, 116, 16155–16158

Sonesson C, Davidson G, Sachs L (2016) Mapping Mining to the Sustainable Development Goals: An Atlas. Available at: https://www.undp.org/content/dam/undp/library/Sustainable%20Development/Extractives/Mapping_Mining_SDGs_An_Atlas_Executive_Summary_FINAL.pdf. Accessed on 1 Oct 2019

Stewart I (2016) Sustainable geoscience. Nat Geosci 9(4):262

Stewart IS, Gill JC (2017) Social geology—integrating sustainability concepts into Earth sciences. Proceedings of the Geologists' Association, 128(2), 165–172

Tullow Oil (2019) Available at: https://www.tullowoil.com/operations/east-africa/kenya. Accessed on 30 Oct 2019

Wood B, Leakey M (2011) The Omo-Turkana Basin fossil hominins and their contribution to our understanding of human evolution in Africa. Evolutionary Anthropology: Issues, News, and Reviews 20(6):264–292

Contents

Editors and Contributors

About the Editors

Joel C. Gill Joel is International Development Geoscientist at the *BritishGeologicalSurvey*, and Founder/Executive Director of the not-for-profit organisation *GeologyforGlobalDevelopment*. Joel has a degree in Natural Sciences (Cambridge, UK), a Masters degree in Engineering Geology (Leeds, UK), and a Ph.D. focused on multi-hazards and disaster risk reduction (King's College London, UK). For the past decade, Joel has worked at the interface of Earth science and international development, and plays a leading role internationally in championing the role of geoscience in delivering the UN Sustainable Development Goals. He has coordinated research, conferences, and workshops on geoscience and sustainable development in the UK, India, Tanzania, Kenya, South Africa, Zambia, and Guatemala. Joel regularly engages in international forums for science and sustainable development, leading an international delegation of Earth scientists to the United Nations in 2019. Joel has prizes from the London School of Economics and Political Science for his teaching related to disaster risk reduction, and Associate Fellowship of the Royal Commonwealth Society for his international development engagement. Joel is a Fellow of the Geological Society of London, and was elected to Council in 2019 and to the position of Secretary (Foreign and External Affairs) in 2020.

Joel is Co-Editor of this book, Lead Author on Chaps. 1 (Zero Poverty), 9 (Infrastructure, Industry, and Innovation), and 16 (Peace, Justice, and Strong Institutions), and Contributing Author on Chaps. 8 (Decent Work and Economic Growth), 10 (Reduced Inequalities), 14 (Life Below Water), and 17 (Partnerships).

 **Martin Smith** Martin is a Science Director with the British Geological Survey and Principle Investigator for the BGS ODA Programme Geoscience for Sustainable Futures (2017–2021). He has a first degree in Geology (Aberdeen) and a Ph. D. on tectonics (Aberystwyth, UK). A survey geologist by training Martin has spent a career studying geology both in the UK and across Africa and India. As Chief Geologist for Scotland and then for the UK he has worked closely with government and industry on numerous applied projects including in the UK on national crises, major infrastructure problems, decarbonisation research and urban geology, and overseas for DFID-funded development projects in Kenya, Egypt, and Central Asia. Martin is a Chartered Geologist and Fellow of the Geological Society of London. He was awarded an MBE for services to geology in 2016.

Martin is Co-Editor of this book, Lead Author on Chap. 11 (Sustainable Cities), and Contributing Author on Chaps. 9 (Infrastructure, Industry, and Innovation), 15 (Life on Land), and 16 (Peace, Justice, and Strong Institutions).

Contributors

Amel Barich Geothermal Research Cluster (GEORG), Reykjavík, Iceland

Nic Bilham Geology for Global Development, Loughborough, UK; University of Exeter Business School, Penryn, Cornwall, UK; Camborne School of Mines, Penryn, Cornwall, UK

Stephanie Bricker British Geological Survey, Environmental Science Centre, Keyworth, Nottingham, UK

Sarah Caven Independent minerals and sustainability consultant, Vancouver, British Columbia, Canada

Lydia M. Chabala Department of Soil Science, School of Agricultural Sciences, University of Zambia, Lusaka, Zambia

Benson H. Chishala Department of Soil Science, School of Agricultural Sciences, University of Zambia, Lusaka, Zambia

Ranjan Kumar Dahal Central Department of Geology, Tribhuvan University, Kirtipur, Kathmandu, Nepal

Marleen de Ruiter Institute for Environmental Studies, Vrije Universiteit Amsterdam, Amsterdam, The Netherlands

Amy Donovan Department of Geography, University of Cambridge, Cambridge, UK

Kim Dowling School of Engineering, Information Technology and Physical Sciences, Federation University Australia, Victoria, Australia;
Department of Geology, University of Johannesburg, Johannesburg, South Africa

Richard Ellison British Geological Survey, Environmental Science Centre, Keyworth, Nottingham, UK

Ezzoura Errami Faculté Polydisciplinaire de Safi, Université Cadi Ayyad, Safi, Morocco;
African Association of Women in Geosciences, Abidjan, Côte d'Ivoire

Singarayer K. Florentine School of Science, Psychology and Sport, Federation University Australia, Victoria, Australia

Joel C. Gill British Geological Survey, Environmental Science Centre, Keyworth, Nottingham, UK;
Geology for Global Development, Loughborough, UK

David Gosselin Environmental Studies, University of Nebraska At Lincoln, Lincoln, NE, USA

Katrien An Heirman Deutsche Gesellschaft für Internationale Zusammenarbeit (GIZ) GmbH, Kigali, Rwanda

Julian Hunt Trinity College, Cambridge, UK;
University College London, London, UK

Ekbal Hussain British Geological Survey, Environmental Science Centre, Keyworth, UK

Hyeon-Ju Kim Seawater Energy Plant Research Center, Marine Renewable Energy Research Division, Korea Research Institute of Ships and Ocean Engineering, Daejeon, South Korea

Elias Kuntashula Department of Agricultural Economics and Extension, School of Agricultural Sciences, University of Zambia, Lusaka, Zambia

Alan MacDonald British Geological Survey, Lyell Centre, Edinburgh, UK

Joseph Mankelow British Geological Survey, Environmental Science Centre, Keyworth, Nottingham, UK

Rachael Martin School of Engineering, Information Technology and Physical Sciences, Federation University Australia, Victoria, Australia

Ellen Metzger Department of Geology, San José State University, San José, CA, USA

Rhoda Mofya-Mukuka Indaba Agricultural Policy Research Institute (IAPRI), Middleway, Kabulonga, Lusaka, Zambia

Melissa Moreano Universidad Andina Simón Bolívar, Toledo, Quito, Ecuador

T. F. Ng Department of Geology, University of Malaya, Kuala Lumpur, Malaysia

Martin Nyakinye Directorate of Geological Surveys, Ministry of Petroleum and Mining, Nairobi, Kenya

Gerel Ochir School of Geology and Mining, Mongolian University of Science and Technology, Ulaanbaatar, Mongolia

Samuel O. Ochola Department of Environmental Studies and Community Development, Kenyatta University, Nairobi, Kenya

Eric O. Odada African Collaborative Centre for Earth System Science (ACCESS), College of Biological and Physical Sciences, University of Nairobi, Chiromo Campus, Nairobi, Kenya

Cailin Huyck Orr Science Education Resource Center, Carleton College, Northfield, MN, USA

Dora C. Pearce School of Engineering, Information Technology and Physical Sciences, Federation University Australia, Victoria, Australia; Melbourne School of Population and Global Health, The University of Melbourne, Melbourne, Australia

Joy Jacqueline Pereira Southeast Asia Disaster Prevention Research Initiative (SEADPRI-UKM), Universiti Kebangsaan Malaysia, Bangi, Malaysia

Silvia Peppoloni Istituto Nazionale di Geofisica e Vulcanologia, Roma, Italy; International Association for Promoting Geoethics, Via di Vigna Murata, Roma, Italy

Evi Petavratzi British Geological Survey, Environmental Science Centre, Keyworth, Nottingham, UK

Michael G. Petterson School of Science, Auckland University of Technology, Auckland, New Zealand

Susanne Sargeant British Geological Survey, The Lyell Centre, Edinburgh, UK

Martin Smith British Geological Survey, The Lyell Centre, Edinburgh, Scotland, UK

Michael H. Stephenson British Geological Survey, Environmental Science Centre, Keyworth, Nottingham, UK

Kirsty Upton British Geological Survey, The Lyell Centre, Edinburgh, UK

Michael Watts British Geological Survey, Environmental Science Centre, Keyworth, Nottingham, UK

Acronyms

10YFP	10-Year Framework of Programmes on Sustainable Consumption and Production
AAAS	American Association for the Advancement of Science
AAWG	African Association of Women in Geosciences
AfDB	African Development Bank Group
AGI	American Geosciences Institute
AGN	African Geoparks Network
AGS	Afghanistan Geological Survey
AGU	American Geophysical Union
AIDS	Acquired Immune Deficiency Syndrome
ALFSIS	Africa Soil Information Service
AMCOW	African Ministers' Council on Water
ANCST	Asian Network on Climate Science and Technology
ANESI	African Network of Earth Science Institutions
ASGM	Artisanal and Small-scale Gold Mining
ASI	Aluminium Stewardship Initiative
ASM	Artisanal and Small-scale Mining
AWG	Association for Women Geoscientists
BGS	British Geological Survey
BIM	Building Information Models
BRI	Belt and Road Initiative (China)
CAES	Compressed Air Energy Storage
CBD	Convention on Biodiversity
CC/CARICOM	Caribbean Community
CCA	Climate Change Adaptation
CCOP	Co-ordinating Committee for Geoscience Programmes in East and Southeast Asia
CCS	Carbon Capture and Storage
CCW	Coupled Carbonate Weathering
CCZ	Clarion Clipperton Zone
CIDA	Canadian International Development Agency
COST	Co-operation in Science and Technology
CRC	Cobalt-Rich Crust
CRIRSCO	Committee for Mineral Reserves International Reporting Standards
CSP	Concentrated Solar Power

DBKL	City Hall of Kuala Lumpur
DDT	Dichlorodiphenyltrichloroethane
DFID	UK Department for International Development (now part of the UK Foreign, Commonwealth, and Development Office)
DMC	Domestic Material Consumption
DNA	Deoxyribonucleic Acid
DRR	Disaster Risk Reduction
DTS	Trans-Saharan fibre-optic backbone
EAGER	East African Geothermal Energy Facility
ECR	Early Career Researcher
EEZs	Exclusive Economic Zones
EFG	European Federation of Geologists
EGU	European Geosciences Union
EITI	Extractives Industry Transparency Initiative
ENSO	El Niño Southern Oscillation
ENVRIplus	Environmental Research Infrastructures Plus
EPA	Environmental Protection Agency
ESD	Education for Sustainable Development
ESIA	Environmental and Social Impacts Assessment
ESWN	Earth Science Women's Network
EU	European Union
FAO	Food and Agriculture Organisation of the United Nations
FIES	Food Insecurity Experience Scale
GAP	Global Action Programme
GCED	Global Citizenship Education
GCRF	Global Challenges Research Fund
GDP	Gross Domestic Product
GEM	Global Earthquake Model
GESAMP	Joint Group of Experts on the Scientific Aspects of Marine Environmental Protection
GEUS	Geological Survey of Denmark
GfGD	Geology for Global Development
GHG	Greenhouse Gas
GHI	Global Hunger Index
GLH/BSU	Ministry of Urban Development and Planning (Hamburg)
GLODAP	Global Ocean Data Analysis Project
GLOSOLAN	Global Soil Laboratory Network
GNH	Gross National Happiness
GNI	Gross National Income
GPS	Global Positioning System
GRACE	Gravity Recovery and Climate Experiment
GSA	Geological Society of America
GSI	Geological Survey of Ireland
GSL	Geological Society of London

GSN (TNO)	Geological Survey of the Netherlands
GSNI	Geological Survey of Northern Ireland
HDI	Human Development Index
HIV	Human Immunodeficiency Virus
HRWS	Human Right to Water and Sanitation
IAGD	International Association for Geoscience Diversity
IAPG	International Association for Promoting Geoethics
ICGLR	International Conference on the Great Lakes Region
ICMM	International Council on Mining and Metals
ICSU	International Council for Science
IEA	International Energy Agency
IFPRI	International Food and Policy Research Institute
IGC	International Geological Congress
IGCP	International Geoscience Programme
IGRAC	International Groundwater Resources Assessment Centre
IGRAC	International Groundwater Resources Assessment Centre
ILO	International Labour Organisation
INASP	International Network for the Availability of Scientific Publications
InSAR	Interferometric Synthetic-Aperture Radar
INSIVUMEH	National Institute for Seismology, Volcanology, Meteorology, and Hydrology (Guatemala)
InTeGrate	Interdisciplinary Teaching About Earth for a Sustainable Future
IODP	International Ocean Discovery Programme
IPCC	Intergovernmental Panel on Climate Change
IP-EEWS	International Platform on Earthquake Early Warning Systems
IRENA	International Renewable Energy Agency
IRMA	Initiative for Responsible Mining Assurance
IRP	International Resource Panel
ISA	International Seabed Authority
ITT	Interagency Task Team
IUCN	International Union for the Conservation of Nature
IUGS	International Union of Geological Sciences
IWM	International Women in Mining
IWRM	Integrated Water Resource Management
IWT	Indus Water Treaty
JMP	Joint Monitoring Programme
JORC	Australasian Code for Reporting of Exploration Results, Mineral Resources and Ore Reserves
KBA	Key Biodiversity Area
KRISO	Korea Research Institute of Ships and Ocean Engineering
L/RBOs	Lake/River Basin Organisations
LDCs	Least Developed Countries

LiDAR	Light Detection and Ranging
MAR	Managed Aquifer Recharge
MDG	Millennium Development Goal
MERS-CoV	Middle East respiratory syndrome coronavirus
MFA	Material Flow Analysis
MODIS	Moderate Resolution Imaging Spectroradiometer
MMA^{III}	Monomethylarsonous acid
M_w	Moment magnitude (*an earthquake magnitude scale*)
MPGG	Mount Paektu Geoscientific Group
MSP	Multi-Stakeholder Partnership
MUST	Mongolian University of Science and Technology
NCGE	National Committee on Gender Equality (Mongolia)
NGDC	National Geoscience Data Centre, British Geological Survey
NGO	Non-Governmental Organisation
NGSPBs	National Geological Societies and Professional Bodies
NGSS	Next-Generation Science Standards
NGU	Geological Survey of Norway
NSET	National Society for Earthquake Technology—Nepal
NUP	National Urban Policy
OCARU	Observatory of Rural Change (Ecuador)
ODA	Official Development Assistance
ODI	Overseas Development Institute
OECD	Organisation for Economic Co-operation and Development
OTEC	Ocean Thermal Energy Conversion (OTEC)
PERC	Pan-European Reserves and Resources Reporting Committee
PEST(LE)	Political, Economic, Social, Technological, Legal and Environmental
$PM_{10, 2.5\,etc.}$	Particulate Matter (with a diameter of <10, 2.5 μm etc.)
PPP	Purchasing Power Parity
PV	Photovoltaic
R&D	Research and Development
REE	Rare-Earth Element
RIGSS	Resilience in Groundwater Supply Systems
RINR	Regional Initiative against the Illegal Exploitation of Natural Resources
RMI	Responsible Mining Index
S4LIDE	Significance of Modern and Ancient Submarine Slope LandSLIDES
SADC	Southern African Development Community
SAGE	Scientific Advisory Group for Emergencies
SDG	Sustainable Development Goal
SEADPRI-UKM	Southeast Asia Disaster Prevention Research Initiative—Universiti Kebangsaan Malaysia

SEGH	Society for Environmental Geochemistry and Health
SESAME	Synchrotron-Light for Experimental Science and Applications in the Middle East
SIDS	Small Island Developing States
SPC	Pacific Community
SPI	Social Progress Index
STEM	Science, Technology, Engineering, and Maths
STI	Science, Technology, and Innovation
SuDS	Sustainable urban Drainages Systems
SWMS	Society for Women in Marine Sciences
TFM	Technology Facilitation Mechanism
TG-GGP	Task Group on Global Geoscience Professionalism
TRIPS	Trade-Related Aspects of Intellectual Property Rights
UN DESA	UN Department of Economic and Social Affairs
UN MGCY	United Nations Major Group for Children and Youth
UN	United Nations
UNDP	United Nations Development Programme
UNDRR	United Nations Office for Disaster Risk Reduction (formerly UNISDR)
UNEP	United Nations Environment Programme
UNESCO	United Nations Educational, Scientific and Cultural Organisation
UNFCCC	United Nations Framework Convention on Climate Change
UNICEF	United Nations Children's Fund (formerly United Nations International Children's Emergency Fund)
UNISDR	United Nations International Strategy for Disaster Reduction
UNWTO	United Nationals World Tourism Organisation
UPA	Agricultural Production Unit (Ecuador)
USGS	United States Geological Survey
VAC	Value Addition Centres (State department for Mining, Government of Kenya)
VDAP	Volcano Disaster Assistance Program (USGS)
WASH	Water, Sanitation and Hygiene
WEEE	Waste Electrical and Electronic Equipment
WEF	World Economic Forum
WGE SIC	Women in Geosciences and Engineering, Special Interest Community
WHO	World Health Organisation
WIM Mongolia	Women in Mining Mongolia
WiPS	Women in Polar Science
WMO	World Meteorological Organisation
WSUD	Water-Sensitive Urban Design
XRF	X-Ray fluorescence spectrometry

End Poverty in All Its Forms Everywhere

1

Joel C. Gill, Sarah Caven, and Ekbal Hussain

J. C. Gill (✉) · E. Hussain
British Geological Survey, Environmental Science
Centre, Nicker Hill, Keyworth NG12 5GG, UK
e-mail: joell@bgs.ac.uk; joel@gfgd.org

J. C. Gill
Geology for Global Development, Loughborough,
UK

S. Caven
Independent minerals and sustainability consultant,
Vancouver, British Columbia, Canada

Abstract

1 NO POVERTY

Poverty is the lack of resources needed to ensure dignity and survival

| Hinders people from reaching their full potential | Reduces life expectancy | Increases vulnerability to economic and environmental shocks | Threatens access to health care and education |

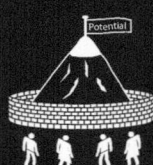

Causes and exacerbators of poverty are complex

| Lack of access to financial services and products | Geographical positions of nations (shaping climate, landscapes, hydrology, and exposure to hazards) | History and past policy decisions within or beyond the country | Differential access to and benefits from natural resources |

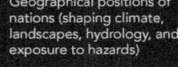

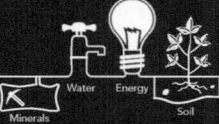

| Cultural norms, practices and beliefs | Weak governance and corruption | Armed conflict | Environmental degradation |

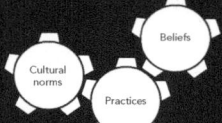

Geoscientists have an opportunity and responsibility to contribute to poverty alleviation

| Improve access to basic services, natural resources, and appropriate technologies | Effective and equitable disaster risk reduction | Ensuring the availability of geoscience for policy, and improving its uptake |

1.1   Introduction

As of 2015, an estimated 736 million people were living in extreme poverty, earning less than $1.90 a day, with almost half the world's population (3.4 billion people) living on less than $5.50 a day/$2000 a year (World Bank 2018). Poverty, however, is multidimensional and extends beyond simply having an adequate income. It encompasses the ability to meet human needs for food, water, sanitation, energy, education, sustainable livelihoods, and empowerment to engage in decision-making (Green 2008; Sachs 2015). Poverty is the 'lack of resources' needed to ensure dignity and survival, be they economic, social, political, or cultural resources.

Poverty hinders individuals and communities from reaching their full potential, reduces life expectancy, and increases vulnerability to epidemics, economic depression, environmental change, and natural hazards. Poverty undermines human rights, threatening the right to work, access to health care and education, freedom of thought and expression, and the right to maintain a cultural identity (Sané 2001; UNESCO 2017). While poverty can drive innovation, pushing people to think beyond the status quo to develop new routes to access financial services or energy resources, it is associated with deeply entrenched deprivation and hardship. Lifting people out of poverty transforms 'the lives and expectations of a nation's inhabitants' to ensure good health, physical safety, meaningful work, and connection to community (Green 2008). While this requires sufficient economic resources, it also needs social, political, natural, and cultural capital.

While extreme poverty is concentrated in sub-Saharan Africa and South Asia (Fig. 1.1), poverty affects every country. In some contexts, people lack sufficient resources to support a

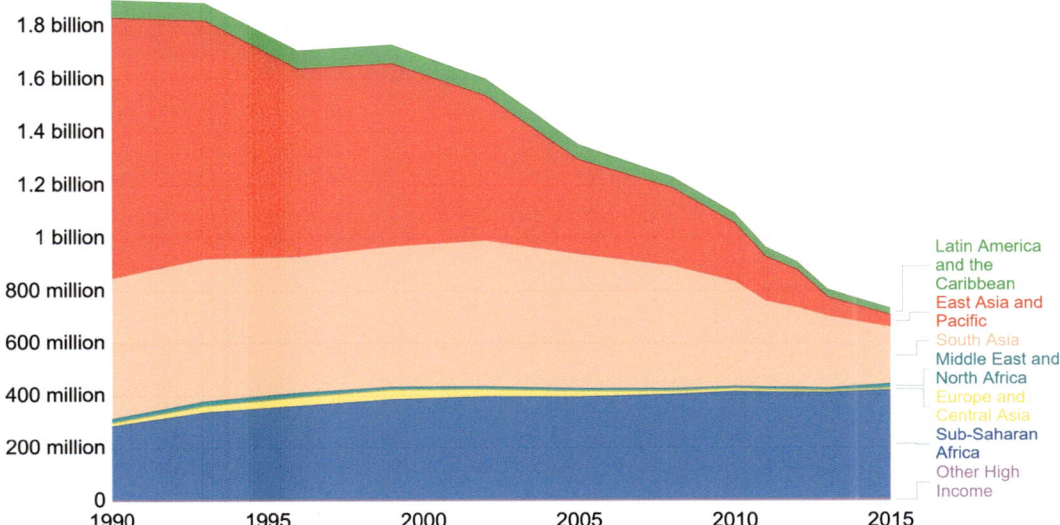

Total population living in extreme poverty, by world region

Numbers are in millions of people. Extreme poverty is defined as living with per capita household consumption below 1.90 international dollars per day (in 2011 PPP prices). International dollars are adjusted for inflation and for price differences across countries.

Our World in Data

Latin America and the Caribbean
East Asia and Pacific
South Asia
Middle East and North Africa
Europe and Central Asia
Sub-Saharan Africa
Other High Income

Source: PovcalNet (World Bank)   OurWorldInData.org/extreme-poverty/ • CC BY
Note: Consumption per capita is the preferred welfare indicator for the World Bank's analysis of global poverty. However, for about 25% of the countries, estimates correspond to income, rather than consumption.

Fig. 1.1   Total population living in extreme poverty, by world region. *Credit* Roser and Ortiz-Ospina (2017), using data from the World Bank (PovcalNet). Reproduced under a CC BY license (https://creativecommons.org/licenses/by/4.0/)

recognised standard of living (relative poverty). In other contexts, including many of the world's least developed countries, those living in poverty face a daily challenge to access the resources they need to meet immediate and basic needs (extreme poverty). The effects of such poverty are widespread, potentially degrading health, hindering education, perpetuating gender inequalities, and reducing political suffrage.

Poverty is not inevitable. The world has made good progress in tackling extreme poverty over the past 25 years (Fig. 1.1). Extreme poverty (those living on $\sim$ \$1.90 a day) fell by more than half between 1990 and 2015, going from 1.9 billion people to 736 million people living in extreme poverty (World Bank 2018). Progress has slowed however; the UN indicate that 6% of the global population will still live in extreme poverty by 2030 (United Nations 2019), and the Covid-19 pandemic will likely exacerbate poverty further. Progress has also been geographically uneven. Reductions in poverty since 1990 have been concentrated in East and South Asia (United Nations 2015). More than half of those

still living in extreme poverty live in sub-Saharan Africa, where estimates suggest the *total* number of people living in extreme poverty is increasing rather than decreasing (World Bank 2018). In sub-Saharan Africa, 84.5% of the population still live on less than \$5.50 a day or \$2000 a year (World Bank 2018). These communities can struggle to meet basic needs, and find themselves particularly vulnerable to falling back into extreme poverty. We must, therefore, take a balanced approach to how we define, measure and interpret the poverty narrative.

When considering actions to address global poverty, we, therefore, have reasons to be positive and reasons to be cautious. **Sustainable Development Goal (SDG) 1—Zero Poverty**—aims to build on this progress and ensure we leave nobody behind, *ending poverty in all its forms everywhere by 2030*. Specific targets (Table 1.1) include ambitions to eradicate extreme poverty, reduce relative poverty, ensure equal rights to basic services and natural resources, and reduce exposure and vulnerability to economic, social, and environmental shocks and disasters.

Table 1.1 SDG 1 targets and means of implementation

Target	Description of Target (1.1 to 1.5) or Means of Implementation (1.A to 1.B)
1.1	By 2030, eradicate extreme poverty for all people everywhere, currently measured as people living on less than \$1.25 a day
1.2	By 2030, reduce at least by half the proportion of men, women and children of all ages living in poverty in all its dimensions according to national definitions
1.3	Implement nationally appropriate social protection systems and measures for all, including floors, and by 2030 achieve substantial coverage of the poor and the vulnerable
1.4	By 2030, ensure that all men and women, in particular the poor and the vulnerable, have equal rights to economic resources, as well as access to basic services, ownership and control over land and other forms of property, inheritance, natural resources, appropriate new technology and financial services, including microfinance
1.5	By 2030, build the resilience of the poor and those in vulnerable situations and reduce their exposure and vulnerability to climate-related extreme events and other economic, social and environmental shocks and disasters
1.A	Ensure significant mobilization of resources from a variety of sources, including through enhanced development cooperation, in order to provide adequate and predictable means for developing countries, in particular least developed countries, to implement programmes and policies to end poverty in all its dimensions
1.B	Create sound policy frameworks at the national, regional and international levels, based on pro-poor and gender-sensitive development strategies, to support accelerated investment in poverty eradication actions

The causes of and solutions to poverty have environmental, social, and governance components. Geoscientists have a role to play in all of these, but particularly around the environmental component, the focus of this chapter. Poverty can be influenced by the physical geography or the underlying geology of a region. Environmental change, degradation, and shocks can threaten development gains and push communities below the poverty line. Poverty cycles can be broken through interventions that enable access and equal rights to clean water, natural resources, and appropriate technologies. Poverty can be a driver of environmental degradation, although it is widely accepted that individuals in wealthier countries typically have the highest ecological footprint. The poorest and other marginalised groups (e.g., indigenous communities) are often the strongest advocates of environmental protection and sustainable consumption, being the most vulnerable to environmental stresses (Broad 1994).

This chapter explores these themes and describes the role of geoscientists, outlining why the geoscience community matters when tackling poverty, and why tackling poverty matters to global security, development and environmental integrity. We characterise poverty and outline its effects (Sect. 1.2), proceeding to describe the progress made in tackling poverty (Sect. 1.3). We discuss the diverse causes of poverty and their relationship to geoscience (Sect. 1.4), and highlight ways that geoscience education, research, and innovation can reduce poverty and help society to end poverty (Sect. 1.5).

1.2 What Is Poverty, and What Are the Effects of Poverty?

1.2.1 Types of Poverty

Two primary definitions of poverty are reflected in the **SDG 1** targets:

- **Absolute** or **Extreme Poverty**: This looks at humanity as one unit and sets a standard (income level) below which humans are considered to be living in extreme poverty. The World Bank currently defines international absolute poverty as living on an income of less than $1.90 a day (although the SDG 1 target includes the measure of 'less than $1.25 a day'). Those living in absolute poverty typically struggle to access immediate needs such as food, safe drinking water, sanitation, health services, shelter, education, and information.
- **Relative Poverty**: This considers differing social contexts, defining a poverty level according to recognised national or regional standards. Relative poverty is, therefore, a measure of income inequality in a given region. In the United Kingdom, for example, relative poverty is currently defined as having an income below 60% of the median household income (UK DWP 2018).

Common to both absolute (extreme) and relative poverty is the idea that there is *a lack of sufficient resources required to meet everyday needs*, although resilience actually requires an ability to not only meet daily needs, but also meet future needs in the event of changing situations. While poverty is traditionally defined in terms of economic resources (e.g., income) or consumption, necessary resources go beyond money. We all need food, water, and shelter to survive. We greatly benefit from access to safe energy, health care, and education if we are to enjoy a reasonable quality of life, and build the skills required to generate an income. The desire to participate in decision-making is shared through diverse societies. Poverty is therefore lack of access to

- *Economic resources*, the financial capacity to purchase essential items.
- *Social resources*, access to basic needs such as food, sanitation, energy, shelter, health care, and education.
- *Natural resources*, access to a healthy natural environment, ecosystem services, and natural assets including geology, soil, water, and biodiversity.

- *Political resources*, a political voice or suffrage.
- *Cultural resources*, cultural opportunities, including access to information and communications.

There are interconnections and reinforcements between these resources. For example, a family lacking economic resources in a context where primary education is not free may be unable to send all of their children to school, and thus limit future opportunities for some of them. A lack of economic resources may also hinder people from exercising political suffrage, as they are unable to take time out of income-generating activities to travel to the nearest polling booth. Lack of economic resources may prevent a family from purchasing a radio and benefitting from weather information that improves their agricultural productivity.

Multidimensional poverty indices attempt to capture this holistic understanding of poverty and ensure actions do not focus solely on relieving economic deprivation (Alkire and Jaha 2018). For example, the 2018 Multidimensional Poverty Index (Table 1.2) incorporates health, education, and standard of living, with 10 indicators weighted to determine an overall poverty index (Alkire and Jahan 2018). The Multidimensional Poverty Index suggests that the percentage of the population living in multidimensional poverty in South Sudan is 91.9%, in Niger is 90.6%, and in Chad is 85.9% (UNDP 2018). These figures are significantly higher than the percentages given when solely focusing on economic deprivation, with the 2006–2016 average percentages of the population living in extreme poverty using the $1.90 definition being 42.7%, 44.5%, and 38.4%, respectively (UNDP 2018).

Table 1.2 2018 multidimensional poverty Index (dimensions, indicators, deprivation cutoffs and weights). Adapted from Alkire and Jahan (2018)

Poverty dimension	Indicator	Deprived if living in the household where…	Weight
Health	Nutrition	An adult under 70 years of age or a child is undernourished	1/6
	Child mortality	Any child has died in the family in the 5-year period preceding the survey	1/6
Education	Years of schooling	No household member aged 10 years or older has completed six years of schooling	1/6
	School attendance	Any school-aged child is not attending school up to the age at which he/she would complete class 8	1/6
Standard of Living	Cooking fuel	The household cooks with dung, wood, charcoal, or coal	1/18
	Sanitation	The household's sanitation facility is not improved (according to SDG guidelines) or it is improved but shared with other households	1/18
	Drinking water	The household does not have access to improved drinking water (according to SDG guidelines) or safe drinking water is at least a 30-minute walk from home, round trip	1/18
	Electricity	The household has no electricity	1/18
	Housing	Housing materials for at least one of roof, walls, and floor are inadequate: the floor is of natural materials and/or the roof and/or walls are of natural or rudimentary materials	1/18
	Assets	The household does not own more than one of these assets: radio, TV, telephone, computer, animal cart, bicycle, motorbike, or refrigerator, and does not own a car or truck	1/18

The Social Progress Index (SPI) is another complimentary metric to Gross Domestic Product (GDP), characerising how countries provide for the social and environmental needs of their citizens (Social Progress Imperative 2018). Some countries (e.g., Bhutan) are experimenting with various forms of Gross National Happiness (GNH) as a measure of national progress beyond simple GDP increases.

1.2.2 Contrasting Poverty Across Settings

The precise financial resources required to meet daily needs will differ from one context to another, as the amount of local currency needed to purchase goods or services may vary. If a household in one country has an income of $500/month but is required to spend an average of $400/month on rent, they will be more vulnerable than a household in another country with an income of $500/month and an average rent of $100/month. Differences in cost of living and inflation affect the purchasing power of a household's income.

Purchasing Power Parity (PPP) enables us to consider these differences and adjust income measures accordingly. We experience this when visiting other countries and find that commodity prices are different to prices at home, once adjusted for currency exchange rates. For example, Table 1.3 shows the nominal *Gross Domestic Product (GDP) Per Capita* of three countries (Tanzania, Guatemala, and the United Kingdom), and contrasts these with the *GDP Per Capita* (adjusted for Purchasing Power Parity). In Tanzania, for example, GDP per capita reflecting PPP is 3.2 times the nominal GDP. This reflects the reduced cost of living in Tanzania.

PPP is embedded into international definitions of extreme poverty. The World Bank's International Poverty Line ($1.90 a day) was determined by contrasting national poverty measures in some of the world's poorest countries, expressed in a common currency using PPP. This means that $1.90 would purchase the exact same basket of goods regardless of the country it is purchased in, which enables a global comparison of absolute poverty.

1.2.3 Poverty Cycles and Traps

The complex and multidimensional nature of poverty can result in poverty cycles or traps, where poverty transmits through multiple generations and continues until there is an adequate intervention. For example, consider the situation of a family who cannot afford to send their children to school, limiting future income-generating opportunities and resulting in future families who cannot afford to send children to school. Cultural factors may also mean that the outcomes of such poverty are discriminatory, with male children getting preferential access to education over female children. Another poverty cycle can exist when pregnant women lack access to good nutrition during foetal development. This can result in babies being born with impaired growth, susceptible to infant mortality, and likely to face challenges of sickness and poverty themselves, throughout their life (Green 2008).

Table 1.3 Contrast between GDP (nominal) per capita and GDP (PPP) per capita in select countries (Data from International Monetary Fund DataMapper 2018)

Country	GDP (Nominal) Per Capita (US$)	GDP (PPP) Per Capita (Int$[a])	Ratio
Tanzania	1160	3680	1 to 3.2
Guatemala	4700	8710	1 to 1.9
United Kingdom	42040	47040	1 to 1.1

[a]International Dollars (Int$) are a hypothetical currency unit that would buy a comparable amount of goods and services in the cited country that a U.S. dollar would buy in the United States at a given point in time

Certain marginalised groups may be more likely to be stuck in cycles of poverty than others may, showing the links between poverty and inequalities (discussed in **SDG 10**). The poverty rates for indigenous peoples in Latin America, for example, are estimated to be twice as high as for others (Calvo-González 2016). Some of this poverty gap can be accounted for by differences in educational levels, sizes of households, access to types of work, and the rural focus of indigenous communities (Calvo-González 2016). The full extent of the poverty gap, however, is likely to be explained by inequality and discrimination (World Bank 2015), with indigenous people earning less than non-indigenous people when they have the same level of education (Calvo-González 2016).

Box 1.1: Examples of Poverty Traps: Artisanal and Small-Scale Mining and Informal Settlements

Artisanal and Small-scale Mining (ASM) is often a more viable economic activity than, for example, subsistence agriculture. The sector is believed to employ over 40 million people globally, and over 150 million indirectly (IGF, 2017) making it a major rural livelihood next to agriculture. It has the potential for higher rewards, and thus artisanal and small-scale miners may determine that there is no economic incentive for them to relocate to agriculture or other types of work or use it as a seasonal substitute to agriculture to augment their incomes (Hilson 2016). Where unemployment is widespread, those working in the ASM sector may be driven to work there with few viable alternatives. This activity is driven by downstream demand and consumption of high-value commodities such as tantalum, gemstones (such as sapphires), and gold. However, artisanal miners often lack access to markets or mainstream finance.

As a result, they may become trapped in debt by taking finance from unscrupulous lenders or, in the case of artisanal gold mining, become indebted to a sole supplier of mercury who dictates their terms of access (Hilson and Pardie 2006) creating financial dependency. Capital on poor terms is compounded by below-market rates for their commodities, and therefore exacerbating rather than relieving poverty. This can hinder their ability to make a sustained, income from ASM or harness natural resources as a catalyst for economic diversification beyond mining dependency.

Informal urban settlements (or slums) can be a form of urban spatial poverty trap (Fig. 1.2), with a lack of natural capital (e.g., clean water) and political capital (e.g., land rights). Informality can result in poor investment and representation, exacerbating issues of poverty and degradation. Slums have defining spatial and social dimensions, with interactions between these, characterised by high population and housing densities, low standards of services and structures, and significant 'squalor' (Grant 2010). The lack of political capital, investment, and policy interest results in ongoing exclusion and the development of an urban spatial poverty trap. Housing density may result in inadequate distances between latrines and water resources (e.g., wells), resulting in contamination and disease (Kimani-Murage and Ngindu 2007). Increased exposure to water-borne diseases could limit income-generating activities and the economic opportunity to move secure environments. Informal settlements are often of substandard building quality, and it is these buildings that are most affected during natural hazards such as earthquakes (e.g., Ahmed 2014).

Once individuals are considered to be living above international poverty measures, sudden or slow-onset changes can occur that push them back into extreme poverty. Interventions to support those living *below* a poverty line are necessary, but it is important to also consider the vulnerability of those living just *above* the poverty line. People may be meeting their daily needs, but unable to set aside resources or access an adequate social security system. When changes in personal circumstances occur (e.g., illness or disability), or environmental shocks (e.-g., earthquakes) and environmental stresses (e.g., climate change), these can impede an individual or community's ability to meet their needs. Tackling poverty and ending cycles of poverty is, therefore, more than ensuring sufficiency for day-to-day needs. It also needs to ensure access to social security, insurances, and an ability to set aside resources to meet future demands.

1.3 Progress in Tackling Poverty

1.3.1 Poverty from 1820 to Today

While recognising that poverty is multi-dimensional, with both monetised and non-monetised approaches to improving well-being, changing income provides an approach to track progress in tackling poverty over time. Bourguignon and Morrisson (2002) estimate that in 1820, 94% of the global population lived in extreme poverty. In 1990, 1.96 billion people were living in extreme poverty, equivalent to 36% of the global population at the time (World Bank 2018). By 2015, the number of people living in extreme poverty had reduced to 736 million (World Bank 2018), or approximately 10% of the global population. Figure 1.3 illustrates this progress in addressing extreme poverty, showing a particularly rapid decrease in the

Fig. 1.2 Informal settlement in Jakarta Indonesia. *Credit* Jonathan McIntosh, reproduced under a CC BY 2.0 license (https://creativecommons.org/licenses/by/2.0/)

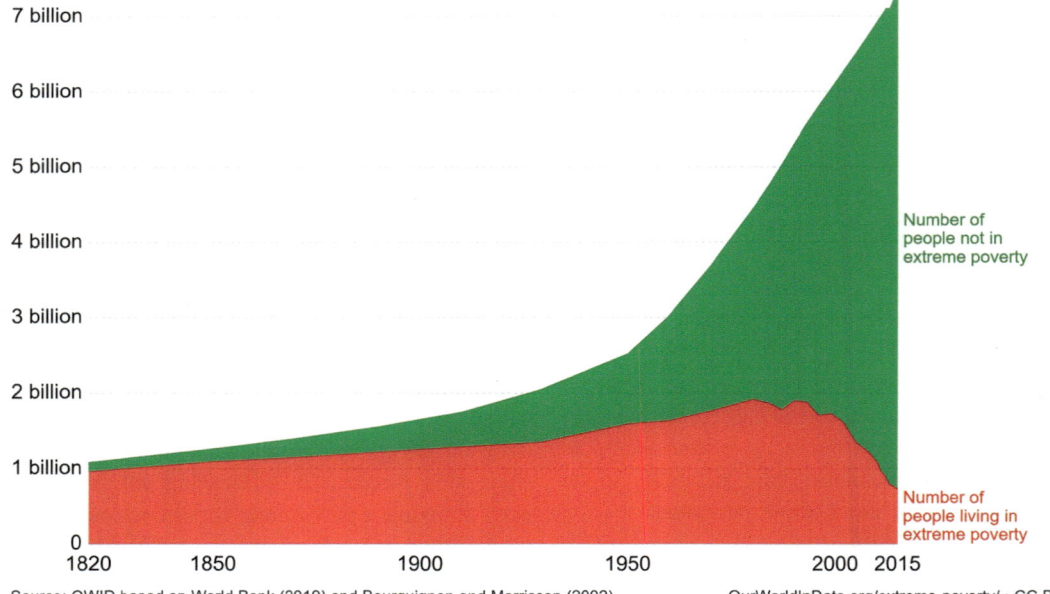

World population living in extreme poverty, 1820-2015
Extreme poverty is defined as living on less than 1.90 international-$ per day. International-$ are adjusted for price differences between countries and for price changes over time (inflation).

Source: OWID based on World Bank (2019) and Bourguignon and Morrisson (2002)   OurWorldInData.org/extreme-poverty/ • CC BY

Fig. 1.3 World population living in extreme poverty, 1820–2015. The line shows the percentage of people in extreme poverty (an income of less than $1.90 per day) while the red portion shows the absolute numbers. *Credit* Roser and Ortiz-Ospina (2017). Reproduced under a CC BY license (https://creativecommons.org/licenses/by/4.0/)

Table 1.4 Poverty rate by region at $5.50 per day. Adapted from World Bank (2018)

Region	1990 (%)	2015 (%)	Percentage Change (1990–2015)
East Asia and Pacific	95.2	34.9	−60.3
Europe and Central Asia	25.3	14.0	−11.3
Latin America and the Caribbean	48.6	26.4	−22.2
Middle East and North Africa	58.8	42.5	−16.3
South Asia	95.3	81.4	−14.0
Sub-Saharan Africa	88.5	84.5	−4.1
World	67.0	46.0	−21.0

past 50 years. More recently this progress is decelerating (United Nations 2019), with estimates that by 2030 6% of the global population will remain in extreme poverty. This does not take into account the effects of the Covid-19 pandemic. Aiming to achieve SDG 1 is therefore ambitious and requires the right support, including a sustained international commitment to eradicating poverty, policies that support

inclusive economic growth, and a focus on rural regions (Chandy and Penciakova 2013).

While the progress in tackling extreme poverty as measured by $1.90 income a day is encouraging, the rate of improvement as measured by $5.50 a day is more subdued (Table 1.4). Between 1990 and 2015, there was a 21% reduction in the number of people living on $5.50 a day, compared to the 26% reduction seen

Share of the population living in extreme poverty, 2017

Extreme poverty is defined as living with per capita household consumption below 1.90 international dollars per day
(in 2011 PPP prices). International dollars are adjusted for inflation and for price differences across countries.

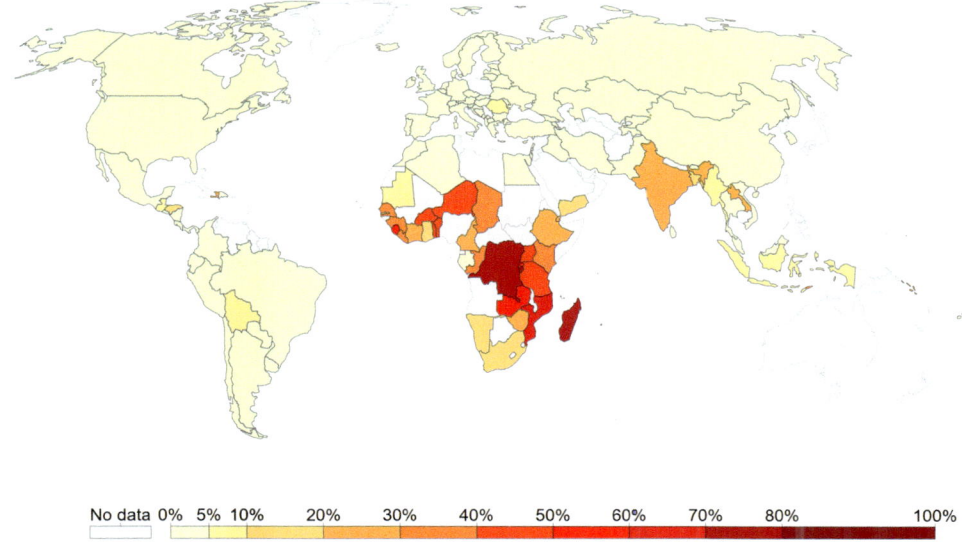

| No data | 0% | 5% | 10% | 20% | 30% | 40% | 50% | 60% | 70% | 80% | 100% |

Source: World Bank   OurWorldInData.org/extreme-poverty/ • CC BY

Fig. 1.4 Share of population living in extreme poverty , 2014 (in 2011 PPP, Int$). *Credit* Roser and Ortiz-Ospina
(2017), using data from the World Bank. Reproduced under a CC BY license (https://creativecommons.org/licenses/by/
4.0/)

for $1.90 a day. This progress has been spatially heterogeneous with dramatic improvements in East Asia and Pacific but relatively little improvement in sub-Saharan Africa (a 4% drop between 1990 and 2015).

1.3.2 Geographic Distribution of Extreme Poverty

The rate of poverty reduction is extremely variable between countries. Today most wealthy nations in Europe, North America, Australia, and Japan have no people living below the World Bank's extreme poverty line. Most of the countries with the highest population share living in extreme poverty are in sub-Saharan Africa (Fig. 1.4). The Democratic Republic of Congo and Madagascar, for example, have more than 70% of their populations living below the

poverty line. When it comes to the absolute number of people living in extreme poverty, India hosts more extremely poor people than any other country, about 29% (210 million people) of the world's total (PovcalNet, 2016). There is also considerable variation in poverty *within* any individual country. Extreme poverty may be concentrated in spatial pockets relating to soil conditions, conflict, or accessibility (see Sect. 1.4.2). While the relationship between urbanisation and poverty reduction is much debated (e.g., Cali and Menon 2013; Imai et al. 2017), it is clear that in most cases the majority of extremely poor people live in rural agricultural environments and most of a country's wealthier individuals live in cities. As cities continue to grow, ensuring low poverty rates in urban environments is dependent on addressing inequalities (see **SDG 10**), and ensuring sustainable and resilient urban environments (see **SDG 11**).

1.4 What Causes Poverty and How Does This Relate to Geoscience?

1.4.1 The Causes and Catalysts of Poverty Are Diverse

In any given location and context, a different combination of factors, with both natural and anthropogenic origins, may contribute to the generation or propagation of poverty. Examples include the following:

- **Access to Financial Services and Products.** A lack of access to the economic capital required to start or grow a business, purchase land, or take innovations to market can exacerbate the poverty of individuals, and stagnate economic growth (see **SDG 8**). Traditional lenders, such as banks, may not be a feasible source of economic resources for many. Innovations (e.g., microfinance and mobile money agents) are helping to address this challenge, however it persists for many.
- **Access to Markets.** Lack of physical access to markets, for example, due to insufficient infrastructure, can sustain poverty (see **SDG 9**). Poor transport infrastructure adds 30–40% to the costs of goods traded among African countries (Ayemba 2018). Approximately two-thirds of the population living in rural Africa are more than 2 km from the nearest road (Ayemba 2018), hindering access to markets.
- **Geography**. The geographical position of a nation shapes its climate, landscapes, hydrology, and exposure to geological and meteorological hazards. These factors can all affect development. Lack of direct access to oceans hinders the ability of many landlocked developing countries to benefit from trading routes, and maximise opportunities from global integration (Arvis et al. 2010).
- **Natural Resources**. Natural resources, including water, soils, and mineral/rock materials underpin social and economic development. Poverty can be triggered by a lack of, or differential access to, natural

resources, or degradation of natural resources. For example, the underlying geology is one critical factor (alongside organic matter, climate, and time) that determines soil type in a given region. Different soil types are suitable for different purposes, with high inputs (therefore potentially high costs) required in some soils to sustain subsistence or commercial agriculture (e.g., grow cash crops). Differential access to natural resources may be a result of poor governance or the legacy of historical decisions (e.g., settlement of communities following conflict), as well as natural variability in the distribution of resources. Travelling long distances to collect water can take time away from education and income generation activities.

- **History**. Poverty today may result from past policy decisions within or beyond the country of interest, including imperialism and colonialism. For example, the decision to exploit hydrological resources for energy generation in Canada resulted in the degradation of fisheries resources for indigenous communities in the province of Manitoba (Hoffman 2008). This contributed to greater poverty in this community, with high welfare dependence and changes to traditional social practices (Hoffman 2008).
- **Culture**. Poverty can be catalysed by cultural norms, practices, and beliefs. For example, cultural practices or biases may determine the division of labour in a household, resulting in women and girls spending a greater proportion of their time on household chores (e.g., fetching water) than men (Blackden and Wodon 2006). This can limit their ability to take part in income-generating opportunities and exacerbate both poverty and inequalities.
- **Governance**. Weak governance, corruption, and poverty are all linked. Chetwynd (2003) found that *'corruption has direct consequences on economic and governance factors, intermediaries that in turn produce poverty'*. The prevalence of corruption and inhibition of democracy has been linked to an abundance of natural resources, particularly

Fig. 1.5 Living in Topographical Extremes, Ladakh. In the Himalaya, Ladakh is a region of environmental extremes, with access often limited during the winter season. Communities have adapted to the environmental extremes, but remain vulnerable due to the difficult geography. *Credit* Joel. C Gill

hydrocarbons and minerals due to the high rents and export income they are associated with (Pegg 2006). This theme is discussed in detail in **SDG 16**.

- **Conflict**. Armed conflict can catalyse poverty and decimate economically productive sectors. For example, Burundi has been affected by chronic violence, contributing to poverty (Brachet and Wolpe 2005).

Each of these factors can act at diverse scales (e.g., national, community, household), with interactions between different factors and the different scales at which each factor could act. For example, the lack of individual access to financial products and services can stagnate national economic growth, and result in lower tax revenues as businesses struggle to develop and flourish. A possible consequence of this is limited ability to invest in the public infrastructure needed to overcome challenges of access,

strengthen governance, or ensure achievement of the maximum social benefits of natural resources. Nations may rely on loans, with unfavourable repayment terms. It is beyond the scope of this chapter to explore all factors contributing to poverty in detail (see the *Further Reading* at the end of this chapter). Through the remainder of this chapter, we focus on those factors that primarily relate to geoscience and the sectors in which geoscientists operate. Tackling poverty, however, requires dialogue across disciplines, and coherent and integrated solutions, a theme we return to in Sect. 1.5.

1.4.2 Geoscience and the Causes of Poverty

Many of the factors contributing to poverty outlined in the previous section have an environmental dimension. Here we discuss

(a) physiography and spatial poverty traps, (b) natural resources, (c) climate and environmental change, and (d) environmental degradation and shocks.

(a) Physiography and Spatial Poverty Traps. Many of those in poverty live in '*spatial poverty traps*', specific regions where physical, natural, social, political, and/or human capital are low, with this resulting in isolation, disadvantage, and marginalisation (Bird et al. 2010). In his book '*Prisoners of Geography*', Tim Marshall outlines how the physical geography (or physiographical) context of nations shapes their political choices and decisions (Marshall 2016). For example, mountains, climate, rivers, and seas can all influence how easy it is to trade, the spread of disease, or the security of a population from external threats. Physiography is, therefore, a form of '*spatial poverty trap*', where particular aspects of the physical geography of a region suppress development efforts and make it harder to overcome poverty. Examples include the following:

- *Landlocked nations,* where lack of access to some types of physical capital (e.g., ports) result in isolation and disadvantage (Gallup et al. 1999). Examples of landlocked developing countries include Afghanistan, Burundi, Central African Republic, Malawi, Mali, Mongolia, Nepal, South Sudan, and Uzbekistan. Transporting goods via land is more than seven times as expensive as via the sea (Venables and Limão 2001). These extra costs, and lack of viable trading routes, can hinder economic growth.
- *Remote communities in mountainous regions* (Fig. 1.5), where there is also a lack of infrastructure or connectivity. Steep topography and a dynamic environment hinder accessibility and infrastructure development. **SDG 9** highlights the challenge of engineering in contexts such as Nepal, where mass movements triggered by earthquakes and heavy rain can result in a difficult construction environment. Remote communities will find it harder to trade and access services, and they are further from economic and political hubs.

Landscapes are an expression of the geological history of a region, shaped by the interplay of tectonics and climate (Allen 2008). The convergence of tectonic plates is responsible for mountain building. Some rock types (e.g., quartzite) are more resistant to erosion than softer rocks and therefore form steep, upland topography that can be difficult to transect. Spatial poverty traps cannot solely be explained in terms of geoscience, but this is a contributing and important factor.

(b) Natural Resources. Differential access to and benefit from natural resources can drive poverty and hinder social and economic development. For example, on a national scale, countries that lack natural resources may struggle to ensure food security or catalyse economic development. At a local scale, access to natural resources drives progress in education, gender equality, decent employment, and health, all of which can help to break poverty cycles. Where resources are prevalent, but wealth from these is not translated into pro-poor sustainable development outcomes or shared equally, poverty can continue or be exacerbated. Key natural resources that have a role in improving the lives and livelihoods in communities or nations facing poverty include soil, mineral, water, and energy:

- **Soil Resources**. Soils are derived from the chemical and physical weathering of rocks, with the chemical composition of these determined by geological processes. For example, the quartzite ridges of the Kagera region of Tanzania are challenging environments to grow crops on, being poor in nutrient-releasing minerals (FAO 2018). Geoscientists, therefore, have a role to play in supporting governments to understand soil geochemistry and the potential health and agricultural implications. Soil quality is particularly important as those living in poverty are often dependent on agriculture, particularly in a rural livelihood context, but also for their nutrient intake (see **SDGs 2** and **3**). Access to land with good quality soil may be competitive. The growth of large, commercial agro-industrial holdings in Romania is linked to low socio-economic development and reduced rural livelihoods opportunities (Popovici et al. 2018). Governments must decide how to balance the need for land for larger scale commercial activities (potentially providing employment opportunities and economic growth, but associated with monoculture or intensive farming practices, which can reduce soil quality) and preserving land to meet the needs of subsistence or small/medium-sized enterprises and future generations. A lack of natural capital (e.g., poor quality soils) may, therefore, also be a form of spatial poverty trap. Subsistence farmers who only have access to poor quality soils may regularly struggle to have a successful harvest and any surplus to trade. A 10% increase in soil quality can lead to a roughly two-percentage point decrease in poverty rates in rural areas, and in sub-Saharan Africa, a nine-percentage point reduction in poverty rates (Heger et al. 2018).

- **Mineral Resources**. A lack of domestic resources can cause a country to rely on imports of raw materials, including basic industrial minerals for construction or fertilisers. This can drive up the cost of infrastructure and agricultural produce. While the presence of mineral resources (e.g., salt, gold, cobalt) is a potential catalyst for economic development, their absence does not necessarily determine poverty. In Costa Rica, a lack of mineral resource definition during the colonial period and a subsequent and continuing ban on mining has led to the development of renewable energy resources, agriculture, human capital, eco-tourism, and high-technology-manufacturing industries. In terms of GDP and social progress, Costa Rica ranks as a high performing outlier (Social Progress Imperative 2018).

- **Water Resources**. A lack of access to clean water can be linked to poverty in many ways, primarily creating health challenges and hindering agricultural productivity which then has implications for education outcomes and economic productivity. Improved irrigation can unlock agricultural productivity and poverty alleviation through increased labour demands, higher crop yields, market-orientated production, creation of growth multipliers through the stimulation of other economic activities, reduced seasonal variability, and improved nutritional outcomes (Hanjra et al. 2009). Resolving access to water can also unlock opportunities by enhancing household and community relationships, thus stimulating economic growth in impoverished communities (Zolnikov and Blodgett-Salafia 2016). Access to clean water and safe sanitation (**SDG 6**) can help to reduce poverty (**SDG 1**) by improving health (**SDG 3**), access to education (**SDG 4**), and gender equality (**SDG 5**).

- **Energy Resources**. The International Energy Agency (IEA) estimates that approximately 1.3 billion people lack access to electricity and 2.7 billion people lack access to clean cooking facilities (IEA 2018). Dependence on solid-fuel (e.g., charcoal) for cooking can result in negative environmental and health implications (Smith 2006). Dependence on charcoal is linked to deforestation and loss of habitat Chidumayo and Gumbo (2012) and in addition charcoal collection absorbs valuable time from other productive activities such as work or education (Hammond et al. 2007). There are an estimated 4 million deaths annually attributed to air pollution (Bruce et al. 2002). Access to clean and affordable energy (**SDG 7**) is, therefore, essential in reducing poverty by improving health (**SDG 3**) and protecting life on land (**SDG 15**).

The relationship between poverty and resources is not necessarily a causal one or inevitable (Lewin 2011). The approaches used to manage soil, mineral, water, and energy resources can influence outcomes either alleviating or exacerbating poverty. A country that is well endowed with natural resources can be subject to what is commonly termed the '*natural resource curse*' where there is a failure to capture and retain value either locally or nationally. It is a global phenomenon, but not inevitable, with many contrasting case studies (McKinley and

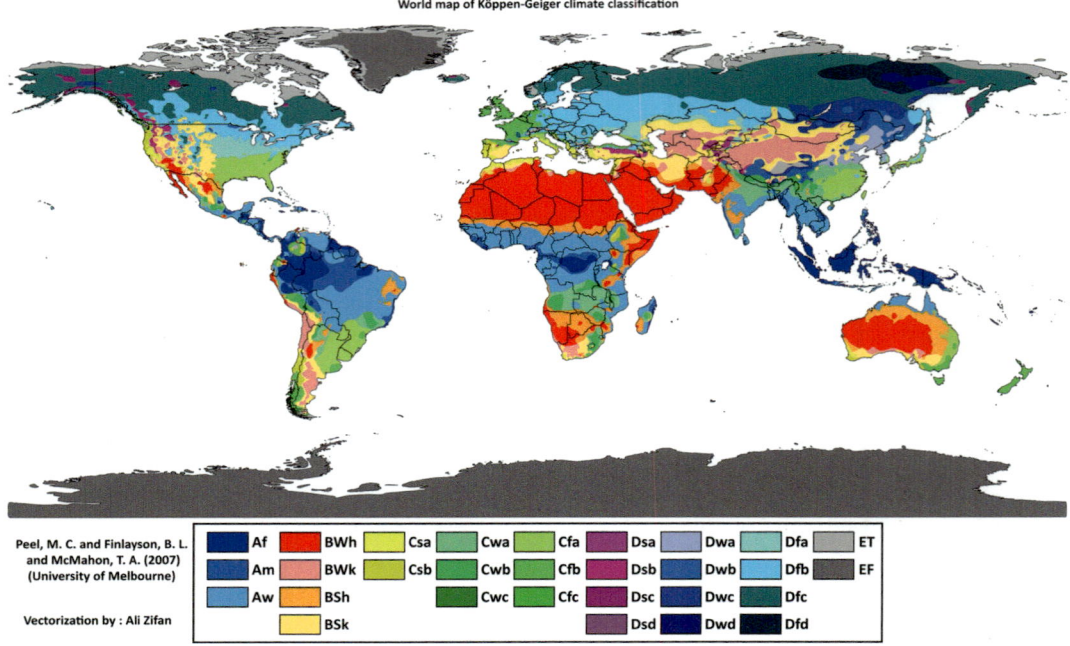

Fig. 1.6 World Map of Köppen-Geiger Climate Classification, with A-Tropical, B-Arid, C-Temperate, D-Cold, and E-Polar. Created by Peel et al. (2016), and used under a CC BY SA 4.0 International license (https://creativecommons.org/licenses/by-sa/4.0/)

Kyrili 2008). For example, while Chile and Zambia are both copper-dependent countries, McKinley and Kyrili (2008) found that Zambia has been less successful in capturing its resource revenues due to poor fiscal policy, and opaque tax systems, royalties, and ownership. In contrast, Chile's state-owned mining company (CODELCO), effective fiscal policy, and Economic and Social Fund and Pensions Reserve Fund have strengthened their ability to avoid the resource curse. There is not one single factor that determines a country's ability to benefit from its resource potential. Rather, it is a complex combination of factors from economics, historical context, fiscal policy, effective governance and institutions, and commodity price shocks.

The social, environmental, and economic challenges surrounding natural resources are diverse and often context specific. While the impacts of poor access to and unequal benefit from natural resources may be most acutely felt at a local and national scale, the actions needed to address resource governance are a global responsibility.

(c) Climate and Environmental Change. Geographic variations in climate can explain differences in crop growth, disease prevalence, and water scarcity and availability (Sachs 2015). Figure 1.6 shows an updated world map of the Köppen-Geiger climate classification (Peel et al. 2007), with tropical, arid, temperate, cold, and polar regions all highlighted. Vector-borne diseases such as malaria and dengue thrive in areas that are typically warm throughout the year (i.e., tropical climates), and therefore some places are more susceptible to the diseases that exacerbate poverty than others. Sub-Saharan African countries bear the heaviest burden of malaria, with 90% of the global cases, and 92% of global malaria deaths occurring there (World Health Organisation 2016). Poverty results in greater exposure to disease, and disease perpetuates this poverty in a reinforcing cycle (Teklehaimanot and Mejia 2008). The costs of malaria fall more heavily on the very poor. In Malawi, for example, the total direct and indirect cost of malaria consumed 32% of the annual household income for very-low-income households, compared with 4.2% of households in low-to-high income categories (Ettling et al. 1994). Climate also affects water availability, influencing precipitation and evaporation. Arid regions (in contrast to temperate regions) need more investment in irrigation and water transport to fulfil domestic and agricultural needs.

Environmental change, for example, due to anthropogenic carbon emissions, may result in new climate patterns and therefore change the spatial viability of disease, crop growth, or water availability. In regions already susceptible, it could exacerbate this. For example, Rangecroft et al. (2013) note the negative effect of climate change on water supplies in the arid mountains of the Bolivian Andes, where poverty means there is limited capacity to adapt. Bolivian glaciers are estimated to have lost 50% of their ice mass over the past 50 years, resulting in concerns regarding the future availability of water resources (Rangecroft et al. 2013).

(d) **Environmental Degradation**. A common, but misguided narrative is that poverty is a major cause of environmental degradation (Durning 1989; Boyce 1994), such as biodiversity loss and reductions in the quality of air, water, and soil. It is asserted that those in poverty degrade resources and the environment in order to ensure their day-to-day survival. This analysis is both simplistic and unjust (Duraiappah 1998; Ravnborg 2003), with degradation largely resulting from the actions of non-poor individuals (Ravnborg 2003), industry and the policies and practices of developed economies. Non-poor actors generally have greater access to land and forest resources, agricultural chemicals, and irrigation (Ravnborg 2003). In contrast, environmental degradation disproportionately affects the poorest in society, exacerbates existing poverty, and increases the proportion of the population living in poverty. Examples include the following:

- Duraiappah (1998) examines select impacts of deforestation (e.g., loss of watershed protection, soil erosion, productivity drop, fuelwood shortage), and notes that each of these will affect low-income groups the most. This is due, in part, to the resultant increase in household expenditure and their lack of resources to meet these needs.
- Broad (1994) describes the onset of commercial logging in the immediate vicinity of agricultural plots of subsistence farmers in the Philippines. This logging changed the availability and quality of water and the vulnerability of soils to erosion, reducing the ability of subsistence farmers to meet their needs.
- Onwuka (2006) highlight environmental challenges associated with hydrocarbon extraction in the Niger Delta (Nigeria). The toxicity of hydrocarbons damages soil fertility, watercourses, and biodiversity in the region, exacerbating poverty due to the elimination of livelihoods (Onwuka 2006; Saliu et al. 2007).

(e) **Natural (Environmental) Hazards.** Disaster risk reduction is embedded into the SDGs. SDG **Target 1.5** highlights the need to build the resilience of those in poverty, helping to reduce the impacts of natural hazards (e.g., earthquakes, tsunamis, landslides, volcanic eruptions, avalanches, floods, droughts, tropical storms, and wildfires) on them. A hazard is a *'process, phenomenon or human activity that may cause loss of life, injury or other health impacts, property damage, social and economic disruption or environmental degradation'* (UNDRR 2017). Hazards are a key component of the risk equation (Fig. 1.7). The risk posed by a hazard is determined by who or what it affects (the exposure), and how susceptible they are to damage or loss because of the hazard (vulnerability). Poverty and inequality both contribute to disaster risk by increasing the vulnerability. In

Fig. 1.7 A common representaton of the risk equation

turn, the occurrence of a hazard will disproportionately affect those in poverty and other marginalised groups. In regions where hazards
regularly occur, they can lead to poverty traps
and threaten national social and economic
development progress. As the economic, social,
and physical impacts of hazards can transcend
national borders, vulnerable communities in one
country can be impacted by a hazard many miles
away. For example, the eruption of the Icelandic
volcano Eyjafjallajökull in 2010 had a significant
economic impact in Kenya, due to the inability of
horticulturalists to transport their cut flowers to
Europe (Waihenya 2010).

While wealthier nations bear the largest fraction
of the total economic burden of disasters, the relative impact on poorer nations is much greater
(Table 1.5). By contrasting total losses from 1998
to 2017 (billions US$) with average annual eco-
nomic losses (billions US$) as a percentage of GDP
over the same timeframe, we note the significant
variation in the top 10 countries listed. Eight out of
the ten highest ranking nations, when considering
the average percentage of GDP lost between 1998
and 2017, are lower income countries.

Poorer nations generally have fewer resources
to cope with disasters and often rely on international financial assistance to recover. The

diversion of time, resources, and personnel
towards rebuilding efforts, and the loss of lives
and livelihoods slows or hinders economic and
human development progress, threatening efforts
to alleviate poverty and promote sustainable
development. For example, in January 2010, a
$M_w = 7$ earthquake struck Haiti, damaging
nearly half of all infrastructure in the epicentral
region, resulting in 100,000–300,000 fatalities,
and displacing more than a million people
(DesRoches et al. 2011). The earthquake is estimated to have resulted in direct economic losses
of US$8–14 billion (Cavallo et al. 2010). For a
country with a 2009 GDP of US$15 billion, the
ability to cope with such losses without international financial assistance is severely limited.
Contrast this with the most expensive earthquake
in the United States, the 1994 $M_w = 6.7$ Northridge earthquake, which costed up to US$40
billion (Petak and Elahi 2000), with GDP in the
United States being approximately US$10.4
trillion at the time. So the Northridge earthquake,
although much costlier, had minimal impact on
the overall US economy. In Haiti, even with
international financial assistance, the impact on
the economy was significant. The 2010 earthquake pushed the country into a recession similar
in magnitude to the 2004 coup d'état.

Table 1.5 Absolute losses versus average annual losses due to natural hazards (1998–2017)[a]

Total losses			Average annual loss relative to GDP		
Country	Loss (billions US$)	OECD classification (2018)	Country	Percentage (%)	OECD classification (2018)
USA	944.8	High Income	Haiti	17.5	Least Developed
China	492.2	Upper Middle Income	Puerto Rico	12.2	High Income
Japan	376.3	High Income	Korea D. P. R.	7.4	Other Low-Income Countries
India	79.5	Lower Middle Income	Honduras	7.0	Lower Middle Income
Puerto Rico	71.7	High Income	Cuba	4.6	Upper Middle Income
Germany	57.9	High income	El Salvador	4.2	Lower Middle Income
Italy	56.6	High Income	Nicaragua	3.6	Lower Middle Income
Thailand	52.4	Upper Middle Income	Georgia	3.5	Lower Middle Income
France	43.3	High Income	Mongolia	2.8	Lower Middle Income

[a] losses adjusted to 2017 US$. Adapted from Wallemacq and House (2018)

1.5 Geoscience Education, Research, and Innovation to Reduce Poverty

Understanding the *causes* of poverty can help inform the interventions required to reduce and eliminate existing poverty, and prevent future generations from suffering this injustice. Interventions may require research, policy development and implementation, and innovation. Many interventions go beyond the scope of the geoscience community, but there are important ways that geoscientists can contribute to ending poverty and meeting the targets of **SDG 1** (Table 1.1). The actions required to end poverty are inextricably linked to the actions required to make progress on other SDGs, such as water and sanitation (see **SDG 6**), reducing inequality (see **SDG 10**), and tackling climate change (see **SDG 13**).

1.5.1 Improving Access to Basic Services, Natural Resources, and Appropriate Technologies

Target 1.4 highlights the need to address differential access to basic services, natural resources and appropriate technologies, recognising their potential to be a catalyst for social and economic development. These three themes are interlinked (e.g., education needs access to energy in evenings to do homework, requiring raw materials to make solar panels). Access to basic services is a necessary step towards true poverty alleviation; ensuring communities are not limited from reaching their full potential. Examples include education (**SDG 4**), health care (**SDG 3**), social welfare, transport infrastructure (**SDG 9**), affordable energy access (**SDG 7**), potable water and improved sanitation (**SDG 6**), and waste management (**SDG 12**). Many of these services are underpinned by access to natural resources (e.g., food, water, energy, minerals, and bio-resources). Given good governance, an ability to access natural resources, and the capture and retention of their full value, there is significant potential for communities or nations to use these to tackle poverty (e.g., through Sovereign Wealth Funds). Appropriate technologies, such as mobile phones, can help to improve access to basic services and management of natural resources. For example, d.light[1] and M-Kopa[2] increase access to solar energy.

Ensuring universal access to these services, resources and technologies are dependent on (i) understanding where resources can be found, (ii) accessing and making use of these resources, and (iii) sustainable management of these resources, understanding and mitigating any negative impacts on lives, livelihoods, and the natural environment (see **SDG 12**). One example is the East Africa Geothermal Energy Facility (EAGER[3]), which was established in 2015 to support governments to address barriers to the advancement of geothermal energy, recognising that this is a frontier market for traditional geothermal energy investors or private investment. It is estimated that geothermal energy could unlock the potential for 10 GW of baseload power across East Africa, a clean energy source for millions (EAGER 2019). In order to develop this sector, however, it must be de-risked, economically viable, and competitive compared to other energy sources (EAGER 2019). Part of EAGER's role is to provide a knowledge hub covering topics such as international best practice business models, exploration of geothermal reserves, and regulatory mandates.

Improving responsible production and consumption of natural resources (**SDG 12**) can contribute to poverty reduction, and ensure continued access to, and availability of, resources to meet the needs of future generations (Fig. 1.8). Is it possible however to promote social and economic development while remaining within the biophysical limits of the planet? As noted by Hickel (2018) and O'Neill et al. (2018), it is typical to observe countries in the Global South living within biophysical boundaries but failing to deliver on social indicators, meanwhile more developed countries typically exhibit

---

[1]www.dlight.com/about/.

[2]www.m-kopa.com/.

[3]https://adamsmithinternational.com/projects/enabling-investment-in-geothermal-power-in-east-africa-2/.

Fig. 1.8 Agriculture in Tanzania. Understanding the underlying geology and the geochemistry of the soil can inform measures to strengthen food security (SDG 2). Geoscience also informs water management for irrigation, ecosystem protection, drinking water, and other domestic services. Image by skeeze from Pixabay

unsustainable levels of consumption while delivering on social indicators. Redistributing resource consumption can provide flexibility for emerging economies to benefit from their natural resources and stimulate the economic growth required to address social indicators, such as access to affordable energy. Certification schemes are one approach that could help catalyse responsible production, but have the potential to exclude marginalised producers unable to participate in these schemes. While recognising limitations to such approaches, there is an opportunity for geoscientists, working alongside other disciplines, to inform schemes encouraging responsible consumption and production.

1.5.2 Effective and Equitable Disaster Risk Reduction

Historically, international efforts to combat disasters have focused on financing disaster response and post-disaster recovery. However, in recent years the emphasis has changed to disaster preparedness and reduction (see Smith and Petley (2009) for a helpful review of different disaster paradigms). Recognising the complex reasons why disasters occur, the Sendai Framework for Disaster Risk Reduction (Box 1.2) is an internationally agreed framework aiming to reduce losses from natural hazards. Agreed at the 3rd UN World Conference on Disaster Risk Reduction in March 2015, it builds on and extends the scope of the Hyogo Framework for Action (2005–2015) to 2030. The Sendai Framework has 4 priorities for action, 7 strategic targets, and 38 indicators for measuring progress on reducing disaster losses (Box 1.2). These indicators align the implementation of the Sendai Framework with the SDGs and the Paris Agreement on climate change. The pursuit of targets in one framework can support the delivery of another.

Geoscientists, with their knowledge of the processes underpinning natural hazards and their

Fig. 1.9 Multi-Hazard Risk in Guatemala City, Guatemala. Multiple natural hazards (e.g., earthquakes, volcanic eruptions, landslides, tropical storms, flooding) affect many of Guatemala's most vulnerable communities. Improved data and research, community engagement, and cooperation across sectors, disciplines and regions can all help to reduce disaster risk. *Credit* Joel C. Gill

impacts on society, have a critical role to play in implementing and monitoring these frameworks (Gill and Bullough 2017). Specifically:

- **Research.** Geoscientists are engaged in research underpinning policies and practice to reduce disaster risk and improve disaster management. Research can include individual PhD projects and large-scale, complex programmes with international consortia. Research can also be within one discipline to increase our understanding of a specific hazard process or landscape evolution, or cross-disciplinary research that integrates advances in the natural and social sciences to understand drivers of risk or the impact of hazards on people (Fig. 1.9). An emerging priority and research focus is understanding the impacts and complex disaster scenarios that can result

from relationships between multiple hazards in a given region (e.g., Kappes et al. 2012; Gill and Malamud 2014; Duncan et al. 2016; AghaKouchak et al. 2018).

- **Global Networks and Cooperation**. Challenges remain in translating new research developments in our understanding of hazards to tools that inform our understanding of potential risk. Cross-disciplinary and international collaborations can aid this innovation. Furthermore, global networks that bring together the international earthquake or volcanic hazard community to support Sendai Framework principles of building effective, meaningful, and strong partnerships. The Global Earthquake Model (GEM[4]), for example, is an organisation of geoscientists

---

[4]https://www.globalquakemodel.org/.

and engineers translating scientific knowledge of active tectonics into an understanding of potential seismic hazards. In 2018, GEM released the first global seismic risk map, which combines seismic hazard with exposure and vulnerability data. Most regional hazard and risk models made by GEM are created in collaboration with local scientists and practitioners who generally have a better understanding of the local distribution of active faults and their relative hazard. Unlike many commercial hazard and risk calculators, the system is open source, which enables better training and in-country capacity building for the long-term sustainability of the maps. Other international, multidisciplinary endeavours are underway to assess and monitor global landslide and volcanic hazards (the Global Landslide Model[5] and the Global Volcano Model[6]).

- **Capacity Strengthening and Community Engagement**. Geoscientists can strengthen capacity and improve community resilience. Through collaboration and cooperation across sectors, local geohazard problems can be addressed and targeted solutions provided. For example, the Great ShakeOut[7] earthquake drill originally started as a means to engage with local schools and businesses on the earthquake hazard posed by the San Andreas Fault in California. However, since its inception, the event has expanded globally and many local ShakeOut drills are conducted around the world to inform, educate, and prepare people for potential future seismic events in their region. In 2018, over 63 million participants took part in ShakeOut drills worldwide, including in many low and lower-middle income countries, including Afghanistan, Iran, the Philippines, and Pakistan.
- **Data.** The explosion of readily available satellite data in near-real time now means that geoscientists can assess rapidly the degree and extent of damage in a disaster. In 2000, the major global space agencies signed up to the *International Charter: Space and Major Disasters*[8], a non-binding agreement that ensures satellite data is freely provided in the event of a disaster or humanitarian emergency. As of December 2018, there have been 593 activations of the Charter for emergencies including earthquake-triggered landslides in Nepal, the Ebola outbreak in West Africa, oil spills in the South China Sea, and wildfires in California.

Box 1.2 Sendai Framework for Disaster Risk Reduction and Paris Agreement on Climate Change The seven *global targets* of the Sendai Framework are:

(a) Substantially reduce global disaster mortality by 2030, aiming to lower average per 100,000 global mortality rate in the decade 2020-2030 compared to the period 2005–2015.

(b) Substantially reduce the number of affected people globally by 2030, aiming to lower average global figure per 100,000 in the decade 2020-2030 compared to the period 2005-2015.

(c) Reduce direct disaster economic loss in relation to GDP by 2030.

(d) Substantially reduce disaster damage to critical infrastructure and disruption of basic services, among them health and educational facilities, including through developing their resilience by 2030.

(e) Substantially increase the number of countries with national and local disaster risk reduction strategies by 2020.

(f) Substantially enhance international cooperation to developing countries through adequate and sustainable support to complement their national actions for implementation of this Framework by 2030.

(g) Substantially increase the availability of and access to multi-hazard early

---

[5]https://pmm.nasa.gov/applications/global-landslide-model.

[6]https://globalvolcanomodel.org/.

[7]https://www.shakeout.org/.

[8]https://disasterscharter.org/.

warning systems and disaster risk information and assessments to the people by 2030.

In order to meet these targets, the Sendai Framework has four *Priorities for Action*:

Priority 1: Understanding disaster risk

Priority 2: Strengthening disaster risk governance to manage disaster risk

Priority 3: Investing in disaster risk reduction for resilience

Priority 4: Enhancing disaster preparedness for effective response and to "Build Back better" in recovery, rehabilitation and reconstruction

Read more: https://www.undrr.org/publication/sendai-framework-disaster-risk-reduction-2015-2030

Paris Climate Change Agreement

The Paris Agreement, also adopted in 2015, aims to limit global average temperature increases to well below 2 °C, while pursuing efforts to limit any increase to 1.5 °C. It also hopes to peak global emissions as soon as possible, provide adequate financing for low-income nations to build a climate-resilient future, develop adaptation strategies for a warming world, and mitigate and minimise losses from the adverse effects of climate change including extreme weather events (e.g., floods, storms) and slow onset events (e.g., droughts).

Read more: https://unfccc.int/process-and-meetings/the-paris-agreement/what-is-the-paris-agreement.

1.5.3 Ensuring the Availability of Geoscience for Policy, and Improve Its Uptake

A primary means of implementation of SDG 1 is the creation of '*sound policy frameworks at national, regional and international levels*' that incorporate '*pro-poor and gender-sensitive*

development strategies to support accelerated investment in poverty eradication actions' (**Target 1.B**). Policy frameworks helping to alleviate poverty could be focused on access to resources or training, or reducing exposure and vulnerability to environmental or economic shocks. These all benefit from a science input, as set out in this chapter.

The Institute of Government (2011) identified seven key characteristics of 'sound' policy: (i) clear goals that are adequately defined, (ii) informed by high-quality and up-to-date evidence, including the evaluation of previous policies, (iii) rigorously tested and determined to be realistic and resilient to adaptation, (iv) external engagement with those affected by the policy, (v) robust assessment of the options, their cost-effectiveness and risks, (vi) roles and accountabilities, and (vii) a realistic plan for obtaining timely feedback. The integration of these characteristics into policy development will help to ensure they are robust and have the intended impact.

In addition, it is essential that different policies are *integrated* and *coherent*. For example, a water policy should not conflict with a minerals governance policy, and these should work together to reinforce (not undermine) the national poverty reduction policy. Underlying many environmental challenges is a segmented policy and regulatory framework, with cross-sectoral policies and institutional partnerships required for effectiveness (UNEP 2015; Getenet and Tefera 2017). Coherent integrated environmental policies are needed to maximise impact and ensure we tackle key causes of poverty, such as environmental degradation.

Box 1.3 Government Policy versus Legislation (details may differ by country)

Government Policy. This is a statement or document outlining the vision and intended actions of a government on a specific theme (e.g., terrorism, education, environmental protection), and how its actions will benefit society. It is not legally binding, with flexibility to evolve in response to additional evidence or changing needs.

In addition to policy being developed by government and the civil service (at the request of government), other stakeholders may also design, propose, and advocate for a particular policy. For example, opposition parties, civil society groups, think tanks, academics, and private sector lobbyists may all try to inform policy (e.g., present evidence that shapes the drafting of government policy) or encourage the uptake of their own policy recommendations (e.g., encourage an elected representative to adopt that policy position as their own, and advocate for it within the national legislature).

Government Legislation. This is enforceable law, approved by a national legislature (e.g., parliament). The necessary legislation to implement a policy is set out in draft form, typically known as a 'bill'. The merits of this bill are scrutinised and debated by the legislature, often with amendments made to increase its quality and reduce any unintended negative consequences. The bill is then voted on, and if approved it goes through a country-specific process to be added to national legislation, becoming a law. *Regulations* may then be required to ensure this law is adhered to, and enforced.

Example A national government may develop a *policy* to reduce the number of people killed during earthquakes, focusing on improvements to building codes and urban planning. Professional bodies representing geotechnical specialists, geologists, urban planners, and the construction industry may submit evidence to help inform this *policy*. Civil servants then prepare draft legislation, in the form of a *Government Bill* (e.g., the Reduction in Earthquake Impacts Bill), which sets out new building codes and punishments for those failing to adhere to these codes. This

Bill is scrutinised and debated by the national legislature. The opposition party introduces an amendment to the *Bill* that says a chartered engineer must also approve designs for new buildings. The national legislature vote on this amendment, and then the *Bill* as a whole. Both votes pass by a substantial majority, and they are then added to the national legislation, becoming enforceable *law*. The national government states that the *regulation* of this *law* is the responsibility of city/municipal governments. They then have the flexibility to develop and implement their own *regulation* procedures to inspect building projects and ensure that they abide by the new *law*.

One of the most significant roles that geoscientists can have in reducing poverty is ensuring the availability of geoscience information to inform policy development. Geoscientists must first increase their understanding of the policy-making process, and national and international priorities in poverty reduction strategies. This can help to guide research questions, as well as increase understanding of how scientists can communicate their science to policymakers. The public may have access to some meetings in national parliaments. In the UK, attending Select Committee meetings provides an opportunity to observe parliamentarians take evidence from expert witnesses (including scientists). Understanding the role of geoscience in delivering the SDGs, as set out in this book, helps us to pre-position ourselves to advocate for geoscience-informed policies to tackle poverty and other sustainable development challenges.

Engaging with the policymaking process can take different forms but is likely to involve communicating in forms beyond the scientific paper (e.g., blogs, policy briefs, videos), and with disciplines outside of our area of study. In some contexts, opportunities exist for fellowships and exchanges with parliamentarians and civil servants, helping to bridge the gap between science and policy. For example, the Royal Society Pairing Scheme brings together research scientists with UK parliamentarians and civil servants. Governments or cross-party parliamentary committees may have public consultations, with the opportunity to submit evidence to inform these. Professional societies are well placed to collect the views of the geoscience community and submit a coordinated response that presents the social value of geoscience on a given topic. A further approach could be to develop enhanced research partnerships with those working in government agencies. This helps to embed the outputs of new geoscience research and innovations into government agencies.

Sound policy frameworks also require strong national scientific institutions and research communities to help inform policy development (see **SDG 16**). Improved scientific capacity in policymaking institutions (e.g., government departments, national and local legislatures) can also enrich the policymaking process. Legislators (e.g., parliamentarians) scrutinise and make decisions on diverse and complex themes, including those directly and indirectly relating to geoscience. Examples include the exploration and extraction of unconventional hydrocarbon resources, the expansion of electric vehicles, and the development of network infrastructure through environmentally sensitive regions. The ability of legislators to make wise decisions and/or hold governments to account will depend on their ability to access and interpret scientific information. This information is typically provided to legislators by their research and support staff. In contexts where required information is highly specialised or of a technical nature, it may be necessary to draw on expertise from specialist staff with scientific backgrounds.

For example, the UK Parliamentary Office of Science and Technology (POST[9]) provides independent, balanced, and accessible advice to UK parliamentarians and analysis of public policy issues related to science and technology. This includes the publication of POSTnotes[10], short summaries of public policy issues based on reviews of the research literature and expert interviews, often co-authored by scientists doing policy fellowships. Relevant examples include science diplomacy (POSTnote 568), environmental Earth observation (POSTnote 566), greenhouse gas removal (549), the water–energy–food nexus (POSTnote 543), access to water and sanitation (POSTnote 521), and deep-sea mining (POSTnote 508). From 2008 to 2012, POST ran a programme to improve parliamentary scrutiny of scientific and technological issues in Uganda. This focused on training parliamentary staff and improving links between parliamentarians and scientists in Uganda.

1.5.4 Capacity Strengthening and Respectful Partnerships

Underpinning all of the interventions in Sects. 1.5.1–1.5.3 is the need to mobilise resources through enhanced development cooperation. This can help to strengthen the capacity of key institutions to implement programmes and policies to end poverty in all its dimensions as indicated in **Target 1.A**. Many national agencies that are involved in supporting economic growth, enhancing human welfare, and strengthening resilience, draw upon the expertise of geoscientists (e.g., geological surveys, water resources, minerals agencies, and civil protection agencies). Development cooperation can take many forms (e.g., public–private partnerships, long-term academic collaborations), but should be characterised by effective and respectful partnerships formed to help achieve priorities identified by Global South stakeholders. See **SDG 17** for a full discussion of this theme.

---

[9]https://www.parliament.uk/mps-lords-and-offices/offices/bicameral/post/.

[10]https://www.parliament.uk/postnotes.

1.6　Key Learning Concepts

- Poverty is a lack of resources (economic, social natural, political, and cultural) to meet everyday needs. Poverty is, therefore, complex and multifaceted. Poverty hinders individuals and communities from reaching their full potential, reduces life expectancy, and increases vulnerability to epidemics, economic depression, environmental change, and natural hazards. There are different ways to measure and define poverty. Many focus on access to economic resources, with others integrating multiple dimensions of poverty.
- Factors contributing to poverty can be either environmental or social. Landlocked countries and remote communities may have limited access to markets. Communities in regions with poor soils, limited water, or unfavourable climate extremes may be more susceptible to food insecurity or diseases. Those living in regions affected by multiple natural hazards may lose development gains due to repeat disasters. Poverty traps and cycles exist, with poverty transmitted through multiple generations until there is an adequate intervention.
- Geoscientists have a unique opportunity and responsibility to contribute to poverty alleviation and have a vital role to play across government, policy, private sector, and NGOs in a range of disciplines from natural resources, hazards, governance. Geoscientists can help improve access to natural resources, while also promoting responsible consumption and production to minimise environmental degradation. Poverty and inequality contribute to disaster risk, and disasters disproportionately affect the poor. Geoscientists working on all aspects of geological hazards and risk reduction can support efforts to reduce poverty.
- Engagement with other disciplines and understanding of context is critical to inform interventions. Geoscientists should increase access to their knowledge and skills, through supporting coherent and comprehensive policies. Partnerships across sectors, disciplines, and regions can help to improve knowledge exchange and technology transfer.

1.7　Educational Ideas

In this section, we provide examples of educational activities that connect geoscience, the material discussed in this chapter, and scenarios that may arise when applying geoscience (e.g., in policy, government, private sector international organisations, NGOs). Consider using these as the basis for presentations, group discussions, essays, or to encourage further reading.

- Identify and contrast a physiographical map of eastern Africa with a map of current poverty levels. Identify and discuss geographical and geological features that may influence the extent of poverty.
- Consider the impact of the arrival of a mineral exploration company on a local community in Zambia. Divide into four stakeholder groups including *host government, the company, the community,* and *migrant artisanal gold miners* working in the region. Debate the potential positive or negative outcomes of the exploration company's presence in the community. Consider themes such as the natural resource curse, infrastructure, knowledge and technology transfer, access to resources, and local procurement. How can geoscientists advocate for positive outcomes?
- Consider you are a senior civil servant in the national government, tasked with reducing all forms of poverty. What steps could help to *increase access to science in Government decision-making,* and how may this influence your strategy for poverty reduction?
- Imagine you are a poor farmer, and your cattle are your primary assets. You live on the slopes of a volcano. What might influence your decision-making process about how to respond to warnings of volcanic unrest, and

requests to evacuate? What measures could support you to protect your life, assets, and livelihood?

Further Reading and Resources

Green D (2008) From poverty to power: how active citizens and effective states can change the world. Oxfam, p 540

OECD (2015) Material resources, productivity and the environment. Available online: https://read.oecd-ilibrary.org/environment/material-resources-productivity-and-the-environment_9789264190504-en (accessed 15 January 2019)

Sachs JD (2015) The age of sustainable development. Columbia University Press. p 543

References

AghaKouchak A, Huning LS, Chiang F, Sadegh M, Vahedifard F, Mazdiyasni O, Moftakhari H, Mallakpour I (2018) How do natural hazards cascade to cause disasters? Nature 561:458–460

Ahmed I (2014) Factors in building resilience in urban slums of Dhaka, Bangladesh. Procedia Econ Financ 18:745–753

Alkire S, Jahan S (2018) The new global MPI 2018: aligning with the sustainable development goals', HDRO Occasional Paper, United Nations Development Programme (UNDP). Available online: http://hdr.undp.org/sites/default/files/2018_mpi_jahan_alkire.pdf (accessed 18 December 2018)

Allen PA (2008) From landscapes into geological history. Nature 451(7176):274

Arvis JF, Marteau JF, Raballand G (2010) The cost of being landlocked: logistics costs and supply chain reliability. The World Bank. Available at: https://elibrary.worldbank.org/doi/abs/10.1596/978-0-8213-8408-4 (accessed 29 October 2019)

Ayemba D (2018) Infrastructure in africa: bridging the gap. Available online: https://constructionreviewonline.com/2018/01/infrastructure-africa-bridging-gap/ (accessed 18 December 2018)

Bird K, Higgins K, Harris D (2010) Spatial poverty traps. ODI Working Paper 321 and CPRC Working Paper 161, p 18

Blackden CM, Wodon Q (eds) (2006) Gender, time use, and poverty in sub-Saharan Africa. The World Bank

Bourguignon F, Morrisson C (2002) Inequality among world citizens: 1820-1992. Am Econ Rev 92(4):727–744

Bourguignon F, Morrison C (2002) Income distribution among world citizens: 1820–1990. Am Econ Rev 92 (September):1113–1132

Boyce JK (1994) Inequality as a cause of environmental degradation. Ecol Econ 11(3):169–178

Brachet J, Wolpe H (2005) Conflict-sensitive development assistance: the case of Burundi. Conflict Prevention & Reconstruction, Environmentally and Socially Sustainable Development Network, World Bank

Broad R (1994) The poor and the environment: friends or foes? World Dev 22(6):811–822

Bruce N, Perez-Padilla R, Albalak R (2002) The health effects of indoor air pollution exposure in developing countries. World Health Organisation. Retrieved from: http://www.bioenergylists.org/stovesdoc/Environment/WHO/OEH02.5.pdf

Cali M, Menon C (2013) Does urbanization affect rural poverty? Evidence from Indian districts, The World Bank

Calvo-González O (2016) Why are indigenous peoples more likely to be poor? Available online: https://blogs.worldbank.org/opendata/why-are-indigenous-peoples-more-likely-be-poor (accessed 25 January 2019)

Cavallo E, Powell A, Becerra O (2010) Estimating the direct economic damages of the earthquake in Haiti. Econ J 120(546):F298–F312

Chandy L, Ledlie N, Penciakova V (2013) The final countdown: prospects for ending extreme poverty by 2030. Brookings Institution. Available online: www.brookings.edu/wp-content/uploads/2016/06/The_Final_Countdown.pdf (accessed 25 January 2019)

Chetwynd E, Chetwynd F, Spector B (2003) Corruption and poverty: a review of recent literature. Manag Syst Int 600:5–16

Chidumayo EN, Gumbo D (2012) The environmental impacts of charcoal production in tropical ecosystems of the world: a synthesis. Retrieved from: http://www.cifor.org/library/3890/the-environmental-impacts-of-charcoalproduction-in-tropical-ecosystems-of-the-world-a-synthesis/ (accessed 18 December 2018)

DesRoches R, Comerio M, Eberhard M, Mooney W, Rix GJ (2011) Overview of the 2010 Haiti earthquake. Earthq Spectra 27(S1):S1-S21

Duraiappah AK (1998) Poverty and environmental degradation: a review and analysis of the nexus. World Dev 26(12):2169–2179

Durning AB (1989) Action at the grassroots: Fighting poverty and environmental decline. Worldwatch Institute, Washington, DC

EAGER (2019) East Africa Geothermal Energy Facility. Website: https://www.adamsmithinternational.com/case-study/enabling-the-development-of-geothermal-energy-in-east-africa (accessed 25 June 2019)

Elias D, Tran P, Nakashima D, Shaw R (2009) Indigenous knowledge, science, and education for sustainable development. In: Indigenous Knowledge and Disaster Risk Reduction. Available online: https://www.researchgate.net/publication/279922682_Indigenous_

knowledge_science_and_education_for_sustainable_ development (accessed 18 December 2018)

Ettling M, McFarland DA, Schultz LJ, Chitsulo L (1994) Economic impact of malaria in Malawian households. Trop Med Parasitol 45:74–79

FAO (2018) Background Information on Natural Resources in the Kagera River Basin. Available online: www.fao.org/fileadmin/templates/nr/kagera/ Documents/Suggested_readings/nr_info_kagera.pdf (accessed 18 December 2018)

Gallup JL, Sachs JD, Mellinger AD (1999) Geography and economic development. International regional science review 22(2):179–232

Getenet B, Tefera B (2017) Institutional analysis of environmental resource management in Lake Tana Sub basin. Soc Ecol Sys Dynam 453–477

Gill JC, Bullough F (2017) Geoscience engagement in global development frameworks. Ann Geophys 60

Gill JC, Malamud BD (2014) Reviewing and visualizing the interactions of natural hazards. Rev Geophys 52 (4):680–722

Grant U (2010) Spatial inequality and urban poverty traps. ODI Working Paper 326 and CPRC Working Paper 166, p 35

Green D (2008) From poverty to power: how active citizens and effective states can change the world. Oxfam

Hammond AL, Kramer WJ, Katz RS, Tran JT, Walker C (2007) The next 4 billion. Market size and business strategy at the base of the pyramid. World Resources Institute International Finance Corporation. Retrieved from: https://wriorg.s3.amazonaws.com/s3fs-public/ pdf/n4b_full_text_lowrez.pdf?_ga=2.68538257. 1846090585.1550552091-1776181383.1543570925

Hanjra MA, Ferede T, Gutta DG (2009) Reducing poverty in sub-Saharan Africa through investments in water and other priorities. Agric Water Manag 96 (2009):1062–1070

Heger M, Zens G, Bangalor M (2018) Does the environment matter for poverty reduction? The role of soil fertility and vegetation vigor in poverty reduction. World Bank Group Policy Research Working Paper 8537, p 39

Hickel J (2018) Is it possible to achieve a good life for all within planetary boundaries? Third World Quarterly. Available online: https://doi.org/10.1080/01436597. 2018.1535895

Hilson G (2016) Artisanal and small-scale mining and agriculture. Exploring their links in rural sub-Saharan Africa. IIED. Available online: http://pubs.iied.org/ pdfs/16617IIED.pdf

Hilson G, Pardie S (2006) Mercury: an agent of poverty in Ghana's small-scale gold-mining sector? Resour Policy 31(2):106–116

Hoffman SM (2008) Engineering poverty: colonialism and hydroelectric development in Northern Manitoba. In: Martin T, Hoffman S (eds) Power Struggles: Hydro Development and First Nations in Manitoba and Quebec. University of Manitoba Press, Winnipeg, MB, Canada, pp 103–128

IEA (2018) World Energy Outlook. Available online: www.iea.org/weo2018/ (accessed 18 January 2019)

Imai KS, Gaiha R, Garbero A (2017) Poverty reduction during the rural–urban transformation: rural development is still more important than urbanisation. J Policy Model 39(6):963–982

International Monetary Fund DataMapper (2018). Available at: https://www.imf.org/external/datamapper/ datasets (accessed 29 October 2019)

IGF - Intergovernmental Forum on Mining, Minerals, Metals and Sustainable Development (2017) Global Trends in Artisanal and Small-Scale Mining (ASM): A review of key numbers and issues. Winnipeg: IISD. Available online: https://www.iisd.org/sites/default/ files/publications/igf-asm-global-trends.pdf (accessed 18 January 2019)

Kappes MS, Keiler M, von Elverfeldt K, Glade T (2012) Challenges of analyzing multi-hazard risk: a review. Nat Hazards 64(2):1925–1958

Kimani-Murage EW, Ngindu AM (2007) Quality of water the slum dwellers use: the case of a Kenyan slum. J Urban Health 84(6):829–838

Lewin M (2011) Botswana's success: good governance, good policies, and good luck. World Bank. Available online: http://siteresources.worldbank.org/AFRICA EXT/Resources/258643-1271798012256/Botswana- success.pdf (accessed 25 January 2019)

Marshall T (2016) Prisoners of Geography. Revised Edition, Elliot and Thompson Limited, p 303

McKinley T, Kyrili K (2008) The resource curse. SOAS Development Digest, No.1. Available online: www. soas.ac.uk/cdpr/publications/dd/file48462.pdf (accessed 25 January 2019)

O'Neill DW, Fanning AL, Lamb WF, Steinberger JK (2018) A good life for all within planetary boundaries. Nat Sustain 1(2)

Onwuka EC (2006) Oil extraction, environmental degradation and poverty in the Niger Delta region of Nigeria: a viewpoint. Int J Environ Stud 62(6):655– 662

Peel MC, Finlayson BL, McMahon TA (2007) Updated world map of the Köppen-Geiger climate classification. Hydrol Earth Syst Sci 11:1633–1644. https://doi. org/10.5194/hess-11-1633-2007

Pegg S (2006) Mining and poverty reduction: transforming rhetoric into reality. J Clean Prod 14(3–4):376– 387

Petak WJ, Elahi S (2000, July) The Northridge earthquake, USA and its economic and social impacts. In: Euro-conference on global change and catastrophe risk management, earthquake risks in Europe, IIASA

Popovici E, Mitrica B, Mocanu I (2018) Land concentration and land grabbing: implications for the socioeconomic development of rural communities in southeastern Romania. Outlook Agric 47(3):204–213

Rangecroft S, Harrison S, Anderson K, Magrath J, Castel AP, Pacheco P (2013) Climate change and water resources in arid mountains: an example from the Bolivian Andes. Ambio, 42(7), 852–863.

Ravnborg HM (2003) Poverty and environmental degradation in the Nicaraguan hillsides. World Dev 31 (11):1933–1946

Riley TA, Kulathunga A (2017) Bringing E-money to the poor: successes and failures. Directions in Development – Finance. World Bank e-Library

Roser M, Ortiz-Ospina E (2019) Global extreme poverty. Available at: https://ourworldindata.org/extreme-poverty (accessed 29 October 2019)

Sachs JD (2015) The age of sustainable development. Columbia University Press

Saliu HA, Luqman S, Abdullahi AA (2007) Environmental degradation, rising poverty and conflict: towards an explanation of the Niger Delta crisis. J Sustain Develop in Africa 9(4):275–290

Sané P (2001) The Role of the Social and Human Sciences in the Fight Against Poverty. MOST-Newsletter, n° 10. Available online: http://digital-library.unesco.org/shs/most/gsdl/collect/most/index/assoc/HASH6991.dir/doc.pdf (accessed 18 December 2018)

Social Progress Imperative. (2018). Social Progress Index 2018 Executive Summary. The Social Progress Imperative. Available online: https://www.socialprogress.org/assets/downloads/resources/2018/2018-Social-Progress-Index-Exec-Summary.pdf (accessed 13 March 2019)

Smith KR (2006) Health impacts of household fuelwood use in developing countries. UNASYLVA-FAO- 57 (2):41

Smith K, Petley D (2009) Environmental hazards. Routledge. p 383

Teklehaimanot A, Mejia P (2008) Malaria and poverty. Ann N Y Acad Sci 1136(1):32–37

UK DWP (2018) Households below average income: an analysis of the UK income distribution: 1994/95-2016/17. Available online: https://assets.publishing.service.gov.uk/government/uploads/system/uploads/attachment_data/file/691917/households-below-average-income-1994-1995-2016-2017.pdf (accessed 13 March 2019)

UNDP (2018) Human Development Indices and Indicators: 2018 Statistical Update. United Nations Development Programme, New York, p 123

UNDRR (2017) DRR Terminology. Available at: https://www.undrr.org/terminology (Accessed 22 October 2020)

UNEP (2015) Global Environment Outlook – Geo 6 Regional Assessment for Africa. United Nations Environment Programme, p 215

UNESCO (2017) Available online: www.unesco.org/new/en/social-and-human-sciences/themes/international-migration/glossary/poverty/ (accessed 18 December 2018)

United Nation (2019) SDG 1. Available at: https://sustainabledevelopment.un.org/sdg1 (accessed 29 October 2019)

United Nations (2015) Millennium Development Goals Report. Available at: https://www.un.org/en/development/desa/publications/mdg-report-2015.html (accessed 29 October 2019)

Venables AJ, Limão N (2001) Infrastructure, geographical disadvantage, transport costs and trade. World Bank Econ Rev 15(3)

Waihenya W (2010) How Iceland's volcano sears Kenya's crops. Available online: www.theguardian.com/commentisfree/2010/apr/20/iceland-volcano-kenya-flight (accessed 25 January 2019)

Wallemacq P, House, R (2018) Economic losses, poverty & disasters: 1998–2017, *UNISDR*. Available at: https://www.unisdr.org/we/inform/publications/61119 (accessed 28 October 2019)

World Bank (2015) Indigenous latin America in the twenty-first century. Available online: http://documents.worldbank.org/curated/en/145891467991974540/pdf/98544-REVISED-WP-P148348-Box394854B-PUBLIC-Indigenous-Latin-America.pdf (accessed 25 January 2019)

World Bank (2018) Poverty and shared prosperity 2018: piecing together the poverty puzzle. World Bank, Washington, DC. License: Creative Commons Attribution CC BY 3.0 IGO

World Bank, PovcalNet. http://iresearch.worldbank.org/PovcalNet/povOnDemand.aspx (accessed 25 January 2019)

World Health Organisation (2016) 10 facts on malaria. Available online: https://www.who.int/features/factfiles/malaria/en/ (accessed 18 December 2018)

Zolnikov TR, Blodgett-Salafia E (2016) Access to water provides economic relief through enhanced relationships in Kenya. J Public Health 39(1):14–19

Joel C. Gill is International Development Geoscientist at the British Geological Survey, and Founder/Executive Director of the not-for-profit organisation Geology for Global Development. Joel has a degree in Natural Sciences (Cambridge, UK), a Masters degree in Engineering Geology (Leeds, UK), and a Ph.D. focused on multi-hazards and disaster risk reduction (King's College London, UK). For the last decade, Joel has worked at the interface of Earth science and international development, and plays a leading role internationally in championing the role of geoscience in delivering the UN Sustainable Development Goals. He has coordinated research, conferences, and workshops on geoscience and sustainable development in the UK, India, Tanzania, Kenya, South Africa, Zambia, and Guatemala. Joel regularly engages in international forums for science and sustainable development, leading an international delegation of Earth scientists to the United Nations in 2019. Joel has prizes from the London School of Economics and Political Science for his teaching related to disaster risk reduction, and Associate Fellowship of the Royal Commonwealth Society for his international development engagement. Joel is a Fellow of the Geological Society of London, and was elected to Council in 2019 and to the position of Secretary (Foreign and External Affairs) in 2020.

Sarah Caven Having started out in mineral exploration, Sarah now draws upon a diversity of global experience spanning private sector, government, social enterprise, and international development. A natural translator across scales and sectors, she contributes to artisanal and small scale mining programs through to regional prospectivity projects. In response to the resourcing future generations challenge, Sarah is passionate about enhancing collaboration, equity and business innovation in mining to unlock development opportunities. Sarah holds a master's in geology (University of Leicester), a Master of Business Administration, (University of British Columbia), and participated in Columbia University's executive education program, Extractive Industries and Sustainable Development.

Ekbal Hussain is a remote sensing geoscientist at the British Geological Survey. He was awarded a Ph.D. in Geophysics and Satellite Geodesy from the University of Leeds and completed his undergraduate studies at the University of Cambridge. Ekbal's research interests are in the use of satellite data to measure ground deformation due to earthquakes, landslides, volcanic activity and subsidence with a focus in low-income nations around the world. He is interested in the disaster risk cycle and the role of geoscience in meeting the Sustainable Development Goals.

Zero Hunger

Benson H. Chishala, Rhoda Mofya-Mukuka, Lydia M. Chabala, and Elias Kuntashula

B. H. Chishala (✉) · L. M. Chabala
Department of Soil Science, School of Agricultural
Sciences, University of Zambia, Lusaka, Zambia
e-mail: bchishala@unza.zm

E. Kuntashula
Department of Agricultural Economics and
Extension, School of Agricultural Sciences,
University of Zambia, Lusaka, Zambia

R. Mofya-Mukuka
Indaba Agricultural Policy Research Institute
(IAPRI), 26A Middleway, Kabulonga, Lusaka,
Zambia

© Springer Nature Switzerland AG 2021
J. C. Gill and M. Smith (eds.), *Geosciences and the Sustainable Development Goals*,
Sustainable Development Goals Series, https://doi.org/10.1007/978-3-030-38815-7_2

Abstract

2 ZERO HUNGER

Overview

Hunger is a state of deprivation where an individual cannot satisfy their basic food needs

842 million people are food insecure, root causes are intertwined with poverty and agricultural production

Most vulnerable are the rural poor in economically depressed and ecologically vulnerable areas

Hunger drives and impacts upon:

Migration and creation of environmental refugees

Urbanisation

Land use – environmental impacts caused by deforestation, land clearing and use of fertiliser

Land rights, justice and equality

Hunger is impacted by climate change

Soil erosion

Rising temperatures

Declining water

Increasing drought

Current status

Current food systems are inefficient and unsustainable accounting for 60% of biodiversity loss and 24% of global greenhouse gas emmisions

Agriculture increasingly competes for land and water supply with forestry, mining and urbanisation

Most global water withdrawal is for agriculture

Better policies and investment in agriculture are needed

Geoscience and hunger

Managing national to regional mineral and soil resources with input of satellite technology, geology and soil maps

Understanding chemical products of rock weathering to develop local sources of natural (mineral) fetilisers

Developing soil data science and digital sensor technologies to monitor soil quality

Improve understanding of groundwater supply, catchment studies and moisture retention in soils

Input to land use and distribution of resources — understanding distribution and quality to optimise the use of natural resources and mitigate potential conflict

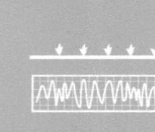 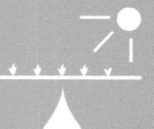 

2.1 Introduction

Over the past century, the global population has quadrupled from about 1.8 billion people in 1915 to about 7.3 billion people, with a projection of reaching 9.7 billion by 2050 (Elferink and Schierhorn 2016). This growth, along with rising incomes in the Global South (which drives dietary changes resulting in greater consumption of protein and meat), means food demand is expected to increase by between 59 and 98% by 2050 (Valin et al. 2013). At the same time as food demand is increasing, we are already failing to meet the needs of many of the world's poorest communities. In 2017, the number of undernourished people around the world reached 821 million, including 151 million children under the age of 5 with stunted growth (GHI 2018). Nearly 45% of deaths of children under the age of 5 were due to starvation (GHI, 2018).

The absolute number of undernourished people increased from 2015 to 2017, by approximately 40 million, with this increase attributed to conflict, especially in regions experiencing climate change (FAO et al. 2017, 2018). Collectively these figures highlight the need for urgent action to *'end hunger, achieve food security and improve nutrition and promote sustainable agriculture'*—as articulated in **SDG 2** (United Nations, 2015). **SDG 2** has five targets (2.1 to 2.5) and three means of implementation (2.A to 2.C) as shown in Table 2.1. These span many dimensions of food security, tackling hunger, and improving agricultural productivity (Fig. 2.1).

Hunger is a multidimensional and complex problem (von Grebmer et al. 2015), and is defined as the distress associated with lack of food and understood as '*a state of deprivation according to which an individual cannot satisfy his/her basic food needs (quantity and quality),*

Table 2.1 SDG 2 targets and means of implementation

Target	Description of target *(2.1 to 2.5)* or means of implementation (2.A to 2.C)
2.1	By 2030, end hunger and ensure access by all people, in particular the poor and people in vulnerable situations, including infants, to safe, nutritious and sufficient food all year round
2.2	By 2030, end all forms of malnutrition, including achieving, by 2025, the internationally agreed targets on stunting and wasting in children under 5 years of age, and address the nutritional needs of adolescent girls, pregnant and lactating women and older persons
2.3	By 2030, double the agricultural productivity and incomes of small-scale food producers, in particular women, indigenous peoples, family farmers, pastoralists and fishers, including through secure and equal access to land, other productive resources and inputs, knowledge, financial services, markets and opportunities for value addition and non-farm employment
2.4	By 2030, ensure sustainable food production systems and implement resilient agricultural practices that increase productivity and production, that help maintain ecosystems, that strengthen capacity for adaptation to climate change, extreme weather, drought, flooding and other disasters and that progressively improve land and soil quality
2.5	By 2020, maintain the genetic diversity of seeds, cultivated plants and farmed and domesticated animals and their related wild species, including through soundly managed and diversified seed and plant banks at the national, regional and international levels, and promote access to and fair and equitable sharing of benefits arising from the utilization of genetic resources and associated traditional knowledge, as internationally agreed
2.A	Increase investment, including through enhanced international cooperation, in rural infrastructure, agricultural research and extension services, technology development and plant and livestock gene banks in order to enhance agricultural productive capacity in developing countries, in particular least developed countries
2.B	Correct and prevent trade restrictions and distortions in world agricultural markets, including through the parallel elimination of all forms of agricultural export subsidies and all export measures with equivalent effect, in accordance with the mandate of the Doha Development Round
2.C	Adopt measures to ensure the proper functioning of food commodity markets and their derivatives and facilitate timely access to market information, including on food reserves, in order to help limit extreme food price volatility

Fig. 2.1 Agriculture **in Hainan Province, China.** With almost 300 million farmers, China is one of the most significant agricultural producers. Its population of approximately 1.4 billion people also makes it the largest consumer of agricultural produce. *Credit* Anna Frodesiak (used under the Creative Commons CC0 1.0 Universal Public Domain Dedication, https://creativecommons.org/publicdomain/zero/1.0/)

required for a healthy and active life' (IRIS and AAH 2017, p. 5). For the Food and Agriculture Organization of the United Nations (FAO), hunger is synonymous with undernourishment which can be defined as the deprivation of food and the consumption of less than 1,800[1] kilocalories per day, the minimum that most people require to live a healthy and productive life (FAO et al. 2014). The ambitions of **SDG 2**, however, extend far beyond ensuring enough calories to also include the complex interactions between food, nutrition, access to food, and resilience of food-producing systems. Hunger, therefore,

includes the '*supply, access, consumption, and intake of food at levels that are insufficient to fulfill human requirements*' (FAO 2018a). Related to hunger are the terms '*undernutrition*' and '*malnutrition*' (Box 2.1) both of which extend beyond calorie consumption and can result from both transitional and chronic situations. For example, acute food shortage leading to famine may be transitional, while long-term systemic food shortage causes chronic undernourishment.

Box 2.1. The Concepts of Hunger (adapted from von Grebmer et al. 2015, and FAO et al. 2017)

Hunger is usually *understood* to refer to the distress associated with lack of food. The FAO defines food deprivation, or

---

[1]This value can range from 1,650 to more than 1,900 kilocalories per person per day for countries in the Global South. Each country's average minimum energy requirement for low physical activity is used to estimate undernourishment (FAO et al. 2014).

undernourishment, as the consumption of fewer than about 1,800 kilocalories a day —the minimum that most people require to live a healthy and productive life.

Undernutrition goes beyond calories and signifies deficiencies in any or all of the following: energy, protein, or essential vitamins and minerals. Undernutrition is the result of inadequate intake of food—in terms of either quantity or quality—poor utilisation of nutrients due to infections or other illnesses, or a combination of these factors. These in turn are caused by a range of factors including household food insecurity; inadequate maternal health or childcare practices; or inadequate access to health services, safe water, and sanitation.

Malnutrition refers more broadly to both undernutrition (problems of deficiencies) and over nutrition (problems of unbalanced diets, such as consuming too many calories in relation to requirements with or without low intake of micronutrient-rich foods).

Food insecurity refers to a lack of 'secure access to sufficient amounts of safe and nutritious food for normal growth and development and an active and healthy life' (FAO et al. 2017).

In this chapter, we reflect on all four aspects above, and use the term *'zero hunger'* to mean access to sufficient calories, adequate intake of food in terms of quality and quantity, and access to a balanced diet that meets the specific requirements of the individual to live an active and healthy life.

SDG 2 (Zero Hunger) recognises the role of agriculture in alleviating hunger especially for those rural households who largely depend on farming for their food provisions. Agriculture is the main primary activity that produces food but when demand outstrips supply, people suffer from starvation and hunger. However, the root causes of hunger are more complex, as we explore in this chapter. Hunger is linked to poverty (**SDG 1**), gender equality (**SDG 5**), inequality (**SDG 10**), responsible consumption and production (**SDG 12**), land degradation (**SDG 15**), and climate change (**SDG 13**). Agricultural management may also be affected by weak marketing policies, priorities for development investment, implementation of sustainable technologies, and governance, together with a lack of political will to develop and implement inclusive policies.

Some key geoscientific inputs to **SDG 2** include the following:

- *Agrogeology* (or agricultural geology). The use of rock and mineral resources can support agriculture through improving soil fertility, water retention, and reducing soil erosion (Van Straaten 2002). Understanding the underlying geology of a region can guide decision-making on what crops may flourish in a region, and what interventions may be needed to support them. Agrogeology can contribute to ending hunger by increasing access to local fertilisers (e.g., from phosphorite), liming materials, and geological resources that improve water retention and reduce soil erosion. Agrogeology can also generate employment in the agro-mineral mining industry, supporting **SDG 8**.
- *Water Resources Management* (including hydrogeology). Identifying, characterising the physical and chemical properties of, and managing groundwater resources in a sustainable manner (see **SDG 6**) can help to support agricultural practice, improve the health of the poor, and support increased productivity.
- *Geochemistry*. Understanding the accumulation and distribution of major and trace elements in agricultural soils can inform decision–making around interventions to protect and improve human health and food safety (Sun et al. 2013).

We explore these themes and examples through this chapter, illustrating the contribution of geoscientists in informing research, practice, and policy to deliver the ambitions of **SDG 2**.

In this chapter, we first examine the key context to **SDG 2**, including the spatial and temporal extent of hunger (Sect. 2.2), and social and environmental factors contributing to hunger (Sect. 2.3). We proceed to examine the role of geoscientists in tackling hunger around the world (Sect. 2.3), focusing on characterising geological resources, groundwater management, and geochemistry to improve health through agriculture. Collectively these contribute to **SDG Targets 2.1** to **2.4.** We finish by synthesising key learning and recommendations (Sect. 2.4).

2.2 The Extent and Distribution of Hunger

It is challenging to measure and supply reliable estimates of hunger to inform policy and progress with development agendas, as it is multi-dimensional, can change rapidly over time, and may vary significantly at very local scales. The FAO annual series of reports called the *State of Food Insecurity and Nutrition in the World*, tracks hunger in the world using the prevalence of undernourishment as a primary indicator, and (as of 2019), also tracking the prevalence of moderate or severe food insecurity. The proportion of undernourished people in developing regions fell from 23.3% in 1990–1992 to 12.9% in 2014–2016 (United Nations 2015b). While there has been a general decline in the share of the population that is undernourished in most regions since 2000, there are some indications that this is changing and an increased share of the population was undernourished in sub-Saharan Africa (for example) in 2014–2016, compared with the previous years (Fig. 2.2).

Another measure used to understand the severity of hunger (in its broadest definition) in a population is the *Food Insecurity Experience Scale (FIES)*, conducted by the FAO of the United Nations. This indicator is assessed by reviewing the answers to eight questions administered at either an individual or household level, and includes the dimension of access to food, making it an improved measure of food insecurity. Collecting this data has, however, proved to be

difficult given the resources required. Figure 2.3 shows the prevalence of severe food insecurity by region, using the FIES global reference scale. While data is only shown for 3 years, Africa in general, but particularly sub-Saharan Africa is more affected by food insecurity, with a greater population affected by food insecurity than undernourishment. Where individuals are just above national and global extreme poverty lines (see **SDG 1**), food insecurity may push people back below the line as they will be particularly susceptible to the impacts of social, economic, and environmental shocks.

The International Food and Policy Research Institute (IFPRI) and partners compute an annual index termed the *Global Hunger Index* (GHI)[2] to monitor the level of hunger in the world, and how it is changing. The GHI reflects undernourishment, child wasting, child stunting, and child mortality (von Grebmer et al. 2015), as defined in Box 2.2. This multidimensional index reflects both the nutritional situation of the whole population, and a particularly vulnerable subset (children). When children lack calories, protein of micronutrients, it can lead to illness, poor development or death (von Grebmer et al. 2015).

> **Box 2.2. Components of the Global Hunger Index, GHI (adapted from von Grebmer et al. 2015)**
>
> **Undernourishment**: the proportion of undernourished people as a percentage of the population (reflecting the share of the population with insufficient caloric intake).
>
> **Child Mortality**: the mortality rate of children under the age of five (partially reflecting the fatal synergy of inadequate nutrition and unhealthy environments).
>
> **Child Undernutrition**: Nutrition targets are measured by stunting and wasting levels in children below the age of five. This includes:

---

[2]https://www.globalhungerindex.org/.

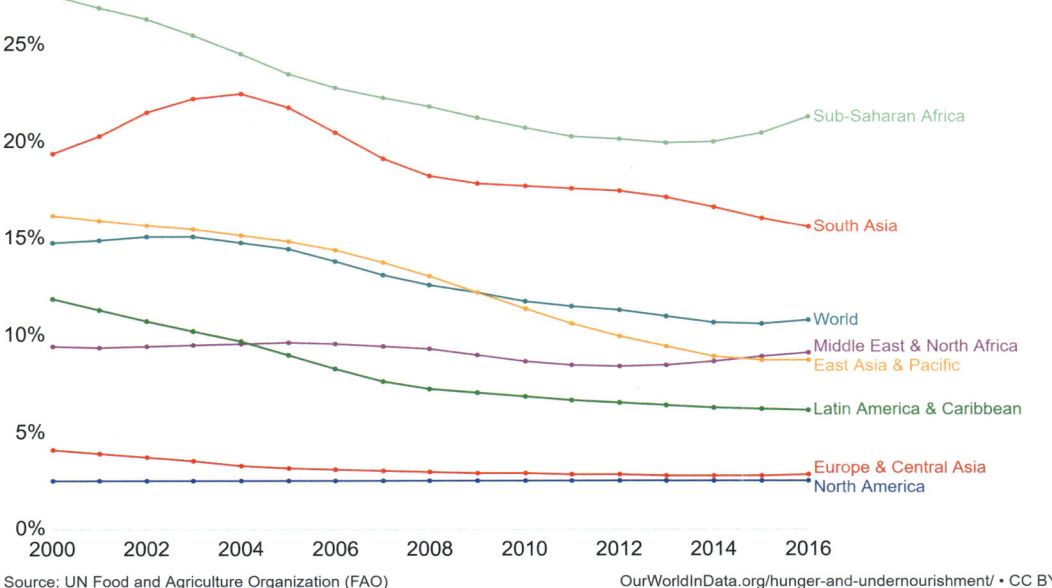

Fig. 2.2 Share of the population that is undernourished, by region. Image from Roser and Ritchie (2019), created using data from the FAO. Used under a CC-BY License (https://creativecommons.org/licenses/by/4.0/)

- **Child Wasting**: the proportion of children under the age of five who suffer from wasting (that is, low weight for their height, reflecting acute undernutrition). Wasting is a result of acute deprivation of nutritious food.
- **Child Stunting**: the proportion of children under the age of five who suffer from stunting (that is, low height for their age, reflecting chronic undernutrition). Stunting indicates long-term nutritional deprivation and may affect mental development, school performance and intellectual capacity.

GHI scores, computed from the above components are then mapped to a severity scale, showing low, moderate, serious, alarming or extremely alarming levels of hunger.

Read more: www.globalhungerindex.org/about.html.

The Global Hunger Index 2019 (von Grebmer et al. 2019) shows declining GHI values in all regions between 2000 and 2019, although there was a slight increase in the GHI of the Near East and North Africa region between 2010 and 2019. Global hunger has a GHI score of 20.0 (moderate to serious), with this being a reduction from 29.0 in 2000 (von Grebmer et al. 2019). This progress reflects improvements in each of the four GHI components (undernourishment, child stunting, child wasting, and child mortality, Box 2.2), although there has been more progress in tackling stunting than wasting in children. The overall trend is, therefore, generally positive, but with more work to do to reduce levels of hunger. Both South Asia and sub-Saharan Africa have GHI scores that indicate serious levels of hunger (von Grebmer et al. 2019). In 2019, the Central African Republic was the only country to have an 'extremely alarming' hunger level, with Yemen, Chad, Madagascar, and Zambia all having 'alarming' levels of hunger. A further 43 countries had

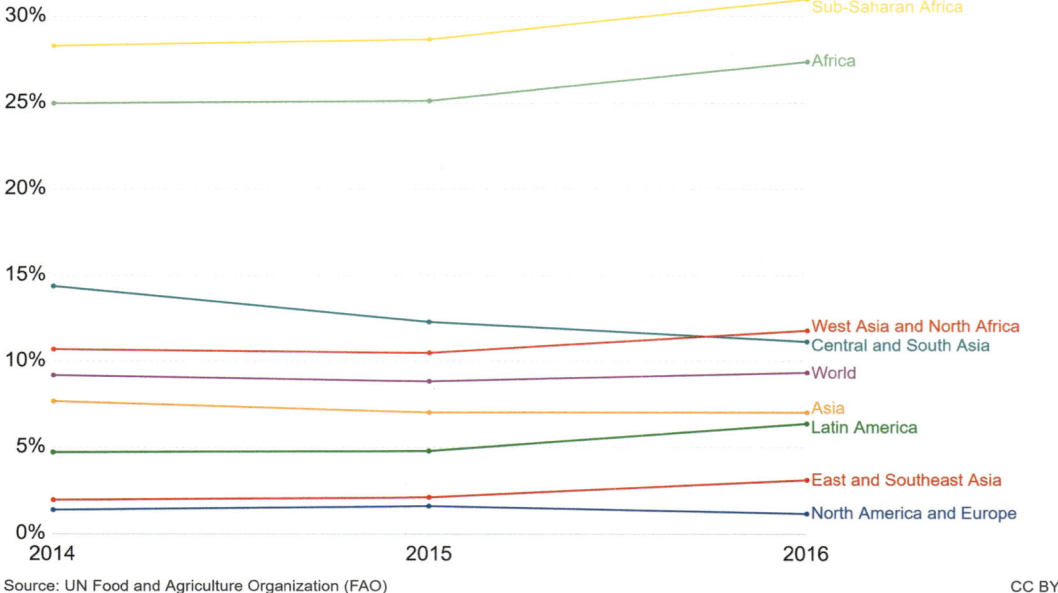

Fig. 2.3 Prevalence of severe food insecurity by region. Image from Roser and Ritchie (2019), created using data from the FAO. Used under a CC-BY License (https://creativecommons.org/licenses/by/4.0/)

'serious' levels of hunger, many of them corresponding with the world's least developed countries. Figure 2.4 shows a map of the 2018 GHI scores for many countries around the world.

Through this section, we see that undernourishment, food insecurity, and multidimensional assessments of hunger all vary spatially, with significant challenges in the Global South. Across the world, we have seen positive steps towards eliminating hunger in the past 20 years, but there are still serious, alarming and extremely alarming levels of hunger in many places. Understanding the complex factors causing this hunger is the first step to determining what actions are needed to tackle hunger and achieve **SDG 2**.

2.3 Hunger Dynamics, Causes, and Catalysts

The root causes of hunger are complex, with links to both human and environmental factors. While our interest is primarily in the links between the natural environment and the challenges and ambitions of **SDG 2**, it is impossible to separate these from other socio-economic processes that drive land-use decisions, land degradation, environmental change, and fluctuations in food demand (the amount of food of the right quality that consumers want at any given time) and production. In this section, we briefly explore two themes that inform the *challenges* around **SDG 2**: (i) social factors contributing to increases in demand for food, and reductions in agricultural productivity, and (ii) environmental impacts of increasing demand for food, the effects of climate change, and water insecurity. In Sect. 2.4, we move from challenges to solutions, particularly examining how geoscientists from across a range of sectors can support these, working in partnership with other disciplines.

2.3.1 Social Factors

An increasing global population is associated with a corresponding increase in demand for

Global Hunger Index, 2018

The Global Hunger Index (GHI) used to track hunger globally and nationally. The index score comprises of four key hunger indicators: prevalence of undernourishment in the total population; childhood wasting; childhood stunting; and child mortality. This calculation results in GHI scores on a 100-point scale where 0 is the best score (no hunger) and 100 the worst. A score >=50 is defined as 'extremely alarming'; 35-50 as 'alarming'; 20-35 as 'serious'; 10-20 as 'moderate' and <10 as 'low'.

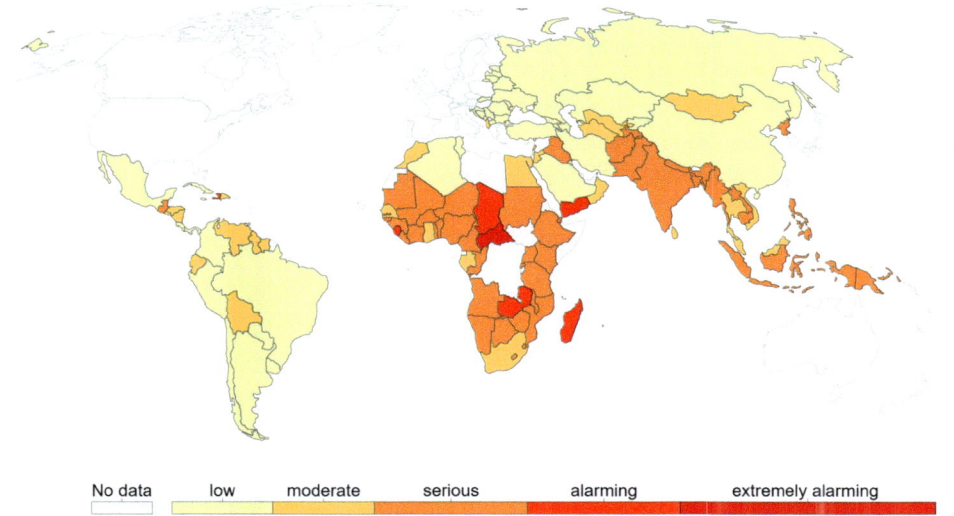

| No data | low | moderate | serious | alarming | extremely alarming |

Source: International Food Policy Research Institute (2018)   CC BY

Fig. 2.4 Global Hunger Index as of 2018. Data is not shown for many countries where the prevalence of hunger is considered to be low (e.g., much of North America and Western Europe). *Credit* Roser and Ritchie (2019), with data from von Grebmer et al. (2018). Figure reproduced under a CC-BY license (https://creativecommons.org/licenses/by/4.0/)

food. The world annual population growth rate has been declining for nearly five decades, but in regions such as Africa and Asia the population growth rate is projected to increase well beyond 2050, and even into the next century (FAO 2017). With technological advancement and progress on other SDGs (e.g., good health and well-being, **SDG 3**), life expectancy at birth has also been increasing. Some countries are currently projected to grow very rapidly, with annual growth rates of more than 2.5% to 2050 projected for Angola, Burundi, Chad, the Democratic Republic of the Congo, Gambia, Malawi, Mali, Senegal, Somalia, the United Republic of Tanzania, Uganda, and Zambia. The share of the population over the age of 65 is also increasing, including in some rural areas of low-income countries. This has an impact on the availability of labour to support agricultural production.

The rural poor, vulnerable, and marginalised groups, and those below or just above the poverty line are particularly susceptible to hunger (see **SDG 1**). They are highly dependent on seasonal rainfall to support agriculture, and often live in economically depressed and ecologically vulnerable areas. Communities living in poverty may have limited access to agricultural information (e.g., meteorological forecasts), services, technologies, and markets. Those in poverty are disproportionately affected by disease and the effects of economic, social, and environmental shocks—reducing productivity through time away from agricultural tasks. Gender inequality (**SDG 5**) can also have significant implications for hunger and poverty. Globally, about 60% of people who go hungry are female (UN Women 2012), amounting to almost half a billion women and girls not having access to the food required to live healthy lives.

Despite agricultural food production being mainly a rural activity, more people now live in cities (see **SDG 11**), and this has important implications for food supply and future in terms of a transition in dietary patterns with significant impact on food systems (FAO 2017). Indeed, the **SDG 2** targets have been criticised as being too limited in scope, taking only an agricultural and rural-centred approach and overlooking growing urban populations and the role of non-agricultural sectors (Burchi and Holzapfel 2015).

Increasing life expectancy and population sizes, the persistence of extreme poverty and inequalities, and urbanisation will have a significant effect on world hunger due to an increase in demand for food, with important repercussions on the agricultural labour force and the socio-economic fabric of rural communities (FAO 2017). These effects are most likely to have the greatest impact in the Global South.

2.3.2 Environmental Factors

There is an ecological and socio-economic trade-off between clearing land to increase the quantity and quality of investment in agriculture and protecting the environment and mitigating climate change in many parts of the world, but particularly in many of the world's least developed countries. Meeting increased food demands entails both land clearing and the intensive use of existing agricultural land to increase crop production. Both activities can have a major impact on the natural environment.

Forests are important for the provision of ecosystem services (e.g., carbon sequestration and biodiversity conservation, see **SDG 15**), but are affected by commercial logging and large-scale land-use change, including the development of palm oil plantations. Deforestation and land clearing drive habitat fragmentation and threaten biodiversity (Dirzo and Raven 2003; Varsha et al. 2016). Palm oil plantations support much fewer species than natural forests, and often also fewer than other tree crops, contributing to habitat fragmentation and pollution (Fitzherbert et al. 2008). Almost all oil palm grows in areas that were once tropical moist forests, and the conversion of these areas, and future expansion, threatens biodiversity and increases greenhouse gas emissions (Varsha et al. 2016). As forest habitat is cleared, endangered species are pushed closer to extinction, and indigenous people who are mostly smallholder farmers who have inhabited and protected the forest for generations are often driven from their land. Many communities engaged in rural agriculture traditionally depend on neighbouring forests for commodities such as fruits and wood to supplement their diet and income.

Land degradation is occurring in almost all world regions, affecting about 20% of the global land area and impacting upon around three billion people (described further in **SDG 15**). This leads to nutrient depletion, especially in Africa, the intensive use of fertilisers, and soil contamination. Continued degradation of land may lead to a rise in rural poverty triggering human conflict, rural instability, and large-scale population migrations crossing borders and regions (von Braun et al. 2017). The annual global cost of land degradation is about US$ 300 billion, with a quarter of this cost relating to degradation in sub-Saharan Africa (Nkonya et al. 2016). This includes both costs to immediate land users due to a reduction in the functioning of the land, and a significant social cost due to loss of local and global ecosystems services (von Braun et al. 2017). Pressures on land use create wider socio-economic effects, including reductions in the size of household farms and reduced household food production, potentially reducing food security, impacting on health, and exacerbating poverty (GRAIN 2014).

While recognising problems of deforestation and land degradation, there is also, however, a growing concern that the amount of crops harvested per unit of land cultivated is not enough to meet the forecasted demand for food. The effects of climate change-driven water scarcity, rising global temperatures, and extreme weather will impact on volumes and distribution of crop yields at both local and national scales, exposing many vulnerable people in rural communities who rely on agriculture. Weather-related events are already affecting food availability in many countries and contributing to a rise in food insecurity (drought and flood) diminishing livestock and productivity, resulting in migration and an expanding the pool of environmental refugees with broadly held grievances (FAO 2019). Many communities across sub-Saharan Africa are affected by an increase in the frequency and intensity of extreme weather conditions and environmental change, as a result of anthropogenic-driven climate change. This includes rising temperatures, declining groundwater tables, changing water flows, droughts, floods, and strong winds which can slow progress toward increasing the productivity of crop and livestock systems and undermine long-term food security (De Pinto and Ulimwengu 2017; von Braun et al. 2017). These effects will particularly impact those regions that are most food

insecure (i.e., many communities in sub-Saharan Africa), together with large food producers such as China and India.

Mining, forestry, and urbanisation all put pressures on water security, which affects agricultural capacity and productivity. Agriculture is estimated to account for 70% of water consumption and 30% of energy consumption and greenhouse gas emissions globally (Janus and Holzapfel 2016). In low-income countries, however, the average share of total freshwater withdrawals used for agriculture is as high as 90%, contrasted with 41% in high-income countries (Ritchie and Roser 2019), illustrated in Fig. 2.5. The agriculture sector is the largest consumer of groundwater resources, with food security therefore inextricably linked to the effective management of groundwater and other freshwater resources around the world (see **SDGs**

Agricultural water as a share of total water withdrawals, 2010
Agricultural water withdrawals as a percentage of total water withdrawals (which is the sum of water used for agriculture, industry and domestic purposes). Agricultural water is defined as the annual quantity of self-supplied water withdrawn for irrigation, livestock and aquaculture purposes.

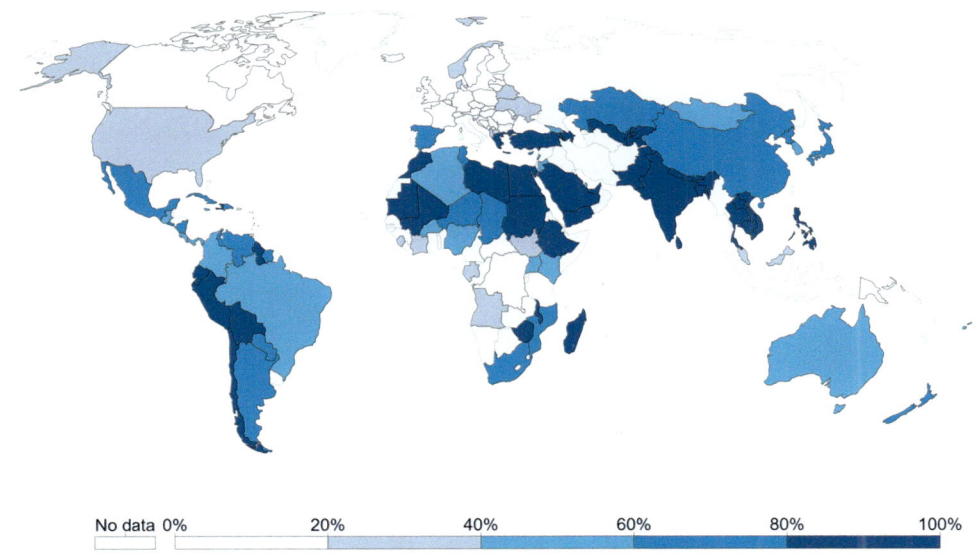

Source: World Bank   OurWorldInData.org/water-access-resources-sanitation/ • CC BY

Fig. 2.5 Agriculture Water as a Share of Total Withdrawals (2010). Share of total water withdrawals (agriculture, industry, domestic), used for agriculture, as of 2010, including water for irrigation, livestock, and aquaculture purposes. *Credit* Ritchie and Roser (2019), with data from the UN Food and Agricultural Organization (FAO) AQUASTAT Database. Figure reproduced under a CC-BY license (https://creativecommons.org/licenses/by/4.0/)

6 and **15**). Pressures on water resources from agriculture will increase with the expansion of irrigation and changes to diets and food consumption. The water requirements for the production of different food types differ enormously (Mekonnen and Hoekstra 2012). One tonne of beef requires more than 47 times the amount of water to produce compared to one tonne of most vegetables. One kilocalorie from beef requires almost 20 times the amount of water to produce compared to one kilocalorie from cereals.

Sustainable land and water use are therefore critical to delivering **SDG 2**, and many other parts of the UN sustainable development agenda, including improved health (**SDG 3**), access to safe drinking water (**SDG 6**), sustainable urbanisation (**SDG 11**), and restoration of ecosystems (**SDG 15**). The UN Environment Programme (UNEP) note that a major overhaul of the global food system is urgently needed if the world is to use natural resources more efficiently and stem environmental damage (UNEP 2016). Food systems are considered to be inefficient and unsustainable, responsible for 60% of global terrestrial biodiversity loss, 24% of global greenhouse gas emissions, overfishing of 29% of commercial fish populations, and overexploitation of 20% of the world's aquifers (UNEP 2016). The concern is also voiced that more than two billion suffer from micronutrient deficiencies —mainly vitamin A, iodine, iron, and zinc—and more than two billion people are overweight or obese (UNEP 2016). The report notes that '*land degradation, the depletion of aquifers and fish stocks and contamination of the environment will lower future food production capacity, thus undermine the food systems upon which our food security depends, as well as cause further degradation of other ecosystem functions*' (UNEP 2016, p. 17). Actions to deliver **SDG 2** must therefore work in coherence with the actions proposed to address **SDGs 6, 14,** and **15**. Geoscientists contributing to these other goals— in the myriad of ways outlined through this book —should have in mind the needs of **SDG 2**, and the ways in which their actions to deliver one or more goals can support or hinder food security and access.

2.4 Delivering SDG 2—The Role of Geoscience in Reducing Hunger

In the previous section we have outlined a range of challenges contributing to hunger and hindering efforts to deliver SDG 2, and demonstrated the need for diverse disciplines and coherence across the SDGs. Environmental and climate change are impacting on agricultural yields, and efforts to drive up agricultural production to meet rising demand are contributing to environmental change. Global and local actions are needed to reduce the environmental impacts of agriculture, manage the human influence on the environment, and achieve both food security and environmental quality (Chen 1990). Geoscience research, knowledge exchange (particularly building partnerships with those developing water, land, health, and agricultural policies), and practice can support efforts to end hunger, achieve food security, improve nutrition and promote sustainable agriculture.

2.4.1 Geological Characterisation to Improve Agriculture

Good management and sustainable use of the Earth's natural resources is essential to eliminating global hunger (addressing SDG **Targets 2.3** and **2.4**), including geological resources such as minerals and rock materials (as well as water resources, discussed in Sect. 2.3.2). Land degradation is a common consequence of poor management of natural resources, affecting food availability. To combat the challenges described in Sect. 2.2, UNEP (2016) recommend a switch to a 'resource-smart' food system, changing the way food is grown, harvested, processed, traded, transported, stored, sold, and consumed. These stages should all have a low environmental impact, use renewable resources sustainably, and use all resources (e.g., soils, fertilisers, water) efficiently. UNEP (2016) recommends approaches to generate higher yields without increasing environmental impacts (e.g., reducing forest loss, making agriculture supply chains carbon neutral), improvements in nutrient efficiency, and reduction of overconsumption and change of unhealthy dietary patterns.

The emphasis on efficiencies in the UNEP (2016) recommendations, provides many opportunities for geoscientists to inform planning and policies. Geological materials and processes influence landscapes and both soil structure and chemistry, and therefore affect the ability to grow crops in any given location. Geological mapping and mineral resource assessment, combined with high-resolution satellite remote sensing data, would therefore provide useful information on national scale soil resources to inform planning. At the community (village or small town) scale, where greatest change is needed, improved understanding of soil chemistry (e.g., mineral depletion and carbon content) can lead to better targeted application of fertilisers.

Geology will also determine the availability of local mineral resources for fertilisers, rock materials for liming and crop cover, and water resources for irrigation. Appleton (1994) notes that in some contexts the direct application of finely ground phosphate rocks (Fig. 2.6) and potassium feldspar rich rocks may provide

nutrients for crops. *Rocks for Crops*: *Agrominerals of Sub-Saharan Africa* by Van Straaten (2002) describes the potential for naturally occurring nutrient-providing rocks and minerals to support agriculture, alongside 'soil amendments' (e.g., sources of lime, and pumice to reduce water evaporation and soil erosion). Van Straaten (2002) summarises the potential role that geological materials can play in sustaining and enhancing soil productivity and biomass production, and provides an inventory of known agricultural mineral resources for 48 countries in sub-Saharan Africa. Understanding local availability of such resources, and utilising these, together with changing tilling practice can all help to improve crop yields, extend the growing season, and mitigate the impacts of drought.

Agricultural geology (or agrogeology) therefore provides information about the state of soil resources and potential means to improve them to increase productivity, particularly of small-scale food producers (**Target 2.3**). There is an urgent need for high-resolution maps of land resources, to help address the challenges of climate change and meeting the ambitions of **SDG 2** in an environmentally sensitive manner. Such

Fig. 2.6 Phosphorite Mine (Oron, Negev, Israel). Phosphorite, or phosphate rock, is a sedimentary rock with large amounts of phosphate minerals. Mining of phosphorite is an important source of fertiliser (as well as animal feed supplements)

information can support reliable crop forecasts in order to project food availability. Despite the importance of crop production, and the associated challenges, some countries in Africa (e.g., Kenya, Senegal, Zimbabwe) lack reliable and timely agricultural production forecasting systems to support decision-making at the national to household levels (FAO 2018b).

Box 2.3 Examples of Soil Mapping and Database Projects

There are global calls to have all soil resources mapped, and this data made freely available. Examples of key initiatives and ongoing projects from national to global scales, include the following:

- A consortium of institutions, coordinated by the FAO, are developing a *Harmonized World Soil Database*[3] with more than 15,000 different soil mapping units (FAO et al. 2009).
- FAO and UNESCO have developed a *Soil Map of the World*[4] at a scale of 1:5,000,000.
- The *Africa Soil Information Service* (ALFSIS)[5], funded by the Bill and Melinda Gates Foundation, works to ensure the application of world-class information technology and data science to Africa's soil and landscape resources.
- The *UK Soil Observatory*[6] is an online archive of UK soils data from nine research bodies, helping people to access soil data, knowledge, and expertise from across a wide range of institutions.
- The European Soil Data Centre have produced a free-to-access *Soil Atlas of Latin America and the Caribbean*[7]

published in English, Portuguese, and Spanish.

Initiatives such as these should be welcomed, supported, and disseminated by geoscientists. They could be complemented with additional knowledge of geological resources to ensure soils are protected, restored, and remain productive for future generations.

2.4.2 Efficient Management of Water Resources

Geoscience underpins our knowledge of both the availability of water resources and the sustainable management of these resources to prevent serious depletion, degradation (e.g., due to salinity), and prolonged water stress. Sustainably managing water resources is key to SDG **Target 2.4**, implementing resilient agricultural practices and strengthening capacity for adaptation to climate change and extreme hydrometeorological events. For example, in some regions, groundwater resources may have a reasonable degree of resilience to climate variability. Understanding location-specific precipitation–recharge relationships can therefore guide decision-making about how to increase climate resilience of agriculture, through improved water management (Cuthbert et al. 2019).

To assess the extent of groundwater resources and how these change over time, there is a need for data. This includes geological and hydrogeological characterisations of the region (i.e., understanding the subsurface, the potential availability of groundwater, aquifer yields, and groundwater chemistry). There is also a critical need for long-term monitoring networks to assess changes to groundwater resources over time. Many places around the world lack hydrological monitoring networks or data is not made available to the public (McNally et al. 2016), making it difficult to monitor and interpret

---

[3]http://www.fao.org/soils-portal/soil-survey/soil-maps-and-databases/harmonized-world-soil-database-v12/en/.

[4]http://www.fao.org/soils-portal/soil-survey/soil-maps-and-databases/faounesco-soil-map-of-the-world/en/.

[5]http://africasoils.net/.

[6]http://www.ukso.org/.

[7]https://esdac.jrc.ec.europa.eu/content/soil-atlas-latin-america.

hydrometeorological variables such as soil moisture (Myeni et al. 2019). Improved understanding of interactions between soil moisture and groundwater could, however, inform preparations for times of water scarcity and adaptation strategies for crop management to improve food security. Soil moisture deficit is the difference between the amount of water actually in the soil and the amount of water that the soil can hold (AMetSoc 2012), and normally results in reduced crop production. Where monitoring networks do not exist, satellite data can provide information on rainfall and vegetation changes (McNally et al. 2016). Improved monitoring and more research is needed, however, to understand long-term changes in seasonal rainfall and accompanying extreme events such as floods and dry spells (Chabala et al. 2013), and their impacts on agricultural production.

Finally, geoscience can ensure water is used efficiently, contributing to coherent and comprehensive water management plans. Section 2.2 illustrated the significance of water withdrawals for agriculture in low-income countries, noting that this forms a major share (compared to domestic and industrial use), and much greater than in high-income countries. Expanding industrialisation (**SDG 9**) and meeting growing domestic demands (**SDG 6**) will both exacerbate water stress, unless significant efficiencies are made in agriculture. This could include: (i) better matching water quality to water use, (ii) more efficient irrigation methods, and (iii) selection of crops and food products that require less water. These measures together with the use of geological resources (e.g., pumice, scoria) to reduce water evaporation from soils could all help.

2.4.3 Geochemistry, Agriculture, and Health

The growth and development of plants is affected by the geochemical characteristics of soil and water, with subsequent impacts on human and livestock health (Thornton 2002; Ma and Li 2019; Rawlins et al. 2012). Elements essential to plant, animal, and human health (micronutrients) are not distributed across all soils evenly. Element abundances being too low can result in nutrient deficiencies and element abundances being too high can result in toxicity, both associated with health problems (Fordyce 1999), as discussed in **SDG 3** Problems may also occur when pollution associated with industry or pesticides has contaminated soils, and this is taken up into plants and then ingested by humans. Regional and national geochemical atlases, developed through a systematic sampling of soils or stream sediments, can inform the optimisation of land-use (Thornton 2002), guiding whether regions are suitable for crop growth or whether they are contaminated by heavy metals, excessive pesticide residues, or organic chemicals (Ma and Li 2019). Geochemists can, therefore, contribute to the delivery of **SDG 2** through understanding the chemistry of soils and the uptake and bioavailability of nutrients (and contaminants) in food crops. This informs interventions to deliver **Target 2.2** (end all forms of malnutrition), as well as other related SDG targets linked to health and well-being (**SDG 3**).

In regions where soils have become depleted of essential micronutrients for plants (e.g., boron, chlorine, copper, iron, manganese, molybdenum, nickel, zinc), biofortification can be used or treatments can be added to soils (Alloway 2008). Understanding the soil geochemistry and structure can help to determine what micronutrient deficiencies may occur, and how to treat these (Fig. 2.7). For example, soils with a high calcium carbonate content, with a low pH, or that are heavily limed may give rise to zinc deficiencies in crops (Alloway 2008). This can be treated using a zinc sulphate or oxide, added to the soil to help improve plant health. How effective this is, and the time needed before a further treatment will again depend on the soil chemistry (e.g., soils being limed to reduce acidity may find treatment ineffective).

Fig. 2.7 Zinc Deficiency in Macadamia Shoots. The youngest leaves, those at the tips of the branches, show yellowing (production of produce insufficient chlorophyll), dwarfing, and malformation. *Credit* Alandmanson (licensed under the CC-BY-SA 4.0 International license, https://creativecommons.org/licenses/by-sa/4.0/)

2.5   Summary and Conclusions

The crucial role that geoscience will play in global food security cannot be overemphasised. This chapter sets out how an understanding of geological resources and processes can help improve soil and water management, contributing to both improved agricultural productivity and agricultural resilience to environmental change. In turn, these contributions will support the global targets of ending hunger and malnutrition.

Geoscience information must be integrated into land-use planning, to ensure conservation and prudent use of resources. Demands for land from diverse sectors and urbanisation will increase. For example, growing biofuels has been cited as contributing to hunger through a reduction in available land for food production (Wahlberg 2008). Geoscientists can help decision-makers at all scales to consider how the subsurface will impact upon the surface activities, helping to appropriately allocate land. Geological materials shape the quality of soils,

and this understanding can guide decisions about what additives are required to improve soils—ensuring this is as efficient as possible. Understanding of locally available geological materials —from groundwater, to nutrient-rich rocks, to soil amendments—can assist in soil management and agriculture. Nations must, therefore, invest in (i) systematic data collection and monitoring networks, (ii) mapping of resources at smaller resolutions, and (iii) research and training of experts, to provide technical advice on how geological resources can be managed to meet the food demands of future generations, while also protecting the environment, and adapting to or indeed mitigating climate change.

We have not covered in this chapter the contribution of geoscientists to developing reliable, resilient, and sustainable infrastructure, essential to getting agricultural produce to markets (see **SDG 9**). We also refer the reader to **SDGs 6** and **15** for a more detailed overview of how geoscientists contribute to water management and protecting ecosystems essential for agricultural productivity. Other innovations such as conservation agriculture and agroforestry are essential for sustainable use of land resources for food production, and we include additional reading associated with these themes at the end of this chapter.

Agricultural and health experts come from many disciplines themselves, and draw on skills ranging from chemistry to statistics, agronomy to meteorology to support sustainable agriculture. Geoscientists must actively build partnerships with these disciplines and listen to their priorities to understand how our science can help to deliver **SDG 2.** Ensuring the end of hunger, food security, and improved nutrition for all demands engagement from many disciplines, working together to increase global food production, with minimum negative impacts on the environment.

2.6   Key Learning Concepts

- Population growth and rising incomes in the Global South (driving dietary changes) are resulting in increased demand for food. We are not meeting the food and nutrition needs of many of the world's poorest communities. In 2017, the number of undernourished people around the world reached 821 million (including 151 million children under the age of 5 with stunted growth), and the absolute number of undernourished people increased by approximately 40 million from 2015 to 2017. **SDG 2** aims to meet this demand and tackle this injustice, ending malnutrition, and ensuring agriculture is efficient and sustainable.
- The Global Hunger Index monitors the level of hunger in the world, and how it is changing, reflecting undernourishment, child wasting, child stunting, and child mortality. Globally, levels of hunger have been falling since 2000, but there remain 48 countries with serious, alarming, or extremely alarming levels of hunger.
- Causes and catalysts of hunger are diverse, with both social and environmental factors and interactions between these. Increasing life expectancy and population sizes, the persistence of extreme poverty and inequalities, a growing middle class, environmental degradation, and poor management of natural resources can all result in production not meeting demands.
- Understanding the underlying geology of a region can inform crop selection, help to understand what additives are required to improve soil performance, and increase efficiency by maximising the use of locally available natural resources for improved soil structure, nutrition and water retention, guiding the extraction of these in a safe and responsible manner.
- Geoscience underpins our knowledge of the availability of water resources and the sustainable management of these resources to prevent serious depletion, degradation, and prolonged water stress. Sustainably managing water resources is key to implementing resilient agricultural practices and strengthening capacity for adaptation to climate change and extreme hydrometeorological events.
- Geochemistry can help inform the optimisation of land use and guide decision-making about where agriculture is or is not suitable. Understanding the bioavailability and uptake of nutrients and contaminants in food crops can inform interventions are appropriate and efficient, contributing to improved nutrition and supporting efforts to deliver **SDG 3**.
- These three contributions require systematic data collection, management, integration, and access to inform decision-making. Investment in (i) systematic data collection and monitoring networks, (ii) mapping of resources, and (iii) research and training, can all support the achievement of **SDG 2**.

2.7 Educational Ideas

In this section, we provide examples of educational activities that connect geoscience, the material discussed in this chapter, and scenarios that may arise when applying geoscience (e.g., in policy, government, private sector international organisations, NGOs). Consider using these as the basis for presentations, group discussions, essays, or to encourage further reading.

- Using the OneGeology Portal[8] explore the rock types in your region and describe their chemistry. Are any of these exploited commercially to support agricultural production (e.g., for fertiliser)? Using van Straaten (2002) as a reference guide, is there any potential for farmers to use geological resources in your region to support agriculture?
- A national Ministry of Water approach you and ask for advice on reducing the share of water withdrawals used for agriculture. Consider the examples mentioned in this chapter to improve water efficiency in agriculture (*e.g., better matching water quality to water use, improved irrigation methods, selection of crops and food products that require less water*) and investigate what each of these may involve and how geoscientists can contribute (you may prefer to have different groups exploring different methods). Present your results to the class.
- What geochemical mapping has been done in your country? Review the scale of mapping, and what exactly has been mapped (e.g., what elements, what regions if not the whole country). How easy is to access this information? Is it available to the public? Discuss with your peers locally appropriate ways that you could improve access to and understanding of geochemical information to inform agriculture.
- What are the components of a healthy diet? How easy is it to access this range of food products in your region? Think about the origins, transport, processing, and marketing of this range of food products. What are the contributions geoscientists can make to ensure access to this food, and to reduce the environmental impact of generating, transporting, processing, and marketing the food?

Further Resources

Combs GF (2013) Geological impacts on nutrition. In: Selinus O, Alloway BJ, Centeno JA, Fuge R, Lindh U, Smedley P (eds) Essentials of Medical Geology. Springer, Dordrecht, pp 179–194

DeVivo, B., Belkin, H., & Lima, A. (Eds.). (2017). Environmental geochemistry: site characterization, data analysis and case histories. Elsevier

Food and Agriculture Organisation of the United Nations (2017). Principles of Conservation Agriculture. http://www.fao.org/publications/card/en/c/981ab2a0-f3c6-4de3-a058-f0df6658e69f/

Food and Agriculture Organisation of the United Nations (2019) The State of the World FAO Flagship Reports. http://www.fao.org/publications/flagships/en/

Geological Society of America (2014) Integrating Geoscience with Sustainable Land-Use Management. GSA position statement. https://www.geosociety.org/documents/gsa/positions/pos13_LandUse.pdf

United Nations (2019) Food security and nutrition and sustainable agriculture. https://sustainabledevelopment.un.org/topics/foodagriculture

Williams, A., Morris, J., Audsley, E., Hess, T., Goglio, P., Burgess, P., Chatterton, J., Pearn, K., Mena, C. and Whitehead, P. (2018). Assessing the environmental impacts of healthier diets. Final report to Defra on project FO0427. https://dspace.lib.cranfield.ac.uk/handle/1826/13574

World Agroforestry Research Centre (2019). http://www.worldagroforestry.org/

References

Alloway BJ (2008) Zinc in soils and crop nutrition. International Zinc Association, Brussels, Belgium, p 139

AMetSoc (2019) Soil Moisture Deficit. Available at: http://glossary.ametsoc.org/wiki/Soil_moisture_deficit (accessed 24 October 2019)

Appleton JD, Mathers SJ, Notholt AJG (1994) Direct-application fertilizers and soil amendments–appropriate technology for developing countries. Industrial minerals in developing countries. AGID Report Series

[8]http://portal.onegeology.org/OnegeologyGlobal/.

Geosciences in International Development 18:223–256

Burchi, F. and Holzapfel, S. (2015) 'End hunger, achieve food security and improved nutrition, and promote sustainable agriculture', in Loewe, M. and Rippin, N. (eds) Translating an Ambitious Vision into Global Transformation The 2030 Agenda for Sustainable Development. Bonn: Deutsches Institut für Entwicklungspolitik gGmbH. Available at: www.die-gdi.de

Chabala LM, Kuntashula E, Kaluba P (2013) Characterization of Temporal Changes in Rainfall, Temperature, Flooding Hazard and Dry Spells over Zambia. Universal Journal of Agricultural Research 1(4):134–144

Chen RS (1990) Global agriculture, environment, and hunger: Past, present, and future links. Environ Impact Assess Rev 10(4):335–358

Cuthbert MO, Taylor RG, Favreau G, Todd MC, Shamsudduha M, Villholth KG, MacDonald AM, Scanlon BR, Kotchoni DV, Vouillamoz JM, Lawson FM (2019) Observed controls on resilience of groundwater to climate variability in sub-Saharan Africa. Nature 572(7768):230–234

De Pinto A, Ulimwengu JM (eds) (2017) 'A Thriving Agricultural Sector in a Changing Climate: Meeting Malabo Declaration Goals through Climate-Smart Agriculture', in ReSAKSS Annual Trends and Outlook Report 2016. International Food Policy Research Institute, Washington, DC

Dirzo, R., & Raven, P. H. (2003). Global state of biodiversity and loss. Annual review of Environment and Resources, 28

Elferink, M., & Schierhorn, F. (2016). Global demand for food is rising. Can we meet it?. Harvard Business Review, 7(04), 2016

FAO (2017). The future of food and agriculture – Trends and challenges. Rome, FAO. Available at: www.fao.org/publications/fofa/en/

FAO (2018a). Monitoring Hunger: Indicators at Global and subnational level. Statistics Division, Working Paper number ESS ESSG/013e. Food and Agricultural Organization of the United Nations

FAO (2018b). Guidelines on the use of Remote Sensing Products to Improve Agricultural Crop Production Forecast Statistics in Sub-Saharan African Countries. Rome. Available at: www.fao.org/publications.enabling environment for food security and nutrition. Rome, FAO

FAO, IFAD, UNICEF, WFP and WHO (2017). The State of Food Security and Nutrition in the World 2017. Building resilience for peace and food security. Rome, FAO. Available at: www.fao.org/policy-support/resources/resources-details/en/1107528/

FAO, IFAD, UNICEF, WFP and WHO (2018). The State of Food Security and Nutrition in the World 2018: Building climate resilience for food security and nutrition. Rome, FAO. Available at: www.wfp.org/publications/2018-state-food-security-and-nutrition-world-sofi-report

FAO, IFAD, and WFP (2014). The state of food insecurity in the world: strengthening the enabling environment for food security and nutrition. Rome, FAO. Available at: www.fao.org/publications/sofi/2014/en

FAO, IIASA, ISRI, ISS-CAS, and JRC (2009). Harmonized World Soil Database (version 1.1). FAO, Rome, Italy and IIASA, Luxemburg, Austria

Fitzherbert EB, Struebig MJ, Morel A, Danielsen F, Brühl CA, Donald PF, Phalan B (2008) How will oil palm expansion affect biodiversity? Trends Ecol Evol 23 (10):538–545

Food and Agriculture Organisation of the United Nations, FAO (2019) The State of the World FAO Flagship Reports. Available at: http://www.fao.org/publications/flagships/en/

Fordyce F (2000) Geochemistry and health, why geoscience information is essential. Geoscience and Development 6:6–8

GRAIN (2014). Hungry for Land. Available at: https://www.grain.org/article/entries/4929 (accessed 25 October 2019)

IRIS and AAH (2017) An Outlook on Hunger: A Scenario Analysis on the Drivers of Hunger Through 2030. Available at: http://www.iris-france.org/wp-content/uploads/2017/10/Hunger-an-outlook-to-2030.compressed.pdf (accessed 1 October 2019)

Janus, H. and Holzapfel, S. (2016) 'Sustainable Development Goals : Pick and choose - or integration at last ?', German Development Institute/ Deutsches Institut für Entwicklungspolitik (DIE) The Current Column, (February). Available at: www.die-gdi.de

Ma, L., & Li, Z. (2019, October). Review of application of geochemistry in agriculture. In SEG 2019 Workshop: Geophysics for Smart City Development, Beijing, China, 29–31 July 2019 (pp. 23-23). Society of Exploration Geophysicists

McNally A, Shukla S, Arsenault KR, Wang S, Peters-Lidard CD, Verdin JP (2016) Evaluating ESA CCI soil moisture in East Africa. Int J Appl Earth Obs Geoinf 48:96–109

Mekonnen MM, Hoekstra AY (2012) A global assessment of the water footprint of farm animal products. Ecosystems 15(3):401–415

Myeni L, Moeletsi ME, Clulow AD (2019) Present status of soil moisture estimation over the African continent. Journal of Hydrology: Regional Studies 21:14–24

Nkonya E, Johnson T, Kwon HY, Kato E (2016) Economics of Land Degradation in Sub-Saharan Africa. In: Nkonya E, Mirzabaev A, von Braun J (eds) Economics of Land Degradation and Improvement – A Global Assessment for Sustainable Development. Springer, Cham

Rawlins BG, McGrath SP, Scheib AJ, Breward N, Cave M, Lister TR, Ingham M, Gowing C, Carter S (2012) The Advanced Soil Geochemical Atlas of England and Wales. British Geological Survey, Keyworth, Nottingham

Ritchie, H. and Roser, M. (2019) Water Use and Stress. https://ourworldindata.org/water-use-stress (accessed 25 October 2019)

Roser, M. and Ritchie, H. (2019) Hunger and Undernourishment. https://ourworldindata.org/hunger-and-undernourishmen (accessed 24 October 2019)

Sun G, Chen Y, Bi X, Yang W, Chen X, Zhang B, Cui Y (2013) Geochemical assessment of agricultural soil: A case study in Songnen-Plain (Northeastern China). CATENA 111:56–63

Thornton I (2002) Geochemistry and the mineral nutrition of agricultural livestock and wildlife. Appl Geochem 17(8):1017–1028

UN Women (2012) Facts and Figures. https://www.unwomen.org/en/news/in-focus/commission-on-the-status-of-women-2012/facts-and-figures (accessed 25 October 2019)

UNEP (2016) Food Systems and Natural Resources. A Report of the Working Group on Food Systems of the International Resource Panel. Westhoek, H, Ingram J., Van Berkum, S., Özay, L., and Hajer M

United Nations (2015): Transforming our world: the 2030 Agenda for Sustainable Development. A/RES/70/1. http://www.un.org/ga/search/view_doc.asp?symbol=A/RES/70/1&Lang=E (accessed 22 October 2019)

United Nations (2015b) Millennium Development Goals Report. Available at: https://www.un.org/millenniumgoals/2015_MDG_Report/pdf/MDG%202015%20rev%20(July%201).pdf (accessed 25 October 2019)

Valin H, Sands RD, Van der Mensbrugghe D, Nelson GC, Ahammad H, Blanc E, Bodirsky B, Fujimori S, Hasegawa T, Havlik P, Heyhoe E (2014) The future of food demand: understanding differences in global economic models. Agricultural Economics, 45(1), 51–67.

van Straaten P (2002) Rocks for Crops: Agrominerals of sub-Saharan Africa. ICRAF, Nairobi, Kenya, p 338

Varsha, V., Stuart, L, P., Clinton, N, J., Sharon, J, S. (2016). 'The Impacts of Oil Palm on Recent Deforestation and Biodiversity Loss', Plos One. https://doi.org/10.1371/journal.pone.0159668 Zeng, J., Yang, J., Zha, Y., and Shi, L.: Capturing soil-water and groundwater interactions with an iterative feedback coupling scheme: new HYDRUS package for MODFLOW, Hydrol. Earth Syst. Sci., 23, 637–655, https://doi.org/10.5194/hess-23-637-2019, 2019

von Braun, J., Gulati, A. and Kharas, H. (2017) Key policy actions for sustainable land and water use to serve people, *Economics*, 11

von Grebmer, K., Bernstein, J., de Waal, A., Prasai, N., Yin, S., and Yohannes, Y. (2015) Global Hunger Index 2015: Armed Conflict and the Challenge of Hunger. Bonn, Washington, DC, and Dublin: Welthungerhilfe, International Food Policy Research Institute, and Concern Worldwide

von Grebmer K, Bernstein J, Hammond L, Patterson F, Sonntag A, Klaus L, Fahlbusch J, Towey O, Foley C, Gitter S, Ekstrom K, Fritschel H (2018) Global Hunger Index 2018: Forced Migration and Hunger. Welthungerhilfe and Concern Worldwide, Bonn and Dublin

von Grebmer, K., Bernstein, J., Mukerji, R., Patterson, F., Wiemers, M., Ní Chéilleachair, R., Foley, C., Gitter, S., Ekstrom, K., and Fritschel, H. (2019). Global Hunger Index 2019: The Challenge of Hunger and Climate Change. Bonn: Welthungerhilfe; and Dublin: Concern Worldwide. 72 p

Wahlberg K (2008) Causes and Strategies on World Hunger: Green Evolution verses Sustainable Agriculture. World Economy and Development in Brief, Global Policy Forum, New York

Benson H. Chishala holds a Ph.D. (Soil Science) from Aberdeen University and a B.Sc degree (Agricultural Science) from the University of Zambia. He worked for the Ministry of Agriculture before joining the University of Zambia where he is Senior Lecturer and currently Dean of the School of Agricultural Sciences. He has served as consultant to different organizations including the Zambia Environmental Management Agency, SGS Zambia, FAO, SADC and the World Bank.

!

Rhoda Mofya-Mukuka is Senior Research Fellow and Head of Research on Agriculture, Food Security and Nutrition at the Indaba Agricultural Policy Research Institute, Zambia. She has more than 20 years of engagement in agricultural and rural development in Zambia. Her current work includes analysis of agricultural policy and implications for food security and nutrition. She has worked as principle investigator on various short term assignments funded by different development partners including the World Bank, UN

Economic Commission for Africa, and US Aid. Rhoda has a Ph.D. in Agricultural Economics from Kiel University and a Masters in Small Business Studies from University of Leipzig, Germany.

Lydia M. Chabala is Head of the Department of Soil Science and Lecturer at the University of Zambia. She is involved in teaching, research, and consultancy and community service. She teaches both undergraduate and postgraduate courses in land use, pedology and pedometrics and the application of GIS in land management. Lydia currently serves as one of the soil experts on the Intergovernmental Technical Panel on Soils (ITPS) of the Global Soil Partnership (GSP - FAO). Prior to joining the University, she worked briefly as Agricultural Officer in the Ministry of Agriculture.

Elias Kuntashula is Senior Lecturer in the Department of Agricultural Economics and Extension at the University of Zambia. He teaches both undergraduate and postgraduate students in a wide range of courses including: agricultural production economics, applied econometrics, quantitative methods in agricultural economics, agricultural policy analysis and international trade. Elias has supervised more than 40 undergraduate and postgraduate students, and has published more than 35 peer reviewed journal and book chapters. He holds a PhD in Environmental Economics and an MSc in Agricultural Economics.

Ensure Healthy Lives and Promote Well-Being for All At All Ages

Kim Dowling, Rachael Martin, Singarayer K. Florentine, and Dora C. Pearce

K. Dowling (✉) · R. Martin · D. C. Pearce
School of Engineering, Information Technology and
Physical Sciences, Federation University Australia,
Victoria, Australia
e-mail: k.dowling@federation.edu.au

K. Dowling
Department of Geology, University of Johannesburg,
Johannesburg, South Africa

S. K. Florentine
School of Science, Psychology and Sport, Federation
University Australia, Victoria, Australia

D. C. Pearce
Melbourne School of Population and Global Health,
The University of Melbourne, Melbourne, Australia

Abstract

3 GOOD HEALTH AND WELL-BEING

Overview

The natural environment (e.g., air, soil, water, rock) has many links to human health

Good wellbeing requires a healthy environment, and allows everyone to reach their full potential

SDG 3 targets include tackling water-borne diseases, promoting mental wellbeing and limiting exposure to toxins in the environment (SDG 3.8)

Snapshot of human health

Health trends

Since 1900, the global average life expectancy has more than doubled

Disasters, conflict and environmental change all add to the burden of disease around the world.

Challenges with human health are not equitably distributed, with extra challenges in low-income countries (e.g., indoor air pollution, poor nutrition)

High Income Countries:

70% of human deaths occur in people aged over 70

Low Income Countries:

20% of human deaths occur in people aged over 70

Geoscience and health

Understanding the origins and movement of pollutants through the environment is critical to sustaining and improving human health

Elevated concentrations of toxic metals and metalloids (e.g., arsenic) in water and soils can increase cancer incidence rates

Pesticides can persist in the environment and produce toxic by-products

Air pollution can be affected by: burning solid fuels and fossil fuels, poor waste management, dust storms, volcanic eruptions.

Inhalation of smoke/particulates contributes to disease and shortens lives

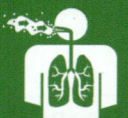

Contact with a healthy natural environment can contribute to physical and mental wellbeing.

Good outdoor space supports lifelong healthy behaviors

Geotourism provides mental physical stimulation.

Connection to place supports sustainability and promotes healthy lives.

3.1 Introduction

Health may be defined as *'a state of complete physical, mental and social wellbeing and not merely the absence of disease'* (WHO 2019a). Wellness is captured in the WHO's definition as encompassing total physical, mental, and social well-being (WHO 2019a). Whereas physical health represents freedom from illness and injury, mental health represents a state of well-being that enables individuals to maximise their potential to participate fully and productively within their social context (WHO 2019b). Social well-being represents a state of positive emotions and a sense of satisfaction and fulfilment in life (CDC 2018). These different contributions to health and well-being are collectively the focus of **SDG 3**.

The concept of wellness encompasses total physical, mental, and social well-being (Fig. 3.1). This is a more complex concept than just the absence of illness. Targeting well-being enhances the possibility for all members of society to reach their full potential, and is critical to true sustainable development. Irrespective of political philosophy, socio-economic status, or cultural background, embracing the aim of this SDG to *ensure healthy lives and promote well-being for all at all ages* appears uncontentious. Good health and well-being allows children to spend more time at school (**SDG 4**), and the wider population to work and generate income (**SDG 8**). Realising the ambitions of **SDG 3** also relies on making progress with other SDGs, such as nutritious food (**SDG 2**), education (**SDG 4**), clean water and sanitation (**SDG 6**), clean energy (**SDG 7**), or high-quality infrastructure (**SDG 9**) and living conditions (**SDG 11**).

Ensuring healthy lives and good well-being for all requires actions to prevent ill-health, and treat both communicable and non-communicable sickness and diseases. *Communicable diseases* include infectious diseases that are transmitted directly to or between individuals. A clear global example is the COVID-19 pandemic, caused by the 2019 novel severe acute respiratory syndrome coronavirus 2 (SARS-CoV-2), and which has drastically affected the lives of millions and the global economy (WHO, 2020a, b). Other infectious diseases include bacterial infections such as leprosy and tuberculosis, viral infections such as various strains of influenza, vector borne parasites such as malaria, and transmission through direct contact with infective bodily fluids such as HIV (WHO, 2017). Communicable diseases also include waterborne diseases, passed from person to person through contaminated water (e.g., cholera, diarrhoea, dysentery). *Non-communicable diseases* may be modifiable by changes to lifestyle factors, including increased physical activity, healthy diet and weight loss, reduction in alcohol and tobacco consumption. Non-communicable diseases are major causes of premature death globally, particularly due to cardiovascular diseases (e.g., heart attack and stroke), cancers, respiratory diseases, and diabetes (WHO 2018b). The targets of **SDG 3** refer to both, and sets this into a broader health and well-being strategy that aims to safeguard the sustainability and to ensure wellness at both individual and the whole-of-population level (Table 3.1).

The targets in Table 3.1 cover maternal mortality (3.1), major epidemics (3.3), and overall mortality (3.4, 3.2), through to limiting exposure to toxins in the environment (3.9) and self-administered toxins (3.5), traffic accidents (3.6), reproductive health (3.7), and accessibility and availability of care (3.8). There are also additional health-related targets in other SDGs, such as the focus on malnutrition in **SDG 2** (discussed in that chapter). While the targets in **SDG 3** articulate desired outcomes (e.g., reduced maternal mortality and less epidemics), they do not set out the means by which these outcomes will be achieved. It falls to health professionals, policy makers, engineers, anthropologists, and geoscientists, among others, to contextualise these targets and associated indicators and to implement solutions that ensure the targets of **SDG 3** are achieved, delivering and protecting sustainable development.

One of the difficulties in delivering human and ecosystem health lies in the identification of appropriate actions (and in-actions) in a complex web of interconnected systems that span

Table 3.1 SDG 3 Targets and Means of Implementation

Target	Description of Target (3.1 to 3.9) or Means of Implementation (3.A to 3D)
3.1	By 2030, reduce the global maternal mortality ratio to less than 70 per 100,000 live births
3.2	By 2030, end preventable deaths of new-borns and children under 5 years of age, with all countries aiming to reduce neonatal mortality to at least as low as 12 per 1,000 live births and under-5 mortality to at least as low as 25 per 1,000 live births
3.3	By 2030, end the epidemics of AIDS, tuberculosis, malaria and neglected tropical diseases and combat hepatitis, water-borne diseases and other communicable diseases
3.4	By 2030, reduce by one-third premature mortality from non-communicable diseases through prevention and treatment and promote mental health and well-being
3.5	Strengthen the prevention and treatment of substance abuse, including narcotic drug abuse and harmful use of alcohol
3.6	By 2020, halve the number of global deaths and injuries from road traffic accidents
3.7	By 2030, ensure universal access to sexual and reproductive health-care services, including for family planning, information and education, and the integration of reproductive health into national strategies and programmes
3.8	Achieve universal health coverage, including financial risk protection, access to quality essential health-care services and access to safe, effective, quality and affordable essential medicines and vaccines for all
3.9	By 2030, substantially reduce the number of deaths and illnesses from hazardous chemicals and air, water and soil pollution and contamination
3.A	Strengthen the implementation of the World Health Organization Framework Convention on Tobacco Control in all countries, as appropriate
3.B	Support the research and development of vaccines and medicines for the communicable and non-communicable diseases that primarily affect developing countries, provide access to affordable essential medicines and vaccines, in accordance with the Doha Declaration on the TRIPS Agreement and Public Health, which affirms the right of developing countries to use to the full the provisions in the Agreement on Trade-Related Aspects of Intellectual Property Rights regarding flexibilities to protect public health, and, in particular, provide access to medicines for all
3C	Substantially increase health financing and the recruitment, development, training and retention of the health workforce in developing countries, especially in least developed countries and small island developing States
3D	Strengthen the capacity of all countries, in particular developing countries, for early warning, risk reduction and management of national and global health risks, including pandemics

geographic, genetic, and temporal boundaries. **SDG 3** requires detailed consideration of the causes of mortality and morbidity. The link between genetic and environmental factors and human illnesses such as cancers, diabetes, immune system, and neurodegenerative disorders, as well as respiratory and cardiovascular diseases has long been established (Lango and Weedon 2008). Environmental factors have a major influence on health outcomes (Alpert 2018). The potential to remove or limit the environmental contributors to disease and other adverse health outcomes, rather than reacting to a medical condition after impact, has both social and financial merit. The synergistic role of the

environment in delivering sustainability and positive health outcomes requires articulation to ensure that it gains appropriate emphasis, funding, and policy commitment. Geoscience offers a way to unpack this complexity. It confirms that sustainability, poverty-elimination, a healthy environment, and human well-being are inextricably linked and provides a unique and valuable lens through which to view the complex interplay of geological, biological, and societal systems.

SDG 3 therefore requires a geogenic and geoscientific lens to evaluate the role the environment plays in both delivering health and enabling sustainable development. The geogenic environment comprises everything that results from our vast geological history (both natural and modified), and includes the environmental compartments of soil, atmosphere, water, and rock. These compartments are interconnected. For example, dust storms in Africa have the potential to impact on water and air quality, and human morbidity thousands of kilometres away in the USA (Crooks et al. 2016). The geogenic environment sustains ecosystem and human health and this interconnectedness of systems means that all changes have consequences which may be both subtle and diverse or catastrophic.

It is easy to argue that bad health outcomes will result from living in any environment with limited or polluted water, where the air is contaminated, where the soil cannot produce healthy crops, where education levels are low and violence is common, and where substance abuse becomes a refuge. This environment exists in both the Global North and Global South, in inner-city low socio-economic districts, in an emerging economy's brownfield zone and in communities where land is contested and climate fluctuations have resulted in drought. For those living in environments where negative environmental factors dominate, maintaining a good diet, taking regular exercise, and living without toxins is an impossibility. Wellness requires a healthy environment.

In this chapter, we explore the links between the geogenic environment and wellness. We discuss health financing and trends (Sect. 3.2), and the geogenic environment and its role in sustaining both ecosystem and human health (Sect. 3.3), with an emphasis on four case studies: arsenic, asbestos, air quality, and pesticides. We proceed to explore the concept of wellness, and how this links with the geogenic environment (Sect. 3.4), and set out geoscience actions to support health and well-being (Sect. 3.5). Through this chapter, we demonstrate the global interconnectedness and significance of the geological environment to human and ecosystem health. The role the environment plays in sustaining all biota on our planet, of which we are only a part, is crucial, and our future is entwined with the fate of the entire ecosystem.

3.2 Human Health—Financing and Trends

3.2.1 Financing Health Improvements

Measures of health and well-being are often reported as ill-health (as a *de facto* measure of the inverse of well-being) and are complex, but human health is recognised as a significant priority given the global health budget is predicted to exceed $US11.5 trillion by 2020, compared with $US4.1 trillion in 1995 (GHDx 2018). Despite this ongoing and growing expenditure, humans and other species die unnecessarily and in greater discomfort than is consistent with a contemporary ethical framework. The role the environment plays in human and environmental health is not always well-articulated and perhaps not properly understood. If this connection was accepted, the financial allocations to both sectors might be more equitable.

When discussing health and health targets, the discourse is often dominated by topics such as vaccine uptake rates, randomised placebo-controlled double-blind clinical trials; availability of medicines; development of interventionist strategies, and research on drug development (Bing et al. 2000). These are all laudable discussions, but it is appropriate to question if this somewhat narrow focus on medical intervention is well placed. There is also a disparity in the amount of money spent on prevention versus treatments. For 2019, the total budget for the US Environmental Protection Agency was approximately US$8.85 billion (EPA 2019). In contrast, in 2018 Americans are estimated to have spent $US3.5 trillion on healthcare services (CMS 2018). There is a general disparity in expenditure for medical interventions versus prevention via environmental protection. Spending too much, too late, is not an effective use of limited funds.

The cost benefits of wellness are rarely discussed. Delivering well-being requires commonality of purpose from governments, corporate/institutional entities and the public. This has proven difficult, as competing needs and objectives, lack of enforceable global environmental regulations, disempowerment and ignorance have hampered success. In 2019, however, the New Zealand government announced their first *'wellness budget'*, framed around five priorities that, while acknowledging the need for economic sustainability, bring focus to environmental sustainability, mental and physical health for all, and education.

3.2.2 Health Trends

Since 1900, the global average life expectancy has more than doubled and although this result is geographically in-homogeneous, the average human lifespan is now 70 years (Roser, 2018). Superficially, these statistics appear pleasing but significant health inequality persists, and a complex picture is emerging. While noting the issues around longevity predictions, Daniels (2006) posits that a current generation of obese children could suffer greater illness and experience diminished lifespan when compared to their parents. In 2018, the Centers for Disease Control and Prevention (USA) reported that for the third straight year there was a downward trend in Americans' average life expectancy. If this trend continues and grows to include other regions, this represents a significant reversal in longevity trends and we assert links to environment quality and lifestyle.

The UN Millennium Development Goals (MDGs), running from 2000 to 2015, had a significant focus on improving global health. Four goals directly related to health objectives, with the United Nations report (2015b) setting out progress and remaining challenges for each of these:

- **Eradicate poverty and hunger (MDG 1)**. Poverty and hunger are inextricably linked with child and maternal mortality, limited educational opportunities, and access to essential medications (United Nations 2015b). The number of people in global poverty (defined as having an income of $2 or less/day) fell from 1.9 billion people in 1990 to 836 million people in 2015. The number of undernourished people living in developing regions fell by almost half since 1990, from 23.3% in 1990–1992 to 12.9% in 2014–2016. Progress has been spatially uneven, with 23% of people in sub-Saharan Africa and 20% of those in the Caribbean being undernourished in 2015, compared with 5% in Latin America. These themes are discussed more in **SDG 1** (Poverty) and **SDG 2** (Zero Hunger).

- **Reduce child mortality (MDG 4)**. Between 1990 and 2015, the global under-five mortality rate has declined by more than half, dropping from 90 to 43 deaths per 1,000 live births. Significant progress has been made in both developed and developing regions, with the mortality rate dropping by 61% and 53%, respectively. A remaining challenge is reducing the deaths of newborn babies. In 2015, 17% of deaths in under-fives occurred within 24 h of being born and 34% of deaths within the first week. Many of these deaths are avoidable through simple interventions such as access to clean water and safe hygiene practices, as well as increased care at the time of delivery. In Southern Asia and sub-Saharan Africa, 48% of all births were *not* attended by skilled health personnel in 2014 (compared with a world average of 29%).

- **Improve maternal health (MDG 5)**. The global maternal mortality ratio reduced by 45% between 1990 and 2015, with reductions of 64% in Southern Asia and 49% in sub-Saharan Africa. At the end of the Millennium Development Goals, the maternal mortality ratio was still approximately 14 times higher in developing regions than in developed regions, with many of these deaths being preventable given access to appropriate care before, during, and after childbirth.

- **Combat HIV/AIDS, malaria and other diseases (MDG 6)**. The number of new HIV infections fell by 40% between 2000 and 2013, from 3.5 million cases to 2.1 million. Between 2000 and 2015, the global malaria incidence rate fell by approximately 37%, and the mortality rate by 58%. It is suggested that interventions helped to avert over 6.2 million deaths from malaria between 2000 and 2015, primarily of young children in sub-Saharan Africa.

Further information on all these statistics, and other Millennium Development Goals, is available in the United Nations report (2015b). While significant progress was made on these four goals, significant challenges remain, and new challenges are emerging. For example, the Ebola epidemic in West Africa (2013–2016) resulted in more than 11,000 deaths and the current epidemic in the Democratic Republic of Congo (2018–) has killed more than 1800 people as of August 2019. This health crisis requires urgent monitoring using real-time geospatial data, and provision of water and sanitation (**SDG 6**) to control its spread. The COVID-19 pandemic has caused global devastation and was preceded by severe acute respiratory syndrome (SARS-CoV) and Middle East respiratory syndrome coronavirus (MERS-CoV), all zoonotic diseases resulting from cross-species transmission of viruses from wild animals, sometimes via domestic animals, to humans (Parrish et al. 2008; Rodriguez-Morales et al. 2020). The vectors vary but human encroachment and destruction of ecosystems are significant contributors. Add to this the impact of climate change on the occurrence and geographical distribution of zoonotic diseases (Sachan & Singh 2010), on habitat contraction and the increased risk of disasters such as extremely destructive wildfires as witnessed across Eastern Australia, California and many parts of Europe.

Prevention is a key focus for all health issues but, it is not possible to anticipate all eventualities, and preparedness (the ability for rapid response to sudden and unexpected events) is required. A reflection on the COVID-19 response has much to teach us about leadership, responsibility, responsiveness to expert advice and information sharing. The effectiveness of any response is amplified by a strong and shared understanding of the strategic terrain in which we must respond, and the geogenic environment critically underpins all aspects of life on the planet.

The WHO (2018c) stated in 2016, that 56.9 million humans died, with more than half attributed to 10 causes of death (Fig. 3.2). The extent to which environmental, mental, and social factors played a role in these deaths was not analysed.

Figure 3.2 provides a global picture, but greater understanding is achieved when we factor in socio-economic status. In low-income countries for this same period, the leading cause of death is respiratory infections and diarrhoeal diseases. People in low-income environments are often dependent on biomass fuels for heating and cooking, with up to 610 deaths per million attributed to indoor air pollution in the worst affected countries (WHO 2018c). Figure 3.3 shows the geographic distribution of premature deaths as a result of using solid fuels for cooking and heating. The high concentrations of particulates arising from the burning of wood and charcoal are associated with an increase in respiratory infections, low birth-weight, and cardiovascular events, which all cause mortality in adults and children (Fullerton et al. 2008). In contrast, the leading causes of death in high-income countries are ischaemic heart disease and stroke (WHO 2018b).

How society responds to these mortality triggers is equally telling. In Bangladesh, tube wells were sunk to gain access to groundwater as an alternative to surface water contaminated with microbial pathogens, successfully reducing deaths from diarrhoeal diseases but increasing the risk of arsenic exposure (Escamilla et al. 2011). Arsenic desorption and dissolution into aquifers (Raessler 2018) has resulted in an estimated 35–77 million Bangladesh people being chronically exposed to arsenic in their drinking water (the impacts of which are described in Sect. 3.3.1). This mass poisoning affected all in the community but the impact was greatest on the young, elderly, and for pregnant women. Pathways to contamination are often complex. Arsenic contaminated groundwater, such as in Bangladesh,

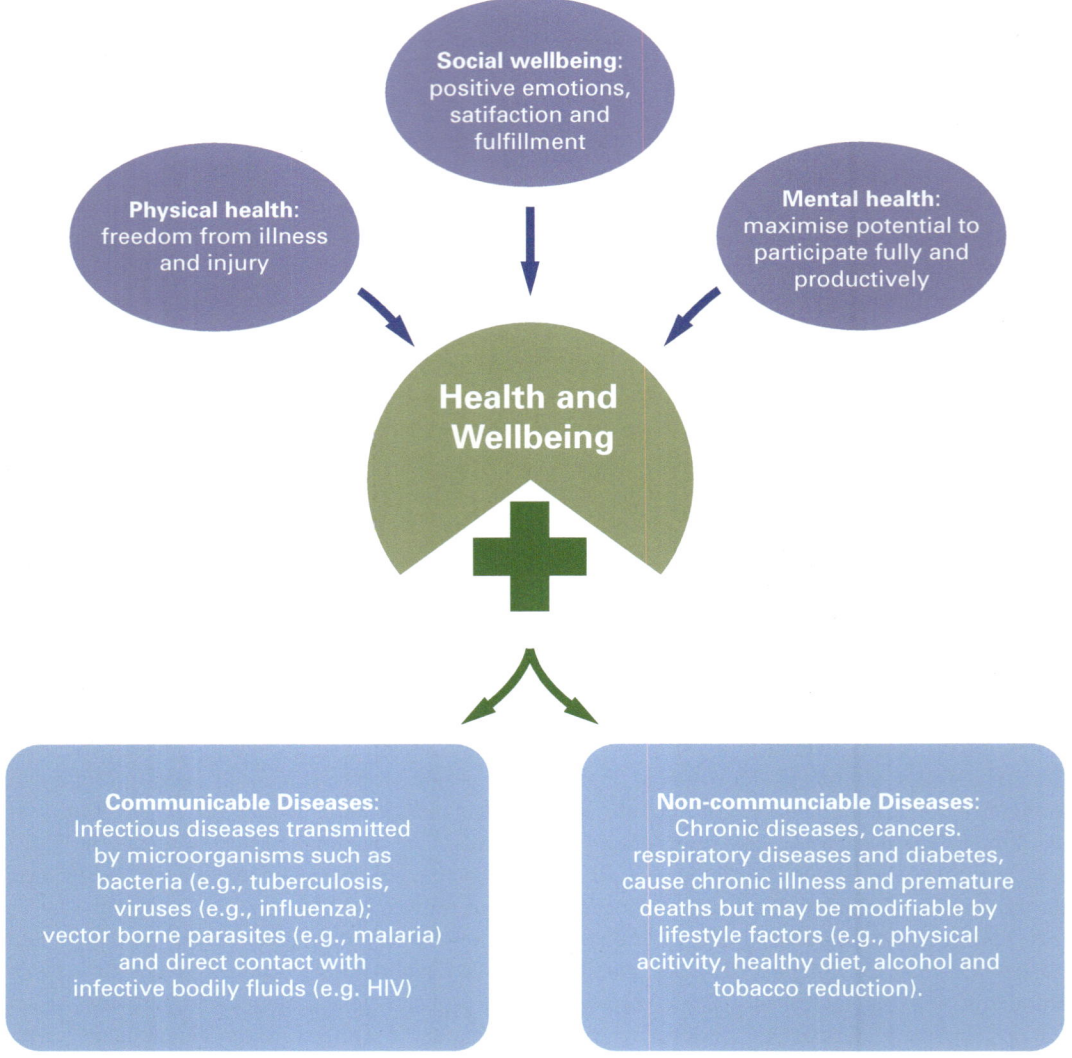

Fig. 3.1 Positive contributions to health and wellbeing and negative contributions made by communicable and non-communicable diseases to the quality of life. Descriptions used within this figure are adapted from WHO (2017, 2018a, 2019b, c) and CDC (2018)

can also result in airborne arsenic that can be inhaled by residents (Joseph et al. 2015). It also affects crops and domestic animals, and as a result, the cow dung cake used as fuel for cooking in unventilated rooms provides another vector for arsenic contamination (Pal et al. 2007).

In the examples above, the link between poor environment and ill-health in low-income households is clear. Can such a case be made

for higher income countries? Heart disease and cancer are major causes of death across the globe and affect people in all income categories from lower-middle to high income (WHO 2018b). While genetic predisposition to diseases is acknowledged, Veronese et al. (2016) report that the risk of premature mortality is lowered with the addition of one or more low-risk lifestyle factors (i.e., healthy eating, high levels of

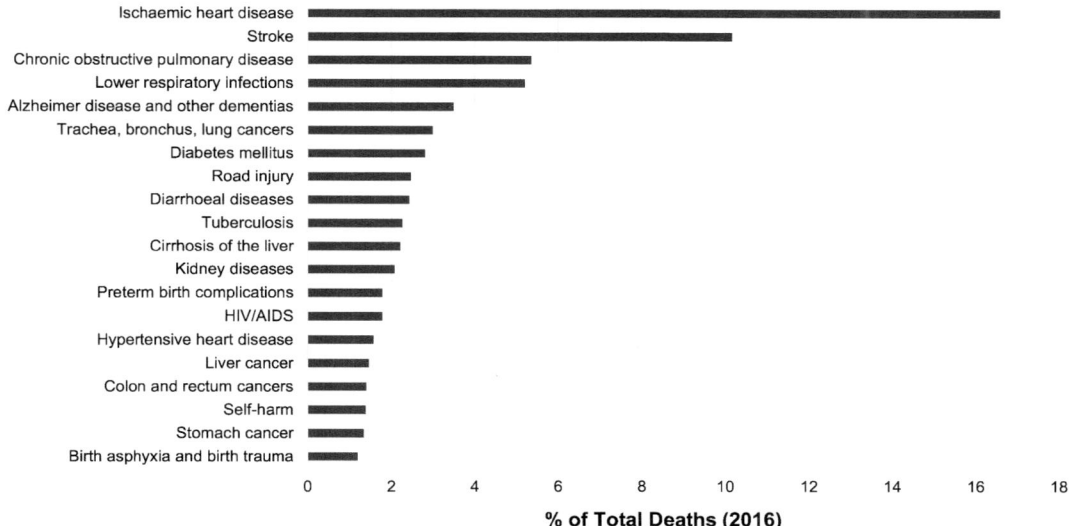

Fig. 3.2 Top 20 causes of death, and the number of deaths as a percentage of total deaths in 2016. Source data from Global Health Estimates 2016: Deaths by cause, age, sex, by country, and by region, 2000–2016. Geneva, World Health Organization; 2018

physical activity, moderate alcohol drinking, and non-smoker). These factors require a healthy environment to deliver access to good food and water, safe places to exercise, education, and a supportive social setting. The environment, built or natural, is critical to encouraging activities that support and nurture our well-being. Humans, if given information and opportunity have the power to intervene in their own health outcomes (Wallerstein 1992). The converse is also true: failure to provide these things to all will result in disparities in well-being outcomes.

Disasters, conflicts, communicable and non-communicable diseases all add to the burden of disease around the world and this burden is not equitably distributed. The likelihood is that escalating climate fluctuations will increase the frequency of disasters, increase conflicts, and increase the necessity to re-home climate-change-refugees. Climate fluctuation and global warming have seen changes in the geographical distribution of vector-borne infectious diseases (Kurane 2010). Even without the escalating stress of climate change, poverty undermines health through poor nutrition and sanitation, and decreases opportunities for education and delivery of safe working environments (Arcaya et al.

2015). Infectious and parasitic diseases persist in some developing countries, and as income levels rise, so too do rates of non-communicable diseases which are often referred to as 'lifestyle' diseases (Amuna and Zotor 2008; Ssewanyana et al. 2018). Globalisation and rapid urbanisation facilitate increases of risky behaviours which are exacerbated by socio-economic inequalities (Hosseinpoor et al. 2012). The picture of disease burden and mortality is complex but there is a link between the environment and the health of both the individual and the society.

3.3 Geogenic Environment

The geogenic environment affects health outcomes and longevity and hence must be central to the achievement of **SDG 3**. Genetic predisposition, social and economic factors clearly play a role in disease development and presentation however if the environment is degraded and uptake of contaminants increases, if poor quality food and water is all that is available and if bleak or limited aesthetic environment frames your vista, these environmental factors have a great potential to impact on longevity, morbidity, and

Number of deaths from household air pollution, 2017
Annual number of premature deaths attributed to illness as a result of household air pollution from the use of solid
fuels for cooking and heating.

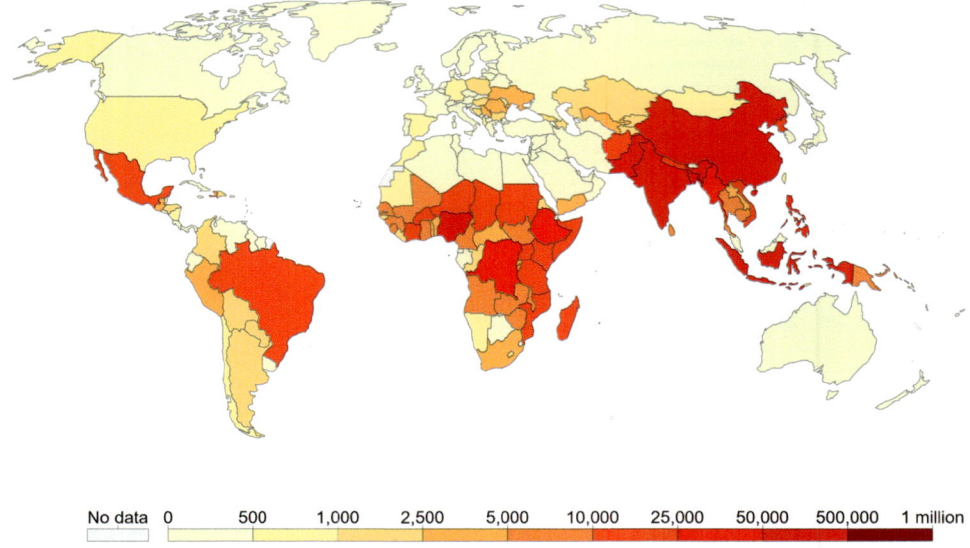

No data   0   500   1,000   2,500   5,000   10,000   25,000   50,000   500,000   1 million

Source: IHME, Global Burden of Disease   OurWorldInData.org/indoor-air-pollution/ • CC BY

Fig. 3.3 Number of premature deaths from household air pollution in 2017, due to the use of solid fuels for cooking and heating. Deaths are inequitably distributed around the world. Image used under a CC BY license (https://creativecommons.org/licenses/by/4.0/), created by Roser and Ritchie (2019), using data from IHME (2018)

quality of life. We investigate the significance of the geogenic environment, and human dependence on these complex systems using four case studies. The following sections look at a variety of environments and consider how the anthropogenic activity has had negative impacts on human and ecosystem consequences.

3.3.1 Arsenic, Soil and Groundwater

The Bangladesh and west Bengal arsenic experience introduced in Sect. 3.2.2 illustrates the need for geoscientists to anticipate issues, recognise unanticipated impacts, and find solutions. The reasons for promoting groundwater use in these regions were clear: surface storage of water led to the proliferation of bacteria and associated illnesses such as cholera, and increased mosquito-breeding centres with a consequent increase in mosquito-borne diseases. Using groundwater,

therefore, seemed prudent 40 years ago, when the extent of naturally occurring arsenic-contaminated groundwater was not realised. As a direct consequence of the use of this groundwater, an estimated 43,000 people in Bangladesh die each year from arsenic-related diseases. These include skin lesions, cancers, and cardiovascular and lung illnesses (WHO nd). 20 million people are still exposed to arsenic levels above the maximum permissible limit of 50 µg/l (WHO nd).

Anthropogenic sources also contribute to arsenic contamination of groundwater with contributions from mining, the burning of fossil fuels and the use of arsenical pesticides and historical wood treatments as well as farming applications (Shankar et al. 2014). Mining and ore processing date back to ancient times, when environmental degradation and adverse human health effects were accepted as inevitable consequences of progress and profit (Dowling et al. 2016). Arsenic contamination of groundwater used for drinking and irrigation, whether from natural or anthropogenic sources, impacts the health of millions of people worldwide, causing death and disability (Shankar et al. 2014).

Although arsenic-contaminated drinking water sources have received much attention, both in the media and by health researchers, airborne arsenic is also of concern in some industrial areas. For example, mining operations, including extraction and smelting processes, may generate arsenic-contaminated airborne particulate matter which can be ingested or inhaled. Of most concern with respect to human exposures are the inhalable size fractions, which can be separated into two broad categories:

- *Coarse particulates*, with aerodynamic diameter of <10 μm (PM_{10}), may penetrate the extra-thoracic region and deposit in the tracheobronchial region, to ultimately be cleared from the lungs by the body's natural defence mechanisms or be ingested;
- *Fine particulates*, with diameter <2.5 μm ($PM_{2.5}$), can travel greater distances from the source and are more likely to be inhaled as deeply as the alveolar region of the lung, where they may be absorbed directly into the pulmonary circulation system, phagocytosed[1] or ultimately cleared by mucociliary transport (Martin et al. 2014).

The bioavailability of arsenic in particulates is of critical importance given the statistically significant inverse relationship between particle size fraction and arsenic concentration (Martin et al. 2016a, 2016b). Living in close proximity to mineralised and mined areas can increase risk of environmental exposures (Martin et al. 2013, 2014, 2016a, 2016b; Pearce et al. 2010) and anthropogenic activities such as wildfires can mobilise sequestered metals from soil organic matter and vegetation (Abraham et al. 2018).

Ongoing systemic absorption of arsenic associated with periodic exposures to contaminated soil by children living in areas of historic

gold mining activity has been demonstrated using synchrotron-based X-ray microprobe techniques (Martin et al. 2013; Pearce et al. 2010). Whereas the dominant oxidised state of arsenic present in mine waste where arsenopyrite (Fig. 3.4) occurs is typically arsenate (As^V) (Arčon et al. 2005), mammalian metabolism of arsenic is thought to involve alternating reduction and oxidative methylation to produce arsenicals of varying toxicity: trivalent species being more toxic than pentavalent species, with arsenite (As^{III}) and Monomethylarsonous acid (MMA^{III}) being considered highly toxic (Vahter 2002; Sattar et al. 2016).

According to WHO (nd):

> Long-term exposure to arsenic from drinking-water and food can cause cancer and skin lesions. It has also been associated with cardiovascular disease and diabetes. In utero and early childhood exposure has been linked to negative impacts on cognitive development and increased deaths in young adults.

So not surprisingly, cancer incidence rates also increase with elevated soil concentrations of toxic metals and metalloids such as arsenic (Núñez et al. 2016; Pearce et al. 2012). Groundwater, dust, food, and the hand to mouth behaviour of children all contribute to the arsenic body burden of humans living in these affected landscapes. The complex interplay between air, water and soil is critical to understanding these impacts.

3.3.2 Asbestos: From "Magic Mineral" to Medical Epidemic

Today's global news headlines are rife with reports of disease epidemics and medical disasters across low to high-income economies (e.g., the Ebola virus; H5N8 bird flu outbreaks; the Flint water contamination crisis in the USA; COVID-19 pandemic). While these crises deserved attention, the health impacts surrounding the asbestos mining and manufacturing industries are less well reported. With an estimated 107,000 annual deaths worldwide (Collegium Ramazzini 2016), asbestos

---

[1]Phagocytosis is the cell engulfing a large particle using the plasma membrane, creating an internal compartment.

Fig. 3.4 Arsenopyrite, an iron arsenic sulphide, is identified as the metallic mineral in this specimen from Talnotry, UK. Quartz veining and granodiorite fragments are also visible. © NERC (used with permission)

has gained notoriety and it invokes fear, anxiety and panic, not only among workers exposed to asbestos dust, but individuals and communities which have potentially been exposed through sources in natural and the built environment. This has particular relevance to **SDG Target 3.9**.

Asbestos is a mineralogical and regulatory term describing a group of silicate minerals that form bundles of long, thin mineral fibres with length to width ratio of at least 3:1 (Table 3.2). While the 'asbestiform' minerals differ in size, shape, and chemical composition, they have in common a fine fibrous structure (Fig. 3.5), and it is this needle-like structure that makes asbestos highly invasive and persistent, especially in the lungs. The pathways by which asbestos causes lung diseases are not fully understood, however it is widely reported that asbestos induces inflammation, tissue damage, and DNA and chromosomal damage (Nymark et al. 2008; Yanamala et al. 2018). Research efforts into the mechanisms for the asbestos induced disease are

ongoing, and key breakthroughs in understanding the pathogenesis of diseases associated with this mineral fibre will allow for important progress in the prevention and treatment of asbestos-related diseases.

Asbestos minerals are significant in work-related cancer through inhalation exposure (Furuya et al. 2018). It is estimated that up to 85% of all occupational lung cancers worldwide can be attributed to asbestos (Takala 2015). Non-occupational and environmental exposures are also a cause for concern and may occur through a variety of pathways such as contact with an exposed worker by family members; exposure during home renovations; environmental exposure by residents proximal to a naturally occurring asbestos deposit or active asbestos mine and incidental contact with asbestos-contaminated materials, wastes and soils (NRC 2007).

With asbestos bans in 55 countries, it seems logical that asbestos-related diseases such as asbestosis, mesothelioma, asbestos-related lung

Table 3.2 Types of asbestos and their use

Series	Name *(common name)*	Commercial Use
Serpentine	Chrysotile (white asbestos)	Construction materials, brake linings, gaskets and boiler seals, insulation for pipes, ducts, and appliances
Amphibole	Actinolite	Rarely used commercially
	Amosite (brown asbestos)	Cement sheets and pipe insulation, insulating board, ceiling tiles and thermal insulation products
	Anthophylite	Used in limited quantities for insulation products and construction materials
	Crocidolite (blue asbestos)	Insulation for steam engines; spray-on coatings; pipe insulation; plastics and cement products
	Tremolite	Rarely used commercially

Fig. 3.5 Images of asbestos clearly showing the fibrous nature: (left) Hand specimen of chrysotile (white asbestos) veins in massive serpentine, Woods Reef New South Wales. (right) Scanning Electron Microscope image of asbestiform riebeckite (crocidolite or blue asbestos), Wittenoom Western Australia. (scale bar 100 microns). © S. McKnight, Federation University Australia (used with permission)

cancer, and asbestos-related pleural disease would be in decline. However, the incidence of mesothelioma (a rare, aggressive cancer of the lining of the lungs, abdomen or heart almost exclusively through asbestos exposure) remains highest in industrialised countries that have already banned asbestos, such as the United Kingdom, the Netherlands, Australia, and New Zealand (Bianchi and Bianchi 2014). In Sweden, for example, since the ban on asbestos in 1982, the total annual number of new mesothelioma cases did not show any evidence of decline until 2014. Similarly, despite asbestos being completely phased out of manufacture in Australia in 2003, the annual number of mesothelioma deaths increased from 403 in 1998 to 656 in 2013 (WHO 2018d).

The protracted latency of mesothelioma, of between 40 and 50 years, means that incidence in many industrialised countries has continued to rise despite the introduction of restrictions and bans on asbestos and materials containing asbestos, with many countries yet to fully realise peak asbestos-related disease figures. The prognosis in countries where a production ban is not in place is for a negative impact on the achievement of **SDG 3**.

Australia has a history of substantial and protracted usage of asbestos products throughout the twentieth century leading to one of the highest incidences of mesothelioma in the world (Allen et al. 2018). As such, it represents a useful case study to examine the potential for future asbestos issues, and as a model for unexpected interactions between soil, air, and human biology. Despite the ban on importation and use in 2003, concern about a 'third wave' of non-occupational asbestos exposure and disease remains. The first and second waves relate to raw asbestos handling and use of asbestos products in industry, respectively. The third wave is associated with short-term and/or low-level exposure to asbestos in the home (Landrigan 1991), such as do-it-yourself home renovators undertaking improvements, and not realising they may be exposing themselves during the process.

With asbestos-containing materials widely distributed in residential buildings, along with an estimated 6,300 tonnes of illegally dumped asbestos waste in Australia (ASEA 2017), asbestos represents an ongoing source of significant concern in Australian urban and rural environments. The incidence and awareness of asbestos-related hazards and diseases has increased the need to more closely regulate procedures for the appropriate management, systematic removal, of asbestos products—especially as they approach the end of their useful life, is clear in the Australian context. This phase will be replicated globally.

The third wave of asbestos contamination is another example of poorly quantified potential health impacts and of interactions between humans and the geogenic environment mediated, in this case, via the built environment. Overall, asbestos illustrates the complexities of the interactions between air, water, and soil in the geogenic environment, the built environment, and humans. The physical properties are the determining factors in its health impacts and many other minerals display these exact same characteristics. Much work is required to address the problems of the past and minimise the negative health impacts into the future.

3.3.3 Air Quality

Air pollution (Fig. 3.6) is a leading global disease risk factor. The meta-analysis in Requia et al.

(2018) demonstrates positive associations between cardiorespiratory diseases and a variety of air pollutants. In 2012, 7 million deaths were linked to ambient air pollution (WHO 2016) and particulate matter ($PM_{2.5}$) is linked to 2 million premature deaths per year (Lozano et al. 2013). Air toxics that are commonly linked with adverse health impacts, and therefore often the subject of national ambient air quality monitoring programs include carbon monoxide, nitrogen dioxide, ozone, sulphur dioxide, lead, and particles less than 10 micron (PM_{10}) and less than 2.5 micron ($PM_{2.5}$).

As with all environmental health impacts, the effects of air pollution are not experienced uniformly across populations. Of great concern is the enhanced effect on young and older members of the population; those with a health-compromised status; poorer communities, and those with limited access to health intervention strategies. Air pollution is asserted to be a developmental neurotoxin and high pollution levels often linked with diminished cognitive performance among children (Chen and Schwartz 2009). In Spain, for example, children from highly polluted schools demonstrated lower growth in cognitive development when compared to children from less polluted school environments (Sunyer et al. 2015). Given that schools across the globe are often conveniently positioned for transport access, their location often coincides with high traffic flow and subsequent elevated motor vehicle emissions. The effects of air pollution are also transgenerational. Perera et al. (2003) provide evidence that environmental pollutants at levels currently encountered in major cities in the USA adversely affect foetal development. The consequences are profound and have important public health implications. The impact of contaminants at times of great developmental growth, including the foetal and early postnatal stages of life, have an increased impact on the individual that may see the effects last a lifetime.

The impact of poor air quality also extends to traditional economic impacts on trade, commerce and tourism. There is a concern in many parts of Asia that poor air quality may influence tourism and business (Jang et al. 2014; Moschino et al.

Fig. 3.6 Air Pollution: Industry and transport are two major sources of pollutants affecting air quality. Image by analogous from Pixabay

2017). Dong et al. (2019), for example, reported that air pollution significantly reduced international inbound tourism in China between 2009 and 2012. The study estimated that for every increase in PM_{10} by 0.1 mg/m³, there would be a corresponding reduction of US$80 million in tourism revenue. Only limited data is available to assess whether air pollution will affect tourists' experiences and participation, but many countries are actively considering the issue.

When considering air quality impact on health it is difficult to look past China as a country with a problem and with significant strategies in place to address impact (Smil 2016). China's approach to air pollution has been organised, consistent, and effective over the past five years (Liu et al. 2014) but it presents an obstinate problem that may take many more years to rectify. Climate change and the global connection of all the environmental segments means that humans and the entire ecosystem will be affected for some time to come, but change is necessary, and China presents an excellent illustration of environmental intervention at an industrial scale.

While the air may seem to be a standalone system in need of restoration, it is precisely the interaction between air, water, rock, and soil (all of which may be disturbed, manufactured, contaminated, or polluted) that is the root cause of all air quality issues. Once again, understanding this interaction is critical to reducing risk efficiently.

3.3.4 Global Agricultural Sustainability: The Tension Between Pesticide Use and Wellbeing

The imperative to achieve and maintain stability in the agricultural sector is a problem as old as agriculture itself. Famine, as embodied in folklore as one of the horsemen of the apocalypse, represents one of the central and continuing fears of civilisation, and the history of all cultures contains examples of failures of crops and the consequent devastating effects on populations.

Toxin-free food, capable of providing for the nutritional needs of a community, is an essential component of **SDG 3** (as well as **SDG 2**).

In understanding the complexities of food production, one must recognise that all agricultural practice is inherently one of environmental disturbance and comes with the usual complexities in the interactions of air, water, and soil. Further, issues such as poor soil, insect attacks, weed infiltration, floods, drought, and, more recently, environmental CO_2 rises, and climate change conditions, all contribute to food production issues. Food, and food quality, have substantial impacts on infant mortality specifically (**Target 3.2**), as well as overall population mortality (**Target 3.1**). Accelerated human population growth necessitates greater food production and pushes agriculture into marginal zones, which in turn leads to increased use of fertilisers and pesticides, with concomitant effects on soil, air, and water at both the point of production and consumption (Fig. 3.7).

The term 'pesticide' covers a wide range of compounds such as insecticides, fungicides, herbicides, and rodenticides (Aktar et al. 2009; Igbedioh 1991). Of particular interest are those pesticides and agrochemicals which are considered essential components of worldwide agriculture systems, and which, together with synthetic fertilisers, have allowed the remarkable increase in crop yields, food production, and improved global food security (Alexandratos and Bruinsma 2012; Igbedioh 1991). Pesticides are a double-edged sword: while increasing productivity, they have a high potential to persist in the environment and to produce toxic by-products. Given the vast quantities of pesticides used globally, large amounts of residual matter are likely to remain in the environment, with many cumulative impacts observed (Samsidar et al. 2018).

Many pesticide degradation products and residues are recognised as being deleterious to human health and the environment (Aktar et al.

Fig. 3.7 Pesticide Spraying. Image by skeeze from Pixabay

2009). This dichotomy is at the heart of the problems faced in any modified environment: they are usually modified for a positive purpose, but the impacts of those changes are often far reaching and complex. Ongoing work is needed to establish robust systems that understand the impacts of agricultural chemicals within the context of their use and production, and which regulate them to adequately protect human and ecosystem health and safety (**Target 3.9**).

Increases in food production, a key requirement to improve well-being, especially in the context of growing populations, must also ensure sustainable food production with better nutritional quality and less contaminants. By definition, sustainable agriculture requires that we meet the food requirements of the current generation without compromising the chances of future generations to meet their requirements and that we acknowledge the needs of environmental systems. Food production has a long history of issues, some of which have inadvertently led to pesticides becoming an environmental contaminant of global concern. These help us to understand the range of factors to be addressed to sustain a healthy agricultural system, and to note how a holistic understanding of a sustainable agricultural system is a relatively recent phenomenon.

Resistance in some crop varieties has rendered pesticides use uneconomic (Maton et al. 2016a). Pesticides and fertilisers were seen initially as essential aids in the growth and protection of food plants, and without these aids, catastrophic food shortages would have resulted (Kumar et al. 2013). During early periods of pesticide application, good harvests and high productivity were celebrated. As demand grew, increasing quantities of pesticides and fertilisers were used to increase crop yields during the so-called 'Green Revolution' (Datta et al. 2016). In the early development phase of these compounds, however, little thought was given to their potential impacts and mobility within air, water, and soil. Over time, targeted pests and plants developed resistance to pesticides, and the immediate

reaction was to increase application rates, but this accelerated the growth of resistance in the target species while at the same time increasing the concentrations of their residues in the environment.

The health of microbial species in the topsoil is an essential element in nutrient production, and this key link in the cycle is compromised with high doses of pesticide. Soil biodiversity plays a critical role in enhancing agricultural sustainability. The destruction of pest predators has also led to increasing virulence of many species of agricultural pests, such as unprecedented mouse plagues. This phenomenon is known as the 'Silent Spring' after the seminal work of Rachel Carson (1962), which documented the use of DDT (dichloro-diphenyl-trichloroethane) and its devastating impact on birds, bees, and other animals.

There is overwhelming evidence that some pesticide chemicals and their degradation products, pose both direct and indirect risks to humans (Forget 1993; Igbedioh 1991). There are no globally accepted health standards for pesticide residue and the diversity and levels for some residues is alarming and hazardous to human and ecosystem health (Handford et al. 2015; Yadav et al. 2015). Those at high risk of exposure include production workers, formulators, sprayers, mixers, loaders, and agricultural farm workers (Aktar et al. 2009). Worldwide, about 25 million agricultural workers experience unintentional pesticide poisonings each year, with approximately 1.8 billion people engaged in agriculture (Carvalho 2017a). Approximately 1 million people per year die or get chronic diseases due to pesticide poisoning (Aktar et al. 2009), with a disproportionate burden observed in people living in developing countries and in close proximity to agricultural activity (WHO 1990). Long-term exposure to pesticides has been shown to be related to cancer, obesity, endocrine disruption and other serious illnesses (WHO 2017; Araújo et al. 2016). At present, there is widespread concern about the herbicide 'glyphosate' which is classified as a carcinogenic agent and currently the most widely applied pesticide worldwide. This substance is clearly a human and ecosystem toxin, with the European Union setting the daily chronic reference dose for glyphosate to 0.5 mg/kg body weight per day (Araújo et al. 2016).

Strategies to limit the negative effects of pesticides should therefore aim to

- Restrict and ultimately prevent the dispersion of pesticide residues into the environment.
- Reverse the unintended reduction of populations of ecologically beneficial species, with particular emphasis on the recovery of bees, birds, amphibians, fish, and small mammals.
- Restore ground water reserves which have been contaminated by leached chemicals.
- Prevent pesticide residues from finding their way into food meant for human and livestock consumption.
- Diminish the possibilities for developed resistance to pesticides that can arise in target pests due to their overuse.
- Reduce the possibilities of poisoning hazards by developing less toxic management techniques.
- Ensure crops and other plant materials are protected from pest insects and plants.
- Remove previous residues from pesticide use from agricultural sites.
- Restore optimum soil properties.

In a functioning ecosystem that incorporates the production of large quantities of healthy foodstuffs, soil biota must be re-established, populations of predator and pollinators must be reinstated, mineral and organic matter must be recycled into the soil, and any leachate must be managed. No other outcome can deliver a sustainable 'systems approach' which addresses the complex interaction of air, water, and soil, required to achieve the SDGs. Agricultural practices must become more efficient while also consuming fewer pesticides and fertiliser, no matter what their current development status or their demand for increasing crops. This apparent

conflict of requirements may have a resolution through the more targeted use of chemicals. For example, direct application of herbicides to weed species or insecticides onto insect larvae will help to reduce the unintended release of pesticides into the environment and protect useful species from overspray. The leaching of chemicals into groundwater is minimised if we reduce the total quantity of pesticide used, which will in turn result in fewer pesticide residues in our food resources. It is also reasonable to assume that, combined with other strategies, it may reduce the risk of the development of 'learned' resistance.

Measures such as moving to production of crops resistant to pests (such as *Bacillus thuringiensis* protein crops), organic farming, development of new cultivars, recuperation of old cultivars, increased use of bio-pesticides, and introducing pheromone traps have already been implemented globally (Carvalho 2017a; Drogui and Lafrance 2012). Such approaches are timely contributions to the development of modern agriculture, and the approaches that they have introduced have already induced a tremendous change to the use of pesticides. Such a change is required for both environmental sustainability and to provide appropriate nutrition for the whole of population wellness.

A sustainable solution needs a continuously healthy ecosystem that contains fertile soil, clean air, clean water, compost, minerals, and natural biological resources (Datta et al. 2016). To achieve this state, immediate reduction of pesticide contamination of surface water is essential to limit toxicological concerns and the adverse effects of pesticides on natural organisms and human health. In addition, efficient treatments for surface waters are needed to remove residual pesticides which have been concentrating for many years. In the short term, the integration of a double-layer filtration unit such as a combined granular activated carbon and sand filtration can constitute an inexpensive and effective pesticide removal device for potable water treatment. It has also been reported that advanced electrolytic-oxidation techniques are promising treatments to remove pesticides from water resources (Drogui and Lafrance 2012).

3.3.5 Further GeoHealth Links

The examples set out in Sects. 3.3.1 to 3.3.4 cover four diverse areas where the work of geoscientists relates to health challenges. In Table 3.3 we note some additional examples, some of which are discussed in other chapters throughout this book. Table 3.3 also illustrates a key tenant of dose response in relation to essential elements and illustrates that elements that are essential to life may also be toxic at an elevated dose.

3.4 Wellness and Longevity Linked to the Geogenic Environment

There is growing evidence that contact with nature and the natural environment positively contributes to the physical and mental health of people (Stokols 1992; Pretty et al. 2005; Ekkel and de Vries 2017). This association is as true in the cities, with appropriate green spaces, as it is in the wild regions. Health-promoting environments span locations as diverse as wilderness zones, parks, and green spaces, to small quiet urban spaces that promote contemplation. The image of this health-promoting environment may not always be '*natural*', but it is arguably clean, and the ecosystem is healthy. Imagine the

Table 3.3 Examples of connections between geoscience and health. N.B. Many health problems can also arise from bacterial, viral, and parasitic organisms, spread by contaminated water. **SDG 6** describes the role of geoscience in providing access to improved water sources

Issue and Link to Health		Further Reading
Fluoride (F-)	Some groundwater has naturally high concentrations of fluoride. Consuming such water can result in dental fluorosis, and if the consumption is long term, skeleton fluorosis. Low concentrations of fluoride are beneficial to dental health. See also **SDG 6**	Yadav et al. (2019); Rasool et al. (2018); Kut et al. (2016); Edmunds and Smedley (2013)
Mercury (Hg)	Historically and now used in artisanal gold mining to extract gold from ore. Impacts of exposure to Hg are neurological, kidney, and possibly immunotoxic/ autoimmune effects	Ha et al. (2017); Gibb and O'Leary (2014)
Micronutrient Deficiencies	As soils become more deficient in micronutrients over time, populations reliant on these soils for subsistence have inadequate mineral and vitamin intakes, leading to poor health. See also **SDG 2**	Knez and Graham (2013)
Selenium (Se)	Essential to human health but harmful at high levels with excessive selenium causing selenosis	Fordyce (2013); Li *et al.* (2012); WHO (1996);
Cadmium (Cd)	Speciation is influential but prolonged exposure via drinking water may result in chronic anaemia amongst other conditions	Burke et al. (2016); Wasana et al. (2016)
Dust	Atmospheric dust particles can be derived from minerals and volcanic eruptions and anthropogenic activities, and constitutes a major influence on human and ecosystem health	Derbyshire (2013); Weinstein et al. (2013)
Radon Gas	A natural radioactive gas formed as a decay product of uranium and is emitted from volcanoes and some rocks/soils. Radon and its decay products emit ionising radiation which is linked to cancer incidence	Linhares et al. (2017); Donald-Appleton (2012)

Scottish Highlands, the African Savana, The Grand Canyon, Uluru, Puerto Natales, and Elephant Island with their breath-taking landscapes as an exemplification of the natural landscape. The interaction of rocks, vegetation, open skies, and perhaps vastness gives a sense of nature and by association, a therapeutic environment. Landscape and wellness are linked (Parsons 1991; Bedimo-Rung et al. 2005; Bowler et al. 2010; Sandifer et al. 2015; Gladwell et al. 2013; Collado et al. 2017).

Health-promoting environments can also improve individual or collective behaviours (e.g., around eating or exercise), to help address many current 'lifestyle' diseases. When addressing obesity, diabetes, and heart disease, the inclusion of exercise is usually part of the journey to better health. A well-designed city, that integrates walking paths and bike paths, that provides outdoor spaces for lunchtime activities, and which promotes human interaction, will contribute to a healthier population. The need for 'green exercise' is important and has public and environmental health consequences (Rogerson et al. 2016).

Work done at Cape Verde, Africa, suggest that there are significant positive benefits from geotourism (Rocha and da Silva 2014). National Parks, Gardens, Geological Parks, and Heritage Parks all provide an experience that promotes well-being (Romagosa et al. 2015). These can provide both mental and physical stimulation and help to engage communities in the larger natural world (see also **SDG 8**). Not only do they promote rural development, but they also promote jobs in traditional as well as novel areas such as pelotherapy (mud therapy) for therapeutic skin and relaxation applications. It also provides educational opportunities for locals and others to understand and appreciate the sustainability of the local environment (Lin and Su 2019). Connection to place is critical to many cultures and this connection supports sustainability and promotes healthy lives (Relph 2016). From Hippocrates (460–370 BCE) to the Aboriginal and Torres Strait Islanders people of Australia today, a connection to place matters.

Kaplan and Kaplan (2003) link health-supportive environments with human behaviour noting that people are more reasonable and satisfied when the natural environment supports them and is explicable to them. In both rural and urban settings, a pleasant environment produced a significantly greater positive effect on self-esteem when compared to a control group (Pretty et al. 2005). This natural and nearby environment, although sometimes neglected, provides an effective resource for health and well-being. A large-scale study of older people found that a predictor of longevity was perceived access to walkable green space, when controlled for age, socio-economic status, gender, and marital status (Takano et al. 2002).

The geogenic environment is clearly more than just a place to visit. It surrounds us; it is where we live and travel. It is also the physical and chemical buffer in the interconnected system of air, soil, and water. It is critical to sustaining life on our planet, but it is also critical to the mental and physical health of humans. Not only can it cause damage to human health when disturbed or contaminated, it also plays a critical role in promoting human health when well maintained. The positive impacts the geogenic environment have on mental and physical health both in urban and regional locations need to have prominence when addressing **SDG 3**.

3.5   Geoscience Actions to Support Health and Wellbeing

Given the direct, indirect, tangible, and intangible benefits of the geogenic environment to human health, it is imperative to outline the ways that geoscientists, environmentalists, and all students of the natural sciences can assist individuals, corporations and governments in delivering **SDG 3**. Our natural resources, water, air, and soil (and the minerals they contain) are essential for life. They enrich our soils and nourish our crops, and they provide energy and the materials for shelter. Negative health outcomes occur when there is insufficient food or where there is contaminated water or air. In contrast, positive well-being flows from a well-maintained therapeutic landscape (Huang and Xu, 2018). This links directly to ensuring healthy lives for all and at all ages.

When air, water, or soil is contaminated, people die or lose years from their lives. When the environment is fragmented, species are lost, biosystems break down and entire ecosystems may be lost, and human health suffers. When the integrity of the geogenic system is lost or the environment undergoes substantive change, new vectors for contamination and disease will be created. Population health is directly linked to patterns of socio-economic disadvantage, and regions of poor environmental quality overlap with low population health outcomes. For society to evolve sustainably, for health outcomes for humans to continue to improve, our environment and its health is a key determinant. Sustainable development is impossible without appropriate attention to our geogenic environment and the maintenance of its health.

Climate fluctuation, with extreme weather events and persistent incremental change, represents a considerable and growing risk to the total environment and will require significant ongoing resources and educational investment to avoid substantial degradation of natural, urban, and rural environments. Everything we eat, drink, and breathe links to the quality of our soil, water, and air and these are in turn deeply affected by climatic fluctuations, which are in turn deeply affected by how we interact with the geogenic environment.

We cannot escape the fact that human and ecosystem health are directly linked, but the complex confounding factors must be understood if we are to achieve **SDG 3**. This requires research, resources (including a boost to funding for environment and wellness research) and intergovernmental commitment. Greater collaborations between geoscientists and health professionals (including those in policy communities) are also important. Box 3.1 outlines some organisations and networks that work to connect geoscience with health. It requires governments to establish and protect national and state parks. It requires planning authorities to mandate green spaces in our cities, it requires corporations to use resources wisely and for our communities to hold them to account, and it requires scientists to understand the deep and complex links between air, water and soil.

Box 3.1 International Geoscience and Health Networks

International Medical Geology Association. The International Medical Geology Association aims to provide a network and a forum to bring together the combined expertise of geologists and earth scientists, environmental scientists, toxicologists, epidemiologists, and medical specialists, in order to characterise the properties of geological processes and agents, the dispersal of geological material and their effects on human populations. Read more: www.medicalgeology.org/

Society for Environmental Geochemistry and Health. SEGH was established in 1971 to provide a forum for scientists from various disciplines to work together in understanding the interaction between the geochemical environment and the health of plants, animals, and humans. We recognise the importance of interdisciplinary research. SEGH members represent expertise in a diverse range of scientific fields, such as biology, engineering, geology, hydrology, epidemiology, chemistry, medicine, nutrition, and toxicology. Read more: www.segh.net/.

AGU GeoHealth Division. The GeoHealth section of AGU aims to nurture transdisciplinary collaborations in order to advance our understanding of the complex interactions between our geospheric environment (including earth, water, soils, and air) and the health, well-being, and continued progress of human populations in concert with all ecosystems. Combining expertise across the geo- and health sciences will facilitate advancement toward a healthier and more sustainable future. GeoHealth is broadly defined to fully encompass the expansive spectrum that covers the earth and climate dynamics, exposure risks, and health impacts. Read more: https://connect.agu.org/geohealthconnect/home.

Understanding the interconnections of systems makes sustainable development possible and is more than just a more efficient system for finding the resources and using them. It requires planning at all stages of development, good access to data analysis tools, development of local policy to provide long-term protection of ecosystems and human health, and global policy designed to assure that poverty, inequality or limited education does not result in unsustainable development. We must engage in capacity building to ensure local and region-specific engagement in sustainable production and resource exploitation and distribution.

Sustainability is not just a goal, it is a requirement for long-term habitability of our planet. It is achievable but requires an understanding of the complexities of our geogenic environment and how humans (and the biosphere in general) interact with that environment. Sustainable development requires that we use our geogenic resources wisely and do so in a socially inclusive manner. Mineral and energy resources have the potential to advance the economic prosperity of many developing countries if resource management practices, laws, and technology evolve to be compliant with environmentally sustainable objectives, protect public health and contribute equitably to the development of local communities (Carvalho 2017b).

The promotion of green space, a respect for our environment and appropriate engagement is the first and most basic step required if we are to promote environmental health and thereby ensure the health of the humans that require that environment to survive and to thrive. We live in a massively interconnected system and understanding that interconnectedness is critical to all aspects of **SDG 3**, and in a larger context, human survival.

3.6 Key Learning Concepts

- Since 1900, the global average life expectancy has more than doubled. Health inequalities persist, however, and actions to prevent ill health, and treat both communicable and non-communicable sickness and diseases, are needed to ensure healthy lives and good well-being for all.
- Good health and well-being allow children to spend more time at school, and the wider population to work and generate income. There are close relationships between good health and well-being and access to nutritious food, education, clean water and sanitation, clean energy, high-quality infrastructure, good living conditions, and protected natural environments.
- Environmental factors have a major influence on health outcomes. The geogenic environment comprises everything that results from our vast geological history (both natural and modified), and includes the environmental compartments of soil, atmosphere, water, and rock. The geogenic environment sustains ecosystems and human health and this interconnectedness of systems means that all changes have consequences. Examples include arsenic contamination of soil and groundwater, exposure to asbestos minerals, poor air quality, and pesticide contamination.
- There is also growing evidence that contact with nature and the natural environment positively contributes to the physical and mental health of people. Green spaces in cities and wider catchments can provide a therapeutic environment, and promote behaviours that address many lifestyle diseases.
- Research, resources (including a boost to funding for environment and wellness research) and intergovernmental commitments

can help to understand the complex links between geoscience and health. Examples of international networks include the *International Medical Geology Association* and the *Society for Environmental Geochemistry and Health.*

3.7 Educational Ideas

In this section, we provide examples of educational activities that connect geoscience, the material discussed in this chapter, and scenarios that may arise when applying geoscience (e.g., in policy, government, private sector international organisations, NGOs). Consider using these as the basis for presentations, group discussions, essays, or to encourage further reading.

- Given the health impacts of using solid fuels for cooking and burning, particularly in the Global South, is a transition to fossil fuels (e.g., gas and oil) justified? Discuss the links between energy poverty and health more widely, reviewing the chapter on SDG 7 to support you.
- How do we manage pesticide use in an overpopulated world? Food production is critical and so is ecosystem health. Is there a balance and how can this move to a sustainable position?
- How could access to sites of geoscientific interest promote positive health and mental well-being? Outline what this looks like at different scales, including (i) access to national parks or geoparks (see SDG 8), and (ii) access to 'green space' in your community.

Further Reading and Resources

Brevik, E.C., Burgess, L.C. (Eds.) (2013). Soils and Human Health. CRC Press, 391 p

Censi, P., Darrah, T., Erel, Y. (Eds.). (2013). Medical Geochemistry: Geological Materials and Health. Springer Science & Business Media

Dobrzhenetskaya, L. (2016) Minerals & Human Health. Cognella Academic press, 322 p

Duffin C.J, Moody R.T.J., Gardner-Thorpe, C. (Eds.) (2013). A History of Geology and Medicine. Special Publication 375. Geological Society, London, 490 p

Mori, I., Ibaraki, H. (Eds.) (2017). Progress in Medical Geology. Cambridge Scholars Publishing. 329 p

Selinus, O., Alloway, B., Centeno, J.A., Finkelman, R. B., Fuge, R., Lindh, U., Smedley, P. (Eds.) (2013). Essentials of Medical Geology, Revised Edition, Springer. 805 p

References

Abraham J, Dowling K, Florentine S (2018) Effects of prescribed fire and post-fire rainfall on mercury mobilization and subsequent contamination assessment in a legacy mine site in Victoria, Australia, Chemosphere: 190. ISSN 144–153:0045–6535. https://doi.org/10.1016/j.chemosphere.2017.09.117

Aktar MW, Sengupta D, Chowdhury A (2009) Review Article: Impact of pesticides use in agriculture: their benefits and hazards. Interdisciplinary Toxicology 2 (1):1–12

Alexandratos N, Bruinsma J (2012) World agriculture towards 2030/2050: The 2012 revision. ESA working paper no. 12–03, Agricultural Development Economics Division, Food and Agriculture Organization of the United Nations. 1–160

Allen LP, Baez J, Stern MEC, Takahashi K, George F (2018) Trends and the Economic Effect of Asbestos Bans and Decline in Asbestos Consumption and Production Worldwide. International Journal of Environmental Research and Public Health 15(3):531

Alpert JS (2018) The role of the environment in health outcomes. https://doi.org/10.1016/j.amjmed.2018.06.001.http://www.who.int/quantifying_ehimpacts/publications/ebd5.pdf

Amuna P, Zotor F (2008) Epidemiological and nutrition transition in developing countries: Impact on human health and development: The epidemiological and nutrition transition in developing countries: Evolving trends and their impact in public health and human development. Proceedings of the Nutrition Society 67 (1):82–90. https://doi.org/10.1017/S0029665108006058

Araújo J, Delgado FI, Paumgartten FJR (2016) Glyphosate and adverse pregnancy outcomes, a systematic review of observational studies. BMC Public Health 16:472

Arcaya MC, Arcaya AL, Subramanian S (2015) Inequalities in health: definitions, concepts, and theories. Global Health Action 8(1):27106–27106

Arčon I, van Elteren JT, Glass HJ, Kodre A, Slejkovec Z (2005) EXAFS and XANES study of arsenic in contaminated soil. X-Ray Spectrom 34:435–8

ASEA, Asbestos Safety and Eradication Agency (2017). National asbestos profile for Australia

Bedimo-Rung AL, Mowen AJ, Cohen DA (2005) The significance of parks to physical activity and public health: a conceptual model. Am J Prev Med 28(2):159–168

Bianchi C, Bianchi T (2014) Global mesothelioma epidemic: Trend and features. Indian journal of occupational and environmental medicine 18(2):82

Bing M, Abel RL, Pendergrass P, Sabharwal K, McCauley C (2000) Data used to improve quality of health care. Tex Med 96(10):75–79

Bowler DE, Buyung-Ali LM, Knight TM, Pullin AS (2010) A systematic review of evidence for the added benefits to health of exposure to natural environments. BMC Public Health 10(1):456

Burke F, Hamza S, Naseem S, Nawaz-ul-Huda S, Azam M, Khan I (2016) Impact of cadmium polluted groundwater on human health: winder. Balochistan. Sage Open 6(1):2158244016634409

Carvalho FP (2017a) Review: Pesticides, environment and food safety. Food and Energy Security 6(2):48–60

Carvalho FP (2017b) Mining industry and sustainable development: Time for change. Food and Energy Security 6(2):61–77

CDC (2018) Centers for Disease Control and Prevention - Well-being concepts. Available at: https://www.cdc.gov/hrqol/wellbeing.htm#three (Accessed 1 October 2019)

Chen JC, Schwartz J (2009) Neurobehavioral effects of ambient air pollution on cognitive performance in US adults. Neurotoxicology 30(2):231–239

CMS (2018) National Health Expenditure Data. Available at: https://www.cms.gov/research-statistics-data-and-systems/statistics-trends-and-reports/nationalhealthexpenddata/nationalhealthaccountshistorical.html (Accessed 19 August 2019)

Collado S, Staats H, Corraliza JA, Hartig T (2017) Restorative environments and health. In Handbook of environmental psychology and quality of life research. Springer, Cham, pp 127–148

Collegium Ramazzini C (2016) The global health dimensions of asbestos and asbestos-related diseases. Journal of Occupational Health 58(2):220–223

Crooks JL, Cascio WE, Percy MS, Reyes J, Neas LM, Hilborn ED (2016) The association between dust storms and daily non-accidental mortality in the United States, 1993–2005. Environ Health Perspect 124(11):1735

Daniels SR (2006) The consequences of childhood overweight and obesity. The future of children 16(1):47–67

Datta S, Singh J, Singh S, Singh J (2016) Earthworms, pesticides and sustainable agriculture: a review. Environ Sci Pollut Res 23(9):8227–8243

Derbyshire E (2013) Natural aerosolic mineral dusts and human health. In Essentials of medical geology (pp. 455–475). Springer, Dordrecht

Dong D, Xu X, Wong YF (2019) Estimating the Impact of Air Pollution on Inbound Tourism in China: An Analysis Based on Regression Discontinuity Design. Sustainability 11(6):1682

Dowling K, Florentine S, Martin R, Pearce D (2016) Chapter 17 Sustainability and Regional Development: When Brownfields Become Playing Fields Sustainability in the Mineral and Energy Sectors

Drogui P, Lafrance P (2012) Pesticides and sustainable agriculture. Farm Food Water Sec 23–55

Edmunds WM, Smedley PL (2013) Fluoride in natural waters. In Essentials of medical geology. Springer, Dordrecht, pp 311–336

Ekkel ED, de Vries S (2017) Nearby green space and human health: Evaluating accessibility metrics. Landscape and urban planning 157:214–220

EPA (2019) EPA Budget. Available online: www.epa.gov/planandbudget/budget (Accessed 19 August 2019)

Escamilla V, Wagner B, Yunus M, Streatfield PK, van Geen A, and Emch M. (2011) Effect of deep tube well use on childhood diarrhoea in Bangladesh. Bulletin of the World Health Organization; 89:21–527. https://doi.org/10.2471/blt.10.085530. Retrieved from http://www.who.int/bulletin/volumes/89/7/10-085530/en/ 30 June 2018

Fordyce FM (2013) Selenium deficiency and toxicity in the environment. In: Essentials of medical geology. Springer, Dordrecht, pp 375–416

Forget G (1993) Balancing the need for pesticides with the risk to human health. In: Forget G, Goodman T and de Villiers A IDRC (eds) Impact of pesticide use on health in developing countries, Ottawa: 2

Fullerton DG, Bruce N, Gordon SB (2008) Indoor air pollution from biomass fuel smoke is a major health concern in the developing world. Trans R Soc Trop Med Hyg 102(9):843–851. https://doi.org/10.1016/j.trstmh.2008.05.028

Furuya S, Chimed-Ochir O, Takahashi K, David A, Takala J (2018) Global Asbestos Disaster. International Journal of Environmental Research and Public Health 15(5):1000

Gibb H, O'Leary KG (2014) Mercury exposure and health impacts among individuals in the artisanal and small-scale gold mining community: a comprehensive review. Environ Health Perspect 122(7):667–672

Gladwell V, Brown DK, Wood C, Sandercock G, Barton J (2013) The Great Outdoors: how a green exercise environment can benefit all. Extreme Physiol Med 2(1):3–3

Ha E, Basu N, Bose-O'Reilly S, Dórea JG, McSorley E, Sakamoto M, Chan HM (2017) Current progress on understanding the impact of mercury on human health. Environ Res 152:419–433

Handford CE, Elliott CT, Campbell K (2015) A review of the global pesticide legislation and the scale of challenge in reaching the global harmonization of food safety standards. Integrat Environ Assess Manag 11(4):525–536

Hosseinpoor AR, Stewart Williams JA, Itani L, Chatterji S (2012) Socioeconomic inequality in domains of health: results from the World Health Surveys. BMC Public Health 12(1):198. https://doi.org/10.1186/1471-2458-12-198

Huang L, Xu H (2018) Therapeutic landscapes and longevity: Wellness tourism in Bama. Soc Sci Med 197:24–32

Igbedioh SO (1991) Effects of Agricultural Pesticides on Humans, Animals, and Higher Plants in Developing Countries. Arch Environ Health: An Int J 46(4): 218–224

Jang YC, Hong S, Lee J, Lee MJ, Shim WJ (2014) Estimation of lost tourism revenue in Geoje Island from the 2011 marine debris pollution event in South Korea. Mar Pollut Bull 81(1):49–54

Joseph, T., Dubey, B., and McBean, E. A. (2015). A critical review of arsenic exposures for Bangladeshi adults. Science of The Total Environment 540–551

Kaplan S, Kaplan R (2003) Health, supportive environments, and the reasonable person model. Am J Public Health 93(9):1484–1489

Knez M, Graham RD (2013) The Impact of Micronutrient Deficiencies in Agricultural Soils and Crops on the Nutritional Health of Humans. In: Essentials of Medical Geology (pp. 517–533). Springer, Dordrecht

Kumar S, Sharma AK, Rawat SS, Jain DK, Ghosh S (2013) Use of pesticides in agriculture and livestock animals and its impact on the environment of India. Asian J Environ Sci 8(1):51–57

Kurane I (2010) The effect of global warming on infectious diseases. Osong Public Health Res Perspect. 1(1):4–9. https://doi.org/10.1016/j.phrp.2010.12.004 Epub 2010 Dec 7

Kut KMK, Sarswat A, Srivastava A, Pittman CU Jr, Mohan D (2016) A review of fluoride in African groundwater and local remediation methods. Groundwater Sustain Develop 2:190–212

Landrigan PJ. (1991) The third wave of asbestos disease: exposure to asbestos in place. Public health control. Introduction. Ann N Y Acad Sci 1991; 643: xv–xv

Lango H, Weedon MN (2008) What will whole genome searches for susceptibility genes for common complex disease offer to clinical practice? J Intern Med 263 (1):16–27

Li S, Xiao T, Zheng B (2012) Medical geology of arsenic, selenium and thallium in China. Sci Total Environ 421:31–40

Lin JC, Su SJ (2019) A New Way of Understanding Geoparks for Society. In Geoparks of Taiwan (pp. 55–67). Springer, Cham

Linhares, D., Garcia, P., & Rodrigues, A. (2017). Radon Exposure and Human Health: What Happens in Volcanic Environments?. Radon, 79

Liu P, Guo Y, Qian X, Tang S, Li Z, Chen L (2014) China's distinctive engagement in global health. The Lancet 384(9945):793–804

Lozano R, Naghavi M, Foreman K (2013) Global and regional mortality from 235 causes of death for 20 age groups in 1990 and 2010 [erratum: Lancet. 381 (9867):628]. Lancet. 2012;380(9859):2095–2128

Martin R, Dowling K, Pearce DC, Florentine S, Bennett JW, Stopic A (2016a). Size-dependent characterisation of historical gold mine wastes to examine human pathways of exposure to arsenic and other potentially toxic elements. Environ Geochem Health 1–18

Martin R, Dowling K, Pearce D, Sillitoe J, Florentine S (2014) Health Effects Associated with Inhalation of Airborne Arsenic Arising from Mining Operations. Geosciences 4(3):128–175

Martin R, Dowling K, Pearce DC, Bennett J, Stopic A (2013) Ongoing soil arsenic exposure of children living in an historical gold mining area in Regional Victoria, Australia: Identifying risk factors associated with uptake. J Asian Earth Sci. ISSN 1367–9120

Martin R, Dowling K, Pearce DC, Florentine S, McKnight S, Stelcer E, Cohen DD, Stopic A, Bennett JW (2016b) Trace metal content in inhalable particulate matter (PM2. 5–10 and PM2. 5) collected from historical mine waste deposits using a laboratory-based approach. Environmental geochemistry and health, 1–15

Moschino V, Schintu M, Marrucci A, Marras B, Nesto N, Da Ros L (2017) An ecotoxicological approach to evaluate the effects of tourism impacts in the Marine Protected Area of La Maddalena (Sardinia, Italy). Mar Pollut Bull 122(1–2):306–315

National Research Council (NRC). (2007). Earth materials and health: Research priorities for earth science and public health. National Academies Press

Núñez O, Fernández-Navarro P, Martín-Méndez I, Bel-Lan A, Locutura JF, López-Abente G (2016). Arsenic and chromium topsoil levels and cancer mortality in Spain. Environmental Science and Pollution Research, 23(17), 17664–17675. https://link.springer.com/article/10.1007/s11356-016-6806-y

Nymark P, Wikman H, Hienonen-Kempas T, Anttila S (2008) Molecular and genetic changes in asbestos-related lung cancer. Cancer Lett 265(1):1–15

Pal A, Nayak B, Das B, Hossain MA, Ahamed S, Chakraborti D (2007) Additional danger of arsenic exposure through inhalation from burning of cow dung cakes laced with arsenic as a fuel in arsenic affected villages in Ganga–Meghna–Brahmaputra plain. J Environ Monit 9(10):1067–1070

Parsons R (1991) The potential influences of environmental perception on human health. J Environ Psychol 11(1):1–23

Parrish CR, Holmes EC, Morens DM, Park EC, Burke DS, Calisher CH, Laughlin CA, Saif LJ and Daszak P. (2008). Cross-species virus transmission and the emergence of new epidemic diseases. Micro Mole Biol Rev 72(3), 457–470

Pearce D, Dowling K, Sim K (2012) Cancer incidence and soil arsenic exposure in a historical gold mining area in Victoria, Australia: a geospatial analysis. J Expo Sci Environ Epidemiol. 22(3):248–57

Pearce D, Dowling K, Gerson A, Sim M, Sutton M, Newville M, Russell R, McOrist G (2010) Arsenic microdistribution and speciation in toenail clippings of children living in a historic gold mining area. Sci Total Environ 408(10):2590–2599

Perera FP, Rauh V, Tsai WY, Kinney P, Camann D, Barr D, Dietrich J (2003) Effects of transplacental exposure to environmental pollutants on birth outcomes in a multiethnic population. Environ Health Perspect 111(2):201–205

Pretty J, Peacock J, Sellens M, Griffin M (2005) The mental and physical health outcomes of green exercise. Int J Environ Health Res 15(5):319–337

Raessler M (2018) The arsenic contamination of drinking and groundwaters in Bangladesh: featuring biogeochemical aspects and implications on public health. Arch Environ Contam Toxicol 75(1):1–7

Rasool A, Farooqi A, Xiao T, Ali W, Noor S, Abiola O, Ali S, Nasim W (2018) A review of global outlook on fluoride contamination in groundwater with prominence on the Pakistan current situation. Environ Geochem Health 40(4):1265–1281

Relph, E. (2016). Senses of place and emerging social and environmental challenges. In *Sense of place, health and quality of life* (pp. 51–64). Routledge

Requia WJ, Adams MD, Arain A, Papatheodorou S, Koutrakis P, Mahmoud M (2018) Global Association of air Pollution and Cardiorespiratory Diseases: a systematic review, meta-analysis, and investigation of modifier variables. Am J Public Health 108(S2):S123–S130

Rocha F, da Silva, EF (2014) Geotourism, Medical Geology and local development: Cape Verde case study. J Afri Earth Sci 735–742

Rodriguez-Morales AJ, Bonilla-Aldana DK, Balbin-Ramon GJ, Rabaan AA, Sah R, Paniz-Mondolfi A, Pagliano P and Esposito S (2020) History is repeating itself: Probable zoonotic spillover as the cause of the 2019 novel Coronavirus Epidemic. Infez Med 28 (1):3–5

Rogerson M, Brown DK, Sandercock G, Wooller JJ, Barton J (2016) A comparison of four typical green exercise environments and prediction of psychological health outcomes. Perspectives Public Health 136 (3):171–180

Romagosa F, Eagles PF, Lemieux CJ (2015) From the inside out to the outside in: Exploring the role of parks and protected areas as providers of human health and well-being. J Outdoor Recreation Tour 10:70–77

Roser M (2018) Life Expectancy. Available at: https:// ourworldindata.org/life-expectancy (Accessed 29 October 2019)

Roser M, Ritchie H (2019) Indoor Air Pollution. https:// ourworldindata.org/indoor-air-pollution (Accessed 1 October 2019)

Sachan N, Singh VP (2010) Effect of climatic changes on the prevalence of zoonotic diseases. Veterinary World 3(11):519–522

Samsidar A, Siddiquee S, Shaarani SM (2018) A review of extraction, analytical and advanced methods for determination of pesticides in environment and foodstuffs. Trends Food Sci Technol 71:188–201

Sandifer PA, Sutton-Grier AE, Ward BP (2015) Exploring connections among nature, biodiversity, ecosystem services, and human health and wellbeing: Opportunities to enhance health and biodiversity conservation. Ecosystem Services 12:1–15

Sattar A, Xie S, Hafeez MA, Wang X, Hussain HI, Iqbal Z, Yuan Z (2016) Metabolism and toxicity of arsenicals in mammals. Environ Toxi Pharma 214–224

Shankar S, Shanker U, Shikha. (2014) Arsenic contamination of groundwater: a review of sources, prevalence, health risks, and strategies for mitigation. The Scientific World Journal, 2014.

Smil V (2016) China's Environmental Crisis: An Enquiry into the Limits of National Development: An Enquiry into the Limits of National Development. Routledge

Ssewanyana D, Mwangala PN, Marsh V, Jao I, van Baar A, Newton CR, Abubakar A (2018) Socioecological determinants of alcohol, tobacco, and drug use behavior of adolescents in Kilifi County at the Kenyan coast. J Health Psyc. https://doi.org/10.1177/1359105318782594

Stokols D (1992) *Establishing and maintaining healthy environments: Toward a social ecology of health promotion.* Am Psychol 47(1):6–22

Sunyer J, Esnaola M, Alvarez-Pedrerol M, Forns J, Rivas I, López-Vicente M, Viana M (2015) Association between traffic-related air pollution in schools and cognitive development in primary school children: a prospective cohort study. PLoS Med 12(3): e1001792

Takala J (2015) Eliminating occupational cancer. Ind Health 53(4):307–309

Takano T, Nakamura K, Watanabe M (2002) Urban residential environments and senior citisens' longevity in megacity areas: the importance of walkable green spaces. J Epidemiol Community Health 56(12):913–918

United Nations (2015b) The Millennium Development Goals Report. Available at: www.un.org/ millenniumgoals/2015_MDG_Report/pdf/MDG% 202015%20rev%20(July%201).pdf (Accessed 13 August 2019)

United Nations (2019). Sustainability Development Goals. Goal 3: Ensure healthy lives and promote well-being for all at all ages. Available from https:// www.un.org/sustainabledevelopment/health/ (Accessed 4 July 2019)

Vahter M (2002) Mechanisms of arsenic biotransformation. Toxicology 181:211–217

Veronese N, Li Y, Manson JE, Willett WC, Fontana L, Hu FB (2016) Combined associations of body weight and lifestyle factors with all cause and cause specific mortality in men and women: prospective cohort study. BMJ, 355, i5855

Wallerstein N (1992) Powerlessness, empowerment, and health: implications for health promotion programs. American journal of health promotion 6(3):197–205

Wasana HM, Aluthpatabendi D, Kularatne WMTD, Wijekoon P, Weerasooriya R, Bandara J (2016) Drinking water quality and chronic kidney disease of unknown etiology (CKDu): synergic effects of fluoride, cadmium and hardness of water. Environ Geochem Health 38(1):157–168

Weinstein P, Horwell CJ, Cook A (2013) Volcanic emissions and health. In Essentials of medical geology (pp. 217–238). Springer, Dordrecht

Global Health Estimates 2016: Deaths by Cause, Age, Sex, by Country and by Region, 2000–2016. Geneva, World Health Organization; 2018

World Health Organization (WHO) (1990) Public health impact of pesticides used in agriculture. World Health Organization, Geneva, p 88

World Health Organization (WHO) (nd). Arsenic Fact Sheet. Available at: https://www.who.int/news-room/fact-sheets/detail/arsenic (accessed 6 September 2019)

World Health Organization (WHO) 2016. 7 million premature deaths annually linked to air pollution. Available at: http://www.who.int/mediacentre/news/releases/2014/air-pollution/en

World Health Organization (WHO) 2017 Regional Office for Africa. Communicable Diseases. https://www.afro.who.int/health-topics/communicable-diseases (Accessed 6 July 2019)

World Health Organization (WHO) 2018a Noncommunicable diseases https://www.who.int/news-room/fact-sheets/detail/noncommunicable-diseases (Accessed 6 July 2019)

World Health Organization (WHO) 2018b, May 24. *The top 10 causes of death*. Retrieved from http://www.who.int/news-room/fact-sheets/detail/the-top-10-causes-of-death (Accessed 30 June 2019)

World Health Organization (WHO) 2018c. *Indoor air pollution and household energy*. Retrieved from http://www.who.int/heli/risks/indoorair/indoorair/en/ (Accessed 30 June 2019)

World Health Organization (WHO) 2018d. Department of Information, Evidence and Research, mortality database http://www-dep.iarc.fr/WHOdb/WHOdb.htm (Accessed 6 June 2019)

World Health Organization (WHO) 2019a. Constitution. Available from https://www.who.int/about/who-we-are/constitution (Accessed 4 July 2019)

World Health Organization (WHO) 2019b. Middle East respiratory syndrome coronavirus (MERS-CoV). Available from https://www.who.int/emergencies/mers-cov/en/ (Accessed 4 July2019)

World Health Organization (WHO) 2019c. Mental health: a state of wellbeing. Available from https://www.who.int/features/factfiles/mental_health/en/ (Accessed 4 July 2019)

World Health Organization (WHO) 2020a. Naming the coronavirus disease (COVID-19) and the virus that causes it. Available from https://www.who.int/emergencies/diseases/novel-coronavirus-2019/technical-guidance/naming-the-coronavirus-disease-(covid-2019)-and-the-virus-that-causes-it. (Accessed 25 October 2020)

World Health Organization (WHO) 2020b. Coronavirus disease 2019 (COVID-19) Situation Report – 94. https://www.who.int/docs/default-source/coronaviruse/situation-reports/20200423-sitrep-94-covid-19.pdf. (Accessed 25 October 2020)

Yadav IC, Devi NL, Syed JH, Cheng Z, Li J, Zhang G, Jones KC (2015) Current status of persistent organic pesticides residues in air, water, and soil, and their possible effect on neighboring countries: a comprehensive review of India. Sci Total Environ 511:123–137

Yadav KK, Kumar S, Pham QB, Gupta N, Rezania S, Kamyab H, Yadav S, Vymazal J, Kumar V, Tri DQ, Talaiekhozani A (2019) Fluoride contamination, health problems and remediation methods in Asian groundwater: A comprehensive review. Ecotoxicol Environ Saf 182:109362

Yanamala N, Kisin ER, Gutkin DW, Shurin MR, Harper M, Shvedova AA (2018) Characterization of pulmonary responses in mice to asbestos/asbestiform fibers using gene expression profiles. Journal of Toxicology and Environmental Health, Part A 81 (4):60–79

Kim Dowling has a long history working in mining and mining affected landscapes. Her research looks specifically at heavy metals and heavy metal mobility in modified and built environments. As leader of the Arsenic Research Group at FedUni, she has engaged in projects investigating air, water, and soil contaminant mobility in both Australia, Bangladesh and South Africa. Her research interests include public health and the meaningful dissemination of environmental information. Kim is a member of the International Medical Geology Association and regularly reports on environmental matters in the media. Her experiences in governance through various leadership roles, and while working in Papua New Guinea as a lecturer and in many remote areas of Australia give her a broad perspective on the complex issues related to the human relationship with the environment.

Singarayer K. Florentine is a Professor in restoration and invasive species ecology, with more than 25 years of experience in research and higher education teaching. Singarayer has led the development of strong collaborative research partnerships with over 30 natural-resource management groups in Australia, as well as with high profile national and international researchers. Singarayer has published over 100 refereed journal articles, and was the 2017 winner of the Australian Federal Government award for University Teaching for the development of innovative curricula in the field of restoration ecology that are linked to high-quality work readiness student outcomes.

Rachael Martin is a multidisciplinary environmental scientist with a keen interest in contaminants and their impacts on human health and the environment. She completed her Ph.D. which investigated the fate and transport of heavy metals in historical gold mine waste deposits in regional Victoria. Since then, she has worked in technical and advisory roles for the New South Wales Government, both in the regulatory and public health sectors.

Dora C. Pearce is interested in environmental epidemiology, focusing on spatial statistical techniques aimed at detecting evidence of adverse health outcomes associated with environmental exposures and infectious disease spread. After gaining experience in organic chemistry and as a hospital scientist, Dora switched to biostatistics and public health research. Dora gained her Ph.D. at the University of Ballarat by modelling the geospatial variation in soil arsenic concentration and cancer incidence at an aggregate population level, and exploring arsenic uptake by investigating the microdistribution and speciation of arsenic in thin sections of children's toenail clippings using synchrotron-based X-ray microprobe techniques.

Quality Education

4

Ellen Metzger, David Gosselin,
and Cailin Huyck Orr

E. Metzger (✉)
Department of Geology, San José State University,
San José, CA, USA
e-mail: ellen.metzger@sjsu.edu

D. Gosselin
Environmental Studies, University of Nebraska At
Lincoln, Lincoln, NE, USA

C. H. Orr
Science Education Resource Center, Carleton
College, Northfield, MN, USA

© Springer Nature Switzerland AG 2021
J. C. Gill and M. Smith (eds.), *Geosciences and the Sustainable Development Goals*,
Sustainable Development Goals Series, https://doi.org/10.1007/978-3-030-38815-7_4

Abstract

4 QUALITY EDUCATION

Geoscience improves education and contributes to the SDGs...

Geoscientists improve access to safe water and reliable energy, allowing children to stay in school longer

Knowledge of natural resources, geohazards and climate change is essential to promoting sustainable development

Geoscientists have innovative ways of thinking across spatial and temporal scales

Enhancing geoscientists' capacity to contribute to the SDGs requires new approaches to geoscience education...

1 More dialogue with other disciplines

2 Incorporation of social and economic sustainability concepts into curricula

3 Development of new skills (e.g., communication, stakeholder mapping)

More public understanding of geoscience will help to foster Earth-literate citizenship, and inspire and inform action towards all the SDGs...

Raise awareness of climate change, and the actions needed to mitigate and adapt to its impacts

Give communities greater understanding of their natural heritage (e.g., through geoparks)

Build understanding of the causes and impacts of hazards, and inspire action to prevent and mitigate disasters

This will foster Earth-literate citizenship, and inspire action towards all the SDGs

Enhancing geoscientists' engagement with the SDGs requires...

More collaboration between Earth scientists

Strengthening pre-college Earth science education & teacher preparation

Incorporating fundamental sustainability concepts and skills into Earth science curricula:

Eradicating poverty

Protecting the natural environment

Ensuring universal access to basic services and ending unsustainable consumption patterns

4.1 Introduction

Education plays a crucial role in delivering the Sustainable Development Goals (SDGs). Quality education is a standalone goal (**SDG 4**), expressed through seven targets and three means of implementation (Table 4.1). It is also, however, an essential catalyst and facilitator for all 17 of the SDGs discussed in this book (UNESCO 2015, 2017a, 2018a). Eradicating poverty, ensuring universal access to basic services, tackling inequality, ending unsustainable consumption patterns, and protecting the natural environment all require new attitudes and ways of living that can only be accomplished through innovative approaches and greater access to education. Quality education, as envisaged in **SDG 4** (*ensure inclusive and equitable quality education and promote lifelong learning*

opportunities for all), provides people with the knowledge, skills, tools, and capacity for lifelong learning that fosters inclusive human well-being while preserving the Earth systems upon which all life ultimately depends (e.g., UNESCO 2014a, 2015, 2017a).

A key message of **SDG 4** is that 'education-as usual will not suffice' (UNESCO 2016a). This has important implications for education across all levels, settings, and disciplines. The solution-resistant challenges to sustainability that arise at the intersection of intertwined and complexly interacting natural and human systems demand innovative thinking regarding the nature and purpose of education. This applies to the education of geoscientists, as much as the general education of the population as a whole.

Transforming Our World: The 2030 Agenda for Sustainable Development presents the

Table 4.1 SDG 4 Targets and Means of Implementation

Target	Description of Target (4.1–4.7) or Means of Implementation (4.A–4.C)
4.1	Ensure that all girls and boys complete free, equitable and quality primary and secondary education leading to relevant and Goal-4 effective learning outcomes
4.2	Ensure that all girls and boys have access to quality early childhood development, care and pre-primary education so that they are ready for primary education
4.3	Ensure equal access for all women and men to affordable and quality technical, vocational and tertiary education, including university
4.4	Substantially increase the number of youth and adults who have relevant skills, including technical and vocational skills, for employment, decent jobs and entrepreneurship
4.5	Eliminate gender disparities in education and ensure equal access to all levels of education and vocational training for the vulnerable, including persons with disabilities, indigenous peoples and children in vulnerable situations
4.6	Ensure that all youth and a substantial proportion of adults, both men and women, achieve literacy and numeracy
4.7	Ensure that all learners acquire the knowledge and skills needed to promote sustainable development, including, among others, through education for sustainable development and sustainable lifestyles, human rights, gender equality, promotion of a culture of peace and non-violence, global citizenship and appreciation of cultural diversity and of culture's contribution to sustainable development
4.A	Build and upgrade education facilities that are child, disability and gender sensitive and provide safe, nonviolent, inclusive and effective learning environments for all
4.B	By 2020, substantially expand globally the number of scholarships available to developing countries, in particular least developed countries, small island developing States and African countries, for enrolment in higher education, including vocational training and information and communications technology, technical, engineering and scientific programmes, in developed countries and other developing countries
4.C	By 2030, substantially increase the supply of qualified teachers, including through international cooperation for teacher training in developing countries, especially least developed countries and small island developing states

Sustainable Development Goals (SDGs) as a plan for 'people and planet'. This captures the idea that human well-being depends not only on peaceful and equitable economic development but also on functioning Earth systems, which are currently in jeopardy. As experts in observing, interpreting, modelling, and monitoring Earth systems, geoscientists are uniquely equipped to address the 'planet' dimension of Agenda 2030. Historically, geoscientists have given less consideration to the 'people' component, including the human actions, institutions, and values that shape how people interact with each other and the planet (Mora 2013; Gill 2016; Stewart and Gill 2017). The thematic content of Earth science education (e.g., natural resources, geohazards; climate change) clearly aligns with the SDGs, but inattention to the social and economic dimensions of sustainability in the formal education and continued professional development of most geoscientists may hinder their ability to engage effectively in international sustainability initiatives. To better integrate geoscience expertise with the SDGs, Earth science education needs to broaden its scope to incorporate key sustainability concepts (Gill 2016; Stewart and Gill 2017; Gill et al. 2018).

Through this chapter, we, therefore, aim to examine the ambitions of **SDG 4** for transformative education and its interconnections with other SDGs. We will highlight the unique knowledge and skills that geoscientists can contribute to sustainable development efforts, and explore how geoscience education can prepare scientists and citizens to act as agents of change for sustainability.

4.2   Education in a Global Development Context

The UN Millennium Development Goals (2000–2015), the predecessor to the SDGs, included one goal focused on education. This aimed, by 2015, to ensure that all children completed a full course of primary schooling. From 2000 to 2015, the number of primary school children not attending school decreased by nearly 50%. However, 57 million children of primary school age were still out of school, with more than half of them living in sub-Saharan Africa (United Nations 2015b). Several key trends in global education underscore additional challenges (all from United Nations 2018):

- Worldwide, an estimated 617 million children and youth of primary and lower secondary school age lack basic mathematics and literacy skills.
- In 2016, only 34% of primary schools in the poorest countries had electricity; less than 40% had basic facilities for handwashing.
- Globally, about 85% of the world's primary school teachers were trained in 2016. This number was 71% in Southern Asia and just 61% in sub-Saharan Africa.

Recent data from the UNESCO Institute for Statistics (2018) highlight wide disparities in access to education based on gender and income:

- 262 million (about one in every five) of the world's children and youth ages 6–17 are out of school; this number rises to approximately one in three in low- and lower-middle-income countries.
- Girls still face obstacles to education, particularly in sub-Saharan Africa, where girls of all ages are more likely than boys to be denied the right to education.
- There are significant gaps in out-of-school rates between the world's poorest and richest countries. Nearly every child of primary age is in school in high-income countries, compared to 80% in low-income countries. The gap widens with age: just 6% of upper-secondary youth are out of school in high-income countries compared to 60% in the poorest countries.

The challenges above and others are addressed by **SDG 4**, which takes a much more holistic approach to education than the Millennium Development Goals, having both a broader agenda and universal applicability (Table 4.1). It has relevance to all nations, from those considered to be least developed to those classified as being high income (UNESCO 2015). **SDG 4** encompasses all levels and types of education (e.g., early childhood, primary, secondary, technical and vocational, higher education, lifelong learning) and different modalities and settings for learning, both formal and informal. It addresses not only the *quantity* of education, but also its *quality,* going beyond access to basic education and enrolment and completion rates, to also consider the quality of learning environments, teacher qualifications, and the effectiveness of curricular frameworks (UNESCO 2015, 2017a).

SDG 4 includes seven targets and three means of implementation (Table 4.1), which we can group into three themes: (1) access to education, (2) learning outcomes, and (3) the educational infrastructure facilitating the implementation of SDG 4 (Table 4.2). Learning outcomes, the focus of this chapter, include not only content knowledge but also development of key competencies, attitudes, and values that empower learners to become change-makers for sustainability (UNESCO 2017a, 2018a).

SDG 4 promotes education that goes beyond the transfer of knowledge to develop core competencies and build capacity for addressing sustainability challenges (UNESCO 2014a, 2015, 2018a). Competencies are broader than skills, described by Wiek et al. (2011) as a constellation of synergistic knowledge, skills, and attitudes that

enable action and problem-solving. UNESCO (2017a) has identified eight overarching sustainability competencies, including cognitive, affective, volitional, and motivational components (Fig. 4.1).

Education that fosters an understanding of the Earth system is a key element of **Target 4.7,** which is subdivided into two complementary and mutually reinforcing components: Education for Sustainable Development and Global Citizenship Education (UNESCO 2015, 2017a).

- **Education for Sustainable Development (ESD)** is interdisciplinary, and addresses both learning content and pedagogy. It includes key sustainability topics (e.g., poverty reduction, sustainable consumption, climate change, and disaster risk reduction) and student-centred, participatory, and action-oriented approaches to teaching and learning (UNESCO 2017a). ESD was the focus of the United Nations Decade of Education for Sustainable Development 2004–2014, which aimed to reorient all aspects and levels of education to include principles and practices of sustainable development (UNESCO 2014a).
- **Global Citizenship Education (GCED)** 'aims to be transformative, building the knowledge, skills, values and attitudes that

Table 4.2 Paraphrased SDG 4 Targets and Means of Implementation, expressed in terms of (1) Access to Education, (2) Learning Outcomes, and (3) Educational Infrastructure

SDG 4 – Paraphrased Targets and Means of Implementation		
Access to Education	Learning Outcomes	Educational Infrastructure
4.1: Equitable and quality primary and secondary education for all **4.2:** Early childhood development and pre-primary education for all **4.3:** Affordable and quality technical, vocational and tertiary education for all **4.5:** Ensure access to all levels of education training regardless of gender, disability, or vulnerability	**4.4:** Increase relevant workforce skills – technical and vocational **4.6:** Increase literacy and numeracy of all **4.7:** Knowledge and skills to promote sustainable development	**4.a:** Quality education facilities **4.b:** Expand access to scholarship resources **4.c:** Expand access to quality educators

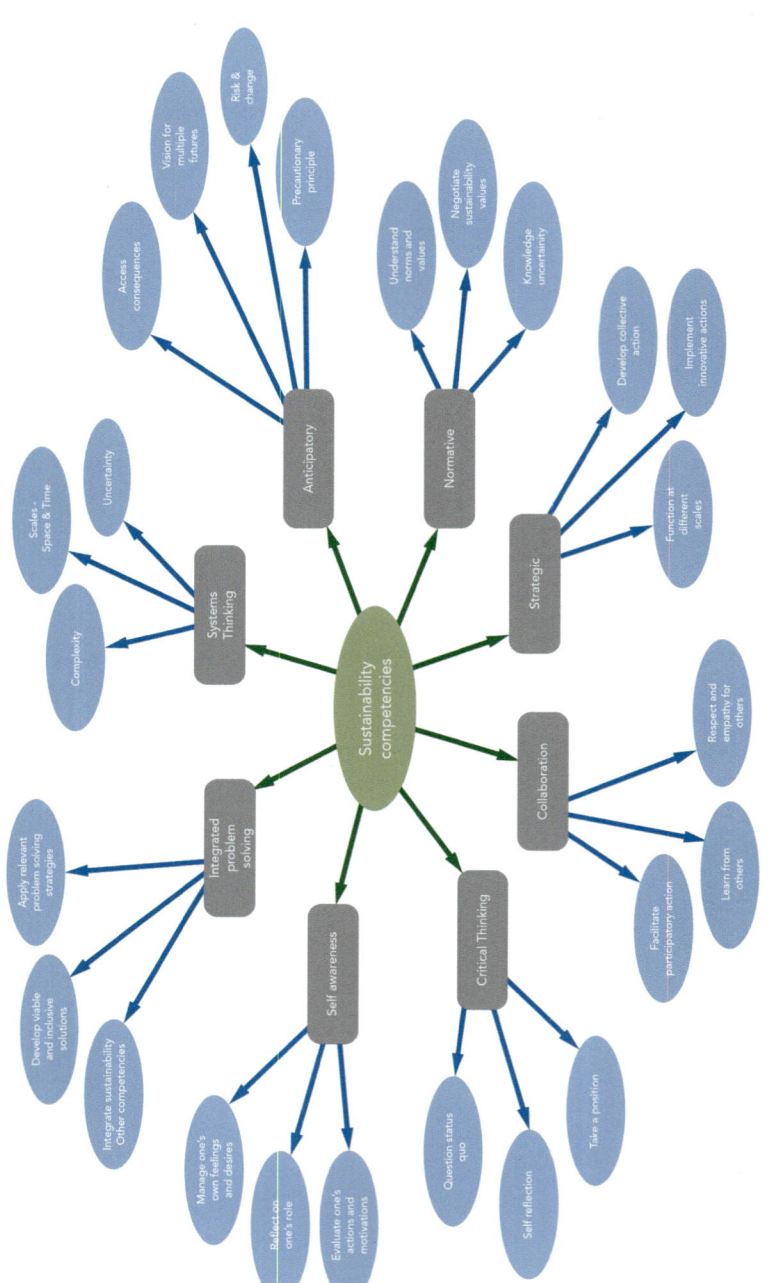

Fig. 4.1 Sustainability Competencies. Credit: Authors' Own

learners need to be able to contribute to a more inclusive, just and peaceful world' (UNESCO 2017b, p. 15). GCED employs a lifelong learning perspective and incorporates concepts and approaches from human rights education, peace education, and Education for Sustainable Development. It is multifaceted, encompassing formal and informal educational approaches, and addresses cognitive, socio-emotional, and behavioural domains of learning (UNSECO 2017b).

The 5-year (2014–2019) Global Action Programme on Education for Sustainable Development (GAP), the official successor to the UN Decade of ESD, builds on its momentum and encompasses all aspects of education (UNESCO 2015, 2017b). A network of about 90 GAP Key Partners, including governments, civil society organisations, academic institutions, and stakeholders from the private sector are working on five interconnected priorities: (1) advancing policy, (2) transforming learning, and training environments, (3) building capacities of educators and trainers, (4) empowering and mobilising youth, and (5) accelerating sustainable solutions at local levels (UNESCO 2014b, 2015). According to a report issued in 2017 midway through the Programme (UNESCO 2017b), GAP Key Partners showed progress in all areas (Table 4.3).

4.3 Geoscience Education and Sustainable Development

4.3.1 Geoscience and the Sustainable Development Goals

Geoscientists possess vital and unique knowledge and skills that are essential to promoting sustainable development (Table 4.4). These include multi-scalar spatial and temporal reasoning, systems thinking, interdisciplinary and collaborative problem-solving, and the ability to cope with the uncertainty inherent in dealing with complex systems and incomplete data (Mora 2013; Gill 2016; Stewart and Gill 2017). Furthermore, an understanding of deep time provides an essential perspective for framing key sustainability issues such as climate change, resource depletion, and loss of biodiversity (Cervato and Frodeman 2012).

With the growing recognition that the problems of sustainability are 'wicked' comes the realisation that science plays a crucial, but not sufficient role in achieving sustainable development. Figure 4.2 summarises the characteristics

Table 4.3 Summary of GAP Key Partners' Achievements 2015–2016. Adapted from UNESCO (2017b)

Advancing policy	Transforming learning environments	Building capacities of educators	Empowering youth	Accelerating solutions at local levels
• 432 strategic documents supported	• Partners helped 73K institutions implement ESD activities	• 1.5 million educators trained by Partners	• More than 1.7 million youth engaged in ESD activities	• Partners worked with local authorities on 1100 ESD activities
• 701 programmes supporting ESD policy development implemented at the country level	• 2.4 million learners involved in ESD activities	• Partners supported 14,000 teacher-training institutions to integrate ESD into teacher education	• 675,000 youth leaders trained	• 754 networks and organisations enabled to conduct ESD activities with local authorities

Table 4.4 Cognitive skills developed through Earth science education (from Gill et al. 2018)

Summary	Description (adapted from King 2008)
1. Interpretive and historical thinking	Earth science involves a range of methodologies, including those required to 'predict' what occurred in the past, interpret what is currently occurring, and review the future. Earth scientists are required to integrate large and incomplete data sets, thinking at both planetary and local scales
2. Systems analysis	Earth scientists are inherently interdisciplinary, combining aspects of physics, chemistry, and biology to understand environmental processes. Earth science supports integrated systems thinking, the impacts of feedback loops, non-linear behaviour, and parameter uncertainty
3. Diverse spatial scales	Earth scientists are trained to think at spatial scales ranging from local to global, and in three dimensions. At a global scale, plate tectonics shapes the surface of the planet. In contrast, chemical weathering of limestone may result in sinkholes affecting localised (1-10 s m^2) areas. Mineralisation at microscopic levels may help to interpret processes contributing to the formation of a rock. Earth scientists work across and integrate these scales. A microscopic analysis of a rock (mm scale) may allow a macroscopic interpretation of a geological unit (km scale)
4. Diverse temporal scales	Earth science requires the ability to think at diverse temporal scales, ranging from seconds to billions of years (so called 'geological time')
5. Field-based applications	Interpreting the geological history of a region requires the collation and integration of diverse field data, such as examining rock strata, fossils, and geochemical sampling. An Earth scientist's training will generally encompass fieldwork, laboratory work, and development of ICT skills (e.g., mapping, modelling, statistical analysis). Earth scientists can therefore connect field data with computer and laboratory simulations

of 'wicked' problems, noting that they are resistant to solutions because they arise from complex, intertwined, and co-evolving socio-ecological systems, are multi-causal, and involve multiple stakeholders with often diverging needs and values (Rittel and Webber 1973). The problems of sustainability do not respect disciplinary boundaries. For example, The Millennium Ecosystem Assessment (2005) assessed how ecosystem change impacts human well-being. This research focused on the interface between natural and social sciences (Brown et al. 2005; Carpenter et al. 2009; Reid and Mooney 2016). However, the education of natural scientists, including Earth scientists, does not typically include attention to the socioeconomic dimensions of sustainability (Mora 2013; Gill 2016; Stewart and Gill 2017; Metzger and Curren 2017). This may limit their ability to communicate scientific knowledge effectively to relevant stakeholders and policy-makers, and to engage effectively in problem-solving at the science–society interface (e.g., Batie 2008; Miller 2015).

Natural hazards mitigation provides an example of the need to rethink how geoscientists engage with communities and decision-makers to address challenges to sustainability. Disaster risk reduction is central to achieving the SDGs and

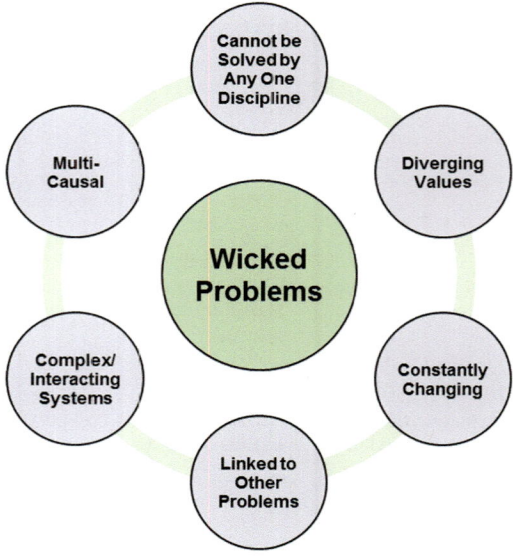

Fig. 4.2 Some characteristics of wicked problems, described by Rittel and Webber (1973) Authors' Own, After Rittel and Webber (1973)

scientific understanding of the physical processes underlying earthquakes, volcanoes, hurricanes, and other natural hazards plays an indispensable role in mitigating risk to vulnerable populations (UNDRR 2015a; UNRISD 2016). Although traditional approaches to reducing risk from natural hazards have been dominated by scientific understanding and technical expertise, it has become increasingly apparent that information alone does not guarantee that risk-reducing action will be taken (e.g., Barclay et al. 2008; Fearnley et al. 2017; García and Mendez-Fajury 2017; Ismail-Zadeh et al. 2017). A tragic example is provided by the 1985 Nevado del Ruiz eruption in Colombia, which claimed 23,000 lives (Voight 1990). This disaster did not result from inadequate scientific information or lack of technology, but rather from the inherent difficulty of conveying scientific uncertainty, ineffective communication, and complex social and economic factors which prevented local authorities and communities from taking action in response scientists' warnings (Voight 1990; Fearnley et al. 2017; García and Mendez-Fajury 2017).

Addressing the social, economic, and political factors that shape responses to risk from natural hazards is as important as understanding the underlying physical processes (Barclay et al. 2008; Fearnley et al. 2017; García and Mendez-Fajury 2017; Ismail-Zadeh et al. 2017). However, the local societal context is often overlooked by geoscientists when communicating risk to policymakers and citizens (Stewart et al. 2017). Social science research shows that people are more likely to take risk-reducing action when they have participated in disaster risk education and communication campaigns that enable the exchange of information and perspectives and development of trust among various stakeholders (Barclay et al. 2008). However, the need to engage with non-scientific audiences presents communication challenges for geoscientists, who seldom have training or experience in sharing scientific knowledge beyond academic journals and professional conferences (Stewart et al. 2017).

Over the past three decades, Earth system science and related global environmental change research programmes have greatly expanded our understanding of the Earth's intertwined and co-evolving biophysical systems while documenting the extensive and accelerating impacts of human activities on these systems (e.g., Rockström et al. 2009; Steffen et al. 2015). There have been numerous and growing calls to more fully integrate scientific investigation of global environmental change with societal needs through the fusion of the natural and human sciences (e.g., Schlosser and Pfirman 2012; Mooney et al. 2013). Just as 'education-as-usual' will not suffice to meet the goals set out in Agenda 2030, 'science-as-usual' will fall far short (Lubchenco et al. 2015). The thematic content of geoscience education intersects with the goals of Education for Sustainable Development and cuts across all the SDGs (as illustrated by this book). However, 'geoscience-education-as-usual' which addresses only the scientific aspects of sustainability will not suffice to help students develop a holistic understanding of sustainability and the actions needed to achieve it.

4.3.2 Geoscience Education in Support of the SDGs

Geoscience has benefited society through exploration for resources, geotechnical engineering, and mitigation of natural hazards, and its societal relevance is growing as humanity grapples with finding ways to balance human well-being now and in the future with respect for planetary limits. Bobrowsky et al. (2017) succinctly state that '*as guardians and developers of geoscience knowledge and given their particular sensitivity towards natural systems, geoscientists must assume the responsibility of promoting new ways of thinking about human lives in relation to Earth systems*' (Bobrowsky et al. 2017, p. 5). The emerging field of geoethics addresses geoscientists' responsibilities to their profession, to the Earth system, and to society (Mogk and Bruckner 2017; International Association for Promoting Geoethics 2018). The fundamental values of geoethics as articulated in the Cape Town Statement on Geoethics (Di Capua et al.

2017) promote geoscience education for all that furthers sustainable economic development, helps to prevent and mitigate natural hazards, and fosters well-being and resilience (Box 4.1).

> **Box 4.1 Fundamental Values of Geoethics, from the Cape Town Statement (Di Capua et al. 2017)**
>
> - Honesty, integrity, transparency and reliability of the geoscientist, including strict adherence to scientific methods;
> - Competence, including regular training and lifelong learning;
> - Sharing knowledge at all levels as a valuable activity, which implies communicating science and results, while taking into account intrinsic limitations such as probabilities and uncertainties;
> - Verifying the sources of information and data, and applying objective, unbiased peer-review processes to technical and scientific publications;
> - Working with a spirit of cooperation and reciprocity, which involves understanding and respect for different ideas and hypotheses;
> - Respecting natural processes and phenomena, where possible, when planning and implementing interventions in the environment;
> - Protecting geodiversity as an essential aspect of the development of life and biodiversity, cultural and social diversity, and the sustainable development of communities;
> - Enhancing geoheritage, which brings together scientific and cultural factors that have intrinsic social and economic value, to strengthen the sense of belonging of people for their environment;
> - Ensuring sustainability of economic and social activities in order to assure future generations' supply of energy and other natural resources.
>
> - Promoting geo-education and outreach for all, to further sustainable economic development, geohazard prevention and mitigation, environmental protection, and increased societal resilience and well-being.
>
> The full Cape Town Statement on Geoethics is available on the website of the International Association for Promoting Geoethics (www.geoethics.org/ctsg).

Gill et al. (2018) outline a multifaceted approach to making better connections between Earth science knowledge and expertise and the international sustainable development agenda. The effort must begin with strengthening pre-college Earth science education, which is either absent or of poor quality in many parts of the world (e.g., King 2008, 2013). Other needs include: (1) incorporating fundamental sustainability concepts and skills, including the ability to engage effectively in cross-disciplinary and cross-cultural communication, into Earth science curricula; (2) more collaboration between Earth scientists and other disciplines, including economics, social sciences, and the humanities; (3) career guidance for Earth science graduates to increase awareness of how they can contribute to sustainable development; and (4) resourcing university-level Earth science education in the Global South (Gill et al. 2018). Teachers are powerful agents of change and another critical need is to infuse sustainability concepts and participatory, learner-centred pedagogies into the preparation of pre-college Earth science teachers (Hale et al. 2017).

4.4 Examples of Geoscience Education in Support of Sustainability

In this section, we present six examples that illustrate how geoscience expertise, education, and outreach can support sustainable development by helping to build conceptual knowledge, foster relevant skills and attitudes, promote greater capacity for adaption and resilience in the face of uncertainty, and inspire action for the SDGs. These examples include the following:

- Three educational initiatives related to natural resources (in this case water), natural hazards, and climate change (Sect. 4.4.1–4.4.3). These topics are closely related to human well-being, involve people–planet interactions, and are included in the thematic content of both Earth science education and Education for Sustainable Development.
- Two UNESCO initiatives, including one supporting Earth science education in Africa, and a global programme using 'geoparks' to promote Earth science education (Sects. 4.4.4 and 4.4.5).
- The *Interdisciplinary Teaching of Geoscience for a Sustainable Future* (InTeGrate) Project (Sect. 4.4.6). This is a United States-based initiative that fosters interdisciplinary teaching about sustainability through the development of instructional materials and programmes that situate geoscience concepts in the context of societal issues and provides a model for cross-disciplinary collaboration and programme design that could be adapted elsewhere (Gosselin et al. 2013, 2016, 2019; Kastens and Manduca 2017).

These six examples range in scale from local to global, representing diverse audiences, types of materials and programmes, and teaching strategies. They also reflect the interconnectedness of Education for Sustainable Development, Disaster Risk Reduction (DRR) education, and climate change education (UNESCO 2010, 2017a), recognising synergies between the SDGs (Sendai Framework for DRR (UNDRR 2015a), and Paris Climate Agreement (United Nations 2015c).

4.4.1 Formal and Informal Education for Water Sustainability

Universal access to safe water and sustainable management of water resources (**SDG 6**) are essential to all aspects of development including poverty reduction, energy and food security, creation of educational and economic opportunities, gender equality, and the health and well-being of people and ecosystems (e.g., UN World Water Assessment Programme 2015). Because management of coupled human-hydrological systems involves feedbacks across multiple scales and constituencies, water is at the centre of some of the most challenging obstacles to sustainability (Silvapalan et al. 2014). Finding solutions requires scientific and technical expertise combined with an understanding of location-specific norms, knowledge, culture, and traditions that shape water use (Kreamer 2016). Well-meaning attempts to improve water access and quality often go wrong due to failures to coordinate with community stakeholders, use of technology that is inappropriate for the setting, a lack of follow up, and an absence of accompanying educational efforts (Kreamer 2016).

A collaboration between Baylor University and Restoration Gateway (RG), a Christian community in rural northern Uganda that includes an orphanage and school, illustrates the key roles of stakeholder engagement, and both formal and informal education in sustainable management of groundwater systems. Wong and Yelderman (2016) describe the outcomes of several visits to RG during which geoscientists and educators from Baylor University combined hydrologic research with both formal and informal education. The Baylor team collaborated with school officials, teachers, and other community members to assess the community's current understanding of hydrogeology and sustainable water development and then worked to increase knowledge through educational interventions. Science classes for students in the RG school, aged approximately 4–10 years, included investigation of the water cycle and aquifers using local examples, an introduction to careers in hydrogeology, and demonstration of relevant scientific tools and methods. Several weeks spent living onsite allowed for daily interactions with the RG community, enhancing the Baylor team's understanding of local cultural perspectives and constraints. Informal conversations with both adults and children allowed for better communication of issues related to the community's water supply and reinforced concepts taught during formal classroom instruction.

4.4.2 Natural Hazards, Disaster Risk Reduction, and Education

Geophysical, meteorological, hydrological, and climatological hazards such as earthquakes, volcanoes, extreme weather, flooding, and drought occur around the world, but not all people are affected equally. For vulnerable populations with insufficient capacity to cope with their impacts, these natural phenomena turn into disasters that bring loss of life and property damage, disrupt social and economic systems, increase poverty, create or worsen conflict, and limit or prevent access to quality education (e.g., UNDRR 2015b). Mitigating risk is thus a key requirement for achieving the SDGs, with education in DRR being a core component of Education for Sustainable Development (UNESCO 2017a). Education, both formal and informal, builds understanding among students (Fig. 4.3), the broader community, government officials, and policy-makers of the causes and impacts of hazards while fostering competencies and skills

that enable and inspire learners to take action to prevent and mitigate disasters (e.g., UNDRR 2012).

Although disasters resulting from natural hazards affect both the Global North and Global South, some areas and populations are at significantly higher risk than others due to environmental, social, and economic factors which often combine in ways that make certain groups, including the poor, children, women, the elderly, and the displaced more susceptible to disasters (PreventionWeb 2015). For example, the Asia–Pacific area experienced nearly half of all global disasters between 2000 and 2017, with the poorest nations suffering the greatest impact (ESCAP 2017; Peters 2018). Additional deaths resulting from extreme weather are expected to move this area from 'high' to 'severe' vulnerability to natural hazards by 2030 (Peters 2018).

The stories of two women, who lent their geoscientific expertise to earthquake education and outreach in hazards-prone Central and South Asia, illustrate how scientists can collaborate with local communities to reduce the risk for those who are most vulnerable. Solmaz Mohadjer, a geohazard researcher and educator at the University of Tübingen and University of Central Asia, uses GPS geodesy and terrestrial remote

Fig. 4.3 Exploring landslide dynamics in Ladakh. In 2014, the University of Jammu partnered with Geology for Global Development to design and deliver a hazards and risk education programme in Ladakh, India. © Geology for Global Development. (used with permission)

Fig. 4.4 Earthquake education in Tajikistan in 2008. Solmaz Mohadjer demonstrates aspects of fault dynamics using simple table-top exercises in a school in Tajikistan © Parsquake (used with permission)

sensing (LiDAR) to quantify hazards such as earthquakes and rockfalls in Central Asia (Roberts-Artel 2017). Mohadjer is also the founder of the ParsQuake initiative,[1] which aims to reduce risk from geohazards in the global Persian community by improving communication between scientists and the public (Fig. 4.4). In collaboration with others, she has developed lesson plans and organised workshops for local teachers (Mohadjer et al. 2010, 2018, 2019; Roberts-Artal 2017). In 2016, Dr. Mohadjer received a *Public Engagement Grant from the European Geosciences Union to* create 10 learning videos that address earthquake science, hazards, and safety.[2] The videos use a paired teaching approach in which questions posed by geoscientists are investigated by students through hands-on activities guided by a classroom teacher (Mohadjer et al. 2018).

Another example of an inspiring geoscientist advancing education for DRR is Anne Sanquini, who has combined seismology with sociology to study what motivates people to act to reduce risk from natural hazards (Frank 2017). Sanquini sits on the Board of Trustees of GeoHazards International, a not-for-profit organisation that works to help people in developing countries take action that addresses safety from natural disasters. After conducting interdisciplinary research into areas prone to strong shaking in the Kathmandu Valley of Nepal, Sanquini concluded that the most vulnerable buildings were schools constructed of bricks or stone with mud mortar. She then investigated how people could be motivated to retrofit the schools (Traer 2015).

Social theory suggests that people will act against hazards when they: (1) know what to do, (2) think it would work, and (3) know someone who did it (Sanquini and Wood 2015; Sanquini et al. 2016). Guided by this theory and in partnership with a technical non-government organisation, the National Society for Earthquake Technology—Nepal (NSET), and the Nepali Department of Education, Sanquini and her team developed a documentary film featuring Nepalese school principals, teachers, parents, and community members who took steps to make their schools more earthquake resistant (Sanquini et al. 2016). The film was tested with community members associated with 16 public schools in Kathmandu, revealing that watching the film increased viewers' knowledge of earthquake-resistant building design and methods and led to

[1]https://parsquake.org/.

[2]https://parsquake.org/resources.php.

increased support for retrofitting schools (Traer 2015). The M 7.8 Gorkha earthquake struck central Nepal just five weeks after testing of the film was completed (Frank 2017). Many of the 16 schools in the research project were heavily damaged. However, the five schools featured in the documentary film suffered no structural damage, and some served as aid stations for their communities in the aftermath. A new ending was created for the film. It featured the five school principals describing how their retrofitted buildings survived the shaking. This film was released to the general public to encourage rebuilding damaged schools using earthquake-resistant methods. (Anne Sanquini, personal communication, February 12, 2019).

4.4.3 Climate Change Education

Climate change, the focus of **SDG 13**, is inextricably linked to sustainable development. It poses significant risks to food and water supplies, disrupts livelihoods, and access to quality education, and disproportionately affects children, the poor, and women (e.g., Bangay and Blum 2010; UNRISD 2016). Climate change is also closely connected to disaster risk, magnifying the effects of natural hazards. 90% of disasters recorded between 1995 and 2015 were caused by floods, storms, heatwaves, and other events linked to climate and weather (Wahlstrom and Guha-Sapir 2015).

Climate change is a complex, interdisciplinary global challenge the potential solutions for which require the navigation of multiple and competing socio-political, cultural, and technical issues (Incopera 2015). Responding to climate change will thus require both science and policy innovations, and fundamental shifts in thinking and ways of living (UNRISD 2016). This is where high-quality, transformative education as called for by **SDG 4** comes in. Education is widely regarded as an essential tool for responding to climate change (e.g., Bangay and Blum 2010; UNESCO 2010; Mochizuki and Bryan 2015) (Fig. 4.5).

The role of education, training, and public awareness in mitigating and adapting to climate change has been highlighted in several international agreements including Article 6 of the UN Framework Convention on Climate Change, the

Fig. 4.5 Encouraging students in Ladakh to explore practical responses to climate change through art and design. © Geology for Global Development (used with permission)

Kyoto Protocol's Article 10e, and Article 12 of the Paris Climate Agreement (UNESCO and UNFCCC 2016). Climate change education is a core element of Education for Sustainable Development and SDG Target 13.3 explicitly refers to education: '*improve education, awareness raising and human and institutional capacity* '. UNESCO has developed instructional materials, online courses, programmes, and initiatives to support climate change education. Examples which address a wide spectrum of target audiences, types of materials, and pedagogical approaches can be found at an online platform[3] developed by the *One UN Climate Change Learning Partnership* (UN CC: Learn 2018).

There are, however, several obstacles to mainstreaming climate change education including an overcrowded curriculum and lack of prepared teachers (Læssøe et al. 2009). Although teaching about climate change is typically compartmentalised into science education, climate change is not solely a scientific phenomenon, but rather a complex socio-scientific issue that cuts across disciplines and does not fit readily into existing curricula and assessments (UNESCO 2010; Stevenson et al. 2017).

No other group of nations is more vulnerable to climate change and rising sea levels than the Small Island Developing States (SIDS). This group of 57 island and coastal countries are distributed across three geographic regions: (1) the Caribbean, (2) the Pacific, and (3) the Atlantic, Indian Ocean, Mediterranean, and South China Sea (Scandurra, 2018). These otherwise heterogeneous countries share distinct social, economic, and environmental vulnerabilities arising from their small size and limited resources, remote locations and isolation from markets, fragile ecosystems and high levels of exposure to natural hazards (UN-OHRLLS 2011; Scandurra et al. 2018).

Crossley and Sprague (2014) note that while SIDS are on the frontline of experiencing the impacts of climate change, they are also at the forefront of developing innovative and creative

approaches for addressing them. One such innovation is Sandwatch, a volunteer network which takes a citizen science approach to climate change adaptation and education and supplies a '*framework for children, youth and adults, with the help of teachers and local communities, to work together to critically evaluate the problems and conflicts facing their beach environments, and to develop sustainable approaches to address these issues*' (Cambers and Diamond 2010, p. 8). Each Sandwatch school or community adopts a local beach and uses simple, readily available equipment to take regular measurements over time of such factors as beach width, currents, waves, and water quality. These data are used to access the stability and health of their beach and coastal environment, identify the nature of any stressors, and take actions to address them (UNESCO Bangkok 2012; Sandwatch Foundation 2018).

Launched in the Caribbean in 1998, Sandwatch is now active in 30 countries worldwide and is supported by a range of online resources (e.g., a manual that guides investigations, training videos, an online course for teachers, and an international database that stores collected data) (The Sandwatch Foundation, 2018). As reflected by the Sandwatch Program, the SIDS countries are well situated to take advantage of place-based approaches to education for sustainable development, which connects scientific investigation to the traditional knowledge that has helped indigenous populations to adapt to location-specific environmental changes through time (Selby and Kagawa, 2018).

4.4.4 The Earth Science Education in Africa Initiative

Africa is rich in natural resources, but poor in the geoscientific expertise needed to sustainably develop them and to mitigate risk from natural hazards and climate change (Martínez-Frías and Mogessie 2012; UNESCO 2012; Jessell et al. 2017). In response to a call from African governments for help in addressing this gap, UNESCO launched the *Earth Science Education*

---

[3]www.uncclearn.org/.

in *Africa Initiative* in 2008 (UNESCO 2012; Mynot 2014). Regional scoping workshops organised across Africa in 2009 and 2010 revealed the need to

1. Raise awareness among the general public and decision-makers of the importance of Earth science as a key contributor to sustainable development;
2. Enhance multidisciplinary approaches to Earth science research and teaching;
3. Include Earth sciences in primary and secondary school curricula;
4. Address the lack of analytical facilities through collaborative exchanges and identification of new funding mechanisms;
5. Build strong connections between industry and universities; and
6. Reinvigorate old and build new networks both within Africa and between African Earth scientists and the global geoscience community (UNESCO, 2012).

Implementation of the *Earth Science Education in Africa Initiative* is facilitated by the African Network of Earth Science Institutions (ANESI), which brings together partners from universities, government agencies, and the private sector to direct projects and plan and host workshops focused on three major themes: (1) geological mapping training, (2) Earth science in schools, and (3) increasing information on the health impacts of mining activities (UNESCO 2012; UNESCO 2017c).

A related initiative is the African Geoparks Network, which was created in 2009 by the African Association of Women in Geosciences (Errami et al. 2015; see **SDG 5**). As described in the next section, UNESCO Geoparks link geoheritage with local socio-economic development through geotourism, which supplies employment opportunities (Errami et al. 2015). Africa has extraordinary geological resources and landscapes. There are currently only two Geoparks in Africa, the M'Goun Global Geopark in Morocco and the Ngorongoro Lengai in Tanzania (Global Geoparks Network 2018). Many projects are underway that aim to increase that number (Errami et al. 2015).

4.4.5 UNESCO's International Geoscience and Geoparks Programme

UNESCO is the only United Nations organisation mandated with supporting research and capacity building in the geosciences (UNESCO 2018d), and they are responsible for the Global Geoparks Network. Geoparks are single, unified geographical regions of international geological value that are holistically managed to protect the Earth's geodiversity, educate the public, and promote sustainable development. (UNESCO 2016b). Together with UNESCO-designated Biosphere Reserves and World Heritage Sites, Geoparks are intended to provide people from across the globe with an opportunity to celebrate the diversity of heritage through the conservation of the world's cultural, biological, and geological diversity and to enhance understanding of societal challenges such as sustainable use of natural resources, mitigation of climate change, and promotion of a culture of sustainable economic development (UNESCO 2016b).

There are currently 140 UNESCO Global Geoparks in 38 countries (UNESCO 2018d). A rigorous process is used to establish and maintain a Geopark designation through a collaborative bottom-up process that involves relevant local and regional stakeholders and authorities including landowners, community groups, tourism providers, indigenous people, and other local organisations. Each UNESCO Global Geopark cooperatively networks with local people, participates in regional networks, and is required to cooperate with other Geoparks as a member of the Global Geoparks Network (UNESCO 2016b). See **SDG 8** for more information.

Geoparks support **SDG 4**, especially **Target 4.7**, and the other SDGs in several ways that also support Earth science education. They serve as outdoor classrooms, implementers of sustainable development and lifestyles, and promote cultural diversity through educational activities for local communities and visitors of all ages. As part of a holistic approach to sustainable development, educational programming is a critical component of Geoparks. The top 10 educational topics among the worldwide network of Geoparks include: Natural Resources; Geological Hazards; Climate Change; Education; Science; Culture; Women; Sustainable Development; Local and indigenous Knowledge; and Geoconservation (UNESCO 2016b).

The Villuercas Ibores Jara UNESCO Global Geopark, located in the south-eastern part of the province of Cáceres in the Extremadura region of Spain, provides an example of how Geoparks foster Earth science literacy for learners of all ages (Villuercas Ibores Jara Geopark 2016; Villuercas Ibores Jara Geoparque Mundial de La UNESCO 2019) (Fig. 4.6). The park's unique landscape consists of quartzite peaks and deep river valleys produced by differential erosion of intensely folded and faulted Precambrian and Paleozoic rocks (Villuercas Ibores Jara Geopark 2016; Villuercas Ibores Jara Geoparque Mundial de La UNESCO 2019). Its fossils record evidence for the " 'Cambrian Explosion' and an abundance of trilobites, brachiopods, and graptolites reflect the rapid diversification of marine life during the Ordovician Period. In addition to its remarkable geomorphology and wealth of paleontological resources, the park is rich in biodiversity and features a number of archaeological sites (Villuercas Ibores Jara Geopark 2016; Villuercas Ibores Jara Geoparque Mundial de La UNESCO 2019).

The 'Geo-schools' project uses the Geopark as an outdoor classroom and aims to foster sustainability through improved geoscience and environmental education (Barrera and Corrales 2011; Villuercas Ibores Jara Geopark 2016; Villuercas Ibores Jara Geoparque Mundial de La UNESCO 2019. The park's education initiative involves rural, primary, and secondary schools as well as adult lifelong education and teacher support centres. It is jointly managed by of the Regional Ministry of Education and Culture, the Cáceres Provincial Council, and the University of Extremadura (Barrera and Corrales 2011; Villuercas Ibores Jara Geopark 2016). Both formal and informal educational approaches are employed to enhance knowledge of the Geopark's geological, biological, and cultural heritage. Examples include interpretive panels, leaflets, and guided tours. Interdisciplinary educational materials are designed by the University in collaboration with teachers and address geology, biology, geography, and history. The book " 'Environmental Awareness of Villuercas- Ibores-

Fig. 4.6 Using art to enhance public understanding of geological time scales © The Villuercas Ibores Jara UNESCO Global Geopark (used with permission of Jose M Barrera, Director)

Jara Geopark' serves as a complementary text for local teachers and students and social media is used to share information, activities, and other resources (Barrera et al. 2015; Villuercas Ibores Jara Geopark 2016).

4.4.6 InTeGrate (Interdisciplinary Teaching About Earth for a Sustainable Future)

The Interdisciplinary Teaching of Geoscience for a Sustainable Future (InTeGrate 2012–2019) STEM Talent Expansion Center, funded by the United States' National Science Foundation, catalysed and supported a higher education community in a broad-based effort to transform undergraduate geoscience education through innovative, learner-centred instruction that connects geoscience to grand societal challenges (Gosselin et al. 2013, 2016, 2019; InTeGrate 2018a). The InTeGrate Project (Box 4.2) explicitly recognised that knowledge of geosciences is necessary, but not sufficient, for the development of effective solutions to the wicked problems of sustainability and focused on the development of interdisciplinary teaching materials and instructional programmes that engage students in understanding geoscience in the context of societal issues (Gosselin et al. 2013, 2019). With the goal of increasing the number of undergraduates from a variety of backgrounds who can act as Earth-literate citizens and to build a sustainability-focused workforce, materials and programmes developed by the InTeGrate Project interweave geoscience concepts and methods with those from other STEM fields, economics, the humanities, and social sciences.

Products of InTeGrate include 32 sets of teaching materials for post-secondary instruction that combine geoscience learning with societal issues and systems thinking (InTeGrate 2018b). InTeGrate materials have also been developed to support the Next Generation Science Standards

Box 4.2 Key features of the InTeGrate Project *Sustainability topics embedded into the resources produced by the InTeGrate project include the following:*

Civil Society/Governance	Food Systems/Agriculture	Risk and Resilience Social/Environmental
Climate Change	Human Health/	Justice
Culture, Ethics and Values	Well-being	Technology
Cycles and Systems	Human Impact/Footprint	Water and Watersheds
Ecosystems	Natural Hazards	
Energy	Natural Resources	
	Pollution and Waste	

(NGSS), the first science standards in the United States to explicitly address climate change, sustainability, and human impacts on the Earth system (InTeGrate 2018b; NGSS Lead States 2013). The InTeGrate modules and courses, which align with the thematic content of Education for Sustainable Development and connect to the SDGs, are freely available online (InTeGrate 2018b). The InTeGrate website also features descriptions of how the InTeGrate materials were developed, implemented, assessed, and disseminated (InTeGrate 2018a).

InTeGrate project participants carry forward into the future their collaborative attitude toward geoscience education, in their individual work, and through relationships built on interactions during the project (Kastens et al. 2014). The structures and tools built to facilitate the development of materials and affect systemic change at multiple levels have proven to be valuable in and of themselves and the resulting curricular materials and approaches serve as examples of how interdisciplinary teaching about sustainability can be brought to geoscience classrooms and programmes (Gosselin et al. 2019, InTeGrate 2018a). The models developed and lessons learned can benefit other communities interested in better integration of geoscience education with **SDG 4** and the other SDGs (Kastens and Manduca 2017).

4.5   Summary and Conclusions

SDG 4 calls for inclusive, high-quality, and transformative education that enables learners to act as agents of change for achieving the SDGs. Geoscientists possess unique knowledge and skills that are essential to promoting sustainable development and the thematic content of geoscience education aligns with Education for Sustainable Development and cuts across all of the SDGs. However, 'geoscience-education-as-usual' which addresses only the scientific aspects of sustainability will not suffice to prepare Earth scientists to contribute to the SDGs and related international sustainability frameworks. Enhancing geoscientists' engagement with the global sustainable development agenda requires a multilevel, multifaceted approach. This includes strengthening pre-college Earth science education and teacher preparation, incorporating fundamental sustainability concepts and skills into Earth science curricula, more collaboration between Earth scientists and other disciplines, career guidance for Earth science graduates to increase awareness of how they can contribute, and providing resources for university-level Earth science education in the Global South.

4.6   Key Learning Concepts

- High-quality, transformative approaches to education are key to communicating and implementing innovative ways of thinking and acting for sustainable development. Education is an essential catalyst and facilitator for all 17 of the SDGs. Quality education provides people with the knowledge, skills, tools, and capacity for lifelong learning that fosters well-being and protects Earth systems.
- Geoscience education needs to expand to include the human elements of challenges to sustainability. The thematic content of Earth science education (e.g., natural resources, geohazards, climate change) clearly aligns with the SDGs, but inattention to the social and economic dimensions of sustainability in the formal education and continued professional development of most geoscientists may hinder their ability to engage effectively in international sustainability initiatives.
- Increasing public understanding of geoscience can help deliver the SDGs, as illustrated with examples of interdisciplinary educational initiatives and resources that address key geoscience themes (natural resources, natural hazards, and climate change) and connect to the environmental and socioeconomic dimensions of sustainability. These examples reflect the interconnectedness of Education for Sustainable Development, disaster risk reduction education, and climate change education.

4.7   Educational Ideas

In this section, we provide examples of educational activities that connect geoscience, the material discussed in this chapter, and scenarios that may arise when applying geoscience (e.g., in policy, government, private sector international organisations, NGOs). Consider using these as the basis for presentations, group discussions, essays, or to encourage further reading.

- Explore what other departments at your institute/university teach courses or modules linking to (i) natural resources, (ii) hazards and disasters, or (iii) climate change. Organise a joint seminar with 15-min presentations from a representative from each department, giving their perspective on the theme. Follow this with an open discussion about what skills are needed to tackle complex development challenges.
- To what extent do you agree with the statement *'a geoscience degree in 2020 will not prepare me for the geoscience jobs in 2030'?* To answer this question, consider how the skills and knowledge themes explored in your degree course may map to the predicted jobs of the future (see **SDG 8** for ideas). What themes and skills would you consider it a high priority to add to your training?

- Review the sustainability competencies in Fig. 4.1, and consider those competencies that are not already integrated into your training. Design a geoscience-themed activity that helps to develop one or more of these competencies.

Further Resources

Earth Science Teachers Association (2019). https://earthscience.org.uk/

European Geosciences Union (2019) Education Resources. www.egu.eu/education/

Geological Society of London (2019) Education and Careers Resources. www.geolsoc.org.uk/education

Higher Education Academy (2019) Education for Sustainable Development. www.advance-he.ac.uk/guidance/teaching-and-learning/education-sustainable-development

InTeGrate (2019) InTeGrate Teaching Materials. Available online: https://serc.carleton.edu/integrate/teaching_materials/index.html

UNESCO (2019) Education for Sustainable Development. https://en.unesco.org/themes/education-sustainable-development

References

Bangay C, Blum N (2010) Education responses to climate change and quality: Two parts of the same agenda? Int J Edu Dev 30(4):359–368

Bangkok UNESCO (2012) Education sector responses to climate change: Background paper with international examples. UNESCO Asia and Pacific Regional Bureau for Education, Bangkok

Barclay J, Haynes K, Mitchell T, Solana C, Teeuw R, Darnell A, Crosweller HS, Cole P, Pyle D, Lowe C, Fearnley C (2008) Framing volcanic risk communication within disaster risk reduction: finding ways for the social and physical sciences to work together. Geol Soc Lond Spec Publ 305(1):163–177

Barrera JM, Corrales JM (2011) Geo-schools: a commitment to education in the territory of the Geopark Villuercas Ibores Jara's project. In Rangnes K (ed) Proceedings of the 10th European Geoparks Conference. European Geoparks Network, Porsgrunn, Norway. https://www.europeangeoparks.org/wp-content/uploads/2015/09/10th-European-Geoparks-Conference-2011-ABSTRACTS.pdf

Barrera JM, Corrales JM, Lopez J, Vazquez J (2015) A key subject called Geopark. In Saari K, Saarinen J, Saastamoinen M (eds) Abstract Book of the 13th European Geopark Conference Rokua Geopark Finland. https://www.europeangeoparks.org/wp-content/uploads/2012/02/Book-of-Abstracts-EGNconference-2015.pdf.

Batie SS (2008) Wicked problems and applied economics. Am J Agr Econ 90(5):1176–1191

Bobrowsky P, Cronin VS, Di Capua G, Kieffer SW, Peppoloni S (2017) The Emerging Field of Geoethics. In: Gundersen L (ed) Scientific integrity and ethics in the geosciences. Special Publication American Geophysical Union, Wiley, New York, pp 175–212

Bralower TJ, Feiss PG, Manduca CA (2008) Preparing a new generation of citizens and scientists to face earth's future. Liberal Edu 94(2):20–23

Brown, K., Viswanathan, K. and Manguiat, M.S. (2005) Integrated Responses. In: Chopra, K, Leemans, R., Kumar, P. and Simons, H. (eds) Ecosystems and Human Well-Being: Policy Responses (The millennium ecosystem assessment series vol. 3) Island Press, Washington D.C., p 425–465

Cambers G, Diamond P (2010) Sandwatch: Adapting to climate change and educating for sustainable development. UNESCO, Paris

Carpenter SR, Mooney HA, Agard J, Capistrano D, Defries RS, Díaz S, Dietz T, Duraiappah AK, Oteng-Yeboah A, Pereira HM, Perrings C, Reid WV, Sarukhan J, Scholes RJ, Whyte A (2009) Science for managing ecosystem services: beyond the Millennium Ecosystem Assessment. Proc Natl Acad Sci USA 106 (5):1305–1312. https://doi.org/10.1073/pnas.0808772106

Cervato C, Frodeman R (2012) The significance of geologic time: Cultural, educational, and economic frameworks. In: Kastens, K.A., and Manduca, C.A (eds.) Earth and mind II: a synthesis of research on thinking and learning in the geosciences: geological society of America special paper 486, p 19–27. https://doi.org/10.1130/2012.2486(03).

Crossley M, Sprague T (2014) Education for sustainable development: Implications for small island developing states (SIDS). Int J Edu Develop 35:86–95

Di Capua G, Peppoloni S, Bobrowsky PT (2017) The cape town statement on geoethics. Ann Geophys 60

Errami E, Schneider G, Ennih N, Randrianaly HN, Bendaoud A, Noubhani A, Al-Wosabi M (2015) Geoheritage and geoparks in Africa and the Middle-East: challenges and perspectives. In: From Geoheritage to Geoparks. Springer International Publishing, Cham, Switzerland, pp 3–23

ESCAP (2017) Leave No One Behind: Disaster Resilience for Sustainable Development, Asia-Pacific Disaster Report 2017. United Nations Economic and Social Commission for Asia and the Pacific, Bangkok

Fearnley C, Winson AEG, Pallister J, Tilling R (2017) Volcano crisis communication: challenges and solutions in the 21st century. Observing the Volcano World. Springer, Cham, pp 3–21

Frank TA (2017) Landing a Second Career: Six Executives Who Got It Right Available online: https://www.drucker.institute/monday-issue/landing-a-second-career-six-executives-who-got-it-right/

García C, Mendez-Fajury R (2017) If I Understand, I Am Understood: Experiences of Volcanic Risk

Communication in Colombia. Observing the Volcano World. Springer, Cham, pp 335–351

Gill JC (2016) Building good foundations: Skills for effective engagement in international development. Geolo Soc Am Spec Pap 520:1–8

Gill JC, Bullough F (2017) Geoscience engagement in global development frameworks. Ann Geophys 60 (Fast Track 7):10

Gill JC, White E, Hartigan J (2018) Enhancing earth science education to support sustainable development. In: Invited submission to the 2nd International Commission on Education for Sustainable Development Practice Report. Geology for Global Development.

Global Geoparks Network (2018) Available online: https://globalgeoparksnetwork.org/?page_id=226 (Accessed 22 Oct 2018)

Gosselin DC, Manduca C, Bralower T, Mogk D (2013) Transforming the teaching of geoscience and sustainability. Eos, Trans Am Geophys Union 94(25):221–222

Gosselin D, Burian S, Lutz T, Maxson J (2016) Integrating geoscience into undergraduate education about environment, society, and sustainability using place-based learning: Three examples. J Environ Stud Sci 6 (3):531–540

Gosselin D, Manduca C, Bralower T (2019) Preparing Students to Address Grand Challenges and Wicked Problems: The InTeGrate Approach. In: Gosselin DC, Egger AE, Taber J (eds) Interdisciplinary teaching about earth and the environment for a sustainable future. Springer International Publishing, Cham, Switzerland, AESS Interdisciplinary Environmental Studies and Science Series

Hale AE, Shelton CC, Richter J, Archambault LM (2017) Integrating geoscience and sustainability: examining socio-techno-ecological relationships within content designed to prepare teachers. J Geosci Educ 65 (2):101–112

Incropera FP (2015) Climate change: A wicked problem: Complexity and uncertainty at the intersection of science, economics, politics, and human behavior. Cambridge University Press, Cambridge, UK

InTeGrate (2018a). About the InTeGrate Project. Available online: https://serc.carleton.edu/integrate/about/index.html (Accessed 22 Oct 2018)

InTeGrate (2018b) InTeGrate Teaching Materials. Available online: https://serc.carleton.edu/integrate/teaching_materials/index.html (Accessed 22 Oct 2018)

International Association for Promoting Geoethics (2018) Available online: https://www.geoethics.org/

Ismail-Zadeh AT, Cutter SL, Takeuchi K, Paton D (2017) Forging a paradigm shift in disaster science. Nat Hazards 86(2):969–988

Jessell M, Baratoux D, Siebenaller L, Hein K, Maduekwe A, Ouedraogo FM, Baratoux L, Diagne M, Cucuzza J, Seymon A, Sow EH (2017) New models for geoscience higher education in West Africa. J Afr Earth Sci https://doi.org/10.1016/j.jafrearsci.2017.12.011

Kastens KA, Manduca CA (2017) Using Systems Thinking in the Design, Implementation, and Evaluation of Complex Educational Innovations, With Examples from the InTeGrate Project. J Geosci Edu 65(3):219–230

Kastens KA, Baldassari C, DeLisi J (2014) InTeGrate Mid-Project Evaluation Report. Available online: https://d32ogoqmya1dw8.cloudfront.net/files/integrate/about/integrate_mid-project_evaluati.pdf (Accessed 22 Oct 2018)

King C (2008) Geoscience education: an overview. Studies in Science Education 44(2):187–222

King C (2013) Geoscience education across the globe-results of the IUGS-COGE/IGEO survey. Episodes 36 (1):19–30

Kreamer DK (2016) Hydrophilanthropy gone wrong - How well-meaning scientists, engineers, and the general public can make the worldwide water and sanitation situation worse. Geolo Soc Am Spec Pap 520:205–219

Læssøe J, Breiting SK, Rolls S (2009) Climate change and sustainable development: The response from education. National reports. The International Alliance of Leading Education Institutes. Available online: https://edu.au.dk/fileadmin/www.dpu.dk/en/research/researchprogrammes/environmentalandhealtheducation/om-dpu_institutter_institut-for-didaktik_20091208102732_cross_national-report_dec09.pdf (Accessed 22 Oct 2018)

Lubchenco J, Barner AK, Cerny-Chipman EB, Reimer JN (2015) Sustainability rooted in science. Nat Geosci 8 (10):741

Manduca CA, Kastens KA (2012) Geoscience and geoscientists: Uniquely equipped to study Earth. Geolo Soc Am Spec Pap 486:1–12

Martinez-Frias J, Mogessie A (2012) The need for a geoscience education roadmap for Africa. Episodes 35 (4):489–492

Metzger EP, Curren RR (2017) Sustainability: Why the language and ethics of sustainability matter in the geoscience classroom. J Geosci Educ 65(2):93–100

Miller TR (2015) Reconstructing sustainability science: knowledge and Action for a Sustainable future. Routledge, New York

Mochizuki Y, Bryan A (2015) Climate change education in the context of education for sustainable development: Rationale and principles. J Edu Sustain Develop 9(1):4–26

Mogk D, Bruckner M (2017) Teaching GeoEthics Across the Geoscience Curriculum. Available online: https://serc.carleton.edu/geoethics/index.html (Accessed 14 Aug 2018)

Mohadjer S, Bendick R, Halvorson SJ, Saydullaev U, Hojiboev O, Stickler C, Adam ZR (2010) Earthquake emergency education in Dushanbe. Tajikistan. J Geosci Edu 58(2):86–94

Mohadjer S, Mutz S, Amey R, Drews R, Kemp M, Kloos P, Mitchell L, Nettesheim M, Gill S, Starke J, Ehlers TA (2018) Using paired teaching for earthquake education in schools. In: EGU General Assembly Conference Abstracts 20:7858

Mohadjer S, Mutz S, Ischuk A, Ehlers TA (2019) Overcoming challenges in earthquake education: a case study from Tajikistan. Geophys Res Abstracts (21), EGU2019–12489

Mooney HA, Duraiappah A, Larigauderie A (2013) Evolution of natural and social science interactions in global change research programs. Proc Natl Acad Sci 110(Supplement 1):3665–3672

Mora G (2013) The need for geologists in sustainable development. GSA Today 23(12):36

Mynott S (2014) Enhancing Earth Science Education in Africa. [Blog] GeoLog. https://blogs.egu.eu/geolog/2014/01/03/enhancing-earth-science-education-in-africa/ (Accessed 26 Sep 2018)

NGSS Lead States (2013) Next Generation Science Standards: For States. The National Academies Press, Washington, DC, By States

ParsQuake—Earthquake Education in the Global Persian Community (2018) Available online: https://parsquake.org/ (Accessed 22 Oct 2018)

Peters K (2018) Accelerating Sendai Framework implementation in Asia: Disaster risk reduction in contexts of violence, conflict and fragility. Overseas Development Report 37. Available online: https://www.odi.org/sites/odi.org.uk/files/resource-documents/12284.pdf. (Accessed 22 Oct 2018)

PreventionWeb (2015) Component of Risk–Vulnerability. Available online: https://www.preventionweb.net/risk/vulnerability (Accessed 22 Oct 2018)

Reid WV, Mooney HA (2016) The Millennium Ecosystem Assessment: testing the limits of interdisciplinary and multi-scale science. Current Opinion Environ Sustain 19:40–46

Rittel HW, Webber MM (1973) Dilemmas in a general theory of planning. Policy Sci 4(2):155–169

Roberts-Artel L (2017) Guest Blog: Earthquake Education in Central Asia. [Blog] GeoTalk. https://blogs.egu.eu/geolog/2017/11/24/geotalk-how-an-egu-public-engagement-grant-contributed-to-video-lessons-on-earthquake-education/ (Accessed 23 Sep 2018)

Rockström J, Steffen W, Noone K, Persson Å, Chapin FS, Lambin E, Lenton TM, Scheffer M, Folke C, Schellnhuber H, Nykvist B, De Wit CA, Hughes T, van der Leeuw S, Rodhe H, Sörlin S, Snyder PK, Costanza R, Svedin U, Falkenmark M, Karlberg L, Corell RW, Fabry VJ, Hansen J, Walker B, Liverman D, Richardson K, Crutzen P, Foley J (2009) Planetary boundaries: exploring the safe operating space for humanity. Ecology and Society 14(2):32. Available online: https://www.ecologyandsociety.org/vol14/iss2/art32/ (Accessed 22 Oct 2018)

Sanquini A (2015) I thought we would probably die. The Nepal Earthquake as Felt from Nepal: Natural Hazards Observer XXXIX (5):9–13

Sanquini AM, Thapaliya SM, Wood MM (2016) A communications intervention to motivate disaster risk reduction. Disaster Prevention and Management 25 (3):345–359

Scandurra G, Romano AA, Ronghi M, Carfora A (2018) On the vulnerability of Small Island Developing States: A dynamic analysis. Ecol Ind 84:382–392

Schlosser P, Pfirman S (2012) Earth science for sustainability. Nat Geosci 5(9):587

Selby D, Kagawa F (2018) Archipelagos of learning: environmental education on islands. Environ Conserv 45(2):137–146

Sivapalan M, Konar M, Srinivasan V, Chhatre A, Wutich A, Scott CA, Wescoat JL, Rodríguez-Iturbe I (2014) Socio-hydrology: Use-inspired water sustainability science for the Anthropocene. Earth's Future 2 (4):225–230

Steffen W, Richardson K, Rockström J, Cornell SE, Fetzer I, Bennett EM, Biggs R, Carpenter SR, De Vries W, De Wit CA, Folke C (2015) Planetary boundaries: guiding human development on a changing planet. Science 347(6223):1259855

Stevenson RB, Nicholls J, Whitehouse H (2017) What Is Climate Change Education? Curriculum Perspectives 37(1):67–71

Stewart IS, Gill JC (2017) Social geology–integrating sustainability concepts into Earth sciences. Proc Geol Assoc 128(2):165–172

Stewart IS, Ickert J, Lacassin R (2017) Communicating seismic risk: the geoethical challenges of a people-centred, participatory approach. Ann Geophys 60, FAST TRACK 7 (2017). https://doi.org/10.4401/AG-7593

The Sandwatch Foundation (2018) Available online: https://www.sandwatchfoundation.org/ (Accessed 22 Oct 2018)

Traer M (2015) Stanford earthquake hazards researcher at center of Nepal quake, Stanford Report, May 22, 2015

UN CC: Learn (2018) Available online: https://www.uncclearn.org/ (Accessed 22 Oct 2018)

UNDRR (2012) Education—a launch pad for disaster resilience. Available online: https://www.unisdr.org/archive/27998

UNESCO (2010) Climate Change Education for Sustainable Development

UNESCO (2012) Earth Science Education Initiative in Africa. Project Flyer. Available online: https://www.unesco.org/new/fileadmin/MULTIMEDIA/HQ/SC/pdf/Earth_Science_Education_in_Africa_brochure-eng.pdf (Accessed 22 Oct 2018)

UNESCO (2014a) Shaping the future, we want: UN Decade of Education for Sustainable Development; final report. UNESCO, Paris.

UNESCO (2014b) Roadmap for Implementing the Global Action Programme on Education for Sustainable Development. Available online: https://unesdoc.unesco.org/images/0023/002305/230514e.pdf

UNESCO (2015) Education 2030: Incheon Declaration and Framework for Action—Towards inclusive and equitable quality education and lifelong learning for all. UNESCO, Paris

UNESCO (2016a) Global education monitoring report, education for people and planet: creating sustainable futures for all. UNESCO, Paris

UNESCO (2016b) UNESCO Global Geoparks Celebrating Earth Heritage, Sustaining local Communities

Available online: https://unesdoc.unesco.org/images/0024/002436/243650e.pdf (Accessed 22 Oct 2018)

UNESCO (2017a) Education for Sustainable Development Goals: Learning Objectives. UNESCO, Paris

UNESCO (2017b) Education for Sustainable Development: Partners in Action: https://unesdoc.unesco.org/ark:/48223/pf0000259719

UNESCO (2017c) Earth Science Education Initiative in Africa. Project Website. Available online: https://www.unesco.org/new/en/natural-sciences/environment/earth-sciences/earth-science-education-in-africa/ (Accessed 22 Oct 2018).

UNESCO (2018a) Issues and trends in Education for Sustainable Development. UNESCO, Paris

UNESCO (2018b) Global Action Programme on Education for Sustainable Development Information Folder. UNESCO, Paris

UNESCO (2018c) Earth Science Education Initiative in Africa. Project Website. Available online: https://www.unesco.org/new/en/natural-sciences/environment/earth-sciences/earth-science-education-in-africa/. (Accessed 22 October 2018)

UNESCO (2018d) International Geoscience and Geoparks Programme (IGGP). Available online: https://www.unesco.org/new/en/natural-sciences/environment/earth-sciences/unesco-global-geoparks/. (Accessed 22 Oct 2018)

UNESCO and UNFCCC (2016) Available online: Action for climate empowerment: Guidelines for accelerating solutions through education, training and public

United Nations (2015a) Transforming Our World: The 2030 Agenda for Sustainable Development: United Nations, Geneva

United Nations, (2015b) The Millennium Development Goals Report

United Nations (2015c) The Paris Agreement. United Nations, Geneva, p 27

United Nations (2018) The Sustainable Development Goals Report. Available online: https://unstats.un.org/sdgs/files/report/2018/TheSustainableDevelopmentGoalsReport2018-EN.pdf

UNESCO Institute of Statistics (2018) New Education Data for SDG 4 and More. https://uis.unesco.org/en/news/new-education-data-sdg-4-and-more. (Accessed 9 March 2019)

UN-OHRLLS (2015) Small island developing states in numbers, climate change edition 2015. Available online: https://unohrlls.org/custom-content/uploads/2015/11/SIDS-IN-NUMBERS-CLIMATE-CHANGE-EDITION_2015.pdf

UNISDR (2015a) Sendai Framework for Disaster Risk Reduction 2015–2030

UNISDR (2015b) Global Assessment Report on Disaster Risk Reduction. The Future of Disaster Risk Management, Making Development Sustainable

UNRISD (2016) Policy Innovations. Transformative Change. Implementing the 2030 Agenda for Sustainable Development. UNRISD Flagship Report. Available online: https://www.unrisd.org/flagship2016-chapter5

Villuercas Ibores Jara Geoparque Mundial de La UNESCO (2019) online in Spanish and English at: https://www.geoparquevilluercas.es/?lang=en

Villuercas Ibores Jara Geopark (2016) Villuercas Ibores Jara Geopark Supplementary Dossier 2011–2105 Revalidation Process: https://www.geoparquevilluercas.es/wp-content/uploads/2016/02/Villuercas-Geopark-supplementary-dossier.pdf

Voight B (1990) The 1985 Nevado del Ruiz volcano catastrophe: anatomy and retrospection. J Volcanol Geoth Res 42(1–2):151–188

Wahlstrom M, Guha-Sapir D (2015) The human cost of weather-related disasters 1995–2015. UNISDR, Geneva

Wiek A, Withycombe L, Redman CL (2011) Key competencies in sustainability: a reference framework for academic program development. Sustain Sci 6(2):203–218

Wong SS, Yelderman JC (2016) Time not wasted: how collaborative research and education help build groundwater sustainability in rural northern Uganda, Africa. Geol Soc Am Spec Pap 520:20–17

World Water Assessment Programme (2015) UN World Water Development Reporte. Available online https://unesdoc.unesco.org/images/0023/002318/231823E.pdf

Ellen Metzger is Professor of Geology and Science Education at San José State University and collaborated with philosopher Randall Curren to write the book Living Well Now and in the Future: Why Sustainability Matters (MIT Press, 2017). She incorporates sustainability concepts into courses for both science students and future teachers and co-directs the Bay Area Earth Environmental STEM Institute (BAESI), a professional development program for teachers founded in 1990.

David Gosselin is Director of Environmental Studies and Earth Systems Scientist at the University of Nebraska-Lincoln, focusing on Earth, environmental, and interdisciplinary education, workforce issues, team development processes, sustainability education, water quality and quantity issues and the application of geochemistry to understand water systems. He has authored or co-authored more than 150 publications that include refereed journal articles, non-refereed abstracts, and contract reports in the field of education and water research.

Cailin Huyck Orr is the Associate Director at the Science Education Resource Center (SERC) at Carleton College. SERC partners with people and institutions across the U.S. to improve the undergraduate experience for students of all backgrounds in STEM and related fields through professional development to promote evidence based teaching, community visioning and bringing science into broader use. Her background is in limnology, particularly lotic systems, and interdisciplinary approaches to environmental problem solving.

Achieve Gender Equality and Empower All Women and Girls

5

Ezzoura Errami, Gerel Ochir,
and Silvia Peppoloni

E. Errami (✉)
Faculté Polydisciplinaire de Safi, Université Cadi
Ayyad, Safi, Morocco
e-mail: errami.e@uca.ac.ma

E. Errami
African Association of Women in Geosciences,
Abidjan, Côte d'Ivoire

G. Ochir
School of Geology and Mining, Mongolian
University of Science and Technology, Ulaanbaatar,
Mongolia

S. Peppoloni
Istituto Nazionale di Geofisica e Vulcanologia,
Roma, Italy

S. Peppoloni
International Association for Promoting Geoethics,
Via di Vigna Murata, Roma, Italy

© Springer Nature Switzerland AG 2021
J. C. Gill and M. Smith (eds.), *Geosciences and the Sustainable Development Goals*,
Sustainable Development Goals Series, https://doi.org/10.1007/978-3-030-38815-7_5

Abstract

5 GENDER EQUALITY

Overview

Gender quality is a fundamental human right, embedded into the United Nations Charter (1945)

Women still face multiple forms of discrimination, harassment and abuse.

Gender equality is foundational to ensuring inclusive sustainable socio-economic development, peace and justice.

Links to other SDGs

Women hold the key to building a world free from poverty (SDG 1) and hunger (SDG 2)

Educational progress underpins changes in attitude and opportunity that enables equality

Improvements to health (SDG 3) through access to clean water and safe sanitation (SDG 6) allows girls to access and stay in school

Improvements to health   Clean water and sanitation

Good governance and justice (SDG 16) is critical to equality

Women are more vunerable than men to the climate change impacts (SDG 13) and hazards (SDG 1)

Lack of equality prevents women from pursuing geoscience training and careers (SDG 8)

Gender equality in geoscience

Tackling barriers that prevent women from pursuing geoscience training and careers

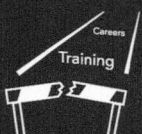

Ensuring equal access to professional development opportunities

Ending harassment, discrimination and bullying

Improving access to leadership positions of geoscience organisations

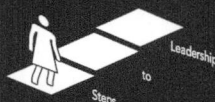

Improving recognition of female scientists

Men advocating for gender equality, as well as women

Everyone should take a personal responsibility for building an inclusive community

Mentors and role models provide motivation, build self-esteem, and incentivise action against challenges.

5.1 Introduction

Equality is a fundamental human right, embedded into the United Nations (UN) Charter (1945). Article 1 states that the UN will facilitate international co-operation *'in promoting and encouraging respect for human rights and for fundamental freedoms for all without distinction as to race, sex, language, or religion'*. The Universal Declaration of Human Rights reiterates this commitment, noting that 'all human beings are born free and equal in dignity and rights' and that no distinction of any kind should be made with regards to the human rights articulated within the Declaration (United Nations 1948).

Women may face multiple and intersecting forms of discrimination, depending on their nationality, sexual orientation, religion, age, socio-economic status, disability, race, ethnicity, marital status, or whether or not they have children (UN Women 2015). Tackling such discriminations and ensuring equal rights, responsibilities, and opportunities for all women and girls is central to the UN Sustainable Development Goals (SDGs), and the specific focus of **SDG 5** (Gender

Equality). Table 5.1 outlines six targets and three means of implementation relating to **SDG 5**. These include ambitions to end all forms of discrimination of, violence against and exploitation of women and girls, creating an environment that ensures their full participation in public life, and equal rights to economic and natural resources.

Gender equality is foundational to ensuring inclusive sustainable socio-economic development, peace, and justice. Partnerships between governments, public and private sectors, civil society and local communities (**SDG 17**) need to be inclusive and recognise the fundamental role of women in local, national, regional, and global human and socio-economic development strategies. Women hold the key to building a world free from poverty (**SDG 1**) and hunger (**SDG 2**) (FAO 2016). Progress in all forms of education (**SDG 4**) underpins changes in attitude and opportunity that enable equality, with improvements to health (**SDG 3**) through access to clean water and safe sanitation (**SDG 6**) allowing girls to access and stay in school. Good and inclusive governance and justice (**SDG 16**) is critical to equality in all contexts. Until gender equality is

Table 5.1 SDG 5 targets and means of implementation

Target	Description of Target *(5.1 to 5.6)* or Means of Implementation *(5.A to 5.C)*
5.1	End all forms of discrimination against all women and girls everywhere
5.2	Eliminate all forms of violence against all women and girls in the public and private spheres, including trafficking and sexual and other types of exploitation
5.3	Eliminate all harmful practices, such as child, early and forced marriage and female genital mutilation
5.4	Recognise and value unpaid care and domestic work through the provision of public services, infrastructure and social protection policies and the promotion of shared responsibility within the household and the family as nationally appropriate
5.5	Ensure women's full and effective participation and equal opportunities for leadership at all levels of decision-making in political, economic and public life
5.6	Ensure universal access to sexual and reproductive health and reproductive rights as agreed in accordance with the Programme of Action of the International Conference on Population and Development and the Beijing Platform for Action and the outcome documents of their review conferences
5.A	Undertake reforms to give women equal rights to economic resources, as well as access to ownership and control over land and other forms of property, financial services, inheritance and natural resources, in accordance with national laws
5.B	Enhance the use of enabling technology, in particular information and communications technology, to promote the empowerment of women
5.C	Adopt and strengthen sound policies and enforceable legislation for the promotion of gender equality and the empowerment of all women and girls at all levels

realised, women will remain more vulnerable than men will to the impacts of climate change (**SDG 13**) and natural hazards (**SDGs 1** and **11**). Lack of equality means that cultural, economic, and political constraints prevent women from receiving access to the resources needed to thrive and build resilient livelihoods.

Achieving **SDG 5** globally by 2030, therefore, requires progress on all of the SDGs, with gender equality embedded into the policies and actions driving these forward. Meaningful and full participation of women at all levels of institutions, and in all partnerships, is critical to these groupings being effective, credible, and having social license to operate. All stakeholders should take action to move towards a more sustainable and equitable world, where women and men have equal rights, shared responsibility towards families, and opportunities to serve their local, national, and global communities.

In this chapter, we aim to show that the geoscience community can help to secure progress in this goal worldwide. We first examine current trends in gender equality, and the scale of the challenge to address (Sect. 5.2). We proceed to explore gender equality in geoscience (Sect. 5.3), including specific challenges faced by women in the geoscience community. We then set out work done by many innovative organisations to address these challenges, with a UK example (Fig. 5.1), a pan-African initiative, and work in Mongolia (Sect. 5.4). We conclude with lessons and recommendations for the broader geoscience community (Sect. 5.5) to advance this critical cause.

Fig. 5.1 Girls into Geoscience Field Trip to Dartmoor (2017). Started by the University of Plymouth in 2014, Girls into Geoscience aims to encourage young women to consider studying geoscience at university, through workshops, conferences, and field trips. © Sarah Boulton/Girls into Geoscience (used with permission)

5.2   Progress in Tackling Gender Equality

There has been progress towards gender equality in recent years, however significant work remains. The United Nations (2015) Millennium Development Goals Report notes that while gender disparity has narrowed in terms of the education of women and girls, they still earn 24% less than men globally. As of 2014, 52 (of 195) countries do not guarantee equality between men and women in their constitutions (UN 2019).

The Global Gender Gap Index aims to assess and monitor gender-based disparities, and increase global awareness about gender gaps. This index (where 0.0 or 0% means total imparity and 1.0 or 100% means perfect parity) considers four sub-indices: (i) economic participation and opportunity, (ii) educational attainment, (iii) health and survival, and (iv) political empowerment, all at a national scale. The mean value of these four sub-indices gives the Global Gender Gap Index. The World Economic Forum (2018) examined the progress made by 144 countries towards gender parity, and determined an average Global Gender Gap Index of 68%. This suggests much room for improvement if we are to achieve universal gender parity, the ambition of SDG 5. Figure 5.2 shows the global average for the four specific thematic dimensions (i–iv) in 2017, with parity in economic participation and opportunity

and political empowerment lagging behind educational attainment and health and survival.

Gender parity varies significantly by region with a regional average of 60% in the Middle East and North Africa, and 75% in Western Europe. While all world regions have narrowed their gender gap during the past 11 years, more efforts are required to accelerate progress. In East Asia and the Pacific, for example, the World Economic Forum (2018) project that it will take 171 years to achieve gender parity (Fig. 5.3). At national levels, the situation is more complex. The 2018 Global Gender Gap Index report shows that the top ten countries are (in order of parity): Iceland, Norway, Sweden, Finland, Nicaragua, Rwanda, New Zealand, Philippines, Ireland, and Namibia. This includes both high-income and low-income countries. While some countries have seen increasing progress on gender parity (e.g., Iceland, Nicaragua), others have seen more variation. Mali has seen an overall decline in gender parity since 2006 (World Economic Forum 2018).

The economic development of countries does not, therefore, control the extent to which there is gender equality, when using this index. Rwanda is classified by the United Nations (UN) as one of the least developed countries (UNDESA 2019); however, it has a global index of 0.804 (as of 2018), with one of the highest number of women parliamentarians in the world (World Economic Forum 2018). In contrast, a lack of gender equality will

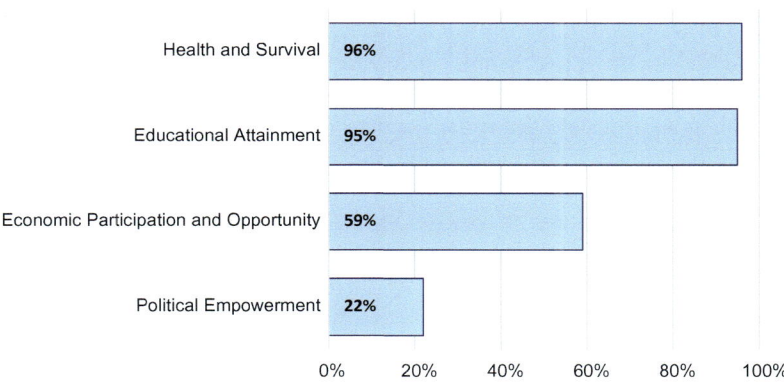

Fig. 5.2 Gender parity (as of 2018) with respect to the four sub-indices of the Global Gender Gap Index. Created using data from the World Economic Forum (2018), covering 149 countries featured in the 2018 index

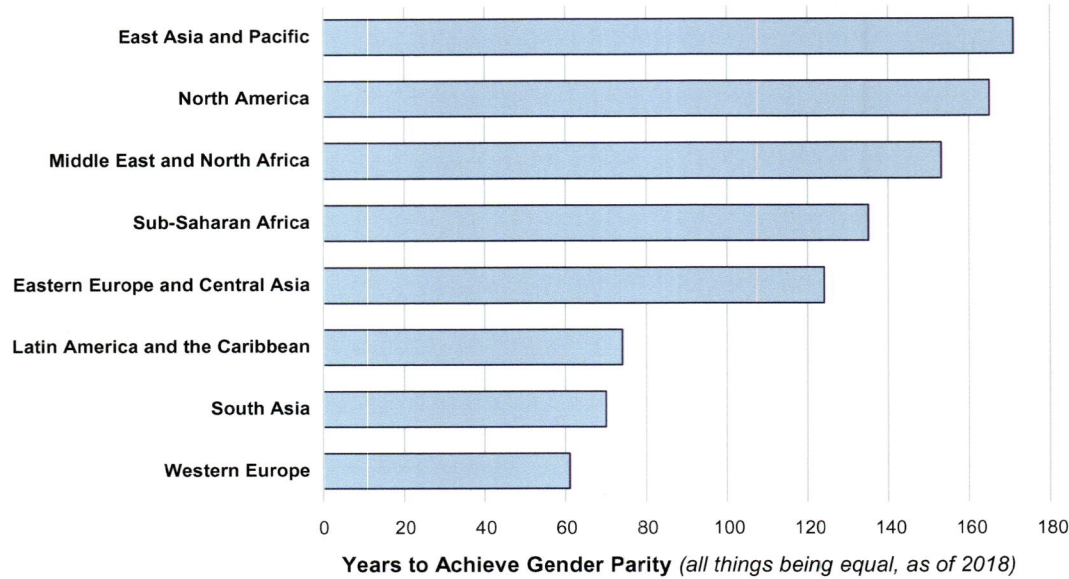

Fig. 5.3 Regional variation in the projected number of years it will take to achieve gender parity (all things being equal). Created by authors, based on data in World Economic Forum, Global Gender Gap Index, 2018

suppress economic growth. The World Bank (2011) suggests that allowing girls to reach the same educational level as boys will increase annual GDP growth rates, with a significant multiplier effect. A country's gender gap closely relates to economic performance, with nations wanting to remain competitive needing to adopt inclusive strategies where women and men have the same opportunities. Countries with equitable sharing of resources act as examples to other countries in their region or income group. Political and public will are critical to driving forward gender parity in diverse contexts. Women's engagement in public life is helping to tackle gender inequality, giving more credibility to institutions and increasing democratic engagement.

A different measure of parity, the *Gender Inequality Index*, also gives a multidimensional view of gender inequalities (UNDP 2019). It includes factors relating to reproductive health, empowerment, and economic status (Ortiz-Ospina and Roser 2018). Figure 5.4 shows country scores from 1995 to 2015, with lower scores in this context meaning lower inequality and higher scores meaning higher inequality.

This index shows considerable variation between countries, and a clearer Global North–Global South divide.

Gender gaps are not uniform across all sectors and industries. The World Economic Forum (2017) found that the largest gender employment gaps are in science, technology, engineering, and mathematics disciplines, and in sectors including software and IT services, manufacturing, energy, and mining.

By collecting and using data to understand the geographical distribution of inequality, how it has changed over time and its prevalence across sectors, it is feasible to identify effective mechanisms to help move towards parity. In the remainder of this chapter, we discuss the pertinence of this theme to the geoscience community.

5.3 Gender Equality in Geoscience

5.3.1 Women in Geoscience

Women have made an important contribution to the historical development of geoscience (Burek

Gender Inequality Index from the Human Development Report

This index covers three dimensions: reproductive health (based on maternal mortality ratio and adolescent birth rates); empowerment (based on proportion of parliamentary seats occupied by females and proportion of adult females aged 25 years and older with at least some secondary education); and economic status (based on labour market participation rates of female and male populations aged 15 years and older). Scores are between 0-1 and higher values indicate higher inequalities.

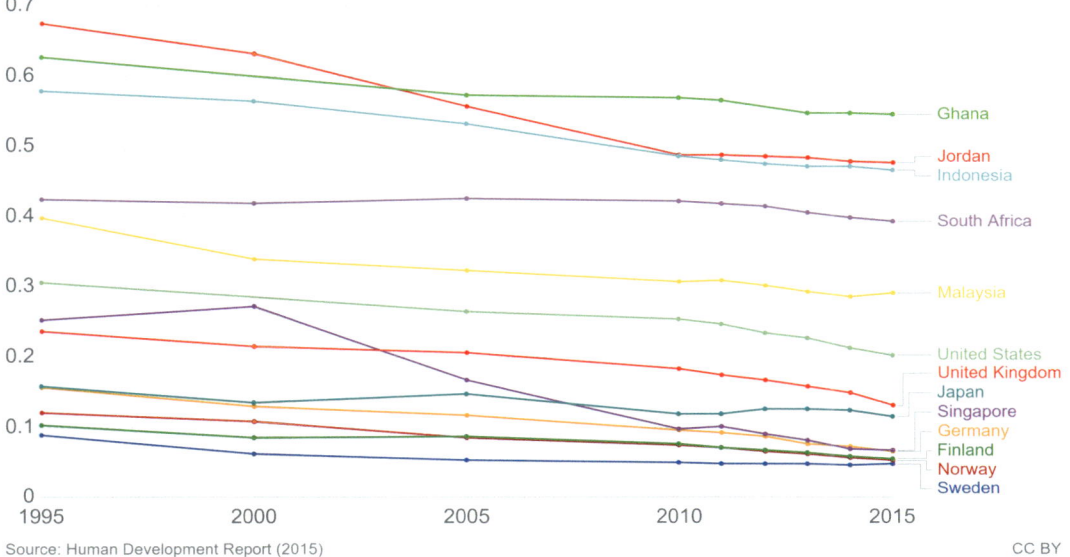

Source: Human Development Report (2015)   CC BY

Fig. 5.4 Gender Inequality Index, 1995–2015. A selection of countries are shown to demonstrate changing trends, and regional differences. Credit: Ortiz-Ospina and Tzvetkova (2018), based on data from the Human Development Report (2015). Reproduced under a CC BY license (https://creativecommons.org/licenses/by/4.0/)

and Higgs 2007; Johnson 2018), but this is not always accurately portrayed by those writing about the subject. The omission of women geoscientists from historical studies of geology is not due to their lack of existence. Kölbl-Ebert (2001) has demonstrated the numerous and important contributions of women. This is a common problem across other broader science, technology, engineering, and mathematics (or STEM) subjects. Recognising this underrepresentation, the British physicist Dr Jessica Wade has written (and encouraged others to write) Wikipedia articles about notable academics to promote women role models in STEM, with similar action for other minority groups (AAAS 2019). We need to remember the historical role that women have played in establishing and nurturing geoscience as a discipline. We need to celebrate their pioneering work across numerous geosciences disciplines, and in different socio-economic and

cultural environments, often when tradition suggested that they should not be engaging in technical and scientific disciplines. Such role models provide inspiration to new generations of women.

Today, the representation of women geoscientists is generally increasing. For example, in Mongolia, by the end of the twentieth century more women than men were graduating as geoscientists (Gerel et al. 2006). In the USA, American Geosciences Institute—AGI (2018a) documented trends from 1975 to 2017 that show an increasing number of females enrolling on to both undergraduate and graduate geoscience programmes, and an increasing number of females granted degrees between 1985 and 2017 (Fig. 5.5). From 2005 to 2017, this increase has been less pronounced, with AGI (2015) suggesting an increase in the number of men enrolling at both undergraduate and graduate levels

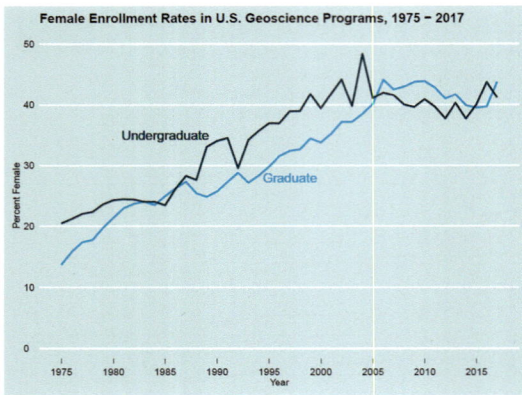
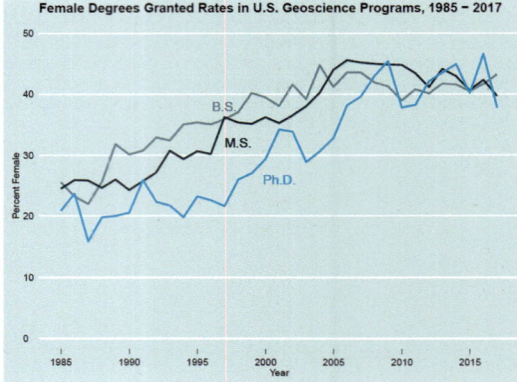

Fig. 5.5 (left) Female enrolment rates in US geoscience programmes, 1975–2017. (right) Percentage of degrees granted to females in US geoscience programmes, 1985– 2017. © American Geosciences Institute (used with permission)

being due to the growing hydrocarbon industry and returning military veterans being encouraged to pursue careers in the resource sector. In Mongolia, by the end of the twentieth century, more women were graduating as geoscientists than men (Gerel et al. 2006), although challenges remain in translating this to equality of opportunity after graduation (see Sect. 5.4.3).

In other countries, although there is a lack of data describing changes in the number of geoscientists, overall statistics for STEM subjects, or the physical sciences (including geology and environmental science courses) exist. In the United Kingdom, for example, 40% of physical science graduates and 23% of those holding Core STEM Occupations in 2017 were women (WISE 2017). Closing gender gaps in access to STEM subjects may require more than tackling gender equality at a high level in society. Stoet and Geary (2018) have highlighted that Finland has high levels of gender equality, education performance, and science literacy of adolescent girls, but one of the world's largest gender gaps in college degrees in STEM fields. Countries that are less equal may have particular pressures to improve quality of life that promote engagement of females in STEM subjects, with STEM professions viewed as bridges to equality, creating greater opportunities to advance socially and economically. In more equal societies, this

perception may be different, and women may have a greater choice to pursue careers that meet their diverse aspirations. Gender equality in geoscience education may, therefore, require careful targeting of interventions, including exploring how perceptions of the value of pursuing one type of career versus another are influenced.

Gender equality is not solely about the number of women enrolling and graduating from geoscience courses, it is also about the way that they are treated throughout their careers (see Sect. 5.3.2 for a discussion of harassment, discrimination, and bullying in the geosciences) and equal access to opportunities to pursue careers in academia, industry, and public life. Kölbl-Ebert (2001) reported, for example, that in 1989 a University of Tübingen student trip to a coal mine was cancelled because the mining company would not allow women underground. Gender diversity is a struggle for the mining sector, at all levels from entry to senior leadership. Mining is traditionally an industry dominated by men, with many areas of employment difficult for women to participate meaningfully. Potential barriers include (*i*) absence of family-friendly workplace policies, (*ii*) sexual harassment and discrimination, (*iii*) a lack of career pathways and leadership roles, and (*iv*) a lack of support and mentoring. Coupled with these barriers, cultural norms and discriminatory legislation may prohibit women from working in certain types of mining activities. Such attitudes persist in many parts of the world, and are hugely detrimental to the encouragement of early-career women. In Sect. 5.4, we outline a range of initiatives that are helping dismantle such prejudices.

In academia, the number of female geoscience tertiary education academic staff is increasing in many countries. For example, in the USA the AGI (2017) estimates 20% of the geoscience academic workforce was female in 2016, up from 14% in 2006. While far from being equal, this increase is predominantly in the junior ranks of academic staff. In the coming years, through retirement of existing staff and promotion of some of these junior staff, it is expected that further progress towards parity will be made. In Mongolia, an analysis of gender equality in the geosciences completed as part of the *Education for Environmental Transition: Mining in Mongolia (2004–2010)* project found that 71% of the Department of Geology and 43% of the Department of Mining staff at the Mongolian University of Science and Technology were female. In Mongolia, the overall proportion of women taking leadership roles as educational/faculty Chairs or Deans is 51%, compared to 21% of industry and government leadership positions being held by women.

Recommendations for further progress towards gender equality include supporting women with families, changing cultural attitudes, and reforming publishing (Cheryan et al. 2017). Equal opportunities to engage in international events, critical for academic networking, research dissemination, and learning and development are also needed. 19% of the 31st International Geological Congress held in 2000 in Rio de Janeiro participants were female (Kölbl-Ebert 2001), and 30% of the Resources for Future Generations conference held in 2018 inVancouver 2018 participants were female. One reason given for this challenge is the lack of consideration given to childcare by academic conference organisers

(Calisi et al. 2018), contributing to the 'baby penalty' that negatively affects women's career mobility (Mason 2013).

Gender equality also requires greater recognition of scientists who are women by national and international geoscience communities, and greater access to leadership positions of influential geoscience organisations. The latter have predominantly been led by men throughout their history, and continue to be so today. For example, at the time of writing, the International Union of Geological Sciences (IUGS) has never had a woman president.

Moving towards gender equality in geoscience requires a multifaceted approach that addresses inequality in education, the workplace, and professional development opportunities. This is a global challenge, but one that requires actions by all members of the geoscience community.

5.3.2 Harassment, Discrimination, and Bullying in the Geosciences

Harassment, discrimination, and bullying have no national, cultural, or ethnic boundaries, although they take different forms depending on the socio-cultural contexts. These are global problems, and are prevalent in both science and academia. Respect is a prerequisite for an ethical working environment, where all persons are equal and have the same opportunities. Harassment, discrimination, and bullying offend the dignity of the person, limit the individual freedom of choice and impede ethical decision-making, compromising the serenity of working environments (Peppoloni and Di Capua 2017). Discrimination, harassment, and bullying unfairly silence many voices, particularly those who are marginalised and already the victims of inequality. For example, young women may be discriminated against if they are (or are considered likely to fall) pregnant and need time off work, or if conscious or unconscious prejudicial discrimination persists in determining some jobs are solely appropriate to men.

Women could meet hostile and discriminating environments in their range of study and work environments, especially during fieldwork. In scientific fields, 71% of women surveyed were sexually harassed while conducting fieldwork and 25% were sexually assaulted (Clancy et al. 2014). Geoscience is not immune to these problems (Marín-Spiotta et al. 2016; Mogk 2017; St. John et al. 2016). A survey conducted by the EU-funded *ENVRIplus* project found that 27% of respondents indicated harassment and 20% gender, racial, or religious discrimination as disgraceful behaviours in their working environments (Peppoloni et al. 2017). These results suggest that harassment and bullying are not marginal problems in the geoscience community. The scientific and professional community has a general tendency, however, to deal with subjects that are strictly technical. Gender issues, ethics, research integrity, and the socio-political implications of geosciences have long been considered 'niche topics', although there are signs that this is changing with some positive initiatives in recent years.

The Statement on Harassment in Geosciences by the *American Geosciences Institute* (AGI 2018b) and the document on Scientific Integrity and Professional Ethics by the *American Geophysical Union* (AGU 2017) have placed harassment, discrimination, and bullying at the centre of discussions in the international geoscience community. The *European Federation of Geologists—EFG* has updated its professional code to include harassment as an unprofessional behaviour that must be reported to relevant authorities for prosecution (EFG 2017). In recent years, the *International Association for Promoting Geoethics—IAPG* has launched initiatives in the scientific community to increase understanding of this problem and negative consequences in ethical and social terms. The IAPG '*zero tolerance towards harassment and discrimination*' campaign gives a clear message to its more than 2200 members across 125 nations, promoting activities that favour strong professionalism and the fight against harassment and discrimination within the scientific and professional community (IAPG 2019).

Furthermore, the *Geological Society of America—GSA* released a statement on removing barriers to career progression for women in geosciences (GSA 2018), and the *National Academies of Sciences, Engineering, and Medicine* (2018) published a report on '*Sexual Harassment of Women: Climate, Culture, and Consequences in Academic Sciences, Engineering, and Medicine*'. This report clarified the problem in quantitative terms, presenting a comprehensive review of research, experiences, and effects of sexual harassment on women and their careers in science, engineering, and medicine. The report provides recommendations for how organisations can prevent and address sexual harassment in academic settings.

Despite the work of these and other organisations, much work remains to remove harassment, discrimination, and bullying in the international scientific community and among the numerous bodies and institutions that compose it. There is a lack of structures (e.g., offices, commissions, working groups) to ensure effective monitoring and reporting, handling of complaints, the development of coordinated actions and initiatives that address problems, raise awareness of fundamental rights, and promotes respect towards all. Geoscientists need access to such structures—as well as broader societal changes in attitude--to hold perpetrators of abuse, harassment, and discrimination to account.

5.3.3 Public Understanding of Geoscience

Many of the themes that geoscientists work on have a gendered dimension, and it is therefore important that geoscientists understand the drivers and impacts of gender inequality. Resource management (e.g., water supply, **SDG 6**), for example, is traditionally the responsibility of women in many contexts (Fig. 5.6). The sustainability of

Fig. 5.6 Water Collection in Tanzania. While water collection is traditionally the responsibility of women in many contexts, their voices are often excluded from discussions on resource management. © Geology for Global Development (used with permission)

these resources may confront women more than other members of the community, being valuable partners in understanding the actions needed to protect and manage the resource. To understand issues of water quality or variation in supply, interviewing women in a community would generally serve geoscientists well.

Initiatives to reduce risk from disasters, such as those triggered by earthquakes, flooding, landslides, or volcanic eruptions, also need to consider gender. Disasters threaten social and economic development, disproportionately affecting the most vulnerable and marginalised in society. Disasters kill more women than men, for example, because of the existing gender inequalities and their lower socio-economic status (Neumayer and Plümper 2007). The 2004 Indian Ocean tsunami killed four times as many women as men in affected areas of Indonesia, Sri Lanka and India (MacDonald 2005). Disaster preparedness efforts often did not consider women, and therefore they were less aware of how to protect themselves. Differences between traditional roles also increased the exposure of women to the tsunami (MacDonald 2005). Formal and informal education, and inclusive participation in disaster risk reduction, community planning, and citizen science, is, therefore, key to minimising impacts of disasters.

5.4 Initiatives to Improve Gender Equality in Geoscience

Numerous local, national, regional, and international initiatives promote the geosciences as a career route for women, increase the visibility of female geoscientists, and break down barriers preventing women from accessing leadership positions of key institutions. In Sect. 5.4.1, we describe a range of organisations—many international—working to improve gender equality in the geosciences. We focus in Sects. 5.4.2 and 5.4.3 on initiatives at two scales, continental

(Africa) and national (Mongolia), reflecting the expertise and leadership of the chapter authors.

5.4.1 Overview of Organisations and Networks

Early-career women in the geosciences need concrete examples of other women who have achieved great results in their professional and academic careers, who have leading roles in the geoscience community, who have been capable to challenge stereotypes and social constraints that assign to women predefined roles. Holmes et al. (2008) argue that the gender gap will be reduced when every geoscience student has access to women geoscientists they can emulate. Mentors and inspiring figures provide personal motivation, build self-esteem, and incentivise action against challenges (Fig. 5.7).

We outline a selection of professional organisations, networks, and projects in Box 5.1. These organisations are fundamental points of aggregation in the scientific and professional community for women and tangible examples of commitment in the working world for early-career geoscientists, forms of self-organisation that very often promote mentoring and inspiring activities. To provide a good and pleasant work environment, all these initiatives are open to constructive discussions and complementarity with others and their engagement is important. The International Women in Mining —IWM have an initiative that engages men in conversations about gender equality, recognising that they are key to change. This initiative celebrates men who advocate for women in mining and help to narrow existing gender gaps (IWM 2019).

> **Box 5.1 Organisations Promoting Gender Equality and Women in Geoscience**
>
> **500 Women Scientists** (https://500women scientists.org/). Founded in 2016, this project aims to give voice to women in

Fig. 5.7 5th Geology for Global Development Annual Conference (2017). The organisers' commitment to excellent representation throughout the history of conferences within this series has helped to attract many young, women Earth scientists. © Geology for Global Development (used with permission)

science (including in geoscience) to empower women to reach their full potential in science, increase scientific literacy through public engagement, and advocate for science and equality. This project provides a resource for journalists, educators, policymakers, and scientists.

100 Women Against Stereotypes (https://100esperte.it/). An Italian project, started in 2016, with the names and CVs of 100 Italian women in science (including 13 geoscientists), aiming to enhance the role of women in contributing to the cultural, scientific, and economic development of society, and promoting excellent scientists (who are female) to the media as opinion makers, consultants, and prominent experts in their disciplines.

African Associations of Women in Geosciences (AAWG) (www.aawg.org/). Aims to encourage women geoscientists to participate in geosciences-related conferences and to inform about or become involved in gender issues related to geosciences.

Association for Women Geoscientists (AWG) (https://awg.org/). Founded in 1977, is an international organisation devoted to enhancing the quality and level of participation of women in geosciences and to introduce girls and young women to geoscience careers, by providing networking and mentoring opportunities to ensure rewarding opportunities for women in the geosciences.

Earth Science Women's Network (ESWN) (https://eswnonline.org/). Dedicated to career development, peer mentoring, and community building for women in the geosciences.

Girls into Geoscience (www.plymouth.ac.uk/research/earth-sciences/girls-into-geoscience). Started by the University of Plymouth in 2014, this aims to encourage young women to consider studying geoscience at university, through workshops, conferences, and field trips.

International Women in Mining (IWM) (https://internationalwim.org/). A network for women in the mining industry, with 10,000 members in over 100 countries, supporting more than 50 groups around the world with 22 in Africa. The association aims to foster women's professional development in mining industry via a global mentoring programme.

International Women in Resources Mentoring Programme (www.iwrmp.com/). Empowers and promotes women working in resources to navigate industry challenges and progress their careers offering the confidence for achievement and leadership to make their mark in the industry.

Organisation for Women in Science for the Developing World (https://owsd.net/). An international forum that gathers women scientists with the objective of strengthening their role in the development process and promoting their representation in scientific and technological leadership.

Society for Women in Marine Science (SWMS) (https://swmsmarinescience.com/). Brings together marine scientists of all career levels to discuss the diverse experiences of women in marine science, celebrate research done by women in the field, and promote the visibility of women in marine science.

Women in Coastal Geoscience and Engineering (https://womenincoastal.org/). A network to help achieve gender equality by inspiring, supporting and celebrating women in coastal geoscience and engineering, working across age groups, career levels, and sectors.

Women in Geoscience and Engineering, Special Interest Community (WGE SIC) (www.eage.org/region/?evp=13076). Part of the European Association of Geoscientists & Engineers, whose main mission is to facilitate the exchange of knowledge and experience and to promote mutual support among their women members.

Women In Mining (www.womeninmining.org.uk/). A non-profit organisation dedicated to promoting and progressing the development of women in the mining and minerals sector.

Women in Polar Science (WiPS) (https://womeninpolarscience.org/). Aims to celebrate and showcase the inspiring work of women in Polar Science, building and strengthening connections, and providing an avenue for mentoring women in polar science at various career stages.

Women Oceanographers (www.womenoceanographers.org/). Aims to encourage young women to pursue careers in science and to remove the mystery that surrounds being a scientist (Fig. 5.8).

Fig. 5.8 Girls into Geoscience Practical Class (2018). Exploring foraminifera (forams) down the microscope at the Girls into Geoscience workshop. © Sarah Boulton/Girls into Geoscience (used with permission)

5.4.2 Tackling Gender Inequality in Africa

Women in Africa are more economically active than women in any other region of the world, at the forefront of agriculture and small businesses, and are better represented in many parliaments than in Europe (African Development Bank 2019). Many challenges remain, however, including reducing the time women spend collecting water and ensuring women receive equal way for equal work. This need for greater gender equity is highlighted in the Africa Mining Vision, adopted in 2009 by African Heads of State and facilitated by the African Union. This sets out how mining can be used to drive development across the continent, emphasising the need to '*initiate empowerment of women through integrating gender equity in mining policies, laws, regulations, standards and codes*' (Africa Mining Vision 2009).

Academia also needs to take similar action on gender equality, with UNESCO (2019) estimating that as of 2016, 24% of tertiary education academic staff in sub-Saharan Africa are women. In some countries, this is much lower: for example, 9% (Burkina Faso), 12% (Cote d'Ivoire), and 6% in Togo. UNESCO (2019) suggests the global average, in 2016, was 42% of tertiary education academic staff being women, reaching 85% in Myanmar in 2017. The African Association of Women in Geosciences (AAWG[1]), founded in 1995, is working to address this in the context of the geosciences, by encouraging women to participate in geoscience conferences, and raising awareness of gender issues related to geoscience (Errami 2009, 2013).

AAWG is an action-orientated organisation, provoking discussion but recognising that this is not sufficient for addressing **SDG 5**. AAWG believes that educating and involving women will benefit the whole of society. AAWG members are involved in mentoring to help women identify and overcome barriers preventing their involvement and progression in the geosciences. AAWG organises meetings with girls and

---

[1]www.aawg.org/.

women in different institutions to determine their problems and reflect on how they can apply learning from other experienced women to help address these challenges. AAWG focuses on three main projects:

Geoscience Conferences. Biannual conferences are coordinated, dedicated to the encouragement of participation of women in the geosciences, and the presentation of their scientific works. These conferences are open to the contribution of both men and women to make them working together. In this framework, men feel that women are giving them opportunities to serve in the activities they are leading, which will push men to give them back the opportunity to be involved in their activities. The conferences are accompanied by short courses, workshops, roundtables, and field trips (Fig. 5.9).

Day of Earth Sciences in Africa and Middle East. This aims to promote geoscience for society, including promoting geoscience to girls in primary and secondary schools, and the public. Since this started, more than 10,000 people have participated in activities, including many women.

The African Geoparks Network (AGN). This network, and three international conferences promoting geoheritage for society, have helped to create new jobs for local women and empowered them through the creation of new products and cooperatives (Errami et al. 2012). This contributes to efforts to alleviate poverty through geotourism (see **SDGs 4** and **8**).

Women lead AAWG activities, which helps to build capacity in organising local, regional, and international scientific events, financial management, and coordinating international scientific

Fig. 5.9 (**a**, **b**) Fifth AAWG preconference meeting entitled *Women and Geosciences for Peace*, Abidjan (Côte d'Ivoire), 4–8 May 2009. Meetings with girls from a local secondary school and women geoscientists working in PETROCI Holding, the main sponsor of AAWG since 2009. (**c**, **d**) Post-AAWG Conference Field Trip (Côte d'Ivoire), with 110 participants. Credit: Authors' Own

teams. It increases their visibility at local, national, and international levels, and builds new and strengthens existing networks. Association with AAWG gives more visibility to women geoscientists at local, national and international levels.

5.4.3 Tackling Gender Inequality in Mongolia

Mongolia currently has one of the highest percentages of women working professionally in the region, with women and men perceived to be relatively equal compared to other countries in the region. Mongolia is ranked 53rd among 144 countries using the Gender Gap index (0.713), and is ranked 4th in the East Asia and Pacific region (Global Gender Gap Index 2017). Mongolia ratified the *United Nations Convention on the Elimination of All Forms of Discrimination Against Women* (UN Women 2009) in 1981, and the subsequent *Optional Protocols* in 2002. A National Law for the Promotion of Gender Equality was approved in 2011, supported by an implementation strategy and action plan for 2013–2016 overseen by the National Committee on Gender Equality (NCGE) and chaired by the Prime Minister.

Mongolia is emerging as a major new frontier for mineral exploration and development, with mining predicted to become the primary industry. The South Gobi province of Mongolia includes the largest mine in Asia, the *Oyu Tolgoi* copper and gold mine, as well as the *Tavan Tolgoi* coal mine (Cane 2014, 2015). There is a strong need for a highly educated, indigenous geoscience, and mining labour pool. It is therefore vital to address gender stereotypes and employment segregation if women are to have equal access to economic opportunities such as employment in mining, mining-related activities, and small to medium enterprises connected to the mining industry. Women's employment at *Oyu Tolgoi* mine does not exceed 38% (recently 40%) of the workforce in any category in mining administration, engineering, environment, geology, processing, information technology, and many

others, and is generally under 20%, suggesting space for improvement (Cane 2015). Generally, female participation in the mining workforce may be more difficult due to organisational culture, and socio-cultural barriers, stereotyping of jobs, discrimination (including through regulations that restrict or prevent women from higher-paying employment as heavy equipment operators or underground workers), sexual harassment, and lack of the mechanisms or will to manage complaints (Cane 2015).

There is also a need to tackle the negative impacts of mining on the lives of women and vulnerable groups (Cane 2015), with growing evidence that gender inequalities are being exacerbated by mining and associated population influx (e.g., through sexual harassment and gender-based violence). It is widely noted that the negative social impacts of mining disproportionately affect women (Macdonald and Rowland 2002; Lahiri-Dutt 2011).

The *Women in Mining Mongolia* (WIM Mongolia[2]) group has engaged with many early-career mining engineers and geologists working in mining companies to explore the contribution of women in Mongolia's extractive industry. Their survey included 16,000 participants (of approximately 37,000 people in Mongolia's mining industry including private- and government-owned companies engaged in various geological, technical, mining, and services fields, and institutions delivering mining education). Their survey found, for example, that almost 17% of the top managers of Mongolian mining companies are women, slightly above the global average of 12%.

Based on this research, WIM Mongolia has developed an action plan to improve diverse and inclusive decision-making practices within Mongolia's mining sector. This aims to strengthen the engagement of women in the mining sector, bringing their understanding and education to the sector, and therefore helping to enhance social license from local communities. WIM Mongolia will initiate dialogue with stakeholders on (i) educating, attracting, and

---

[2]www.wimmongolia.org/.

recruiting women into the mining sector; (ii) helping advance women in a male-dominated culture; and (iii) coaching and mentoring to develop leadership and accountability skills.

The *Education for Environmental Transition: Mining in Mongolia (2004–10)* project, funded by the Canadian International Development Agency (CIDA), has worked to build the capacity of the Mongolian University of Science and Technology (MUST) to train geologists, government officials, and other professionals in the skills necessary to implement environmentally sustainable mining practices. A specific project activity during this time included an analysis of gender equality in the geosciences in Mongolia, tracking differences between gender equality in academia and industry. At the end of the twentieth century, more women than men were graduating as geoscientists but this was not translating into career opportunities for women (Gerel et al. 2006), especially in mining companies.

The *Association of Mongolian Women Geoscientists* was established to increase the awareness of gender-related issues among relevant individuals and institutions, including mining companies, and to develop strategies for the further integration of women into all spheres of geoscience-related research and careers (Gerel et al. 2006). Activities help women to enhance their professional skills, share experiences, and promote the implementation of government policies on geology and the mining sector. Specific projects have included the publication of books to raise the profile of Mongolian female geoscientists, the development of a geopark, and coordinating an exhibition with the Museum of Geology and Mineral Resources.

Mongolia is improving aspects of gender equality compared to neighbouring countries (Cane 2015), however, pervasive gender discrimination and harassment continue to create barriers to women's economic empowerment and leadership. The maintenance and growth of a healthy and robust geoscience community in Mongolia requires careful attention to the way in which gender roles are changing. For educated early-career female geoscientists, a primary challenge is gaining access to what have traditionally been male roles, particularly in the mining sector. The rapid growth of mining in Mongolia has the potential to have a positive impact on the lives of women and other vulnerable groups, if managed appropriately.

5.5 Conclusions and Recommendations

Despite global progress towards gender equality and the empowerment of women worldwide, discrimination continues, depriving women and girls of their basic human rights and the same opportunities as men. Achieving gender equality is the responsibility of all (including both men and women), and should be a primary consideration when developing actions to address all 17 of the SDGs. It requires diverse changes in policy and practice in many domains. Legislation is necessary, but alone will not tackle the social, cultural, and historical obstacles that hinder women from accessing equality. UN Women (2015) notes that equality written in the law does not always translate into equality in practice. To address discrimination, we must also eliminate gender-based stereotypes (e.g., *'science is not for girls'*) and those social norms and practices (e.g., *'a girls education is prioritised less than a boys education)* that result in discrimination persisting (UN Women 2015).

Women and girls need equal access to the tools and services (e.g., education and training, health care, capacity building, suffrage) to help them fight for and defend their universal human rights. Each member of the geoscience community should take a personal responsibility for ensuring that equal rights and opportunities are at the heart of the study and practice of geoscience, building inclusive communities free of discrimination and abuse. Limiting inclusion means we are not harnessing the full collective intellect and creativity of the geoscience community to address pressing scientific and societal challenges. It does significant damage to individuals, to the geoscience profession, and to society as a whole.

In Box 5.2, we outline our recommendations, summarising key reflections from this chapter, to provoke discussion and action in the geoscience community about what more we can do to achieve **SDG 5** in our spheres of influence.

Box 5.2 Summary of Recommendations

1. Gender equality needs to be acknowledged as a primary concern by all in the geoscience community (i.e., it is not just an issue for women, and it is not a specialist interest theme for a small subset of the community).

2. Gender equality cuts across many of the SDGs, and therefore needs to be in the forefront of thinking when designing policies and programmes, or developing partnerships and networks, to address the full spectrum of SDGs, alongside tackling broader inequality (**SDG 10**) and ensuring actions are pro-poor (**SDG 1**).

3. Increase policymakers and communities' awareness of how gender equality in the geosciences can contribute to peaceful, healthy, and more prosperous communities.

4. Listen to the needs of both women in geoscience and communities, ensuring a bottom-up approach to tackling gender inequality.

5. Put systems in place to ensure equitable access to formal and informal education and training, including continued professional development and mentoring programmes. This will help to empower women and build their capacity. Mentorship encourages women, and helps to build professional networks.

6. Design educational and workplace environments to ensure the specific needs of females are met without them being exposed to any kind of abuse or discrimination. For example, access to safe water and sanitation is critical if girls are to complete their education (see **SDG 6**).

7. Increase the visibility of women in the geosciences to encourage children and youth considering careers in science, technology, engineering, and mathematics. Initiatives such as the *Girls into Geoscience* programme (Box 5.1) make a valuable contribution to this aim, and should be encouraged and resourced by the broader geoscience community.

8. Employers need to provide flexible work environments and more childcare support to help staff balance work and family commitments.

9. Develop toolkits focused on practical ways to facilitate and assess progress towards gender equality in geoscience training and employment environments could help drive forward action on gender equality. These would require engagement by all stakeholders, to ensure these toolkits are used.

10. Encourage women to stand for local, national, regional, and global leadership positions in the geosciences, with individual organisations determining the barriers that may prevent women from applying for these roles and taking action to remove these.

11. The geoscience community (both men and women) should be vocal about changes needed in attitudes, beliefs, and behaviours that result in gender-based discrimination and abuse (e.g., gender-based violence, stereotyped gender roles, early marriage) within geoscience institutions and society as a whole.

12. Strengthen local, national, regional, and global geoscience networks (such as those in Box 5.1) to facilitate the career enhancement of females in geoscience.

5.6 Key Learning Concepts

- Gender equality is a fundamental human right, embedded into the United Nations Charter, Universal Declaration of Human Rights, and UN Sustainable Development Goals. **SDG 5** aims to end all forms of discrimination of, violence against and exploitation of women and girls, creating an environment that ensures their full participation in public life, and equal rights to economic and natural resources. Significant work remains to ensure gender equality, with projections suggesting it will take anything from 60 to 170 years if we continue with business as usual.

- Women have made an important contribution to establishing and nurturing geoscience as a discipline, around the world. While there is some evidence to suggest an increase in the representation of women in geoscience programmes, challenges remain in translating this to equality of opportunity after graduation. Moving towards gender equality in geosciences requires a multifaceted approach that addresses inequality in education, the workplace, and professional development opportunities. This includes supporting women with families, changing cultural attitudes, and considering the services provided at international events to make it easier for those with children to participate.

- Harassment, discrimination, and bullying are global problems, and are prevalent in both science and academia. While not limited to targeting women, evidence suggests that they are currently likely to meet hostile and discriminating environments in their range of study and work environments. Creating a safe working environment is the responsibility of all geoscientists, supported by access to formal structures to ensure effective monitoring and reporting, handling of complaints, the development of coordinated actions and initiatives that address problems, raise awareness of fundamental rights, and promotes respect towards all.

- Many of the themes that geoscientists work on (e.g., resource management, disaster risk reduction) have a gendered dimension, and it is, therefore, important that geoscientists understand the drivers and impacts of gender inequality.

- Numerous local, national, regional, and international initiatives promote the geosciences as a career route for women, increase the visibility of women geoscientists, and break down barriers preventing women from accessing leadership positions of key institutions. Examples include Girls into Geoscience, the African Associations of Women in Geosciences, and Women in Mining Mongolia. These initiatives increase the visibility of diverse role models in the geosciences, and should be encouraged and resourced by the broader geoscience community.

5.7 Educational Ideas

In this section, we provide examples of educational activities that connect geoscience, the material discussed in this chapter, and scenarios that may arise when applying geoscience (e.g., in policy, government, private sector international organisations, NGOs). Consider using these as the basis for presentations, group discussions, essays, or to encourage further reading.

- Take part in a Wikipedia edit-a-thon, to enrich the information available online about women in the geosciences. The current list of geologists[3] omits many notable women geoscientists from around the world. Help to change this.

- Organise a *Question and Answer* session with your Head of Department to ask about the actions in place to increase gender equality in your department. What can you do to support these?

- Carefully consider what steps you think are needed to improve gender equality in the global geoscience community. Write a letter expressing both your concerns and ideas for

---

[3]https://en.wikipedia.org/wiki/List_of_geologists.

improvement, and send this to your local geological society, or international union.

Further Resources

Burek CV, Higgs B (Eds) (2007) The role of women in the history of geology. Geological Society of London Special Publications 281

Holmes MA, OConnell S, Dutt K (Eds) (2015) Women in the geosciences: practical, positive practices toward parity. John Wiley & Sons

Johnson BA (Ed) (2018) Women and geology: who are we, where have we come from, and where are we going? (Vol 214). Geological Society of America

UN Women (2015) United Nations entity dedicated to gender equality and the empowerment of women. https://www.unwomen.org/en. Accessed 14 Feb 2019

References

AAAS (2019) Profile of Dr Jessica Wade BEM. https://www.aaas.org/membership/member-spotlight/jessica-wade-physicist-and-author-hundreds-wikipedia-entries-about. Accessed 25 June 2019

Africa Mining Vision (2009) https://www.africaminingvision.org/amv_resources/AMV/Africa_Mining_Vision_English.pdf. Accessed 14 Feb 2019

African Development Bank (2019) Gender equality index. https://www.afdb.org/en/topics-and-sectors/topics/quality-assurance-results/gender-equality-index. Accessed 3 Oct 2019

AGI (2015) U.S. Female Geoscience Enrollments Level Off. https://www.americangeosciences.org/workforce/currents/us-female-geoscience-enrollments-level. Accessed 19 Feb 2019

AGI (2017) Female geoscience faculty representation grew steadily between 2006–2016. https://www.americangeosciences.org/workforce/currents/female-geoscience-faculty-representation-grew-steadily-between-2006-2016. Accessed 19 Feb 2019

AGI (2018a) U.S. Female geoscience enrollments show variability in 2017. https://www.americangeosciences.org/sites/default/files/currents/Currents-133-GenderEnrollmentsDegrees.pdf. Accessed 19 Feb 2019

AGI (2018b) AGI statement on harassment in the geosciences. AGI—American Geosciences Institute, https://www.americangeosciences.org/content/agi-statement-harassment-geosciences

AGU (2017) AGU Scientific Integrity and Professional Ethics. AGU—American Geophysical Union, https://ethics.agu.org/files/2013/03/Scientific-Integrity-and-Professional-Ethics.pdf

Calisi, RM and a Working Group of Mothers in Science (2018) Opinion: How to tackle the childcare–conference conundrum. https://www.pnas.org/content/115/12/2845. Accessed 19 Feb 2019

Cane I (2014) Community and company development discourses in mining: the case of gender in Mongolia. (Doctorate of Philosophy), University of Queensland, Brisbane

Cane I (2015) Social and gendered Impacts related to mining, Mongolia. Prepared by: Adam Smith International. https://dfat.gov.au/about.../mongolia-social-gendered-impacts-related-to-mining.docx

Cheryan S, Ziegler SA, Montoya AK, Jiang L (2017) Why are some STEM fields more gender balanced than others? Psychol Bull 143(1):1–35. https://doi.org/10.1037/bul0000052

Clancy KBH, Nelson RG, Rutherford JN, Hinde K (2014) Survey of academic field experiences (SAFE): trainees report harassment and assault. PLoS ONE 9(7): e102172. https://doi.org/10.1371/journal.pone.0102172

EFG (2017) EFG Code of Ethics. EFG—European Federation of Geologists. https://eurogeologists.eu/wp-content/uploads/2017/07/Code_Ethics_June2017.pdf

Errami E (2009) Pre-congress meeting of the fifth conference of the African Association of women in geosciences entitled "Women and Geosciences for Peace." Episodes 32(3):210–211

Errami E, Andrianaivo L, Ennih N, Gauly M (2012) The first international conference on African and Arabian Geoparks—Aspiring Geoparks in Africa and Arab World. El Jadida, Morocco, 20–28 November 2011. Episodes 35, 2. 349–351

Errami E (2013) African Association of Women in Geosciences "AAWG": Past achievements and future challenges. Il Ruolo Femminile Nelle Science Della Terra: Esperienze a confronto e prospettive future. 27–30. ISPRA, Atti 2013. ISBN 978-88-448-0585-2

FAO (2016) Women hold the key to building a world free from hunger and poverty, https://www.fao.org/news/story/en/item/460267/icode/. Accessed 19 Feb 2019

Gerel O, Batkhyshig B, Myagmarsuren S (2006) Mongolian women in geosciences: the appearance and reality of changing gender roles'. Atlantic Geol 42:1

GSA (2018) GSA statement: removing barriers to career progression for women in the geosciences. Geological Society of America. https://www.geosociety.org/GSA/Science_Policy/Position_Statements/Current_Statements/gsa/positions/position26.aspx. Accessed 27 March 2019

Holmes MA, O'Connell S, Frey C, Ongley L (2008) Gender imbalance in US geoscience academia. Nat Geosci 1:79–82. https://doi.org/10.1038/ngeo113

Human Development Report (2015) Human development data. https://hdr.undp.org/en/data. Accessed 8 Oct 2019

IAPG (2019) Zero tolerance by IAPG towards harassment and discrimination. https://www.geoethics.org/harassment-discrimination. Accessed 27 March 2019

IWM (2019) Engaging men. https://internationalwim.org/library/engaging-men. Accessed 27 March 2019

Kölbl-Ebert M (2001) On the origin of women geologists by means of social selection: German and British comparison. Episodes 24(3):182–193

Lahiri-Dutt K (2011) Introduction: gendering the masculine field of mining for sustainable community livelihoods. Gendering the field: Towards sustainable livelihoods for mining communities. In: Lahiri-Dutt K (Ed). Canberra, ANU E Press, pp 1–19. Asia Pacific Environment Mongograph 6

MacDonald R (2005) How women were affected by the Tsunami: a perspective from Oxfam. PLoS Med 2(6): e178. https://doi.org/10.1371/journal.pmed.0020178

Macdonald I, Rowland C Eds (2002) Tunnel vision—women, mining and communities. Melbourne, Oxfam Community Aid Abroad

Marín-Spiotta E, Schneider B, Holmes MA (2016) Steps to building a notolerance culture for sexual harassment, EOS, 97. https://doi.org/10.1029/2016EO044859

Mason MA (2013) The baby penalty. https://www.chronicle.com/article/The-Baby-Penalty/140813. Accessed 5 Feb 2019

Mogk D (2017) Geoethics and professionalism: the responsible conduct of scientists. In: Peppoloni S, Di Capua G, Bobrowsky PT, Cronin V (eds) Geoethics at the heart of all geoscience. Annals of Geophysics, Vol 60, Fast Track 7. https://doi.org/10.4401/ag-7584. https://www.annalsofgeophysics.eu/index.php/annals/issue/view/537.

National Academies of Sciences, Engineering, and Medicine (2018) Sexual Harassment of Women: Climate, Culture, and Consequences in Academic Sciences, Engineering, and Medicine. https://sites.nationalacademies.org/SHStudy/index.htm

Neumayer E, Plümper T (2007) The gendered nature of natural disasters: the impact of catastrophic events on the gender gap in life expectancy, 1981–2002. Annals of the Association of American Geographers 97 (3):551–566. ISSN 0004-5608. https://doi.org/10.1111/j.1467-8306.2007.00563.x

Ortiz-Ospina E, Tzvetkova S (2018) Women's employment. Published online at https://ourworldindata.org, Retrieved from:https://ourworldindata.org/female-labor-supply [Online Resource]

Ortiz-Ospina E, Roser M (2018) Economic inequality by gender. https://ourworldindata.org/economic-inequality-by-gender. Accessed 19 February 2019

Peppoloni S, Di Capua G (2017) Geoethics: ethical, social and cultural implications in geosciences. In: Peppoloni S, Di Capua G, Bobrowsky PT, Cronin VS (Eds) Geoethics at the heart of all geoscience. Annals of Geophysics, Vol. 60, Fast Track 7. https://doi.org/10.4401/ag-7473. https://www.annalsofgeophysics.eu/index.php/annals/issue/view/537.

Peppoloni S, Di Capua G, Haslinger F (2017) D13.1—Questionnaire to analyse the ethical and social issues and assessment report on questionnaire answers. ENVRIplus project deliverable. https://www.envriplus.eu/wp-content/uploads/2015/08/D13.1.pdf. Accessed 27 March 2019

St. John K, Riggs E, Mogk D (2016) Sexual harassment in the sciences: a call to geoscience faculty and researchers to respond. J Geosci Educ November 2016 64(4):255–257. https://nagt-jge.org/doi/full/10.5408/1089-9995-64.4.255

Stoet G, Geary DC (2018) The gender-equality paradox in science, technology, engineering, and mathematics education. Psychol Sci 29(4):581–593. https://doi.org/10.1177/0956797617741719

UN Women (2009) The convention on the elimination of all forms of discrimination against women (CEDAW). https://www.un.org/womenwatch/daw/cedaw/. Accessed 27 March 2019

UN Women (2015) Human Rights of Women. https://www.unwomen.org/en/digital-library/multimedia/2015/12/infographic-human-rights-women. Accessed 5 February 2019

UNDP (2019) Gender Inequality Index. https://hdr.undp.org/en/content/gender-inequality-index-gii. Accessed 3 Oct 2019

UNDESA (2019) Least developed countries. https://www.un.org/development/desa/dpad/least-developed-country-category/ldcs-at-a-glance.html. Accessed 3 Oct 2019

UNESCO (2019) Tertiary education, academic staff (% female) https://data.worldbank.org/indicator/SE.TER.TCHR.FE.ZS?view=chart&year_high_desc=true. Accessed 19 February 2019

United Nations (2015) Millennium development goals report. https://www.un.org/millenniumgoals/2015_MDG_Report/pdf/MDG%202015%20rev%20(July%201).pdf. Accessed 27 March 2019

United Nations (1945) Charter. https://www.un.org/en/charter-united-nations/. Accessed 27 March 2019

United Nations (1948) Universal declaration of human rights. https://www.un.org/en/universal-declaration-human-rights/. Accessed 27 March 2019

UN (2019) Sustainable Development Goals Report 2019. https://unstats.un.org/sdgs/report/2019/The-Sustainable-Development-Goals-Report-2019.pdf.

World Economic Forum (2017) The Global Gender Gap Report 2017. World Economic Forum

World Economic Forum (2018) The Global Gender Gap Report 2018. World Economic Forum

WISE (2017) Women in science and engineering statistics, https://www.wisecampaign.org.uk/statistics/. Accessed 5 Feb 2019

Silvia Peppoloni is a researcher at the Italian Institute of Geophysics and Volcanology, focused on geohazards, engineering geology, geomorphology and geoeducation. She is an international leader of geoethics, founding member and Secretary General of the International Association for Promoting Geoethics, Councillor 2018–2022 of the International Union of Geological Sciences, and sits on the boards of several scientific organisations. Silvia has authored and edited a range of scientific books and articles, as well as pieces for the Italian media. She has been awarded with Italian prizes for science communication and natural literature.

Ezzoura Errami is a lecturer and researcher with a strong academic grounding in Earth Sciences and over 15 years' experience in international geoscientific professional organizations. Her current major research interest is geoheritage, geoeducation, sustainable development and gender related studies in geosciences. Ezzoura is President of the African Association of Women in Geosciences, Africa coordinator of the International Association for Promoting Geoethics (IAPG) and Coordinator of the African Geoparks Network. She is also Vice-President of the ArabGU, a Founding member of the South Asian Association of Women Geoscientists, and the regional focal point for the (Arabic countries and Africa, International Union for the Conservation of Nature (Geoheritage Specialist Group). Ezzoura has previously served as a Council member for the International Union of Geological Sciences.

Gerel Ochir is a Professor and Director of the Geoscience Center at the Mongolian University of Science and Technology. She developed geoscience education in Mongolia, with 30 years leading the Department of Geology. She has published many papers and books, supervised Masters and Ph.D. students, and led 20 international research projects. She has deployed her research skills to understand gender in the geosciences in Mongolia, editing and publishing a book on '*Mongolian Women in Geoscience*'. She was elected International Union of Geological Sciences Vice-President (2008–2012) and International Geoscience Programme Working Group lead and member (2004–2012). Gerel is active in many professional societies, including the Society of Mongolian Women Geologists.

Clean Water and Sanitation

Kirsty Upton and Alan MacDonald

K. Upton (✉) · A. MacDonald
British Geological Survey, Lyell Centre, Research
Avenue South, Edinburgh E14 4AP, UK
e-mail: kirlto@bgs.ac.uk

© Springer Nature Switzerland AG 2021
J. C. Gill and M. Smith (eds.), *Geosciences and the Sustainable Development Goals*,
Sustainable Development Goals Series, https://doi.org/10.1007/978-3-030-38815-7_6

Abstract

6 CLEAN WATER AND SANITATION

Overview

Access to clean water and safe sanitation are human rights - essential for social well-being, health, education and livelihoods

Water resources are unevenly distributed around the globe

Climate change and population growth present significant challenges to understanding and managing resilient water supplies

Groundwater is a key element of climate resilient water supplies

Sustainable groundwater management requires cooperation in use, monitoring and regulation

Current status

Since 2000, more people have access to basic and safely managed drinking water across Asia and Sub Saharan Africa.

Progress towards SDG targets is uneven: in sub Saharan Africa more than 50% of the rural population lack access to even basic water services.

Reduced access to water services or the degradation of water resources impacts gender equality and dispro-portionately affects the poorest communities.

Financing of the water sector is reported as inadequate in 80% of countries, hindering SDG 6 progress.

Role of geoscience and groundwater

Provision of geological data and understanding the groundwater potential and aquifer architecture.

Protection of water-related ecosystems by understanding interactions between surface and sub-surface water.

Input to groundwater management through quantifying abstraction, recharge and environmental flows, and mapping aquifer characteristics.

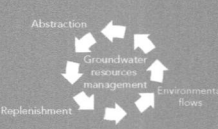

Monitoring water quality and protecting groundwater from contamination.

6.1 Introduction

Water and sanitation are inalienable rights of humanity as enshrined in the Human Right to Water and Sanitation (HRWS) by the United Nations General Assembly and adopted by all Member States in 2010. Access to safe water and sanitation is essential for social well-being, supporting outcomes in health (**SDG 3**), education (**SDG 4**), livelihoods (**SDG 8**), and gender equality (**SDG 5**) for urban and rural populations globally (Bartram and Cairncross 2010). Water and sanitation-related diseases, particularly diarrhoeal diseases, remain one of the major causes of death in children under five (Wang et al. 2016). The health burden associated with poor water and sanitation services, along with the burden placed on women and children to collect water when services are located away from the home, impacts on levels and equality of education, as well as economic productivity (Hutton et al. 2007). The role of water in the agricultural sector, particularly where irrigation supports agricultural production and development, contributes to economic growth through revenue and employment and increases food security at a

household, national, and even global scale. Through hydropower and renewables, water can also contribute to improved access to affordable and clean energy (**SDG 7**).

Realisation of the social and economic benefits of safe water and sanitation requires an increase in service levels, particularly across sub-Saharan Africa, and the sustainable management and protection of water resources across the globe. This is the focus of **_SDG 6: Ensure availability and sustainable management of water and sanitation for all_**, which strives to achieve universal access to safe and affordable drinking water and sanitation and aims for efficient use, integrated management, and protection of freshwater resources and water-related ecosystems, as summarised by the targets and associated indicators in Tables 6.1 and 6.2.

The SDGs build on decades of work aimed at improving lives, reducing poverty, and protecting the environment at a national and global level. Preceding the SDGs, the Millennium Development Goals (MDGs), which were adopted by the United Nations General Assembly in 2000, aimed (under Goal 7: Ensure Environmental Sustainability) to halve the proportion of

Table 6.1 SDG 6 targets and means of implementation

Target	Description of Target (_6.1_ to _6.6_) or Means of Implementation (_6.A_ to _6.B_)
6.1	By 2030, achieve universal and equitable access to safe and affordable drinking water for all
6.2	By 2030, achieve access to adequate and equitable sanitation and hygiene for all and end open defecation, paying special attention to the needs of women and girls and those in vulnerable situations
6.3	By 2030, improve water quality by reducing pollution, eliminating dumping, and minimising release of hazardous chemicals and materials, halving the proportion of untreated wastewater and substantially increasing recycling and safe reuse globally
6.4	By 2030, substantially increase water-use efficiency across all sectors and ensure sustainable withdrawals and supply of freshwater to address water scarcity and substantially reduce the number of people suffering from water scarcity
6.5	By 2030, implement integrated water resources management (IWRM) at all levels, including through transboundary cooperation as appropriate
6.6	By 2020, protect and restore water-related ecosystems, including mountains, forests, wetlands, rivers, aquifers, and lakes
6.A	By 2030, expand international cooperation and capacity-building support to developing countries in water- and sanitation-related activities and programmes, including water harvesting, desalination, water efficiency, wastewater treatment, recycling, and reuse technologies
6.B	Support and strengthen the participation of local communities in improving water and sanitation management

Table 6.2 SDG 6 indicators

Indicator	Description of indicator
6.1.1	Proportion of population using safely managed drinking water services (see Table 6.3 for definition)
6.2.1	Proportion of population using safely managed sanitation services (see Table 6.3 for definitions), including a hand-washing facility with soap and water
6.3.1	Proportion of wastewater safely treated
6.3.2	Proportion of bodies of water with good ambient water quality
6.4.1	Change in water-use efficiency over time
6.4.2	Level of water stress
6.5.1	Degree of IWRM implementation (0–100)
6.5.2	Proportion of transboundary basin area with an operational arrangement for water cooperation
6.6.1	Change in the extent of water-related ecosystems over time
6.A.1	Amount of water- and sanitation-related official development assistance that is part of a government-coordinated spending plan
6.B.1	Proportion of local administrative units with established and operational policies and procedures for participation of local communities in water and sanitation management

the global population without sustainable access to an improved drinking water source and sanitation facility by 2015.

- An *improved drinking water source* has the potential to provide safe water as it is protected from contamination through its design and construction; this includes piped water, boreholes or tubewells, protected dug wells, protected springs, rainwater, and packaged or delivered water.
- An *improved sanitation facility* is designed to separate excreta from human contact, including flush or pour flush to piped sewer systems, septic tanks or pit latrines, ventilated improved pit latrines, composting toilets, or pit latrines with slabs.

As outlined in the final MDG Report (United Nations 2015), the drinking water target was achieved in 2010, and by the end of the MDG period in 2015, almost 90% of the global population had access to an improved drinking water source. However, significant inequalities persisted across the globe: sub-Saharan Africa, for example, missed the drinking water target completely with only around 68% of the population accessing an improved source in 2015. While globally, urban dwellers achieved a higher level of access to improved sources than those in rural areas (96% compared to 84%). The MDG target for sanitation was not achieved with around one-third of the global population still using unimproved sanitation facilities in 2015, with a starker contrast between urban and rural access to improved facilities (82% compared to 50%).

Targets 6.1 and **6.2** of the SDGs go beyond the aims of the MDGs for access to improved services, introducing a service ladder (Table 6.3), which ultimately aims for the much more ambitious goal of safely managed services (note that the MDG for improved services equates to a limited level of service under the SDGs). Moving to safely managed services is a considerable challenge, with less than 30% of the population in sub-Saharan Africa, and 71% of the global population, estimated as having a safely managed water source in 2017 (Joint Monitoring Programme (JMP) 2019c). The SDGs also go beyond the focus of the MDGs and incorporate the sustainable management and protection of all water resources. This is necessary not only to achieve the drinking water target, but to balance multiple competing demands for water while maintaining the resilience and biodiversity of water-related ecosystems (see **SDG 15**), which provide many other services upon which humans depend, such as carbon sequestration and

Table 6.3 Service ladder for drinking water and sanitation. Outlined by the Joint Monitoring Programme (JMP) of the World Health Organisation (WHO) and UNICEF

Service Level	Drinking water definition	Sanitation definition
Safely Managed	Drinking water from an improved source that is located on premises, available when needed, and free from faecal and priority chemical contamination	Use of improved facilities which are not shared with other households and where excreta are safely disposed in situ or transported and treated off-site
Basic	Drinking water from an improved source, provided collection time is not more than 30 min for a round trip, including queuing	Use of improved facilities which are not shared with other households
Limited	Drinking water from an improved source for which collection time exceeds 30 min for a round trip, including queuing	Use of improved facilities shared between two or more households
Unimproved	Drinking water from an unprotected dug well or unprotected spring	Use of pit latrines without a slab or platform, hanging latrines, or bucket latrines
Surface Water (6.1) / Open Defecation (6.2)	Unsafe or unimproved drinking water directly from a river, dam, lake, pond, stream, canal or irrigation canal (example in Fig. 6.1)	Disposal of human faeces in fields, forests, bushes, open bodies of water, beaches, and other open spaces or with solid waste

Notes (1) an improved drinking water source has the potential to provide safe water as it is protected from contamination through its design and construction; these include piped water, boreholes or tubewells, protected dug wells, protected springs, rainwater, and packaged or delivered water. (2) An improved sanitation facility is designed to separate excreta from human contact, including flush or pour flush to piped sewer system, septic tanks or pit latrines, ventilated improved pit latrines, composting toilets, or pit latrines with slabs

storage, air and water pollution control, nutrient cycling, erosion prevention, food, medicine, livelihoods, recreation opportunities, and spiritual health (Wood et al. 2018).

Target 6.3 aims to improve ambient water quality to protect both ecosystems and humans from harmful pollutants, including hazardous substances. Progress towards this target is measured by the percentage of wastewater treatment, including wastewater derived from households, commercial and industrial activities, urban run-off, and agriculture, and the percentage of water bodies in a country with good ambient water quality. Water quality is measured by a core set of parameters: dissolved oxygen, electrical conductivity, pH, nitrogen, and phosphorous for surface water, and electrical conductivity, pH, and nitrate for groundwater.

Target 6.4 addresses water scarcity by aiming for sustainable withdrawals (withdrawals defined as freshwater taken from surface or groundwater sources, either permanently or temporarily, for agricultural, industrial or domestic use) and increased water use efficiency. Water use efficiency is measured as a productivity metric, defined as a country's total gross domestic product (GDP) per

unit of freshwater withdrawal, where a high GDP per unit of freshwater withdrawal indicates a water-efficient economy. Water scarcity is indicated by the level of water stress at a national scale, defined as the ratio between total freshwater withdrawal and total renewable freshwater resources, after taking into account environmental water requirements. A country would be considered water-stressed if 25–60% of renewable water resources are withdrawn; if this proportion is higher at 60–75% or > 75%, a country would be considered water scarce or severely water scarce, respectively. It should be noted that water scarcity can also be considered in terms of economic or institutional water scarcity—where water shortages are caused, not by a lack of water availability, but by poor accessibility due to inadequate investment or capacity to develop and supply secure water sources.

Integrated water resources management (IWRM), **Target 6.5**, seeks to bring together stakeholders representing different sectors or geographical regions to ensure collaborative, cooperative, and coordinated management of water resources at the scale of individual basins, which may cross national borders. The degree of

Fig. 6.1 Surface water in Tanzania. Example of 'surface water' (see Table 6.3 for definition) used for drinking and watering animals. © Joel Gill (used with permission)

implementation is assessed through the four components of IWRM: enabling environment, institutions and participation, management instruments, and financing. **Target 6.6** aims to protect water-related ecosystems (Fig. 6.2), by halting degradation and destruction of ecosystems, or regenerating those already degraded. Water-related ecosystems include vegetated wetlands, rivers, lakes, reservoirs, and groundwater, with special mention of those occurring in mountains and forests (linking to **SDG 15**). The indicator for this target tracks changes over time in the spatial extent of water-related ecosystems and inland open waters, and the quantity and quality of water in these ecosystems (overlapping with indicator 6.3.2).

Means of Implementation 6.A and **6.B** recognise that international and local cooperation is needed to achieve SDG 6, aiming for increased funding for water and sanitation, particularly as official development assistance to developing countries, and increased involvement of local communities in water and sanitation management

to ensure the needs of all people are being met. Equality is a core principle of the SDGs, particularly achieving gender equality and the empowerment of women and girls to enjoy equal access to education, economic resources, employment, and political participation (helping to deliver **SDG 5**). This has particular relevance for **SDG 6** due to the unequal burden put on women and children to collect water when sources are located off-site.

Achieving **SDG 6** requires an understanding of the interlinkages between targets within the Goal, not simply consideration of the targets in isolation (Fig. 6.3). For example, increased sanitation must be accompanied by wastewater treatment to ensure water quality is maintained for both drinking water and ecosystem services. Likewise, water resources must be managed sustainably to ensure sufficient quantity for all services, including drinking water and ecosystems, but also other economic uses such as agriculture, industry, and energy. IWRM links all targets of **SDG 6**, providing a management framework for addressing these linkages (both

Fig. 6.2 Freshwater resources in Iceland. The targets of SDG 6 emphasise both provision of safe and affordable drinking water, and the protection of freshwater ecosystems. Integrated Water Resources Management promotes a coordinated approach to the management of water, land, and related resources. Image by Free-Photos from Pixabay

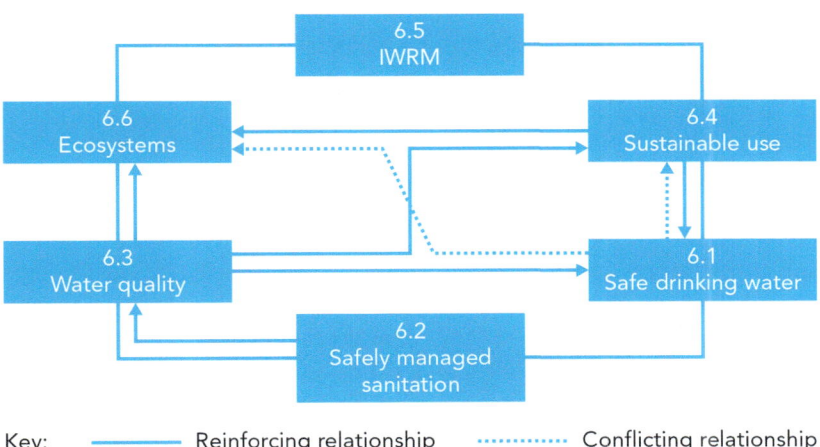

Fig. 6.3 Interlinkages between targets of SDG 6. IWRM refers to Integrated Water Resources Management. Solid lines refer to reinforcing relationships, and dashed lines are potentially conflicting relationships

synergetic and conflicting) to balance competing demands on water resources.

Understanding the linkages between **SDG 6** and the other goals within the development framework is also crucial for supporting decision-making to achieve long-lasting development outcomes. The SDGs are, by design, an integrated set of goals and there are multiple intersection points where individual goals, or targets within them, act to reinforce, or in some cases conflict, with others. There are multiple interlinkages between **SDG 6** and the other 16 SDGs. Water and sanitation underpin many areas of development and poverty reduction—from health and well-being (**SDG 3**) to economic growth (**SDG 8**) and food security (**SDG 2**). Schools have an important role to play in improving WASH outcomes through education and access to services, while the health benefits of improved WASH lead to improved school attendance, particularly for girls (see **SDG 4**). Agricultural productivity can be increased by expanding access to irrigation and increasing the use of fertilisers and pesticides (see **SDG 2**), but this increases the demand for water and potentially pollutes freshwater resources. The strength and nature of the interlinkages often depend on the context, and therefore, vary geographically. A detailed exploration of the interlinkages for **SDG 6** can be found in *'A Guide to SDG Interactions: from Science to Implementation'* (International Council for Science (ICSU, 2017) and *'Water and Sanitation Interlinkages across the 2030 Agenda for Sustainable Development'* (UN-Water 2016).

In this chapter, we will look at global progress towards the targets of **SDG 6** in more detail and introduce some of the key challenges for achieving this goal. We will then focus on groundwater and the crucial role that it can, and is playing in achieving **SDG 6**. We will explore the role that geoscientists can play in improving groundwater management and development so that the potential socio-economic benefits of groundwater are realised without significant environmental degradation and risk to future water resources.

6.2 Challenges and Progress Towards SDG 6

6.2.1 Challenges to Achieving SDG 6: Climate Change, Population Growth, and Conflict

The SDGs represent an ambitious set of targets for sustainable economic, social, and environmental development. For water and sanitation, these targets are set within the context of a changing climate (see **SDG 13**) and rapidly growing population, which puts pressure on global water resources both in terms of supply and demand. On top of these pressures are challenges such as rising inequality, environmental degradation, urbanisation, industrial production, agricultural intensification, conflict and migration, and a lack of investment and adequate governance, which affect the availability, accessibility, and quality of water resources globally.

Water resources are not spread evenly across the globe. Not all areas have access to frequent rainfall throughout the year to replenish reservoirs, rivers, and aquifers and sustain aquatic ecosystems (Fig. 6.4). The availability of year-round water, or the ability to store and transfer water has a direct impact on a nation's economic development (Grey and Sadoff 2007). Much of Africa and South Asia, are challenged by long dry seasons or low annual rainfall. This uneven global distribution is being further affected by climate change.

The most recent climate change synthesis report from the Intergovernmental Panel on Climate Change (IPCC 2014) states that global warming is unequivocal and summarises the impacts that are already being seen in the global climate system. Multi-decadal globally averaged land and sea surface temperatures increased between 1880 and 2012. Precipitation over mid-latitude land areas in the northern hemisphere has increased since 1901 (there is low confidence in precipitation trends at other latitudes). Glaciers have continued to shrink worldwide. Global mean sea level rose by 0.19 m between 1901 and 2010. An increase in extreme events has been observed since 1950, with an increase in the frequency of heatwaves across Europe, Asia, and Australia, and more areas experiencing an increase in heavy rainfall events compared to those seeing a decrease in extreme precipitation. Looking to the future, projected changes in temperature and precipitation remain uncertain and vary geographically, but it is very likely that heatwaves will occur more often and last longer, and that extreme precipitation will become more

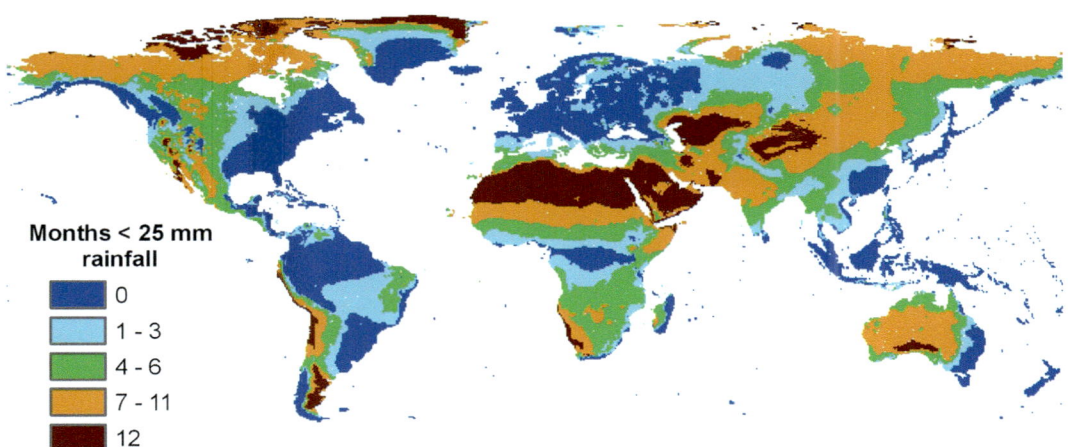

Fig. 6.4 Global distribution of rainfall showing the number of months with limited (<25 mm) rainfall. From: Hunter et al. (2010) Water Supply and Health. PLoS Med 7(11): e1000361. https://doi.org/10.1371/journal.pmed.1000361. Reproduced under a CC BY license (https://creativecommons.org/licenses/by/4.0/)

intense and occur more often in many areas. Changes in average precipitation are more variable, with some areas likely to experience an increase in mean annual precipitation and others likely to see a decrease. This has consequences for global water resources, with an increased risk of flood and drought, and in some areas (particularly dry subtropical regions), a reduction in renewable water resources. Risks related to climate change disproportionately affect the poor in part, because most developing countries are in tropical or arid regions where the effects of climate change are likely to be most severe, but also because poorer populations have less capacity to adapt to, withstand, and recover from climate-related risks such as flood and drought.

In 2015, the global population reached 7.3 billion. This is expected to increase to 8.5 billion by 2030, 9.7 billion by 2050, and 11.2 billion by 2100, with more than half the growth occurring in Africa (United Nations Department of Economic and Social Affairs Population Division 2019). This puts obvious pressure on water resources in terms of demand for drinking water (Fig. 6.5), but will also increase the amount of water required for food production and other resources to ensure continued economic and social development. Superimposed on the global trend of population growth, is an increase in the proportion of the population living in urban areas, which is expected to increase from around 55% in 2018 to 68% in 2050 (United Nations Department of Economic and Social Affairs Population Division 2018). This puts particular pressure on water and sanitation services in urban areas, which are already struggling to cope with rapid population growth in many developing countries.

Whether caused by a lack of availability or accessibility, the implications of water scarcity are potentially significant and wide-ranging. In addition to hindering socio-economic development, water scarcity, in extreme cases, can be a contributing factor to migration, conflict, and humanitarian crises, like that witnessed in 2015/16, in East Africa, during the El Nino-related drought (Box 6.1). Even if not the primary cause, water scarcity is often one of many complex environmental, social, economic, and political factors leading to unrest and conflict. One of the most well-known examples of this is in the Middle East, with access to water a critical component of the ongoing conflict in the West Bank and Gaza. Much of the recharge to aquifers exploited in Israel, occurs upstream in the mountains of the West Bank, where abstraction is strictly controlled to protect downstream flows. Similar tension between upstream and downstream users occurs in many river basins with the Nile Basin, Indus, and Mekong river basins all sources of potential conflict. Considering the increasing pressures on global water resources, particularly related to climate and land use change and population growth, water scarcity is likely to become a more widespread and significant issue, making the need for sustainable management and protection of water resources ever more critical.

Box 6.1. Impacts of El Niño in Eastern and Southern Africa

What is El Niño? The El Niño Southern Oscillation (ENSO) is a global climate phenomenon that influences interannual temperature and precipitation patterns across the globe, most significantly in the tropics. The ENSO has a neutral phase and two opposite phases—El Niño and La Niña —driven by changes in the sea surface temperature gradient and atmospheric pressure gradient over the tropical Pacific Ocean (Met Office 2019). The impacts of ENSO are felt beyond the Pacific region. In Africa, El Niño episodes are generally associated with drought conditions in Southern Africa and the horn of Africa, with extreme rainfall often occurring in Tanzania, Uganda, and Kenya.

The 2015–16 El Niño event was one of the strongest on record (Siderius et al. 2018). Rainfall perturbations, occurring on top of multiple preceding dry years, resulted in drought conditions across southern Africa, as well as parts of Ethiopia, Somalia, and Kenya. The hydrological effects of this drought included reduced river flows, unusually low lake levels, exceptional soil moisture deficits, reduced groundwater storage and reduced spring flows across the region (Philip et al. 2018; Siderius et al. 2018; Kolusu et al. 2019; MacDonald et al. 2019).

Impacts of the 2015–16 El Niño event were felt across southern and eastern Africa. There was significant disruption to the urban water supply in Gaborone, Botswana, and hydroelectric load shedding in Zambia (Siderius et al. 2018); severe water shortages and water collection times of more than 12 hours were experienced in the Ethiopian Highlands (MacDonald et al. 2019); and crop failures caused food shortages for millions of people across the region. In Ethiopia, the government, along with the United Nations, released a Humanitarian Response Document in 2015, asking for emergency assistance for over 10 million people. Continued below-average rainfall means the region is still experiencing a humanitarian crisis several years later (ReliefWeb 2019). However, people that had access to groundwater through boreholes were much less severely impacted, and many of the boreholes continued to function through the drought (MacDonald et al. 2019).

6.2.2 Monitoring Global Progress

The monitoring framework for tracking progress towards the SDGs is global, however, the review process is voluntary and country-led, often supported by regional or sub-regional commissions or organisations. In some cases, national baseline data, against which progress is monitored, does not exist and the SDGs call for increased support for data collection at a national level to inform the measurement of progress. In 2018, less than half of Member States had comparable data on progress towards meeting the targets of **SDG 6**; just over 40% had data available for more than four indicators, and only 6% had data available on more than eight indicators (United Nations 2018). Targets for water, sanitation, and hygiene (**6.1** and **6.2**), have a long history of data collection under the MDGs, i.e., since 2000, but the others generally have data available over much shorter time periods, if at all.

Fig. 6.5 Queuing for water at a hand dug well in Northern Nigeria. Photo by Alan MacDonald. © UKRI/British Geological Survey

6.2.3   Global Progress: Drinking Water and Sanitation

Despite the pressures described above, global progress has been made towards achieving the targets of **SDG 6**. The Joint Monitoring Programme (JMP) of the World Health Organisation (WHO) and UNICEF use the service ladders shown in Table 6.3, to monitor progress towards **Targets 6.1** and **6.2**. Continued use of the MDG definitions of improved and unimproved services allows comparison across the MDG and SDG periods, showing a significant increase in the percentage of the total population with access to basic and safely managed drinking water services since 2000, particularly in rural areas, across Asia and sub-Saharan Africa (Table 6.4). A similar trend is seen for sanitation services across Asia, Latin America, and the Caribbean, while progress has been less significant in sub-Saharan Africa (Table 6.5).

There is still some way to go if we are to meet these targets by 2030 (see Figs. 6.6 and 6.7). In 2015, 30% of the global population still lacked access to safely managed drinking water services and 12% lacked access to even basic services—most of these in sub-Saharan Africa and Oceania. More than 60% of the global population lacked access to safely managed sanitation services, while 32% lacked access to basic services—again, mostly in sub-Saharan Africa and Oceania, although central and southern Asia also has some way to go. Oceania is the only region to have experienced a decrease in service levels, which has occurred for sanitation across rural and urban areas. Within these regions, levels of access are significantly lower in fragile states. For example, in sub-Saharan Africa, some of the lowest service levels for drinking water and/or sanitation are found in Chad, South Sudan, Democratic Republic of the Congo, and Somalia, which in 2019, were ranked in the top ten most fragile states in the world (The Fund for Peace[1]). Equally, Yemen and Afghanistan, are amongst the most fragile states and have the lowest service levels for drinking water and sanitation in Western and Central Asia.

6.2.4   Global Progress: Sustainable Management

The United Nations Synthesis Report (2018), summarises progress towards all targets of **SDG 6**. In 2014, levels of water stress were highest in Northern Africa and Western, Central and

---

[1]https://fundforpeace.org/.

Table 6.4 Global progress towards Target 6.1 for drinking water services (Joint Monitoring Programme (JMP) 2019a)

Region	% population with at least basic drinking water services (safely managed services)					
	Total		Urban		Rural	
	2000	2017	2000	2017	2000	2017
Australia and New Zealand	100 (-)	100 (-)	100 (92)	100 (97)	99 (-)	100 (-)
Central and Southern Asia	81 (41)	93 (60)	93 (66)	96 (62)	76 (31)	91 (60)
Eastern and South-Eastern Asia	81 (-)	93 (-)	97 (91)	98 (91)	71 (-)	86 (-)
Europe and North America	98 (90)	99 (95)	100 (97)	99 (97)	96 (-)	98 (-)
Latin America and the Caribbean	90 (56)	96 (74)	96 (82)	99 (82)	71 (-)	88 (42)
Northern Africa and Western Asia	91 (-)	92 (-)	94 (-)	97 (-)	71 (-)	84 (-)
Oceania (not Aus/NZ)	54 (-)	55 (-)	91 (-)	92 (-)	40 (-)	44 (-)
Sub-Saharan Africa	46 (18)	61 (27)	78 (42)	84 (50)	30 (6)	46 (12)

Table 6.5 Global progress towards Target 6.2 for sanitation services (Joint Monitoring Programme (JMP) 2019a)

Region	% population with at least basic sanitation services (safely managed services)					
	Total		Urban		Rural	
	2000	2017	2000	2017	2000	2017
Australia and New Zealand	100 (61)	100 (72)				
Central and Southern Asia	25	61	57 (-)	74 (-)	12 (7)	55 (40)
Eastern and South-Eastern Asia	61 (32)	85 (64)	81 (28)	91 (72)	47 (27)	75 (52)
Europe and North America	95 (69)	97 (76)	98 (79)	99 (85)	89 (-)	94 (48)
Latin America and the Caribbean	74 (12)	87 (31)	82 (15)	91 (37)	47 (-)	70 (-)
Northern Africa and Western Asia	77 (26)	88 (38)	88 (40)	95 (49)	64 (-)	76 (-)
Oceania (not Aus/NZ)	38 (-)	30 (-)	75 (-)	70 (-)	26 (-)	18 (-)
Sub-Saharan Africa	23 (15)	30 (18)	37 (17)	45 (20)	17 (14)	22 (18)

Southern Asia, and lowest in Oceania, sub-Saharan Africa, Latin America, and the Caribbean. In Northern Africa and Western Asia, 79% of available freshwater is withdrawn, while in Central and Southern Asia, the proportion is slightly lower at 66%. Twenty-two countries are defined as water-stressed, indicating a high probability of future water scarcity, with 15 countries already withdrawing more than 100% of their renewable water resources. Most countries need to accelerate their implementation of IWRM to achieve the 2030 target. Levels of implementation are highest in Australia, New Zealand, Europe, and North America, and the lowest in Latin America and the Caribbean. However, even in regions with low overall implementation, there are examples of countries with high levels of IWRM implementation, highlighting that levels of development are not always prohibitive. Levels of cooperation for managing transboundary water resources are generally higher for surface water than groundwater, with around 59% of transboundary basins covered by an operational agreement in 2017. The highest levels of cooperation are seen in Europe, North America, and sub-Saharan Africa, again indicating that levels of development do not have to prohibit effective water governance.

Share of the population with access to improved drinking water, 2015

An improved drinking water source includes piped water on premises (piped household water connection located inside the user's dwelling, plot or yard), and other improved drinking water sources (public taps or standpipes, tube wells or boreholes, protected dug wells, protected springs, and rainwater collection).

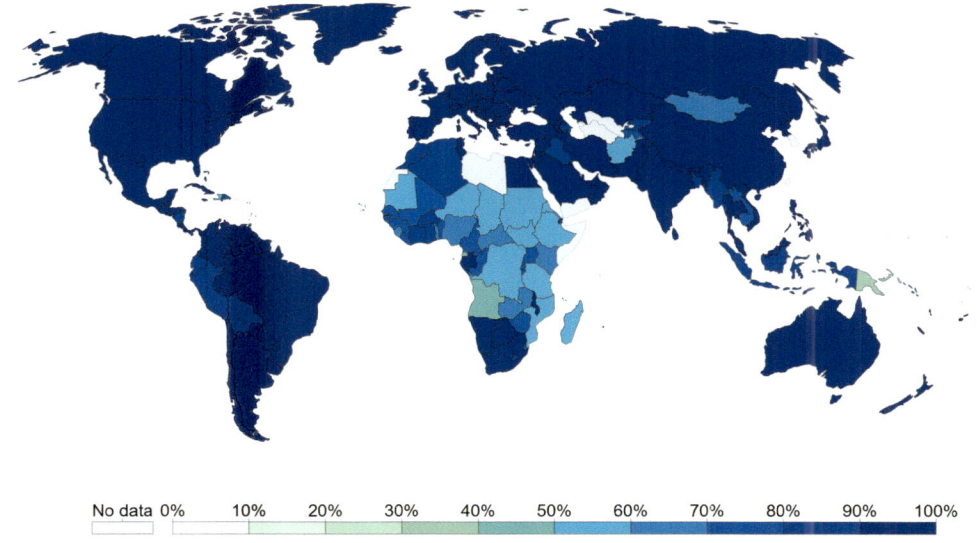

No data 0%   10%   20%   30%   40%   50%   60%   70%   80%   90%   100%

Fig. 6.6 Share of the population with access to improved drinking water (as of 2015). An improved water source includes safely managed, basic and limited services under the SDG service ladder (Table 6.3). *Credit* Ritchie and Roser (2019b), using data from World Bank, World Development Indicators. Reproduced under a CC BY License (https://creativecommons.org/licenses/by/4.0/)

Achieving all targets within **SDG 6** will require sufficient financing of the water sector and between 2012 and 2016, funding to the water sector dropped globally by more than 25%. In 2017, 80% of countries reported inadequate financing to meet the targets of **SDG 6**.

6.2.5 Global Progress: The Role of Groundwater

Groundwater makes a significant contribution to water supplies for domestic, agricultural, and industrial use globally. Reliable estimates of groundwater abstraction are not readily available at a global scale due to lack of monitoring, however, in 2010, global withdrawals were estimated to provide around 36% of domestic water supply, 42% of irrigation water for agriculture, and 27% of industrial water supply (Döll

et al. 2012). In parts of the southern and eastern UK, groundwater accounted for 100% of the total public water supply in 2015 (British Geological Survey 2019). In the USA, California is the state most reliant on groundwater, which in 2015, accounted for 21% of total freshwater withdrawals (United States Geological Survey 2019). India is the largest user of groundwater in the world, estimated to use more than 25% of the global total, with 60% of irrigated agriculture and 85% of drinking water reliant on groundwater (World Bank 2010). Although incomplete for Africa, data from the JMP in 2015, indicated that over 50% of the rural population in Africa, is reliant on groundwater as a primary source of drinking water (UPGro 2017).

Groundwater has an important role to play in achieving **SDG 6**, as will be discussed further in Sect. 6.3, but it is also relevant to other targets through several reinforcing and conflicting

Share of population with improved sanitation faciltities, 2015

Improved sanitation facilities are designed to ensure hygienic separation of human excreta from human contact. Improved sanitation facilities include flush/pour flush (to piped sewer system, septic tank, pit latrine), ventilated improved pit (VIP) latrine, pit latrine with slab, and composting toilet.

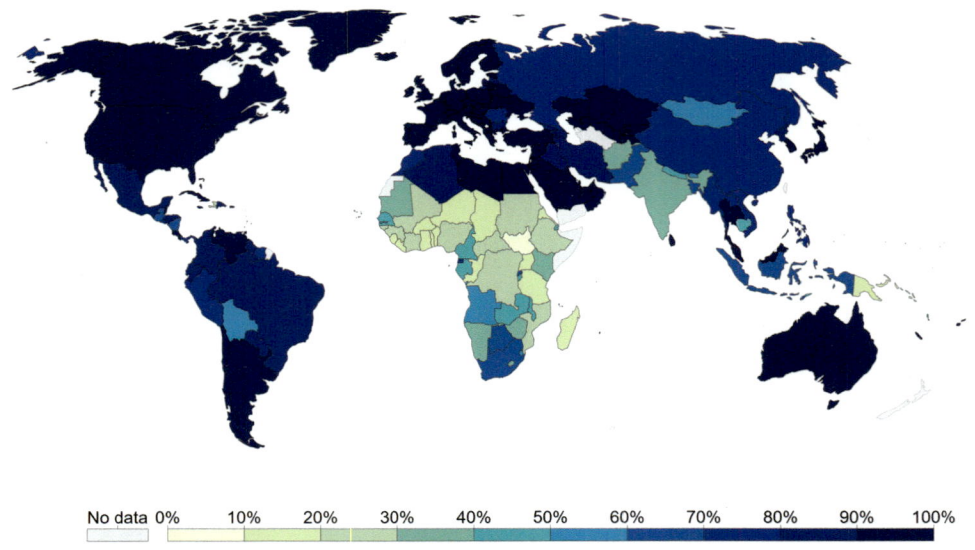

No data 0%   10%   20%   30%   40%   50%   60%   70%   80%   90%   100%

Source: World Bank – WDI   OurWorldInData.org/water-access-resources-sanitation/ • CC BY

Fig. 6.7 Share of the population with access to improved sanitation facilities (as of 2015). An improved sanitation facility includes safely managed, basic and limited services under the SDG service ladder (Table 6.3).

Credit Ritchie and Roser (2019b), using data from World Bank, World Development Indicators. Reproduced under a CC BY License (https://creativecommons.org/licenses/by/4.0/)

linkages (Guppy et al. 2018). Groundwater has the potential to increase resilience to water-related disasters (namely floods and droughts) and climate change as targeted by **SDGs 1.5**, **2.4**, and **13.1**. Through environmentally sound waste management, as targeted by **SDG 12.4**, improvements to groundwater quality will also be achieved. Similarly, achieving sustainable management and efficient use of natural resources, as targeted by **SDG 12.2**, will have positive outcomes for groundwater, and water resources more generally. As mentioned above, increased agricultural productivity may have negative implications for groundwater through increased demand for groundwater-fed irrigation and pollution by the use of fertilisers and pesticides.

6.2.6   Equity and Leaving no One Behind

The SDGs are based on the principle of leaving no one behind, paying particular attention to the least developed countries, in particular African countries, and to the most vulnerable members of society, including children and youth, those with disabilities, those living in extreme poverty, those living with HIV/AIDS, older people, indigenous peoples, refugees, and internally displaced persons. While geoscience undoubtedly plays a critical role in achieving the SDGs, and particularly **SDG 6**, it is important to recognise and understand the complex issues of equality and the challenges associated with addressing inequality in the effort to achieve the SDGs (see **SDG 10**). In the case of **SDG 6**, this predominantly concerns access to water services, which is highly unequal across the globe. Addressing these inequalities has long been an issue for academics and practitioners alike, with many past failures in progressing towards universal access to safe and affordable water attributed to errors or misjudgements by those in power (Chambers 1997). Understanding the realities and prioritising the needs of the most vulnerable members of society is essential to achieving **SDG 6**. For this reason, geoscientists are increasingly working alongside social scientists with the skills and methods to ensure that engineering or environmental solutions to water supply are centred on the needs of the most vulnerable.

6.3 Geology and SDG 6

6.3.1 Groundwater and the Water Cycle

Science, and earth science, in particular, has an important role to play in achieving **SDG 6**, with each of the four main branches of study—lithosphere, hydrosphere, atmosphere, and biosphere—contributing vital knowledge and understanding for addressing one or more of the targets within this goal. Of particular importance is an understanding of the water cycle (Fig. 6.8): how different components of the water cycle interact with one another, and with people, to determine the quantity and quality of water available and how this varies over time and space. Geoscientists can help answer critical questions such as: (1) how much rainfall is lost to evapotranspiration, how much becomes run-off to enter surface water stores such as rivers, lakes, and reservoirs, and how much infiltrates into the ground to enter groundwater stores or aquifers? (2) What is the nature of the subsurface and what does this mean for groundwater flow and storage? (3) How much water can be removed from an aquifer without causing long-term depletion or environmental degradation? (4) What is the natural quality of water stored on the surface or underground, and

how is this affected by human activity? (5) How often do extreme climatic events, such as heavy rainfall or prolonged dry periods occur, and what impact does this have on surface and groundwater in terms of flood and drought? Answering these questions to achieve the targets of **SDG 6** requires expertise from many disciplines within the geosciences—climate science, hydrology, hydrogeology, hydrochemistry—as well as other disciplines, such as engineering and the social sciences, to address the technological, environmental, social, and economic aspects of water service delivery.

Groundwater plays a key role in achieving **SDG 6**, particularly **Target 6.1**, because it is widely distributed, resilient to drought, and generally of good natural quality. The widespread distribution of groundwater across the globe (Fig. 6.9), means it can often be accessed close to the point of use where other sources, e.g., rainwater or surface water, are absent or insufficient. This is particularly relevant for dispersed rural communities that are distant from large-scale water supply infrastructure. Groundwater sources are generally more resilient to drought than surface water sources due to the significant amount of water that can be stored in aquifers compared to rivers, lakes, and reservoirs. This storage provides a buffer against short-term rainfall variability, often allowing a reliable supply of water when other sources fail during prolonged dry periods. The quality of groundwater is generally very good due to the natural filtration process that occurs when water infiltrates into the ground and flows through the pore spaces in a rock. Being underground also provides a level of protection from potentially polluting activities at the surface, meaning that groundwater often requires less treatment to achieve safety standards for drinking than surface water.

Exploiting groundwater for water supply, whether for domestic, agricultural, or industrial use, is not, however, always straightforward. The groundwater environment is complex and needs to be properly understood to ensure that aquifers are exploited appropriately and sustainably, without risk to the long-term quality or

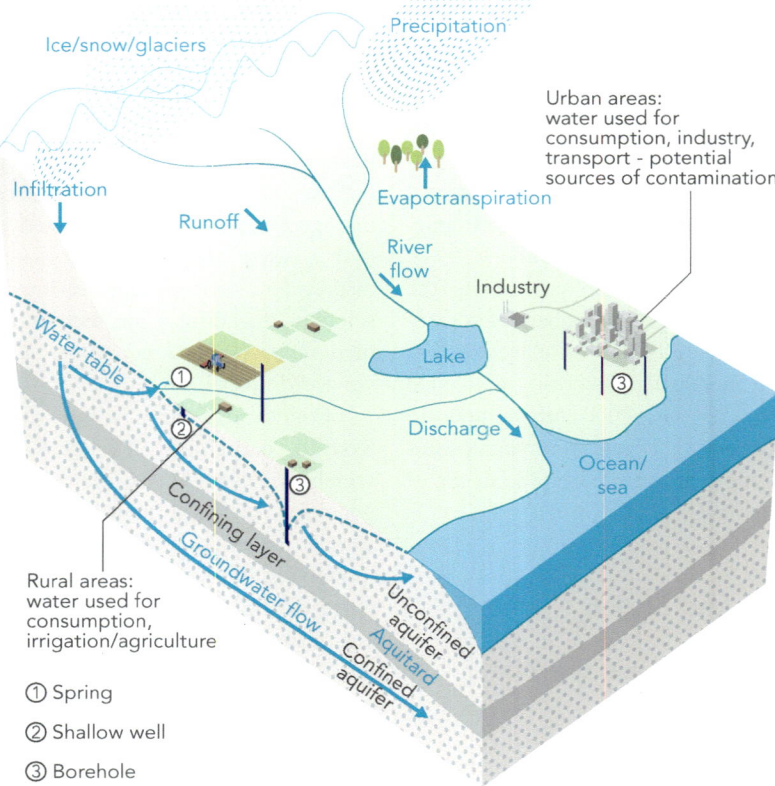

Fig. 6.8 The Water Cycle

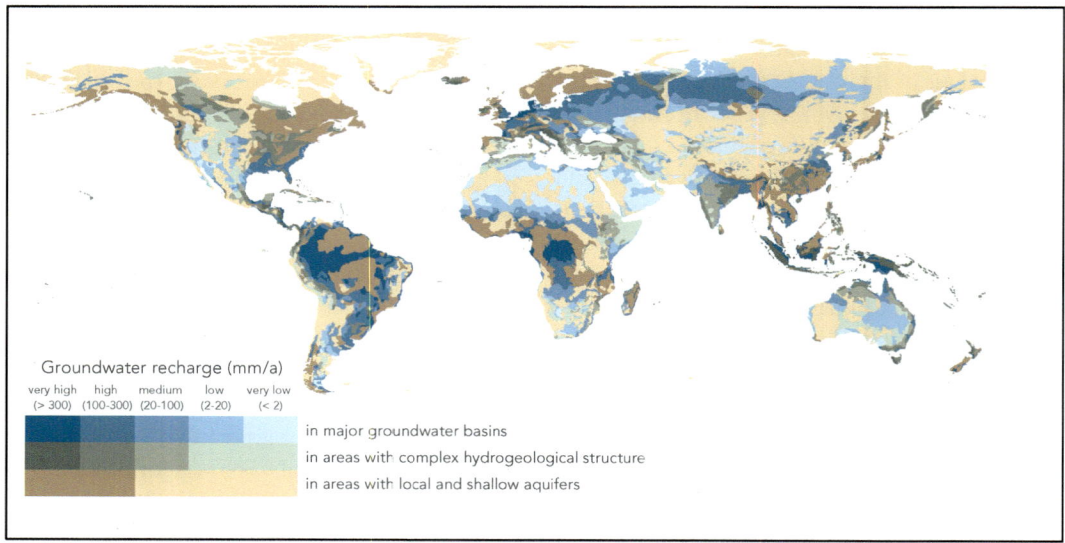

Fig. 6.9 Groundwater resources of the world. © BGR & UNESCO (2010), used with permission

availability of the resource. A sound understanding of the groundwater environment is also necessary for protection, integrated management, and efficient use of groundwater resources, as targeted by **SDGs 6.3** to **6.6**.

6.3.2 Key Groundwater Concepts

Groundwater—the freshwater stored in rocks and sediments beneath the ground surface—accounts for 30% of the total freshwater on Earth. Considering almost 70% of this freshwater is locked up in ice caps, glaciers, permanent snow, and permafrost, the majority (>98%) of accessible freshwater exists as groundwater (Gleick 1996). Hydrogeology, meaning water in rocks, is the discipline within the geosciences concerned with the study of groundwater. Groundwater can be found, to some extent, in almost all rock types but its potential usefulness as a resource is dependent on the quantity, quality, and sustainability of available water.

The amount of groundwater present at any given location will largely depend on the porosity and permeability of the rock and the amount of water entering the ground as recharge (see Box 6.2 for definitions). If the porosity, permeability, and recharge are high enough, water will accumulate in the pore spaces and fractures in a rock, usually above an impermeable base layer. If the rock becomes fully saturated, this forms an aquifer (Fig. 6.8). Groundwater flows naturally through an aquifer from the point of recharge to a point of discharge —usually a spring, river or the sea. Where groundwater is exploited for human use, wells or boreholes also act as points of discharge. The quantity of groundwater that can be stored and transmitted through an aquifer to a discharge point is dependent on the characteristics of the aquifer: mainly the transmissivity, storage, and 3D architecture (Box 6.2). These characteristics are largely controlled by geology. The depth and lithology of an aquifer also determine how easily accessible the groundwater is and what technology is required to exploit it. Rocks that do not transmit water easily are called aquitards.

Box 6.2. Basic Hydrogeological Concepts

Porosity (%) is the total void space within a rock and therefore defines the total amount of groundwater stored within an aquifer. Primary porosity refers to the pore space between grains, while secondary porosity refers to the space within fractures.

Permeability (measured in m^2) describes the ability of a porous media to allow fluids to pass through it.

Hydraulic conductivity (m/day) describes the ease with which a fluid would flow through a rock; it is dependent on the permeability of the rock and the properties of the fluid.

Transmissivity (m^2/day) describes the ability of an aquifer to transmit volumes of water; it is calculated by multiplying the hydraulic conductivity of an aquifer by its saturated thickness.

Yield (m^3/day or litres per second) describes the average volume of water that can be abstracted from an aquifer from a borehole, well or spring.

Storativity (dimensionless) describes the volume of water released from an aquifer per unit drop in groundwater head per unit area.

Depth to groundwater is the depth to the water table (or to the top of a confining layer where the water table has risen above the top of an aquifer and is therefore under pressure).

3D architecture describes the way in which the properties of the aquifer (i.e., permeability and storativity) vary with depth.

Piezometric level is a way of expressing the pressure in a confined aquifer. It is the level at which water would rise in a borehole drilled into the confined aquifer.

Water table is the upper surface of a groundwater body in an unconfined aquifer. It can be measured by the static water level in a well or borehole in an unconfined aquifer.

Recharge describes the amount of water that replenishes an aquifer, usually from precipitation, but also from seepage from rivers, lakes, or canals.

Discharge describes the amount of water removed from an aquifer, either by natural discharge to the environment (e.g., rivers, springs, lakes, wetlands), or through abstraction for human consumption.

Aquifers are generally classified or mapped according to the dominant groundwater flow mechanism—whether flow occurs mainly through the pore space or fractures in a rock—often combined with a measure of the productivity of an aquifer, lithology, or average recharge to an aquifer. As for geology, the hydrogeology of any region is complex and spatially variable, both laterally and vertically. However, the main types of aquifer found across the globe can be summarised into just a few key hydrogeological environments, which are described in Table 6.6, and illustrated in Fig. 6.10. In some hydrogeological environments, for example, an alluvial plain that is homogeneous, laterally extensive, permeable, and receives significant recharge, groundwater is readily available and easily accessed by a shallow hand dug well or manually drilled borehole. In more complex hydrogeological environments, such as deep, fractured basement rocks with low primary porosity, developing a successful groundwater source is more challenging. However, even relatively low permeability rocks can be capable of providing sufficient flow to a well to support an individual household or community water supply, or small-scale irrigation scheme.

6.3.3 Water Supply

In those parts of the world with most work to do to achieve **SDG 6.1**, the challenges of groundwater development for water supply are different in urban and rural contexts. In many urban areas,

the public water supply infrastructure cannot expand fast enough to provide a piped water supply to the rapidly growing population. As a result, urban populations often obtain water from multiple sources according to availability and cost. Sources may include private water vendors, utility stand-posts and kiosks (Fig. 6.11), and unimproved shallow wells and surface water, with many individuals drilling their own private wells or boreholes to ensure they have a reliable source of water for drinking and other domestic uses (Box 6.3). Private borehole development is, however, often completely unregulated resulting in issues of over abstraction and contamination, as documented in parts of Asia and Africa (Foster and Vairavamoorthy 2013). Although in many rapidly expanding urban areas, private wells or boreholes are helping to bridge the gap between supply and demand, there are equity issues in terms of access as low-income households often lack the resources, both in terms of land ownership and capital, to instal a private well. Private borehole development may also ultimately lead to a reduction in revenue for water utilities, further reducing their ability to expand piped water infrastructure and provide lower tariffs to poorer households.

Box 6.3. Informal urban water supply and sanitation in Lusaka, Zambia

In Lusaka, Zambia, repeated cholera outbreaks during the rainy season, are linked to contaminated drinking water. During an outbreak in 2017–18, one of worst in recent years, more than 5000 cases were reported in Lusaka, eliciting an emergency response (International Federation of Red Cross and Red Crescent Societies 2018) and requiring a multifaceted public health response including increased chlorination of municipal water supplies, provision of emergency water supplies, a vaccination campaign, and rapid training for health care workers.

Lusaka sits on carbonate rocks that are overlain by permeable superficial deposits

Table 6.6 Hydrogeological Environments. Adapted from MacDonald et al. (2005)

Hydrogeological Environment	Lithology	Flow Mechanism	Productivity	Description
Crystalline basement aquifers	Highly weathered/fractured metamorphic or magmatic rocks	Fracture flow	Moderate	Groundwater can be found in well-developed fracture networks and/or a thick weathered zone
	Poorly weathered/fractured metamorphic or magmatic rocks	Fracture flow	Low	Groundwater can exist in small fractures and may be locally important, but is difficult to find
Consolidated sedimentary aquifers	Sandstones	Intergranular or fracture flow	Moderate to high	Groundwater can be found in pore spaces and fractures; productivity will increase with coarseness and degree of fracturing
	Limestones	Fracture flow	Moderate to high	Groundwater can be found in fractures, which may be enhanced by dissolution; limestones have low primary permeability
	Mudstones	Fracture flow	Low	Groundwater can be found in fractures in hard, consolidated mudstones; often interbedded with sandstone or limestone layers
Unconsolidated sedimentary aquifers	Major alluvial or coastal sands and gravels	Intergranular flow	High	Groundwater can be found in thick unconsolidated sands and gravels deposited in major rivers basins or shallow seas
	Valley and coastal dune sands and gravels	Intergranular flow	Moderate	Groundwater can be found in smaller, dispersed sand and gravel deposits found in many modern-day river valleys and coastal dune environments
Volcanic	Lava, ash, and pyroclastic deposits	Fracture flow	Low to high	Groundwater often found along fractured contacts between lava flows in complex layered aquifer systems

of varying thickness (Nkhuwa et al. 2018). Groundwater in the karstic aquifer flows through a system of well-developed conduits and channels, making it a highly productive aquifer, which satisfies more than half the city's water requirements. However, its high permeability and limited protection also means that contaminants can easily infiltrate and be transmitted through the aquifer. This, combined with poor sanitation and waste management, results in the aquifer being extremely vulnerable to contamination.

As occurs in many rapidly expanding African cities, inadequate water supply and sewerage service provision has led many residents across Lusaka to instal their own private water supplies and on-site sanitation facilities. These are largely unregulated, often resulting in inadequately protected pit latrines being located very close to wells or boreholes (Fig. 6.12). This can result in untreated sewage leaking or discharging to the underlying aquifer, which residents then use for water supply (FRACTAL and LuWSI 2018).

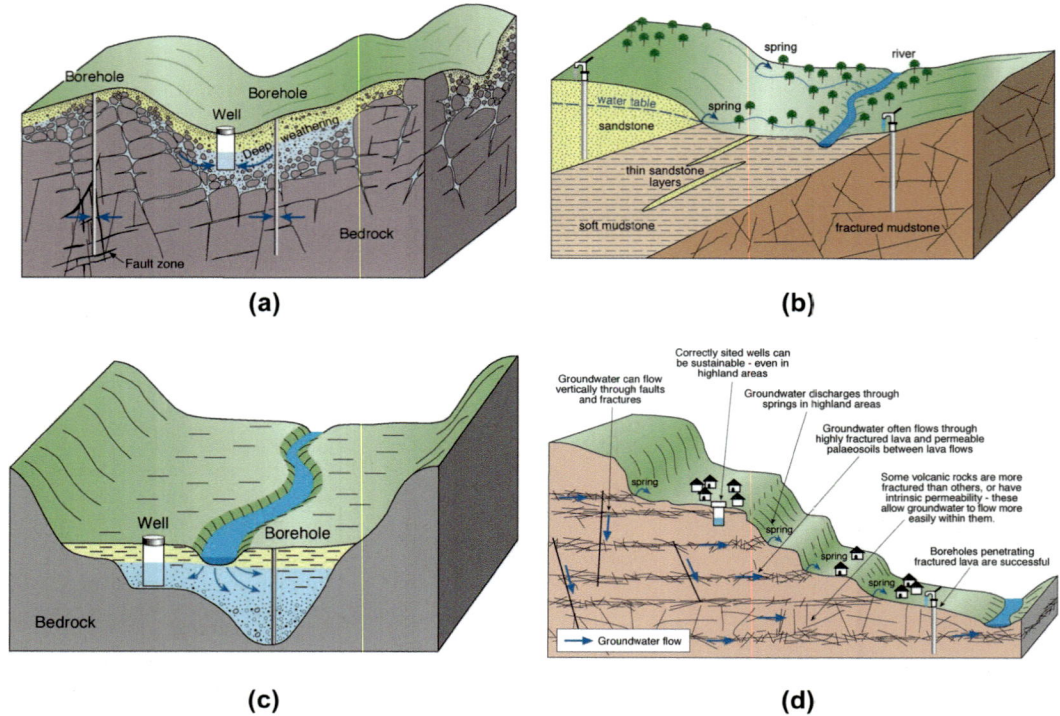

Fig. 6.10 Hydrogeological Environments. a weathered basement aquifer; **b** sandstone and mudstone sedimentary aquifers; **c** major alluvial aquifer; **d** volcanic aquifer.

From MacDonald et al. (2005), used with permission from ITDG publishing. © NERC

Low-income, high-density peri-urban areas are most vulnerable to issues of groundwater contamination as service provision is lower and inhabitants can often only afford to access shallow groundwater through unprotected wells, which are highly susceptible to contamination (Nkhuwa 2006). However, groundwater contamination due to inadequately maintained septic tanks has also been observed in high-income, low-density parts of the city (Nkhuwa et al. 2015).

If the water quality and water supply targets of **SDG 6** are to be met, these issues need to be addressed through increased service provision, regulation, source protection, and water treatment.

In rural settings, where water supply infrastructure is sparse or non-existent, groundwater often represents the only viable option for safe and reliable water supply through either household or community wells or boreholes. In sub-Saharan Africa, the majority of the rural population source their drinking water from groundwater through wells, boreholes, and springs. Properly sited and constructed boreholes, equipped with handpumps, have proved an excellent method for increasing access to safe drinking water, and have revolutionised rural water supply over the past 50 years. However, questions still remain about the best methods to maintain and manage these supplies over the long-term and how to increase their current low levels of functionality. It should be noted that community water points are considered a basic level of

Fig. 6.11 Water Kiosk in Chipata, Zambia. *Credit* GIZ Rahul Ingle (reproduced under a CC BY SA 2.0 License, https://creativecommons.org/licenses/by-sa/2.0/)

service under the SDG indicators because water is not available on an individual's premises, but it is likely that many rural populations will be reliant on these for decades to come, particularly in sub-Saharan Africa. As in urban settings, achieving equitable access is a challenge in rural areas, with the possibility of the location of a community water point privileging some members of the community over others.

Geoscientists have a key role to play in improving access to safe drinking water. Expertise is required in: planning and designing water supply programmes; siting and commissioning individual water points; mapping the location, quantity, quality, and renewability of available groundwater resources; and carrying out research into the reliability and sustainability of supply. In many areas, groundwater resources are relatively easy to find and standard techniques and methods

can be used to develop sustainable supplies (MacDonald et al. 2005), however, in other areas, groundwater resources can be much more difficult to develop. Geoscientists, therefore, have a vital role in helping to design appropriate drilling programmes, ensuring the correct techniques and methods are employed. Geophysics is often used to site individual boreholes, and pumping tests and water quality sampling undertaken on individual sources. These methods require qualified geoscientists to correctly apply the methods and interpret the results. In many parts of the world, groundwater resources are yet to be mapped at a sufficient scale to be useful for helping to design drilling programmes, with a particular gap in water quality mapping. There are still many unanswered questions for research to address—particularly around the sustainability of groundwater as demand for water increases—

Fig. 6.12 Close proximity of groundwater well and pit latrine in peri-urban area of Lusaka, Zambia. *Credit* Kenedy Mayumbelo (reproduced under a CC BY 2.0 license, https://creativecommons.org/licenses/by/2.0/)

and in the successful management of water services, which requires geoscientists to work with other disciplines to make progress.

6.3.4 Groundwater Quality

The natural, or baseline, quality of groundwater is generally very good, but varies considerably in different hydrogeological environments due to reactions between the water and rock. Groundwater naturally contains many dissolved constituents, which at certain concentrations are not harmful, and in fact, in many cases are essential for human health. However, groundwater quality can be affected by both naturally occurring and human-induced contaminants, which at elevated concentrations can have serious implications for human and ecosystem health. The World Health Organisation provides guidelines and standards for drinking water, which set recommended limits for microbial, chemical, and radiological aspects of water quality (World Health Organisation (WHO) 2017). Some of the major contaminants of concern for groundwater globally are summarised below and in Table 6.7.

Chemical contaminants, which are naturally occurring in the environment, can be introduced to a groundwater system by natural and anthropogenic processes. The main natural contaminants (also referred to as geogenic contaminants) of concern globally are fluoride and arsenic.

Table 6.7 Summary of common chemical and biological contaminants in groundwater

Inorganic chemical constituents		Pathogens	Organic compounds	Others
Major Elements[a] [b]: Sodium[c] Sulphate[c] Nitrate[c] Magnesium[c] Potassium[c]	Trace Elements[a]: Fluoride[c] Iron[c] Manganese[c] Arsenic Selenium[c] Cadmium Nickel[c] Chromium[c] Lead Aluminium	Coxsackievirus Echovirus Norovirus Hepatitis Rotavirus E. Coli Salmonella Shigella Campylobacter jejuni Yersinia Legionella Cryptosporidium parvum Giardia lamblia	Chlorinated solvents Aromatic hydrocarbons Pesticides	Pharmaceuticals Radionuclides Salinity

[a]Naturally occurring in groundwater
[b]The other major chemical constituents in groundwater, also considered essential for human health, are Bicarbonate (HCO_3), Calcium (Ca), Chloride (Cl), and Silicon (Si).
[c]Essential for human health at certain concentrations.

Fluoride occurs in groundwater where it dissolves fluorine-bearing minerals such as fluorite, apatite, and micas, which are particularly common in crystalline rocks such as granites. Elevated fluoride is more likely to occur where groundwater has a long residence time in an aquifer as this provides more time for water-rock interactions to occur. In active volcanic regions, elevated fluoride in groundwater can also occur due to mixing with hydrothermal fluids or gases. Fluoride is an issue across many parts of the world, particularly arid parts of northern China, India, Sri Lanka, North Africa, the East African Rift System, and Argentina (Box 6.4).

The occurrence of arsenic in groundwater is complex and can be related to a number of natural and anthropogenic processes. It can occur naturally where groundwater interacts with arsenic-bearing minerals such as sulphide minerals precipitated from hydrothermal fluids in volcanic environments, and pyrite and iron oxides that often accumulate in sedimentary environments. Human activities such as mining (particularly for coal and sulphide minerals), industry, and the use of certain arsenic-bearing pesticides, can also be sources of arsenic in groundwater. High arsenic concentrations tend to occur in strongly reducing (low oxygen) groundwaters or oxidising groundwaters with high pH, which inhibit adsorption of arsenic onto sediments and soils. Arsenic is a well-documented issue in anaerobic alluvial and deltaic aquifers in Bangladesh, West Bengal (eastern India), Nepal, northern China, Vietnam, and Cambodia, and in aerobic but high pH loess (wind-blown sediment) aquifers in Argentina and Chile.

Long-term exposure to elevated concentrations of these elements can cause dental and skeletal fluorosis in the case of fluoride, and a vast number of dermatological, cardiovascular, neurological, and respiratory issues, as well as several cancers, in the case of arsenic.

Box 6.4. Health Impacts of Elevated Fluoride in Groundwater, India

More than 200 people worldwide are believed to be drinking water with fluoride in excess of the WHO guideline of 1.5 mg/L (Edmunds and Smedley 2013). India is one of the worst affected countries (Podgorski et al. 2018), with parts of Sri Lanka, China, Mexico, and East Africa also significantly impacted.

Groundwater normally contains low concentrations of fluoride (<1.5 mg/l), which we require to maintain good dental health. However, high fluoride concentrations in drinking water can lead to health complications when consumed over long periods of time (BGS and WaterAid 2000; Edmunds and Smedley 2013). Long-term exposure to concentrations of 1.5–4 mg/l can lead to dental fluorosis, the most common issue associated with excessive fluoride consumption, which in extreme cases causes the tooth enamel to become pitted and discoloured. Higher concentrations (>4 mg/l) can cause skeletal fluorosis —a bone disease causing painful damage to bones and joints—or, in the worst cases crippling fluorosis which can ultimately lead to paralysis. Children under the age of seven, whose teeth and are still developing, are most vulnerable to dental fluorosis, which can be exacerbated by calcium and vitamin C deficiency.

Endemic fluorosis affects at least 17 States in India, with Andhra Pradesh, Rajasthan, Haryana, and Gujarat being the worst affected (BGS and WaterAid 2004). Much of India is underlain by Precambrian basement rocks, which mainly comprise gneisses and granites, with lesser amounts of metasedimentary rocks. In some areas the basement is overlain by younger sedimentary rocks and about half the land area of non-peninsular India is covered by Quaternary alluvial deposits. The alluvial deposits form the most productive aquifers, but Tertiary sediments and the Precambrian basement are also widely used for water supply. Elevated fluoride is most commonly (but not exclusively) associated with groundwater circulation in granitic basement rocks in arid and semi-arid areas of the country.

Fluoride can be removed from the water, but many individuals or countries lack the resources to treat water adequately. One of the best-known methods— the Nalgonda technique—was developed in India. This involves adding a combination of alum, lime, and bleaching powder to contaminated water, which is stirred and left to settle, allowing fluoride to be removed through the process of flocculation, sedimentation and filtration (BGS and WaterAid 2000). This method can be applied at the household level in a bucket, and at the community level in defluoridation plants.

Nitrate, although naturally occurring, is generally elevated in groundwater by human activities. The most common sources of nitrate in groundwater are nitrogen fertilisers, sewage, and wastewater. The use of nitrogen fertilisers to increase crop yields has grown significantly since the 1970s. Intensive application of fertilisers, particularly where double or triple cropping is practiced alongside poorly controlled irrigation, can lead to leaching of nitrate from the soil to an underlying aquifer. This occurs in agricultural areas across the world (Box 6.5). Intensive livestock farming, through manure and slurry pit leachate and effluent, is another potential source of nitrate contamination in groundwater, along with untreated sewage and wastewater. This is a particular problem in urban areas where sanitation infrastructure, much like water supply infrastructure, cannot expand fast enough to meet the needs of a growing population. In these circumstances, many households instal their own private waste disposal facilities—usually a pit latrine or septic tank—that can leak if not properly constructed and maintained. This poses a potential threat to an aquifer, and ultimately human health, particularly where unimproved sanitation facilities are combined or co-located with unimproved drinking water services (as described previously in Box 6.3).

Box 6.5. The Nitrate Time-Bomb When nitrate is leached from the soil it travels through the unsaturated zone before reaching the water table below. The travel time will depend on the geology and thickness of the unsaturated zone, and it can take as long as 100 years for nitrate to travel from the soil to an underlying aquifer. This large delay is sometimes referred to as the Nitrate Time-Bomb since the full impact of nitrate contamination from the use of nitrogen-based fertilisers, may not be observed for many years to come.

In areas with a history of intensive agriculture, such as Europe, North America and China, a significant amount of nitrate has built up in the unsaturated zone. This may cause groundwater contamination issues for decades to come, despite the introduction of legislation to control the use of fertilisers (Ascott et al. 2017). While this is a more significant problem in agriculturally intense countries, it is an issue that could become more severe in less developed countries as agriculture intensifies to meet the growing food demand (Fig. 6.13).

Elevated nitrate in groundwater, which ultimately discharges to rivers, lakes and coastal areas, can cause significant damage to ecosystems and increase the cost of water treatment. There are also health issues associated with high concentrations of nitrate—most notably a rare condition referred to as 'blue-baby syndrome', whereby nitrate reduces to nitrite in the stomach of young children, oxidising haemoglobin to methaemoglobin, which is unable to transport oxygen around the body. There are no reliable estimates of the extent of the problem worldwide (WHO 2019).

Poor sanitation practices are also the primary source of microbiological contaminants, particularly in shallow aquifers in urban and peri-urban areas (Lapworth et al. 2017). Pathogens that are easily transported in groundwater and potentially very harmful to human health include *Norovirus*, *Hepatitis*, *E. Coli*, *Salmonella*, and *Legionella*.

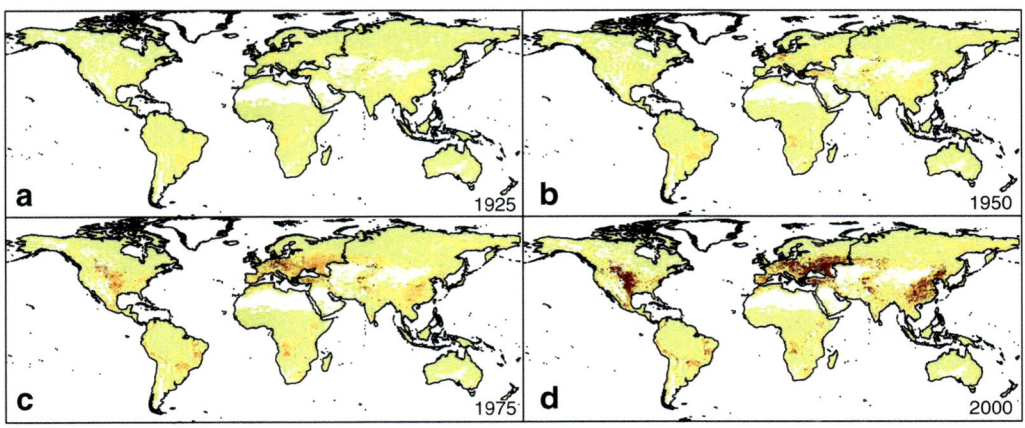

Vadose zone N storage (kg N ha^{-1})
High : 4334
Low : 0

Fig. 6.13 Build-up of nitrate in the unsaturated (vadose) zone over time. From Ascott et al. (2017). Reproduced under a CC BY 4.0 License (https://creativecommons.org/licenses/by/4.0/)

Other contaminants, such as heavy metals, synthetic organic compounds, and a range of emerging contaminants including food additives, caffeine, pharmaceuticals, and synthetic hormones, can be introduced to groundwater systems from industrial, agricultural, and domestic sources, posing a potential threat to vulnerable aquifers. Emerging contaminants in particular are mostly unregulated and not effectively removed by conventional treatment practices (Stuart et al. 2011), requiring an improved understanding of how they behave in the environment.

Groundwater vulnerability is often considered in the context of source-pathway-receptor. The vulnerability of a receptor (this may be an aquifer, well, borehole, spring, or river), will depend on the pathways that exist to transport a contaminant from its source to the receptor. The vulnerability of an aquifer to contamination from any of the sources discussed above is dependent on the properties of the soil and unsaturated zone, through which contaminants have to travel before reaching an aquifer, and the properties of the aquifer itself. In high permeability rocks, contamination can move quickly from the surface to an aquifer, then through an aquifer to a receptor (e.g., a borehole, spring, river, or wetland). Wells or boreholes in shallow fractured aquifers located close to the source of contamination will be highly vulnerable as there is little opportunity for attenuation, either in the unsaturated or saturated zone. Deep aquifers with low permeability will provide greater opportunity for attenuation between the source and receptor and are therefore less vulnerable to contamination.

Groundwater salinity is a widespread problem, which at shallow depths can be a major constraint on the development of groundwater resources. Elevated mineral concentrations have health impacts when water is routinely used for drinking and can reduce the value of water for industry and agriculture, causing damage to the soil if used for irrigation. The processes that lead to groundwater becoming saline are complex and can be divided into two broad categories: natural and those that are exacerbated by human activity.

The source of naturally occurring saline groundwater can be marine, where seawater enters coastal aquifers, or terrestrial, associated with low rainfall, shallow water tables and high rates of evaporation. Some aquifers have also become increasingly saline due to irrigation, either from leaching of salts in the soil, or waterlogging and subsequent salinization. Pakistan and the Indus valley have seen some of the worst increases in groundwater and soil salinization due to a long history of irrigation.

Geoscience has a role to play in addressing **Targets 6.2** and **6.3** through groundwater and source protection as part of IWRM, waste management, and groundwater remediation. As described above, there are many potential sources of contamination than can negatively impact groundwater systems, including human excreta and sewage from poor sanitation practices, wastewater from domestic, agricultural or industrial activities, solid waste, and hazardous waste from industry. Management of any type of waste requires capture, storage, transport, treatment, and disposal or reuse, which may involve simple domestic-scale systems such as pit latrines up to large-scale infrastructures such as centralised sewer systems, wastewater treatment plants, and landfills. Whether considering small-scale storage of human excreta and sewage in a pit latrine or large-scale storage of solid waste in a municipal landfill, an understanding is required of how this waste behaves in the environment and what mitigation measures are required to minimise any negative impacts on both the environment and people. Geoscientists can help answer questions such as.

1. What are the potential sources of contamination and how close are these to environmental or human receptors such as drinking water sources, ecological sites, or areas used for recreation?
2. Is there potential for contaminants to be mobilised by water infiltrating at the surface?
3. If mobilised, how easily could contaminants move through the subsurface?
4. Is the waste well contained given the nature of the subsurface and what additional measures are required for safe storage?

Again, these questions require inputs from various disciplines within the geosciences, along with others involved in the waste management process.

6.3.5 Sustainable Groundwater Management

Groundwater is not an easy resource to manage. It is out of sight, and therefore, often overlooked by both the public and governments. This can lead to the misconception that wells, boreholes, and springs will continue to supply high quality water indefinitely, irrespective of how much water is abstracted or polluting activities occurring in the surrounding area (Smith et al. 2016). At the catchment scale, groundwater can have many users with competing demands: drinking water, industrial production, and agriculture. Trade-offs also develop between urban and rural users, and between groundwater abstraction and the ecological functioning of wetlands or baseflow to rivers. Balancing these abstraction demands, along with the environmental requirements for groundwater, is a challenge, but is essential if all targets within **SDG 6** (and other linked goals, e.g., **SDG 2**, **SDG 8**, **SDG 15**) are to be met. Achieving sustainable groundwater management requires local groundwater users, technical experts (including geoscientists) and policymakers to work together to develop understanding, drive change, and develop and implement appropriate tools (Smith et al. 2016).

Pressures on groundwater are increasing from both abstraction and pollution, and resources need to be protected and managed. High abstraction in parts of the world have led to rapidly falling water tables, sometimes accompanied by land subsidence or degradation of water quality through saline intrusion. Parts of India, Pakistan, the USA, Iran, Saudi Arabia, and China have been identified as experiencing severe overexploitation of groundwater (Gleeson et al. 2012). In other areas, such as parts of sub-Saharan Africa, groundwater resources remain less developed, and opportunities exist to develop groundwater for social, economic and health benefits (Cobbing and Hiller 2019). Changing land use—and in particular intensive agriculture and urbanisation—have led to widespread groundwater contamination (Morris et al. 2003). Nitrate concentrations are high in many aquifers in agricultural areas; and beneath many cities, groundwater has been polluted by a cocktail of different organic and inorganic chemicals. Because of the long residence times of groundwater, it can take many years, decades or centuries for contaminants to be flushed out of an aquifer. Management of groundwater is important not just for today but for future generations.

Groundwater systems vary considerably—based on the geology, climate, links to surface water systems, and land use—which means they respond differently to pressures and require different management solutions. The starting point for groundwater management is, therefore, to characterise how groundwater systems work: what is the geological and hydrogeological environment; how much recharge does the system receive; how much groundwater is naturally discharged, and where; and what is the vulnerability of an aquifer to pollution? Using this knowledge, effective monitoring systems can be designed to bring to light the impact on groundwater from abstraction and land use. Given the nature of groundwater as a common pool resource many different stakeholders then need to be involved to develop reasonable visions and plans for groundwater governance that leave no one behind (Villholth et al. 2017). As well as considering groundwater as a source for human consumption, the role of groundwater in maintaining ecosystems, which provide many services to both humans and the environment, is also of concern. Integrated Water Resources Management (IWRM) provides a framework to help manage water resources across catchments, taking into account the uses of water from all parts of the water cycle. This paradigm shift in management approach moved the emphasis away from individual well fields or aquifers to entire water systems. The European Union has been at the forefront of applying the principles of IWRM to groundwater and are set out in the Water

Framework Directive of 2000, and supplemented by the Groundwater Protection Directive of 2006 (Quevauviller 2007). In summary, these approaches manage the balance of abstraction from groundwater with the recharge and unwanted impact to others and the environment and protect groundwater quality through groundwater friendly rural land use, regulation to penalise point source pollution, and the development of precautionary engineering structures to contain point source pollution such as landfill sites.

To achieve sustainable groundwater management various methodologies have been developed and proved useful, for example, detailed 3D mapping of aquifers and groundwater systems; monitoring systems with in situ monitoring of water levels and chemistry and the use of satellite data such as InSAR, and GRACE; sophisticated land zoning methods based on the vulnerability of groundwater to contamination, or travel times to abstraction boreholes; the development of numerical groundwater models to test possible future scenarios or track sources of pollution. Some technical engineered interventions are also sometimes used, such as rainwater harvesting and managed aquifer recharge (MAR) to increase the natural recharge to the system (Box 6.6); the use of scavenger wells to control pollution particularly in saline areas; and the construction of engineered structures to control pollution or flooding. Geoscientists are fundamental to developing and adapting these methods and technologies.

Box 6.6. Managed Aquifer Recharge (MAR)

MAR involves artificially recharging aquifers with excess surface water during wet periods, or in some cases treated wastewater, which is stored underground and can be accessed during dry periods when surface water is scarce. MAR is gaining increased attention as an adaptation measure to improve water security and resilience to climate variability. It is increasingly important as a management strategy in conjunction with demand management to maintain stressed groundwater systems (Dillon et al. 2019). However, there are limitations to the applicability of MAR, which always need to be fully considered when assessing the viability of this solution.

The International Groundwater Resources Assessment Centre (IGRAC[2]) document over 1000 examples of MAR schemes worldwide, which use different methods and technologies to artificially recharge an aquifer (Stefan and Ansems 2018). The application of MAR has grown rapidly since the 1960s, with an estimated capacity of 10 km^3 per year in 2018 (Dillon et al. 2019). However, with estimated annual global groundwater abstraction of 800 km^3, there is still room for growth. Natural groundwater recharge through rainfall and river and lake leakage remains the overwhelming method by which groundwater is renewed.

Techniques to enhance groundwater recharge range in scale and sophistication (Dillon et al. 2019). Enhanced recharge from rivers is widely used across India, where hundreds of thousands of constructed dams create ponds within the river channel to increase infiltration. Recharge can be further induced from the river by drilling abstraction boreholes close to the banks of the river. This pulls water from the river into the aquifer and naturally filters the water through the aquifer material. Water spreading is a method used to capture floodwater and spread it over a larger area to increase soil moisture and promote infiltration to an aquifer. Some schemes involve dedicated recharge boreholes which pump treated surface water directly into the aquifer. All methods come with risks of increasing contamination of the groundwater and need to be monitored carefully.

---

[2]https://www.un-igrac.org/.

Although groundwater is essentially a local resource with a flow rarely more than one metre per day, aquifers do not respect international borders. Large aquifers crossing international or state borders (referred to as transboundary aquifers) require some level of cooperation to be successfully and sustainably managed. The level of cooperation could extend from a shared understanding of the extent and nature of an aquifer to joint monitoring and agreed regulation. Given the slow nature of groundwater movement, transboundary aquifers can be viewed more as a vehicle and opportunity for technical cooperation, rather than a source of conflict.

6.4 Conclusions

Groundwater has an important role to play in achieving the SDGs, particularly through meeting the targets of **SDG 6**. Geoscientists have a critical role to play in achieving safely managed drinking water and sanitation for all (**Targets 6.1 and 6.2**), protecting the quality of the globe's water resources (**Target 6.3**), ensuring sustainable water use and reduction of water scarcity (**Target 6.4**), achieving integrated water resources management (**Target 6.5**), and protecting water-related ecosystems (**Target 6.6**). Understanding, characterising, monitoring, forecasting, and communicating groundwater dynamics and the connections with the wider ecosystem are not straightforward. In addressing these targets, geoscientists are required to work alongside policymakers, and often water users, to ensure the best evidence informs decisions about water resource development and allocation. This may happen from the local scale—where scientists work alongside communities or local authorities to inform water resources management in small basins or catchments—up to the regional or continental scale—where scientific evidence is used by national governments to inform the development and management of large transboundary water resources. With an increasing rate of global environmental change, the demand for groundwater as a reliable source of water will only increase.

6.5 Key Learning Concepts

- Water and sanitation are key components of economic and social development
- Progress towards the targets of SDG 6 is highly unequal across the globe and often, but not always, related to levels of development
- There are significant challenges to achieving SDG 6, such as climate change and population growth, the effects of which are also unequal across the globe
- Groundwater has a key role to play in achieving SDG 6, particularly through the provision of sustainable and climate resilient water supplies
- Groundwater resources are out of sight and often difficult to understand, requiring expertise across a range of disciplines
- Overexploitation and pollution of groundwater is a global issue, but can be addressed through IWRM and sound management and governance strategies

6.6 Educational Ideas

In this section, we provide examples of educational activities that connect geoscience, the material discussed in this chapter, and scenarios that may arise when applying geoscience (e.g., in policy, government, private sector international organisations, NGOs). Consider using these as the basis for presentations, group discussions, essays, or to encourage further reading.

- From 1990 to 2015 (25 years), access to improved drinking water in Tanzania has gone from 53.90% to 55.60% of the population.[3] At this rate of progress, it will be 2667, before Tanzania has 100% access to improved drinking water. Explore the reasons for this rate of progress and the actions (from

---

[3]https://ourworldindata.org/grapher/share-of-the-population-with-access-to-improved-drinking-water?tab=chart&time=1990..2015&country=OWID_WRL+IND+KEN+BRA+TZA.

geoscientists and others) that may help catalyse action towards 100% access to improved
drinking water in Tanzania.

- Review the information in this chapter on
groundwater and fluoride. Prepare an information sheet for NGOs drilling boreholes,
summarising key geological environments
associated with elevated fluoride.
- Integrated water resources management aims
to bring different stakeholders together to
ensure collaborative, cooperative, and coordinated management of water resources.
Reflecting across the SDGs, and how demand
for water may change by 2030, consider the
range of stakeholder this may include, and
what priorities each may have in terms of the
quantity and quality of water required to fulfil
their needs. As a class, discuss what recommendations you would make to resolve conflicting demands on water resources in an
equitable way, leaving no one behind, while
protecting resources for future generations.

Further Reading and Resources

Books and Articles

United Nations (2018) Sustainable Development Goal 6
Synthesis Report on Water and Sanitation. United
Nations, New York. https://www.unwater.org/
publication_categories/sdg-6-synthesis-report-2018-
on-water-and-sanitation/
MacDonald A, Davies J, Calow R, Chilton J (2005)
Developing groundwater: a guide for rural water
supply. ITDG publishing. https://www.
developmentbookshelf.com/doi/book/10.3362/
9781780441290
IAH Strategic Overview Series. https://iah.org/education/
professionals/strategic-overview-series
WaterAid/British Geological Survey Groundwater Quality Fact Sheets: Country. https://www.bgs.ac.uk/
downloads/browse.cfm?sec=9andcat=115. Element
https://www.bgs.ac.uk/downloads/browse.cfm?sec=
9andcat=116
GW-MATE Briefing Note Series, Case Profile Collection,
Book Contributions and Strategic Overview Series.
available via IGRAC. https://www.un-igrac.org/
special-project/gw-mate

Videos

IGRAC Groundwater, the Hidden Resource. https://www.
youtube.com/watch?v=tzkBvLXa8jsandt=13s
RWSN A borehole that lasts a lifetime. https://vimeo.
com/128478995

Tools and Online Resources

Africa Groundwater Atlas. https://www.bgs.ac.uk/
africagroundwateratlas/index.cfm
USGS Water Science School. https://www.usgs.gov/
special-topic/water-science-school
UK Groundwater Forum. https://www.groundwateruk.
org/
IGRAC Global Groundwater Information System. https://
www.un-igrac.org/global-groundwater-information-
system-ggis/
World-wide Hydrogeological Mapping and Assessment
Programme (WHYMAP). https://www.whymap.org/
whymap/EN/About/about_node_en.html

References

Ascott MJ, Gooddy DC, Wang L, Stuart ME, Lewis MA,
Ward RS, Binley AM (2017) Global patterns of nitrate
storage in the vadose zone. Nature Commun 8(1):1416
Bartram J, Cairncross S (2010) Hygiene, sanitation, and
water: forgotten foundations of health. Plos Med 7
(11):e1000367. https://doi.org/10.1371/journal.pmed.
1000367
BGR and UNESCO (2010) Groundwater Resources of the
World (WHYMAP GWR)
BGS and WaterAid (2000) Water Quality Fact Sheet:
Fluoride
BGS and WaterAid (2004) Groundwater Quality: North-
ern India
British Geological Survey (2019) Current UK groundwa-
ter use. https://www.bgs.ac.uk/research/groundwater/
waterResources/GroundwaterInUK/2015.html.
Accessed July 2019
Chambers R (1997) Whose Reality Counts?, Practical
Action Publishing. https://www.developmentbook-
shelf.com/doi/abs/10.3362/9781780440453
Cobbing J, Hiller B (2019) Waking a sleeping giant:
realizing the potential of groundwater in Sub-Saharan
Africa. World Dev 122:597–613. https://doi.org/10.
1016/j.worlddev.2019.06.024
Dillon P, Stuyfzand P, Grischek T, Lluria M, Pyne RDG,
Jain RC, Bear J, Schwarz J, Wang W, Fernandez E,
Stefan C (2019) Sixty years of global progress in
managed aquifer recharge. Hydrogeol J 27(1):1–30

Döll P, Hoffmann-Dobrev H, Portmann FT, Siebert S, Eicker A, Rodell M, Strassberg G, Scanlon BR (2012) Impact of water withdrawals from groundwater and surface water on continental water storage variations. J Geodyn 59:143–156

Edmunds WM, Smedley PL (2013) Fluoride in fluoride in natural waters. Essentials of medical geology. Springer, Dordrecht, pp 311–336

Foster S, Vairavamoorthy K (2013) Urban groundwater—policies and institutions for integrated management. Global Water Partnership

FRACTAL and LuWSI (2018) Policy Brief, Lusaka. Groundwater Pollution: Key threat to water security and health, Lusaka, Zambia

Gleeson T, Wada Y, Bierkens MF, van Beek LP (2012) Water balance of global aquifers revealed by groundwater footprint. Nature 488(7410):197

Gleick PH (1996) Water resources. Encyclopedia of climate and weather. In: Schneider SH. New York, Oxford University Press. 2, pp 817–823

Grey D, Sadoff CW (2007) Sink or swim? Water security for growth and development. Water Policy 9(6):545–571. https://doi.org/10.2166/wp.2007.021

Guppy L, Uyttendaele P, Villholth KG, Smakhtin V (2018) Groundwater and Sustainable development goals: analysis of interlinkages. UNU-INWEH Report Series, Issue 04, Hamilton, Canada. https://www.nature.com/articles/nature11295#supplementary-information

Hunter PR, MacDonald AM, Carter RC (2010) Water supply and health. PLoS Med 7(11):e1000361

Hutton G, Haller L, Bartram J (2007) Global cost-benefit analysis of water supply and sanitation interventions. J Water Health 5(4):481–502

ICSU, International Council for Science (2017) A guide to SDG interactions: from science to implementation. I. C. f. Science, Paris

International Federation of Red Cross and Red Crescent Societies (2018) Zambia: Cholera Outbreak Lusaka - Emergency Plan of Action Final Report. https://reliefweb.int/sites/reliefweb.int/files/resources/MDRZM011dfr.pdf

IPCC (2014) Climate change 2014: synthesis report. contribution of working groups i, ii and iii to the fifth assessment report of the intergovernmental panel on climate change [Core Writing Team, R.L. Pachauri and L.A. Meyer (eds.)]. IPCC, Geneva, Switzerland

Joint Monitoring Programme (JMP) (2019a) Estimates on the use of water, sanitation and hygiene by country (2000–2017); Countries, areas and territories; Updated July 2019, United Nations Children's Fund (UNICEF) and World Health Organisation (WHO). https://www.washdata.org

Joint Monitoring Programme (JMP) (2019c) Progress on household drinking water, sanitation and hygiene 2000–2017. Special focus on inequalities. United Nations Children's Fund (UNICEF) and World Health Organisation (WHO), New York

Kolusu SR, Shamsudduha M, Todd MC, Taylor RG, Seddon D, Kashaigili JJ, Ebrahim GY, Cuthbert MO,

Sorensen JP, Villholth KG, MacDonald AM (2019) The El Nino event of 2015–2016: climate anomalies and their impact on groundwater resources in East and Southern Africa. Hydrol Earth Syst Sci 23(3):1751–1762

Lapworth DJ, Nkhuwa DCW, Okotto-Okotto J, Pedley S, Stuart ME, Tijani MN, Wright J (2017) Urban groundwater quality in sub-Saharan Africa: current status and implications for water security and public health. Hydrogeol J 25(4):1093–1116

MacDonald A, Davies J, Calow R, Chilton J (2005) Developing groundwater: a guide for rural water supply. ITDG publishing

MacDonald AM, Bell RA, Kebede S, Azagegn T, Yehualaeshet T, Pichon F, Young M, McKenzie AA, Lapworth DJ, Black E, Calow RC (2019) Groundwater and resilience to drought in the Ethiopian Highlands. Environ Res Lett 14(9):095003

Met Office What are El Niño and La Niña? (2019). https://www.metoffice.gov.uk/weather/learn-about/weather/oceans/el-nino. Accessed 29 Oct 2019

Morris BL, Lawrence AR, Chilton PJC, Adams B, Calow RC, Klinck BA (2003) Groundwater and its susceptibility to degradation: a global assessment of the problem and options for management (Vol. 3). United Nations Environment Programme

Nkhuwa DCW (2006) Groundwater quality assessments in the John Laing and Misisi areas of Lusaka. In: Xu Y, Usher B (eds) Groundwater pollution in Africa. Taylor and Francis/Balkema, Netherlands, pp 239–251

Nkhuwa DCW et al (2015) Groundwater resource management in the st. Bonaventure township, lusaka, Delft

Nkhuwa DCW et al (2018) Africa Groundwater Atlas: Hydrogeology of Zambia. August 2019. https://earthwise.bgs.ac.uk/index.php/Hydrogeology_of_Zambia

Philip S, Kew SF, Jan van Oldenborgh G, Otto F, O'Keefe S, Haustein K, King A, Zegeye A, Eshetu Z, Hailemariam K, Singh R (2018) Attribution analysis of the Ethiopian drought of 2015. J Clim 31(6):2465–2486

Podgorski JE, Labhasetwar P, Saha D, Berg M (2018) Prediction modeling and mapping of groundwater fluoride contamination throughout India. Environ Sci Technol 52(17):9889–9898

Quevauviller P (2007) European Union Groundwater Policy. An International Overview. P. Quevauviller, Royal Society of Chemistry, Groundwater Science and Policy

ReliefWeb (2019) Humanitarian Crises in Southern and Eastern Africa. https://reliefweb.int/topics/humanitarian-crises-southern-and-eastern-africa

Ritchie H, Roser R (2019a) Clean water access. https://ourworldindata.org/water-access. Accessed 27 Oct 2019

Ritchie H, Roser R (2019b) Sanitation Access. https://ourworldindata.org/sanitation. Accessed 27 Oct 2019

Siderius C, Gannon KE, Ndiyoi M, Opere A, Batisani N, Olago D, Pardoe J, Conway D (2018) Hydrological response and complex impact pathways of the 2015/2016 El Niño in Eastern and Southern Africa. Earth's Future 6(1):2–22

Smith M, Cross K, Paden M, Laban P (2016) Spring—Managing groundwater sustainably. IUCN, Gland Switzerland

Stefan C, Ansems N (2018) Web-based global inventory of managed aquifer recharge applications. Sustain Water Resour Manag 4(2):153–162. https://doi.org/10.1007/s40899-017-0212-6

Stuart ME, Manamsa K, Talbot JC, Crane EJ (2011) Emerging contaminants in groundwater. BGS Open Report (OR/11/013)

United Nations (2015) The millennium development goals report. United Nations, New York

United Nations (2018) Sustainable development goal 6 synthesis report on water and sanitation. United Nations, New York

United Nations Department of Economic and Social Affairs Population Division (2018) 2018 Revision of World Urbanisation Prospects, New York

United Nations Department of Economic and Social Affairs Population Division (2019) World Population Prospects 2019. United Nations, Online Edition

United States Geological Survey (2019) Groundwater Use. https://www.usgs.gov/mission-areas/water-resources/science/groundwater-use. Accessed July 2019

UN-Water (2016) Water and sanitation interlinkages across the 2030 Agenda for Sustainable Development, Geneva

UPGro (2017) Groundwater and poverty in sub-Saharan Africa. UPGro Working Paper, St. Gallen

Villholth KG, Lopez-Gunn E, Conti K, Garrido A, Van Der Gun J (Eds) (2017) Advances in groundwater governance. CRC Press

Wang H, Naghavi M, Allen C, Barber RM, Bhutta ZA, Carter A, Casey DC, Charlson FJ, Chen AZ, Coates MM, Coggeshall M (2016) Global, regional, and national life expectancy, all-cause mortality, and cause-specific mortality for 249 causes of death, 1980–2015: a systematic analysis for the Global Burden of Disease Study 2015. The Lancet 388 (10053):1459–1544

WHO (2019) Water-related diseases: Methaemoglobinemia. https://www.who.int/water_sanitation_health/diseases-risks/diseases/methaemoglob/en/

Wood SL, Jones SK, Johnson JA, Brauman KA, Chaplin-Kramer R, Fremier A, Girvetz E, Gordon LJ, Kappel CV, Mandle L, Mulligan M (2018) Distilling the role of ecosystem services in the Sustainable Development Goals. Ecosys Serv 29:70–82

World Bank (2010) Deep wells and prudence: towards pragmatic action for addressing groundwater overexploitation in India. The International Bank for Reconstruction and Development/The World Bank, Washington, USA

World Health Organisation (WHO) (2017) Guidelines for drinking-water quality: fourth edition incorporating the first addendum. W. H. O. (WHO), Geneva

Kirsty Upton is a hydrogeologist with the British Geological Survey, focussing on applied and interdisciplinary groundwater research in the UK and sub-Saharan Africa. Her research focuses on methods for groundwater resource assessment and understanding groundwater availability during droughts. She is a lead author of the Africa Groundwater Atlas and has worked with a variety of stakeholders, both in the UK and Africa, to understand ways in which groundwater research can be translated to inform policy and practice.

Alan MacDonald leads international groundwater research at BGS and is an honorary professor at the University of Dundee. He started work as a hydrogeologist in southern England working mainly on the chalk aquifer before spending 3 years in rural Nigeria with WaterAid. For the last 20 years he has focused on water security and supply issues in Africa and South Asia, publishing 3 books and > 100 papers. He recently led two major studies for the UK Department for International Development in Africa and South Asia, leading to new understanding about the resilience of water supplies to changing climate and abstraction.

Affordable and Clean Energy

Energy Geoscience and Human Capacity

Michael H. Stephenson

M. H. Stephenson (✉)
British Geological Survey, Environmental Science
Centre, Nicker Hill, Keyworth, Nottingham NG12
5GG, UK
e-mail: mhste@bgs.ac.uk

© Springer Nature Switzerland AG 2021
J. C. Gill and M. Smith (eds.), *Geosciences and the Sustainable Development Goals*,
Sustainable Development Goals Series, https://doi.org/10.1007/978-3-030-38815-7_7

M. H. Stephenson

Abstract

7 AFFORDABLE AND CLEAN ENERGY

Overview

Distribution of fossil fuels greatly influences the economic development and energy mix in the global South

There is a conflict between energy production contributing to greenhouse gas emissions and supporting sustainable food and water supplies for economic development

Many developing countries lack power grid infrastructure and regulation to develop their energy potential

Governance and integrated approaches are essential to transitioning to clean energy resources

Current status

Energy demand is projected to grow to 2040, with 30% of the growth in the developing world

There is likely to be a continuing reliance on fossil fuels, including coal

There is a significant opportunity in Africa for growth in renewable energy resources (e.g., solar, geothermal, hydropower)

Indigenous knowledge and research can help to understand national resources and capabilities, underpinning regulation and environmental safeguards

Role of geoscience: to provide strategic knowledge that helps decarbonise

Identification and monitoring of suitable sites for Capture and Storage of CO_2

Resource mapping of geothermal potential

Provision of geological solutions to the long term storage of nuclear waste.

Characterisation of underground storage in caverns for hydrogen and compressed air

Utilisation of underground aquifers and geothermal gradient to provide cooling and heat

Rigorous science to underpin regulation and support public engagement and understanding

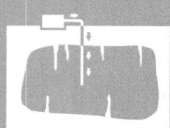  

7.1 Introduction

Before the industrial revolution, global energy demand was limited and supplied essentially by traditional renewable sources. The evolution of simple steam engines accelerated in the seventeenth and eighteenth centuries and improvements by Thomas Newcomen and James Watt in the mid-1700s produced the modern steam engine powered by coal, providing energy for locomotives, factories, and farm implements. Coal was also used for heating buildings and smelting iron into steel. In 1880, coal powered a steam engine attached to the world's first electric generator leading to the development of thermal power stations which still provide most of the world's electricity and in the late 1800s, petroleum began to be processed into gasoline (petrol) for firing internal combustion engines.

With the advent of cheap cars in the early 1900s, and the spread of electricity, energy demand increased and by the 1950s, nuclear power joined coal and petroleum to help satisfy that demand. However, geopolitical concerns over petroleum affected the pattern of energy supply in the latter part of the 1900s, as well as concerns over the safety of nuclear electricity generation (Fig. 7.1). In addition, increased evidence of anthropogenic climate change (IPCC Fifth Assessment Report 2014), recognised that the largest human influence has been the emission of greenhouse gases such as carbon dioxide, mostly related to the burning of energy fuels in transport and electricity production. For a detailed discussion of climate change, see **SDG 13**.

Despite the huge rise in demand and supply of energy for electricity and transport, its global distribution remains very uneven. In 2016, sub-

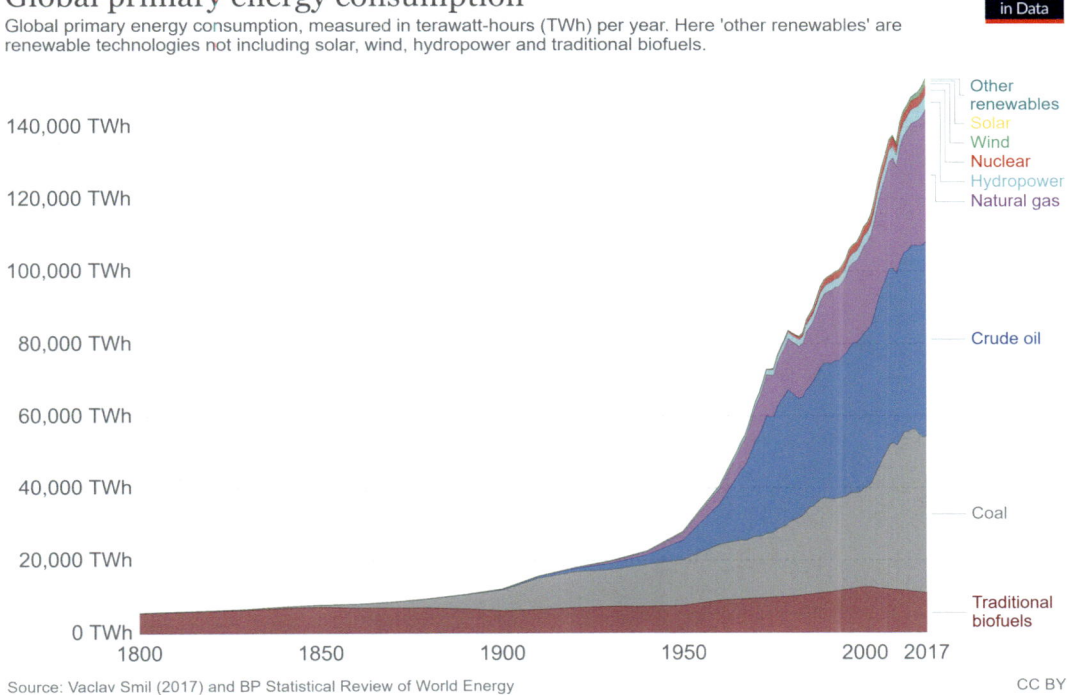

Fig. 7.1 Global Primary energy Consumption (measured in terawatt-hours, TWh). *Credit* Ritchie and Roser (2019a). Reproduced under a CC BY License (https://creativecommons.org/licenses/by/4.0/)

Saharan Africa and South Asia had approximately 600 million and 200 million people, respectively, with no access to electricity (Ritchie and Roser 2019a), while in the Global North this figure was negligible. Approximately 800 million Indians and 600 million sub-Saharan Africans, use traditional biomass as their primary cooking fuel (Kaygusuz 2012). Thus, a central industrial and social challenge of the twenty-first Century is to satisfy growing energy demand while reducing emissions related to energy production, but also to ensure that energy is available to all (Fig. 7.2).

The Sustainable Development Goals (SDGs) are a collection of 17 global goals set by the United Nations General Assembly in 2015, for the year 2030. **SDG 7: 'Ensure access to affordable, reliable, sustainable, and modern energy'** aims at improving energy access, increasing renewables in the energy mix, energy efficiency and integration, and international cooperation, and has targets to 2030, and indicators of progress. Many of the targets are closely associated with geoscience, for example, in exploration and feasibility studies for subsurface renewables such as geothermal, as well as sustainable use of fossil fuels within strict carbon budgets (Table 7.1).

Energy in its broadest sense enables business, industry, agriculture, transport, communications, and modern services such as health care; but it also enables improvements in living standards. **SDG 7** is, therefore,intimately connected with most of the other 17 SDGs (Fig. 7.3), mainly through providing improved living standards, economic growth and activity, and improved environmental protection. The services that energy provides improve human, social, economic, and environmental conditions; and final energy use and the Human Development Index (HDI) are correlated (Steckel et al. 2013), the correlation implying early rapid gains in HDI

Fig. 7.2 Wind turbines in rural India. India is one of the largest producers of renewable energy, currently accounting for approximately 35% of energy production *Credit* Vestas (CC BY 2.0, https://creativecommons.org/licenses/by/2.0/)

Table 7.1 SDG 7 Targets by 2030

Target	Description of Target (7.1 to 7.3) or Means of Implementation (7.A to 7.B)
7.1	By 2030, ensure universal access to affordable, reliable, and modern energy services
7.2	By 2030, increase substantially the share of renewable energy in the global energy mix
7.3	By 2030, double the global rate of improvement in energy efficiency
7.A	By 2030, enhance international cooperation to facilitate access to clean energy research and technology, including renewable energy, energy efficiency, and advanced and cleaner fossil fuel technology, and promote investment in energy infrastructure and clean energy technology planning
7.B	By 2030, expand infrastructure and upgrade technology for supplying modern and sustainable energy services for all in developing countries, in particular, least developed countries, small island developing States, and land-locked developing countries, in accordance with their respective programmes of support

Fig. 7.3 Relationship of SDG 7 to other SDGs

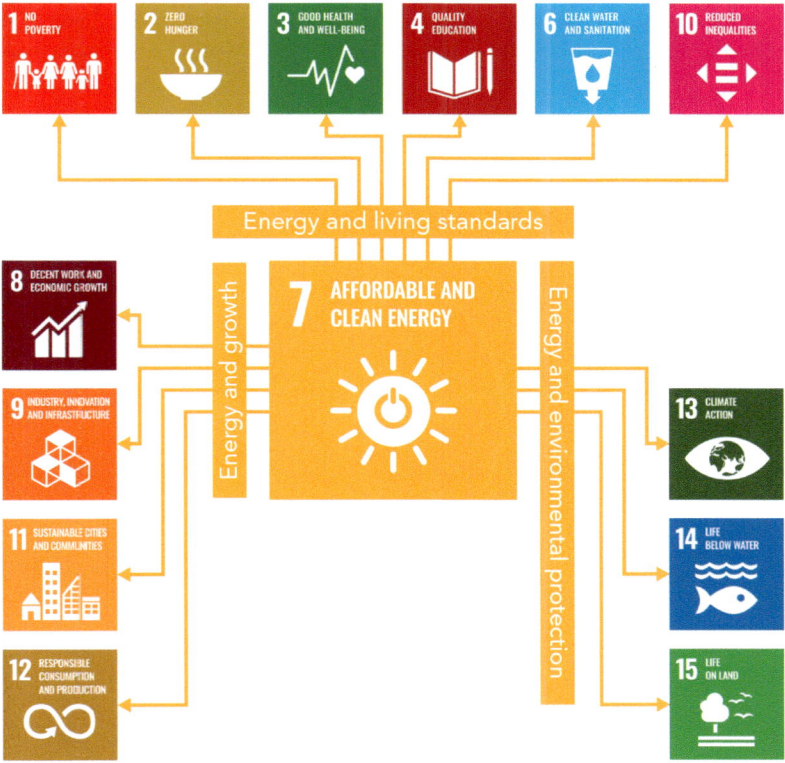

with relatively small gains in energy usage, with HDI levelling off at levels of energy usage around 75 GJ/yr per capita.

On the negative side, energy (for example, fossil fuel power and hydropower), can be produced and deployed in ways that pollute the environment, affect land use, and increase greenhouse gas emissions. Similarly, energy is an element of the food-energy-water nexus and thus its sustainability is tensioned against water (**SDG 6**) and food (**SDG 2**). Finally, the financial value that energy can release, can also be syphoned into the ruling elites of kleptocracies and autocracies, rather than be cascaded down to benefit society at large. Thus, the benefits of energy for sustainable development are strongly dependent on ethical governance and strong institutions (**SDG 16**).

In general, the effectiveness of energy systems to supply sustainable development depends on a number of factors (illustrated in Fig. 7.4). These include

- Availability, affordability, security, reliability, and safety of energy supplies
- Environmental sustainability of the energy supply
- Planning, design, construction, operation, financing and pricing of energy-using buildings, industrial processes, and transport systems in end-use sectors
- Social and cultural norms of the use of energy

- Access to alternative technologies and energy sources
- Investment assistance to develop and deploy energy service
- Government policies that ensure energy systems develop in a way that best supports and accords with sustainable development.

Geoscience has a direct role in several of these areas including in establishing the geographical distribution, geological habitat, geotechnical feasibility of construction and infrastructure, and environmental sustainability, of energy supply (Table 7.2). Though it discusses affordable and clean energy, which is a requirement across the world to a greater and lesser extent, this chapter

Fig. 7.4 Energy for sustainable development. Reprinted from Renewable and Sustainable Energy Reviews, 16 (2), Kaygusuz, K., Energy for sustainable development: A case of developing countries, 1116–1126, Copyright (2012), with permission from Elsevier

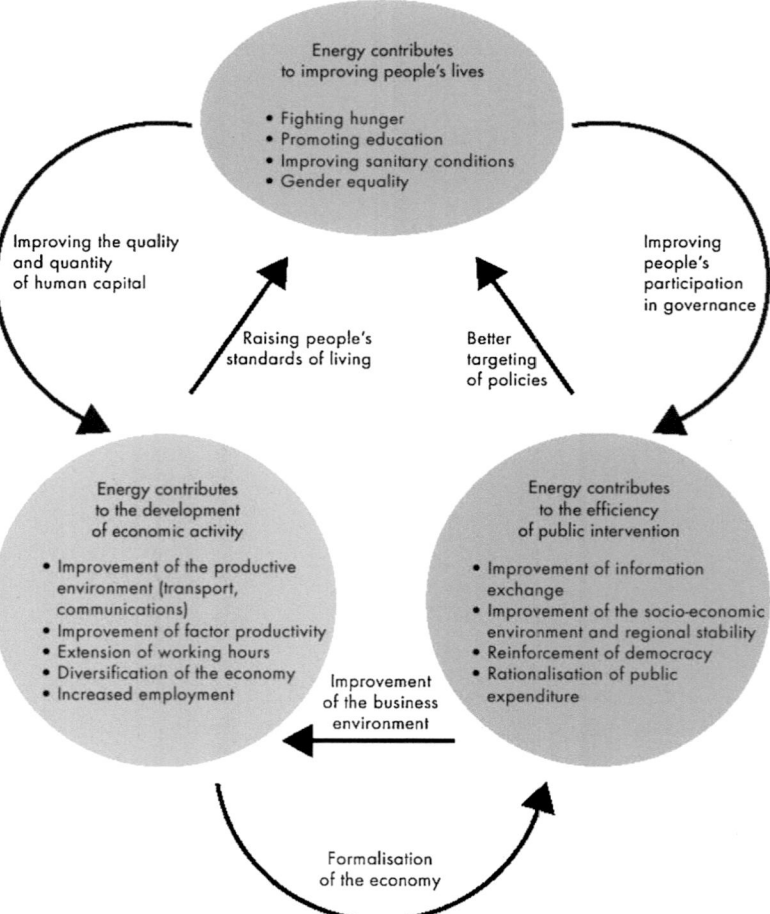

Table 7.2 SDG 7 Indicators and geoscience relevance

Indicators	Geoscience link
7.1.1 Proportion of population with access to electricity	Geoscience for exploration and sustainability of renewable and appropriately used fossil energy sources
7.1.2 Proportion of population with primary reliance on clean fuels and technology	
7.2.1 Renewable energy share in the total final energy consumption	Geoscience to support the expansion of renewables, e.g., geothermal, wind turbine ground conditions
7.3.1 Energy intensity measured in terms of primary energy and GDP	Holistic planning involving the subsurface
7.A.1 Mobilised amount of United States dollars per year starting in 2020, accountable towards the $100 billion commitment	Improved links between geoscientists/geoscience institutions and other energy specialists
7.B.1 Investments in energy efficiency as a percentage of GDP and the amount of foreign direct investment in financial transfer for infrastructure and technology to sustainable development services	Improved links between geoscientists/geoscience institutions and energy system specialists including energy distribution specialists and finance sector

concentrates on affordable and clean energy in the Global South (so-called, developing countries), which has the most to develop in energy and the most to gain economically. It also has the most difficult challenges. In this chapter, Sect. 7.2 will examine the distribution of present-day energy resources and forecasts for future supply and demand, and Sect. 7.3 the geoscience implications of the main options for affordable and clean energy, including research and development needed, as well as training needs.

7.2 Energy Resource Distribution and Use

7.2.1 Fossil Fuels in the Global South

The distribution of fossil fuels has a bearing on the way that nations develop and the energy mix that they develop, as well as on the human capacity needed to develop and maintain sustainable supplies. Amongst oil (Fig. 7.5), proven reserves are concentrated in well-explored parts of the Middle East, North America, Africa, Northern Asia and South America. The occurrence of significant resources of unconventional oil (from low permeability reservoirs) is notable in North America, Eastern Europe and Eurasia,

and South America. Gas has a similar pattern (Fig. 7.6). North America, Eastern Europe and Eurasia, South Asia, and Asia-Pacific all have large coal reserves (Fig. 7.7). Of relevance to this chapter is that several of the largest developing countries: India, Indonesia, and South Africa, have very large coal resources.

The International Energy Agency (IEA), forecasts using three 'scenarios' which contain predictions of energy infrastructure investment and energy demand and supply, based primarily on the need to reduce CO_2 emissions and on common assumptions of economic conditions and population growth. The most relevant is the New Policies Scenario which takes account of broad policy commitments and plans that have been announced by countries and their governments, including national pledges to reduce greenhouse gas emissions and plans to phase out fossil energy subsidies, even if the measures to implement these commitments have yet to be identified or announced. This might be regarded as the most realistic and widely quoted of the IEA's scenarios.

The IEA's 2016 World Energy Outlook New Policies Scenario (IEA 2016), predicts an increase in energy demand between now and 2040. 30% of this increase will be from the Global South, particularly in Asia and Africa.

Oil Proved Reserves, 2015

Total proved oil reserves, measured in barrels. Proved reserves is generally taken to be those quantities that geological and engineering information indicates with reasonable certainty can be recovered in the future from known reservoirs under existing economic and operating conditions.

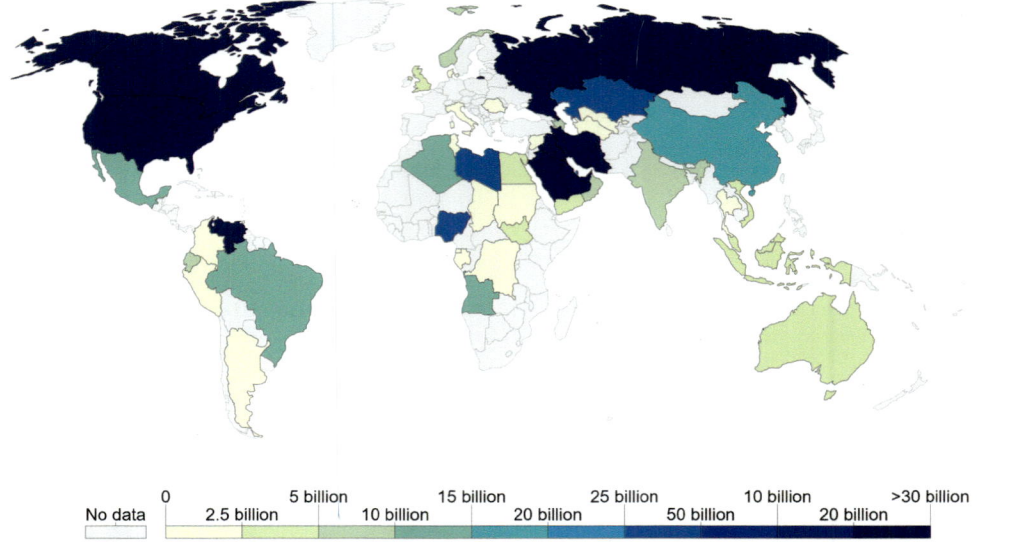

Source: BP Statistical Review

OurWorldInData.org/fossil-fuels/ • CC BY

Fig. 7.5 Oil Proven Reserves (as of 2015), measured in barrels *Credit* Ritchie and Roser (2019b), using data from BP Statistical Review of World Energy 2016. Reproduced under a CC BY License (https://creativecommons.org/licenses/by/4.0/)

Natural Gas Proved Reserves, 2015

Total proved gas reserves, measured in trillion cubic metres. Proved reserves is generally taken to be those quantities that geological and engineering information indicates with reasonable certainty can be recovered in the future from known reservoirs under existing economic and operating conditions.

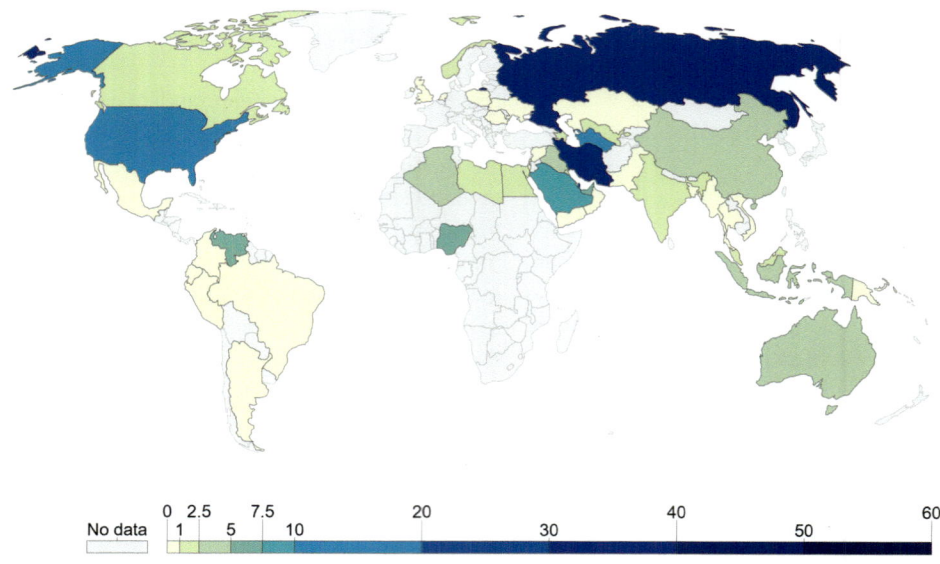

Source: BP Statistical Review

OurWorldInData.org/fossil-fuels/ • CC BY

Fig. 7.6 Natural Gas Proven Reserves (as of 2015), measured in trillion cubic metres. *Credit* Ritchie and Roser (2019b), using data from BP Statistical Review of World Energy 2016. Reproduced under a CC BY License (https://creativecommons.org/licenses/by/4.0/)

Coal Proved Reserves, 2015

Total proved coal reserves, measured in tonnes. Proved reserves is generally taken to be those quantities that geological and engineering information indicates with reasonable certainty can be recovered in the future from known reservoirs under existing economic and operating conditions.

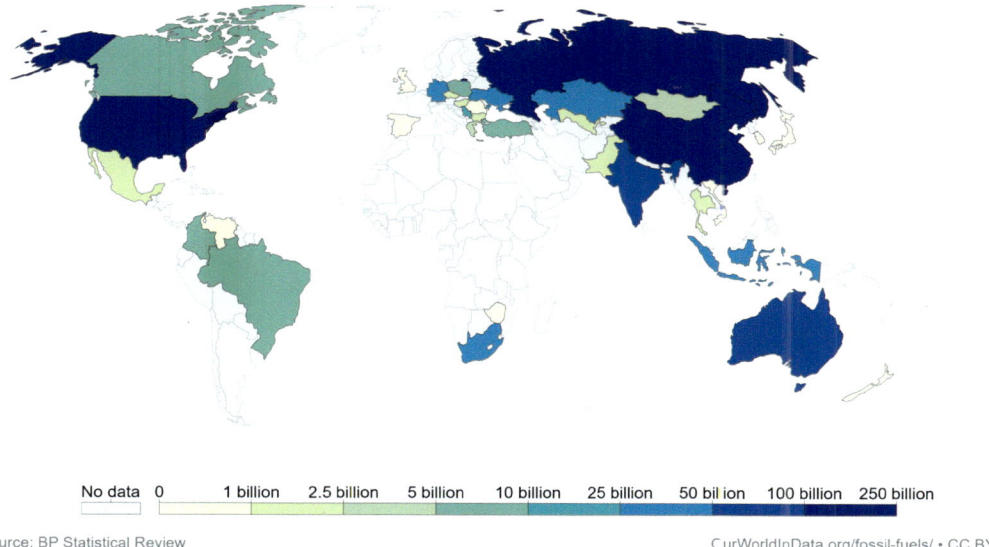

| No data | 0 | 1 billion | 2.5 billion | 5 billion | 10 billion | 25 billion | 50 billion | 100 billion | 250 billion |

Source: BP Statistical Review   OurWorldInData.org/fossil-fuels/ • CC BY

Fig. 7.7 Coal Proven Reserves (as of 2015), measured in tonnes. *Credit* Ritchie and Roser (2019b), using data from BP Statistical Review of World Energy 2016. Reproduced under a CC BY License (https://creativecommons.org/licenses/by/4.0/)

According to the IEA, breaking this down into types of energy including the three major fossil fuels shows little change between now and 2040. Important factors to note include the marked increase in renewables, but also that all three of the main fossil fuels do not decline but in fact increase. Focussing on coal alone in the IEA's New Policies Scenario suggests a shift from the developed to the developing world.

The drive for development, poverty alleviation (**SDG 1**), and improved health (**SDG 3**) are all connected with greater energy requirements. Countries in the Global South that contain significant fossil fuel resources will be able, through conventional forms of commercialisation, to realise those resources for electrical power and other energy requirements (e.g., transport, heat or air conditioning). Coal has particular appeal to countries with large reserves and problems of rural development, poverty alleviation, and health.

Box 7.1. Coal in India and South Africa

India has very large coal resources and problems of rural development, poverty alleviation, and health. Much domestic energy in India, comes from biomass (firewood, crop residue, dung) and India consumes 200 Mtoe (million tonnes of oil equivalent) of biomass each year. One hundred million Indian households still use firewood to cook food, mainly in rural areas (Kaygusuz 2012). Cooking with firewood takes its toll on the health of Indians with an estimated 50,000 deaths per year (household fires, accidents, and ill health). India's rural electrification programme aims to introduce healthier fuel in households, to improve agricultural production (for example, through better irrigation pumps), and to develop business and trade in agriculture. Significant inroads into rural electrification have been made with small scale solar, particularly to provide domestic lighting in rural areas (e.g., Kamalapur and Udaykumar 2011).

However, the forecasts of the IEA (2016), suggest that at least some of India's future electricity supply will come from coal (Fig. 7.8). At present coal provides about 70% of India's electricity but about 243 GW of coal-fired power is planned in India, with 65 GW actually being constructed and an extra 178 GW proposed. Shearer et al. (2017) surveyed this proposed 'fleet' of coal power stations to forecast the amounts of power that could be provided should these power stations be completed. Their survey shows average annual capacity additions beginning in 1960, as well as future additions based on proposed new plants. For the future, Shearer's survey shows that coal plants under development could be producing

435 GW of coal power by 2025, and assuming an average lifetime of 40 years, the coal plants could be operating as far ahead as 2065. South Africa, like India, has a large number of rural people without access to electricity (roughly 60% of South African households), but also a strong demand for electricity, particularly for the mining industry. South Africa's coal reserves are large—28 billion tonnes—which would allow 100 more years of mining at current rates. According to the IEA New Policies Scenario (IEA 2016), South African coal production will be driven mainly by domestic demand for coal for power supply. At present, more than 90% of electricity is generated by coal in South Africa, and this will remain the case well into the next decades. Coal production is predicted to rise to a peak of around 230 Mtce by 2020, and then fall to 210 Mtce by 2035 (IEA 2011).

Fig. 7.8 Coal mining in India. India's coal plants could be operating as far ahead as 2065. *Credit* Nitin Kirloskar (CC BY 2.0, https://creativecommons.org/licenses/by/2.0/)

Gas and oil resources in the developing world are less certain than those of coal because there have been fewer systematic surveys in these areas. Having said this, there are well-established producing areas in Africa and southeast Asia, for example, Nigeria, Angola, Gabon, and Sarawak. The IEA New Policies Scenario (IEA 2016), forecasts gas and oil demand to grow to 2040 in Africa and India, and a new gas province to emerge offshore Mozambique and Tanzania. Egypt's gas production is expected to grow. The potential for shale gas in Africa and Asia is not known in detail but South Africa, India, Indonesia, and Pakistan are believed to have significant resources (EIA 2013). The continued extraction and use of gas and oil, like coal, depends on the adherence to climate policies and more locally to increasing standards of environmental assurance.

7.2.2 Renewable Energy Resources in the Developing World

Solar energy uses concentrated solar power (CSP) systems and photovoltaic (PV) systems. Global horizontal irradiation data (Shahsavari and Akbari 2018), indicate that much of the developing world is suited to solar power (Fig. 7.9). Global wind potential modelled with wind climate data with high-resolution terrain information shows high potential in coastal areas and high latitude areas with rather lower potential in the tropics and subtropics. Wind and solar development require site-specific information to aid investment decisions though suffer the same need for site-specific information to aid investment decisions (Gies 2016).

Miketa and Saadi (2015) and the Africa Progress Panel (2015), note the challenges to realise solar and wind as bankable technologies. The locations of wind and solar resources in Africa are not known in enough detail at present to stimulate private investment by companies hoping to select sites for projects. Another problem is that Africa and the developing world lack big electricity grids and transmission lines to move large amounts of power within countries and across regions (Gies 2016).

Hydropower plants are highly site-specific, but can be broadly categorised into three. Storage hydropower uses a dam to impound river water, which is then stored for release when needed. Storage hydropower can be operated to provide base-load power, as well as peak-load through its

Fig. 7.9 Solar Power Plant Telangana II in the state of Telangana, India. *Credit* Thomas Lloyd Group (CC BY SA 2.0, https://creativecommons.org/licenses/by-sa/2.0/)

ability to be shut down and started up at short notice according to the demands of the system. It can offer enough storage capacity to operate independently of the hydrological inflow for many weeks, or even up to months or years.

Run-of-river hydropower channels flow water from a river through a canal or penstock to drive a turbine. Typically, a run-of-river project will have short term water storage and result in little or no land inundation relative to its natural state. Run-of-river hydro plants provide a continuous supply of electricity and are generally installed to provide base load power to the electrical grid. Pumped-storage hydropower provides peak load supply, harnessing water which is cycled between a lower and upper reservoir by pumps, which use surplus energy from the system at times of low demand. When electricity demand is high, water is released back to the lower reservoir through turbines to produce electricity, and thus a zero-sum electricity producer.

Africa has abundant hydropower resources. It is estimated that around 92% of technically feasible potential has not yet been developed. Central Africa has about 40% of the continent's hydro resources. At the end of 2014, there was 28 GW of hydro capacity installed in Africa (IRENA 2015). Of the resources available, the Congo River has the largest discharge of African rivers, followed by the Zambezi, the Niger, and the Nile.

India's economically exploitable and viable hydroelectric potential is estimated to be 148,701 MW (Govt. of India 2018), but south Asia hydropower is cross-border in nature due to the size of catchments and so its development involves geopolitical factors (Box 7.2).

Box 7.2. Indus hydropower cross-border issues.

The large discharges that are needed for hydropower are sustained best by very large catchments which often span several countries so that the building of hydropower dams have effects on downstream countries. The Indus River is an example, being one of the longest rivers in Asia. It originates in the Tibetan Plateau and flows through Ladakh, India, towards the Gilgit-Baltistan region of Pakistan and the Hindukush ranges, and south through Pakistan to the Arabian Sea near Karachi. The river's catchment is more than $1,165,000$ km^2 and its annual flow of 243 km^3 is one of the largest in the world. The river and its catchment spans four countries and supports 215 m people. India and Pakistan, the two main countries in the basin, divided up rights to the various tributaries under the Indus Water Treaty (IWT) of 1960.

India has approved major dams on the Chenab, and Jhelum rivers, and the Indus itself. However, the Nimoo Bazgo hydro plant, situated at Alchi village (Fig. 7.10), is under dispute. Pakistan says that the 57-metre high 45 MW Nimoo Bazgo dam will substantially reduce downstream water flows in the Indus River, because the project is designed to store 120 million cubic metres of water. This Pakistan says will allow India to regulate the water of Indus, a situation which is "not acceptable to Pakistan." Because the IWT treaty does not provide a definitive solution, the two countries have been in dispute. Downstream in the Punjab, India and Pakistan share the alluvial Indo-Gangetic aquifer (recharged partly by the Indus River) which helps support the huge population of the Indus region, accounting for 48% of all water withdrawals in the basin.

Geothermal is an important renewable source of energy with a strong geoscience aspect. It can be divided into two types: heat that is sufficient to generate electricity, and heat that is sufficient only for supplementing heating systems in buildings or for industrial processes. By 2017, about 13.4 GW of geothermal electricity was being produced from power stations globally; but a much larger amount of power, about 28 GW, is provided for direct heating of houses and public buildings, spas, industrial processes, desalination, and glass

Fig. 7.10 Nimoo Bazgo Power Project, situated at Alchi village, on the Indus River in Ladakh. *Credit* Mehrajmir13 (CC BY SA 4.0, https://creativecommons.org/licenses/by-sa/2.0/)

houses (Dickson and Fanelli 2003). Conventional electric power production is commonly limited to fluid temperatures above 180 °C, but with binary fluid technology, lower temperatures can also be used to generate electricity down to about 70 °C. For direct district heating, useful temperatures range from 80 °C to just a few degrees above the ambient temperature. At least 90 countries have potential geothermal resources though only about 70 tap this potential. Electricity is produced from geothermal energy in only 24 countries (Dickson and Fanelli 2003).

In the developing world, geothermal potential is considered high in East Africa, the Philippines, and Indonesia. In the East African rift valley, geothermal potential is considered greater than 20,000 MW of electricity, though currently, only Kenya has operational geothermal power stations (Omenda 2018). Slow progress in East Africa relates to high start-up costs (including costs of

drilling), inability to secure finance and lack of trained human capacity (Omenda 2018).

Tidal range resource potential varies considerably across the globe and is amplified by basin resonances and coastline bathymetry. Tidal energy has a relatively high cost but limited availability of sites with sufficiently high tidal ranges or flow velocities, but technological developments and improvements, both in design (e.g., dynamic tidal power, tidal lagoons) and turbine technology (e.g., new axial turbines, cross-flow turbines), extend the suitable locations and bring down costs. Historically, 'tide mills' have been used in Europe and on the Atlantic coast of North America. Tidal is not well developed in developing countries, though India is reported to have tidal energy potential of around 8,000 MW (Energy World 2018). **SDG 14** discusses Ocean Thermal Energy Conversion Technologies.

7.2.3 Urbanisation and Climate Change

Cities presently cover only approximately 3% of the land surface but they account for >70% of energy consumption and 75% of carbon emissions and are, therefore, major contributors to climate change (see **SDG 11**). Most future urbanisation will take place in the developing world, having strong effects on supply and generation of energy.

A recent study in China (Zhao and Zhang 2018), showed that for every 1% increase in the urban population relative to the total population, national energy consumption rose 1.4% and also that urbanisation increased energy consumption through urban spatial expansion, urban motorisation, and increase in energy-intensive lifestyles. Urban households consume 50% more energy than rural households per capita, which indicates that continued urbanisation in China, will increase national energy consumption. This is likely across the developing world. Urban policies are required, therefore, to encourage compact urban growth, green buildings, and low energy vehicles.

Smart grids will likely also improve energy response to urbanisation. These are electricity transmission networks that use digital technology to allow two-way communication between supplier and customers allowing the grid to respond digitally to quickly-changing electricity demand. This will also allow more efficient transmission of electricity, quicker restoration of electricity after power disturbances, lower power costs for consumers, less energy wastage, and better integration of large-scale renewable energy systems.

Climate change (see **SDG 13**) adds another level of complexity to urbanisation with implications for energy supply and usage including the requirement for more cooling/heating, and for power supply reliability.

7.2.4 Transport

Transport worldwide consumes about one-fifth of global primary energy and increasing transport demand is expected in the rapidly growing economies of the developing world because economic growth is strongly correlated with growth in transport volumes (Aßmann and Sieber 2005). After the energy sector, transport is the most important producer of carbon dioxide so it is likely that an increasing proportion of global emissions will come from transport. Transport is also in many cases harder to decarbonise than fixed infrastructure such as power stations or cement works.

Thus, a sustainable transport strategy has to take into account the growing transport demands in developing countries and reduce emissions at the same time. Technical solutions could include (1) more efficient conventional engines, better designed vehicles, improved inspection and maintenance, and fuel quality; (2) renewable fuels in transport, such as ethanol and biogas; (3) better transport demand management, land-use planning and fuel pricing; (4) lower carbon natural gas vehicles; and (5) electric vehicles.

Of these options, only electric vehicles has implications for geological science research or human capacity, mainly because an understanding of the primary resources used to make batteries is crucial. These include deposits of lithium, sodium, vanadium, copper, cobalt, and nickel. Estimates based on electric vehicles being 30% of the global vehicle fleet by 2030 suggest that an extra 2 million tonnes of copper, 1.2 million tonnes of nickel, and 260,000 tonnes of cobalt will have to be mined per year into the future (Financial Times 2017). These are considerable increases on present production levels and suggest that more resources will have to be found, and that recycling of materials will have to be improved.

7.3 Geoscience Research for Affordable and Clean Energy in Developing Countries

7.3.1 Fossil Fuels

IEA forecasts (IEA 2016), indicate that fossil fuels will have a role to play in future energy systems in the developing world and these are likely to be used as feedstock for chemical industry and fertilisers, but also combusted in transport, heating and in electricity generation (with or without CCS, see below). As such, geoscience research into fossil fuels will contribute to **SDG 7.1** '*Proportion of population with access to electricity*'.

Coal, oil, and gas need similar geological research and knowledge for their exploration and extraction, for example, 3D seismic, resource and basin analysis, and structural geology. Coal, oil, and gas are also related in that research and knowledge are commonly provided by commercial, often multinational, companies following earlier pre-competitive surveys done by in-country geological survey organisations. However 'home-grown' knowledge and research, in exploration techniques, for example, are needed to be able to provide data on potential new resources for inward investment, to ensure a development pipeline of resources. Regulatory bodies also need the technical capability to maintain optimal environmental and sustainability safeguards, and to ensure that commercial negotiations over the development of resources are done on an equal footing to ensure equitable distribution of value between developer and government.

7.3.2 Electrification

If electrification is seen as a solution in primary energy in transport (electric vehicles) and heating and cooling in the developing world, then it is likely that demand for electricity will be very high and that the production of electricity will have to be fundamentally decarbonised. Geoscience research that enables the decarbonisation of electrification would contribute to **SDG 7.1**. In a scenario with high renewables or nuclear electricity, this would not be an issue, but for countries in the developing world with large coal or hydrocarbon reserves, for example, South Africa, Indonesia, and India, decarbonising electricity at source would require carbon capture and storage on fossil fuel power stations (see also **SDG 13**).

Geological storage of CO_2 relies on the ability to demonstrate that the storage operator can predict the future evolution of the CO_2 plume within known limits of certainty. Doing this requires robust and reliable observations of the site behaviour before, during, and after injection of CO_2 (Holloway 2007). Geological CO_2 storage must also lead to the permanent containment of the CO_2. Fundamental to the safety of achieving this reduction in atmospheric CO_2 emissions is the need to select and characterise geological sites that are expected to enable permanent containment. The largest stores are saline aquifers, which require the displacement of the in situ pore waters during CO_2 injection. The rate of injection and ultimately the mass of CO_2 that can be injected can be limited by the pressure increases that can occur during injection. Research is still needed to better understand the limits on pressure increases, improved methods for improving injectivity and managing pressure increases in saline aquifers. In contrast, depleted oil and gas fields that are subsequently used for CO_2 storage may require careful management during injection due to a number of processes that might limit injectivity, including Joule–Thomson cooling, well integrity, and seal integrity issues. Key challenges in CCS for the developed and developing world include de-risking the economic model for CCS and de-risking the full supply chain, as well as looking into public attitudes to CCS. In many cases, the solution to these problems will involve the development of industry-scale CO_2 storage pilots (Stephenson 2013; Holloway 2007).

Renewable energy that can feed directly into electricity production includes high enthalpy geothermal, solar and wind. Geological research and knowledge input into solar is focused on the raw materials needed for their production, and for wind consists mainly of the provision of geotechnical data for construction offshore and onshore. However, to support high enthalpy geothermal, geological studies will involve accurate resource mapping, as well as a detailed understanding of the fracture systems, geochemistry, hydrogeological systems, and thermal properties of the potential source rocks.

Geological studies on geothermal would, therefore, contribute to **SDG 7.1** and **7.2**.

In the case of a greatly increased need for low carbon nuclear electricity, geological considerations are very important. Long-term, safe management of highly radioactive waste is a significant challenge for countries with developed nuclear industries and will continue to be as nuclear energy plays a role in the future energy mix. Deep geological disposal is a key solution to managing waste for the long term, but it requires understanding and validation of complex subsurface processes and their interactions and feedbacks for up to one million years into the future taking into account seismicity and volcanism, as well as climate-related processes such as permafrost and ice loading and unloading (e.g., McEvoy et al. 2016).

The isolation of waste from the geosphere over long timescales requires fundamental knowledge of flow paths from the waste canister, through natural and induced discontinuities in the engineered barriers and surrounding host rock, to the surface environment. Geomechanics also play an important role in the long-term evolution of a repository and can strongly influence flow. Key science questions include the influence of stress state, burial history, and the generation and behaviour of faults and fractures. The long-term integrity of a repository and its surrounding geological and surface environment is central to developing safety arguments. Understanding near-field (geological characteristics, hydrogeological regime) and far-field (plate tectonics, climate) processes is required to build an integrated understanding of the evolution of the subsurface. Studies of deep geological disposal would, therefore, contribute to **SDG 7.1** and **7.2**.

7.3.3   The Hydrogen Economy

The hydrogen economy encompasses fuel for transport (road vehicles and shipping), stationary power generation (for heating and power in buildings), and an energy storage medium feeding from off-peak excess electricity. A system for hydrogen generation, salt cavern storage and electricity generation can begin with wind and solar energy. At times of excessive wind or solar electricity production, electrolysers can use this electricity to produce hydrogen and oxygen from water. The hydrogen is stored below the plant in a salt cavern. A gas combustion power plant using hydrogen alone or combined with natural gas can generate electricity. Excess renewable electricity can also be used to produce hydrogen from natural gas, through steam reforming (Ozarslan 2012).

An important geological aspect of the hydrogen economy is the need for the large-scale, long-term, and intermittent storage. The technology of compressed hydrogen gas storage in salt caverns is similar to that of natural gas, however, hydrogen energy density by volume is only one-third of that of natural gas, and so gaseous hydrogen energy storage is more expensive. For an integrated hydrogen economy, geological survey of salt beds including detailed facies mapping would be required mainly because salt cavern construction and performance are strongly impacted by salt heterogeneity. Studies are also required of the response to salt of repeated pressurisation cycles over long periods (Ozarslan 2012). Studies of geological hydrogen storage would contribute to **SDG 7.1** and **7.2**.

7.3.4   Energy Storage

With both electrification and hydrogen decarbonisation strategies, grid-scale energy storage will be needed, including compressed air energy storage (CAES). In CAES a storage pressure of about 70 bar is envisaged. Salt caverns are favoured because, being impermeable, there are no pressure losses, and because there is no reaction between the oxygen in the air and salt. Again studies of the facies variation and mechanical properties of the salt will be required (Evans et al. 2009). For the siting of other grid-scale storage options, for example, pumped hydro storage schemes, detailed geotechnical and seismic risk studies are required for dam building and deeper geological site characterisation for tunnels and underground installations. Studies of subsurface energy storage would contribute to **SDG 7.1** and **7.2**. Holistic planning involving the subsurface in relation to energy storage coupled with smart grid electricity distribution could contribute to **SDG 7.3**.

7.3.5 Ground Source Heat and Cold

Global energy demand from air conditioning is expected to triple by 2050, with climate change and developing country growth, requiring large new electricity capacity (IEA 2018). Air conditioning use is expected to be the second-largest source of global electricity demand growth after the industry sector, and the strongest driver for buildings by 2050 (IEA 2018). Although electricity is likely to power many air conditioners, a geothermal heat pump or ground source heat pump that transfers heat to or from the ground can also be used to cool and provide heat to buildings. This is achieved by using the shallow subsurface as a heat source (in winter) or a heat sink in the summer (contributing to **SDGs 7.1** and **7.2**). Although the use of ground source heat pumps is growing, common scientific and technical uncertainties that impede private investment include accurate estimates of the potential of the subsurface and rates of natural replenishment of extracted heat. In addition, the frequent lack of regulation of ground source heat and cold discourages investment.

7.3.6 Regulation and Compliance

Geological monitoring of the integrity and efficiency of subsurface energy installations will be important, as will mathematical concepts of risk and uncertainty. In areas such as disposal (CCS) and extraction (geothermal), geoscience and society (including engagement and communication with the public, Government, industry, and other stakeholders) will be important to secure a social licence.

In data science and infrastructure, the monitoring for compliance of subsurface energy installations will particularly demand more

capability in telemetry, data streaming techniques and visualisation, as well as a greater ability to store and manage very large amounts of data. To understand the change in data (for example, change points and anomalies in production or containment behaviour), and in order to forecast better, new statistical and artificial intelligence techniques will be needed. To manage subsurface energy installations a full suite of modelling techniques will be needed from conceptual static modelling to dynamic modelling, to forward modelling, to simulation.

It is worth noting that improved energy efficiency in buildings, industrial processes and transportation could reduce the world's energy needs in 2050 by one third (IEA 2016), however, at present geoscience and geological materials have a little direct role in energy efficiency, improvements in domestic appliances and building design being more important.

Box 7.3 Study of the effects of geothermal development on the Maasai community of the Kenyan Rift Valley

The study carried out by the Kenya Electricity Generating Company Ltd, Olkaria Geothermal Project (Mariita 2002) examined (1) the beneficial impacts of the project (e.g., employment, provision of water and infrastructure); and (2) the negative impacts of the project (e.g., displacement, noise, and pollution). On whether the geothermal project has had any impact on their lives, many respondents mentioned the positive benefits such as water, shops, and school. Most said that the noise or gas emissions did not discomfort them in any way, nor have any of their livestock been hurt by the project facilities. Negative comments included resentment due to resettlement, reduction in their land for grazing and Maasai cultural values being eroded by outsiders. Perhaps most telling, some respondents complained that they did not receive any of the energy generated at the site (Fig. 7.11).

Fig. 7.11 Olkaria Geothermal Project, Kenya © Chris Rochelle (used with permission)

7.3.7 Energy Governance

For many developing countries, rich resources have often paradoxically lead to low economic growth, environmental degradation, deepening poverty, and in some cases, violent conflict (Pegg 2006; Fischer-Kowalski et al. 2019). Primary resources such as oil and gas, apart from supplying energy for a nation's infrastructure, also constitute a source of revenue. For oil and gas revenue to contribute to a nation's wealth, particularly for its poorer citizens, depends on a number of factors including the manner in which resource income is spent, system of government, institutional quality and governance, type of resources, and stage of industrialization (Torvik 2009). Similarly, for the benefits of rural electrification or other affordable and clean energy to be made available to the wider population, good governance is needed. As an example, Van Alstine et al. (2014), indicated that within the emerging 'petro-state' of Uganda, four significant governance gaps might allow a lack of equitable development: (1) lack of coherence amongst civil society organisations; (2) limited civil society access to communities and the deliberate centralisation of oil governance; (3) industry-driven interaction at the local level; and (4) weak local government capacity.

Improvements can be made in public sector institutions like geological surveys, and government departments such as mines, energy, and water ministries through capacity building programmes. These programmes aim at understanding the business of the organisation within the context of government and regulation and can initiate training needs analysis to improve the qualifications and skills of staff (Box 7.4). Programmes can also advise on the functions of organisations and parts thereof. The ways that donor funds and projects are used can also be optimised so that development projects are not primarily organised to reflect donor agendas rather than the needs of the recipient institution (Stephenson and Penn 2005).

7.4 Geoscience Training for Affordable and Clean Energy in Developing Countries

An analysis of the training needed for modern energy geoscience is shown in Table 7.3. The table divides the geoscience energy disciplines into 5 major categories: geothermal and renewables, energy storage, radioactive waste disposal, CCS, and hydrocarbon systems. Each of these also has subtopics.

In general, the main skills needed include rock volume characterisation and process understanding in order to establish the geological feasibility of different solutions to energy, decarbonisation,

Table 7.3 Training needed for modern energy geoscience

Subtopic	Highly radioactive Waste Disposal		CCS		Energy Storage		Geothermal and Renewables		Hydrocarbon Systems	
	Description	Required training and level	Description	Required training and level	Description	Required training and level	Description	Required training and level	Description	Required training and Level
	Containment (fluid processes)	Geophysics, hydrogeology, geochemistry, computing, telemetry; BSc, MSc, PhD	Developing a plan for CO2 storage pilot	Geophysics, geomechanics hydrogeology, geochemistry, computing, economics, law; BSc, MSc, PhD	Thermal storage	Geophysics, hydrogeology, geochemistry, computing; BSc, MSc, PhD	Geothermal shallow	Geophysics, hydrogeology, geochemistry, computing; engineering geology; BSc, MSc, PhD	Conventional	Geophysics, structural geology, stratigraphy, biostratigraphy, organic geochemistry
	Siting — geological context	Geophysics, structural geology, stratigraphy; BSc, MSc, Ph.D.	Developing and maintaining technologies and methodologies	Computing; BSc, MSc, Ph.D.	Cavern storage	Geophysics, geomechanics, computing, telemetry; BSc, MSc	Geothermal deep	Geophysics, hydrogeology, geochemistry, computing; BSc, MSc, PhD	Unconventional	Geophysics, structural geology, stratigraphy, biostratigraphy, organic geochemistry
			Containment: safety, site characterisation	Geophysics, petroleum engineering; hydrogeology, geochemistry, computing, telemetry; BSc, MSc, PhD	Formation storage	Geophysics, structural geology, sedimentology, stratigraphy; BSc, MSc	Offshore siting (wind, barrage)	Geophysics, geomorphology, sedimentology, stratigraphy		
			Injectivity, pressure management, storage optimisation	Geophysics, petroleum engineering hydrogeology, geochemistry, computing, telemetry; BSc, MSc, PhD	Pumping storage	Geophysics, geomorphology, structural geology, stratigraphy; BSc, MSc				
			Planning and licensing regulation	Geology and policy; BSc, MSc						

and low carbon industry, all of which have a strong relationship to appropriate regional and site-specific geology, and the processes associated with that type of geology. For example, in radioactive waste disposal and CCS, the site-specific characterisation of rock masses is vital to understand the feasibility of containment; similarly, process understanding in relation to that rock mass is vital to understand long term change in the subsurface. In general, these areas of research and activity will need a full range of qualification level in the tertiary education sphere from BSc to Ph.D. In rock volume characterisation and process understanding the quantitative disciplines of geoscience will be most important, including geophysics, geochemistry, geomechanics, and in some cases petroleum engineering.

Beyond rock volume characterisation, existing and novel methods of monitoring will be become important, as geological energy and decarbonisation options develop—as will concepts of risk and uncertainty, thus requiring mathematical and statistical training. Training in science and society will be needed to understand the social licence and public engagement (see **SDG 4**). Again for these areas of research and study, the full range of qualification level in the geoscience tertiary education sphere from BSc to Ph.D. will be needed.

As well as developing home-grown training, skills, and knowledge can be transferred between the developed and developing world through shared geoscience courses, integration of energy industry expertise and training (e.g., visiting professorships in the developing world) and international cooperation. An example of such cooperation is included in Box 7.4.

Box 7.4. Capacity Building in Afghanistan.

Using experience in a number of developing country and post-conflict contexts, a methodology for Business Needs Analysis was developed and used at the Afghanistan Geological Survey in Kabul between 2003 and 2004 (Stephenson and Penn 2005). The main aim of the analysis was to help the AGS to function better in the post-conflict context, providing independent information on sustainable use of resources to the Afghanistan government. Extensive stakeholder analysis carried out as part of the Business Needs Analysis gauged the organization's strengths and weaknesses and took account of the local social, political, and business context. Training was designed to be tuned to business need, including appropriate IT and communication skills, technical and scientific (Fig. 7.12).

A vital part of the training was to foster a corporate understanding of the private sector in the AGS, so that it could interact successfully with business and commerce.

Fig. 7.12 Capacity Building in Afghanistan. Left: The Afghanistan Geological Survey, a crucial public sector science institution providing independent information on sustainable use of resources to the Afghanistan government. Right: female Afghan students learning about geological maps. Photos by author (M Stephenson)

The analysis also established a system allowing regular cyclical business/training review, so that the AGS could adapt to further change.

Following feedback from stakeholders, it was considered important to help to understand how donor projects (often from very different donors with different priorities and agendas) could be better organised, coordinated and tuned to the business need of the institutions. Further information on this collaboration is included in the chapter exploring **SDG 17**.

7.5 Discussion and Conclusions

The concept of energy transition (e.g., Sovacool 2016), concerns the wholesale change of energy supply from one source to another, for example, from wood biomass to coal in the Industrial Revolution of the eighteenth Century. According to Sovacool (2016), amongst the stages that are experienced are a period of extended experimentation with small scale technology and a diversity of design, followed by scale up of technologies as designs improve and economies of scale emerge, and finally by scaling up at the industry level. As industry structure becomes standardised and core markets become saturated, further industry growth is driven by globalisation and the diffusion of a successful design from the innovation core to rim and periphery markets.

The speed of energy transition in developing countries will be governed by the rate at which new technology becomes available, knowledge of the opportunities for new energy technologies including the locations for suitable developments, as well as the opportunities for scale up beyond the small scale which involves commercialisation, addressing market failure and strategic investment. In many ways, geoscience can be expected to provide the strategic knowledge that addresses market failure and encourages investment, rather like the way, for example, that the way that strategic public sector science

research investment in the 1980s and 1990s in the US paved the way for a successful shale gas industry (Stephenson 2015).

What kind of energy system will evolve in developing countries? Will it be electricity-dominated with primary decarbonised and centralised electricity sources (e.g., large fossil fuel power stations with CCS), or dominated by more distributed renewable resources, or based on hydrogen as a fuel. How much will ground source geothermal provide heat and air conditioning to buildings? As in the developed world, these questions are not easy to answer. The forecasts of the IEA suggest that fossil fuels will continue to be used for several decades in the developing world while renewables gain ground. Facilities and technologies for storing grid-scale electricity as well as transporting it are mostly inadequate in the developing world and will be needed whichever system is adopted, even if (most likely) it is a hybrid.

It is also likely that much development in subsurface energy will take place along development or trade corridors in the developing world, for example, the Nacala and Northern corridors in East Africa (Stephenson 2018), and so targeted regional geological studies will be needed to support integrated decarbonisation and resource management technologies (including integrated hydrogen and CCS). Also, geotechnical studies will be needed to support new railways, roads, pipelines and tunnels.

The present underinvestment in 'home-grown' education and research will tend to concentrate expertise in the commercial sector, which often being multinational, will not necessarily encourage local expertise, and will take away some of the ability of the government institutions to deal with multinational companies on an equal footing. So it seems clear that increased investment in energy geoscience training and research that has a clear application in developing country energy challenges and opportunities is needed.

In the compiling of this chapter, it became clear early on that data on energy resources and energy geoscience for developing countries is nowhere conveniently stored or collated, being concentrated more often on particular areas of interest, for example, Africa and India, as geographical categories. This means that the similarities and differences between developing countries cannot easily be ascertained, so that useful generalisations are difficult. This hampers planning and better understanding. For example, IEA and BP energy statistics do not routinely contain sections on the developing world as a category. It is also true to say that sectoral differences in energy—for example, between the oil and gas industry and the geothermal industry, or other renewables—makes a generalised view difficult. Thus, it might be wise to institute better developing country energy geoscience data collation and storage.

7.6 Key Learning Points

- Energy in its broadest sense enables business, industry, agriculture, transport, communications, and modern services such as health care; but it also enables improvements in living standards.
- Geoscience has a direct role in several of these areas including in establishing the geographical distribution, geological habitat, geotechnical feasibility of construction and infrastructure, and environmental sustainability, of energy supply.
- The geographical distribution of energy resources in the developing world indicates large potential in fossil fuels, including oil and gas but particularly coal, in several large key developing nations. Whether these fuels will be developed will depend on local needs and emissions policies. Many renewable energy resources are abundant in the developing world but their development depends on market conditions, often in tension with fossil fuels.
- Underpinning geological activities for affordable, reliable, sustainable, and modern energy will need to gain an understanding of the resource, but also how it can be used within limits that do not damage the environment, locally or through emissions, but still also be affordable. Electricity will clearly be key and decarbonisation of electricity could involve carbon capture and storage on fossil fuel power stations, which will involve in-depth geological studies in resource and containment.
- Geological studies will be needed for other low carbon electricity, for example, wind turbines, geothermal, and nuclear power. For the hydrogen economy, similar in-depth geological studies will be required.
- Geological studies will also have to feed into appropriate regulation that manifestly protects people and property, and regulations will need to be enforced by strong, independent local institutions. Facilities and technologies for storing grid-scale energy (electricity), as well as transporting it will be required, as will be an understanding of the development corridors where the most activity will take place.
- It is clear that increased investment in energy geoscience training and research is needed; and better data on energy geoscience research and training needs for the developing world would allow for better analysis and planning.

7.7 Educational Resources

In this section, we provide examples of educational activities that connect geoscience, the material discussed in this chapter, and scenarios that may arise when applying geoscience (e.g., in policy, government, private sector international organisations, NGOs). Consider using these as the basis for presentations, group discussions, essays, or to encourage further reading.

- Think about some of the rocks you've recently described as part of a petrology practical class. What contribution could they make to delivering SDG 7? What types of geological environments (think of both rock types and geodynamics) could be suitable for (i) carbon capture and storage, (ii) containment of radioactive waste, and (iii) hosting minerals used in solar panels?
- Prepare a review of information on energy access, energy consumption, and energy generation in (i) Zambia, (ii) Fiji, and

(iii) Canada. How easy is to access information for the three countries, and what may the reasons be for any differences in the availability of statistics? With the information you have collated, consider what steps each country could take to tackle climate change?

- How may implementation of the 16 other SDGs increase/decrease demand for energy? Are these changes likely to be the same everywhere, or affect particular regions? What are the implications of your findings on geoscience training?

- Explore the energy use per capita for different locations around the world (e.g., Tanzania, India, Vanuatu, and Australia)? Debate the statement *'it is unreasonable to prevent countries with very low energy use per capita from increasing their use of fossil fuels'.*

Further Reading

Fouquet, R. (2015). Handbook on energy and climate change. Edward Elgar Publishing Ltd, 752 pp

Helm, D. (2015) Natural capital: valuing the planet. Yale University Press, 320 pp

Kuzemko C, Goldthau A, Keating M (2016) The global energy challenge: environment, development and security paperback – 30 Sep 2015. Palgrave, Dev Secur, 264 pp

Letcher T (ed) (2013) Future energy: improved, sustainable and clean options for our planet, 2nd Edition, Elsevier Science, 738 pp

MacKay D (2009) Sustainable energy without the hot air. UIT

References

Africa Progress Panel (2015) Power people planet seizing Africa's energy and climate opportunities, 182 pp, Geneva

Aßmann D, Sieber N (2005) Transport in developing countries: renewable energy versus energy reduction? Transp Rev 25(6):719–738

Dickson MH, Fanelli M (2003) Geothermal energy: utilization and technology. UNESCO, Paris, 221 pp

Energy Information Administration (EIA) (2013) Technically Recoverable Shale Oil and Shale Gas Resources: An Assessment of 137 Shale Formations in 41 Countries Outside the United States

Energy World (2018) https://energy.economictimes.indiatimes.com/news/power/india-possesses-tidal-energy-potential-of-around-8000-mw-r-k-singh/62349353

Evans D, Stephenson M, Shaw R (2009) The present and future use of 'land' below ground. Land Use Policy 134:34–58

Financial Times (2017) https://www.ft.com/content/82158952-7da3-11e7-9108-edda0bcbc928

Fischer-Kowalski M, Rovenskaya E, Krausmann F, Pallua I, Mc Neill JR (2019) Energy transitions and social revolutions. Technol Forecast Soc Change 138:69–77

Gies E (2016) Can wind and solar fuel Africa's future? Nature 539:20–22

Govt. of India (2018) Reports of ministry of power. http://www.cea.nic.in/monthlyhpi.html

Holloway S (2007) Carbon dioxide capture and geological storage. Philosop Trans R Soc Lond A 365:1095–1107

International Energy Agency (2011) World Energy Outlook, 450 pp

International Energy Agency (2013) Resources to Reserves, Paris 272 pp

International Energy Agency (2016) World Energy Outlook, 684 pp

International Energy Agency (2018) World Energy Outlook, 714 pp

IPCC (2014) Climate Change 2014: Synthesis Report. Contribution of Working Groups I, II and III to the Fifth Assessment Report of the Intergovernmental Panel on Climate Change [Core Writing Team, R.K. Pachauri and L.A. Meyer (eds.)]. IPCC, Geneva, Switzerland, 151 pp

IRENA (2015) Africa 2030: roadmap for a renewable energy future. In: IRENA, Abu Dhabi, 72 pp

Kamalapur GD, Udaykumar RY (2011) Rural electrification in India and feasibility of photovoltaic solar home systems. Electric Power Energy Syst 33:594–599

Kaygusuz K (2012) Energy for sustainable development: a case of developing countries. Renew Sustain Energy Rev 16:1116–1126

Mariita NO (2002) The impact of large renewable energy development on the poor: environmental and socioeconomic impact of a geothermal power plant on a poor rural community in Kenya geothermal energy and the rural poor in Kenya. Energy Policy 30:1119–1128

McEvoy FM, Schofield DI, Shaw RP, Norris S (2016) Tectonic and climatic considerations for deep geological disposal of radioactive waste: a UK perspective. Sci Total Environ 571:507–521

Miketa A, Saadi N (2015) Africa power sector: planning and prospects for renewable energy. In: IRENA, 44p

Omenda P (2018) Geothermal outlook in East Africa. In: IRENA—International Geothermal Association. Presentation, 37 pp

Our World in Data (2019) https://ourworldindata.org/

Ozarslan A (2012) Large-scale hydrogen energy storage in salt caverns. Int J Hydrogen Energy 37:14265–14277

Pegg S (2006) Mining and poverty reduction: transforming rhetoric into reality. J Clean Prod 14:376–387

Ritchie H, Roser M (2019a) Energy Access. https://ourworldindata.org/energy-access. Accessed 27 Oct 2019

Ritchie H, Roser M (2019b) Fossil Fuels. https://ourworldindata.org/fossil-fuels. Accessed 27 Oct 2019

Shahsavari A, Akbari M (2018) Potential of solar energy in developing countries for reducing energy-related emissions. Renew Sustain Energy Rev 90:275–291

Shearer C, Fofrich R, Davis SJ (2017) Future CO_2 emissions and electricity generation from proposed coal-fired power plants in India. Earth's Future 5 (4):408–416

Sovacool BK (2016) How long will it take? Conceptualizing the temporal dynamics of energy transitions. Energy Res Soc Sci 13:202–215

Steckel JC, Brecha RJ, Jakob M, Strefler J, Luderer G (2013) Development without energy? Assessing future scenarios of energy consumption in developing countries. Ecol Econ 90:53–67

Stephenson MH (2013) Returning carbon to nature: coal, carbon capture, and storage. Elsevier, Amsterdam, Netherlands, 143 pp

Stephenson MH (2015) Shale gas and fracking: the science behind the controversy. Elsevier, Amsterdam, Netherlands, 153 pp

Stephenson MH (2018) Energy and climate change: an introduction to geological controls interventions and mitigations. Elsevier, Amsterdam, Netherlands, 206 pp

Stephenson MH, Penn IE (2005) Capacity building of developing country public sector institutions in the natural resource sector. In: Marker B, Petterson MG, Stephenson MH, McEvoy F (eds) Sustainable minerals for a developing world. Geological Society Special Publication, vol 250, pp 185–194

Torvik R (2009) Why do some resource-abundant countries succeed while others do not?. Oxford Rev Econ Policy 25:241–256

Van Alstine J, Manyindo J, Smith L, Dixon J, Amaniga Ruhanga I (2014) Resource governance dynamics: the challenge of 'new oil'in Uganda. Resour Polic 40:48–58

Zhao P, Zhang M (2018) The impact of urbanisation on energy consumption: a 30-year review in China. Urban Clim 24:940–953

Michael H. Stephenson is Executive Chief Scientist at the British Geological Survey. He has done research in the Middle East and Asia, including highlights in Oman, Saudi Arabia, Jordan, Pakistan, Iran, Israel and Iraq. He has honorary professorships at Nottingham and Leicester universities in the UK and is a visiting professor at the University of Milan, Italy and the University of Nanjing, China. His most recent book '*Energy and Climate Change: An Introduction to Geological Controls, Interventions and Mitigations*' examines the Earth system science context of the formation and use of fossil fuel resources, and the implications for climate change.

Decent Work and Economic Growth

Katrien An Heirman, Joel C. Gill,
and Sarah Caven

This article/publication reflects the personal opinion of
the author.

K. A. Heirman (✉)
Deutsche Gesellschaft für Internationale
Zusammenarbeit (GIZ) GmbH, P.O. Box 59, Career
Center Building, KG 541 Str, Kigali, Rwanda
e-mail: katrien.heirman@giz.de

J. C. Gill
British Geological Survey, Environmental Science
Centre, Nicker Hill, Keyworth, Nottingham NG12
5GG, UK

J. C. Gill
Geology for Global Development, Loughborough,
UK

S. Caven
Independent minerals and sustainability consultant,
Vancouver, British Columbia, Canada

Abstract

8 DECENT WORK AND ECONOMIC GROWTH

Decent work and economic growth helps to facilitate sustainable development

Economic growth and employment enables investment in healthcare, education and infrastructure

Decoupling economic growth from environmental degradation provides new opportunities

Decent, well-paid and fulfilling jobs boosts development and well-being

Sustainable and resilient economies are achieved through:

Reduced consumption of natural resources (e.g., minerals, water)

Improved understanding of the complex interactions between the environnment and society

Economic diversification, reducing the impacts of economic and environmental shocks

Technological upgrading, knowledge exchange and research to improve economic productivity

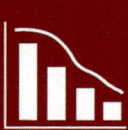

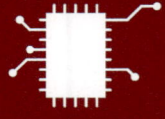

1900 1950 2000 2050

Retail Agriculture Tourism Finance Energy Transportation

Decent work boosts economic growth, which helps to sustain livelihoods:

Job creation is critical to meet growing demand, particularly in the Global South

More jobs will support environmental sustainability, decarbonisation and the circular economy

Sustainable geotourism also provides new jobs

Workplaces must be safe and secure, and free of violence and harassment of all kinds

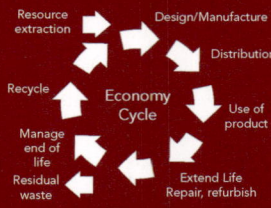

Resource extraction Design/Manufacture Distribution Recycle Economy Cycle Use of product Manage end of life Residual waste Extend Life Repair, refurbish

8.1 Introduction

Strong, sustainable, and resilient economies are key to sustainable development, together with inclusive social development and environmental integrity. Strong and productive economies enable increased investment in healthcare, education, and infrastructure. Sustainable and resilient economies ensure that growth does not come at a high environmental cost, and can withstand environmental, social and economic changes, and shocks. Meeting the aims of **SDG 8** will, therefore, enable progress across a suite of SDGs. This progress is secured through improved access to economic services, economic diversification (not relying on one income source), the creation of decent and fulfilling jobs, and the valuing and implementation of good practice regarding labour standards and workplace safety. These factors are represented in the targets and means of implementation of **SDG 8**, in Table 8.1, which aims to '*promote sustained, inclusive and sustainable economic growth, full and productive employment and decent work for all.*'

The geological environment and geoscience profession contributes both directly and indirectly to economic growth. For example, geoscientists help societies to access the natural resources—including groundwater, minerals, and hydrocarbons—that underpin our economies, and are essential to the manufacturing of goods

Table 8.1 SDG 8 targets and means of implementation

Target	Description of target (8.1 to 8.10) or Means of implementation (8.A to 8.B)
8.1	Sustain per capita economic growth in accordance with national circumstances, and in particular, at least 7% gross domestic product growth per annum in the least developed countries
8.2	Achieve higher levels of economic productivity through diversification, technological upgrading, and innovation, including through a focus on high-value added and labour-intensive sectors
8.3	Promote development-oriented policies that support productive activities, decent job creation, entrepreneurship, creativity and innovation, and encourage the formalisation and growth of micro-, small-, and medium-sized enterprises, including through access to financial services
8.4	Improve progressively, through 2030, global resource efficiency in consumption and production and endeavour to decouple economic growth from environmental degradation, in accordance with the 10 Year Framework of Programmes on Sustainable Consumption and Production, with developed countries taking the lead
8.5	By 2030, achieve full and productive employment and decent work for all women and men, including for young people and persons with disabilities, and equal pay for work of equal value
8.6	By 2020, substantially reduce the proportion of youth not in employment, education or training
8.7	Take immediate and effective measures to eradicate forced labour, end modern slavery and human trafficking and secure the prohibition and elimination of the worst forms of child labour, including recruitment and use of child soldiers, and by 2025, end child labour in all its forms
8.8	Protect labour rights and promote safe and secure working environments for all workers, including migrant workers, in particular, women migrants, and those in precarious employment
8.9	By 2030, devise and implement policies to promote sustainable tourism that creates jobs and promotes local culture and products
8.10	Strengthen the capacity of domestic financial institutions to encourage and expand access to banking, insurance and financial services for all
8.A	Increase Aid for Trade support for developing countries, in particular, least developed countries, including through the Enhanced Integrated Framework for Trade-related Technical Assistance to Least Developed Countries
8.B	By 2020, develop and operationalize a global strategy for youth employment and implement the Global Jobs Pact of the International Labour Organisation

and provision of services that contribute to GDP. With the increasing demand for energy (including 'green technologies' such as solar panels) and ongoing industrialisation in the Global South (see **SDG 7** and **9,** respectively), the demand for natural resources is likely to continue and increase. The transition to electric vehicles, for example, will increase demand for many raw materials used in lithium-ion batteries (e.g., lithium, cobalt, nickel, manganese, and graphite) and in power generation, grid storage, and charging infrastructure (e.g., copper, cobalt, and nickel) (BGS 2018). As progress is made towards a circular economy (see **SDG 12**), but geoscientists will be critical to guiding this process and improving resource use efficiency. The subsurface will play a more significant role in society given restrictions on urban expansion. By 2030, approximately 60% of the world population will live in an urban setting (see **SDG 11**), with cities needing to expand downwards and increasing numbers of activities moved underground (e.g., transportation, storage, parking, shopping). The subsurface will also be crucial in the energy transition, providing solutions for large and long-term subsurface energy and heat storage (see **SDG 7**).

Other aspects of the natural environment contribute indirectly to economic activity through ecosystem and geosystem services, for example, helping to sequester carbon or cycle nutrients (Everett et al. 2010; Van Ree and van Beukering 2016). A good understanding and sound assessment of geohazards are essential to take measures of risk reduction to create a more resilient economic system. Consideration of geodiversity helps inform the development of nature-based solutions to environmental challenges and increasing demand for resources (Schrodt et al. 2019). Geoscientists in research, industry, and the public sector play a fundamental role in understanding geological systems and their relationships with ecosystems, as well as the approaches needed to manage these to support economic activity that is decoupled from environmental degradation. Natural landscapes, shaped by geological processes, are often fundamental to tourism, particularly geotourism. Research, education, and outreach that connect tourists to the geological environment can enhance the understanding of Earth systems and planetary boundaries.

This chapter explores these themes. We describe the role of geoscientists in supporting economic growth and examine economic diversification from the perspective of natural resources (Sect. 8.2). We explore the future of work, opportunities for geoscientists in ensuring environmental sustainability, and the opportunities that geoheritage and geotourism offer to broader job creation (Sect. 8.3). We conclude by discussing safe and secure work environments, profiling particular challenges faced by geoscience sectors, and recognise the responsibilities of geoscientists in ensuring workplaces free of harassment, discrimination, and modern slavery (Sect. 8.4).

8.2 Strong, Sustainable, and Resilient Economies

The first target of **SDG 8** advocates for per capita economic growth to be sustained in accordance with national circumstances, with a pre- COVID-19 ambition of sustaining 7% GDP growth per year in the world's least developed countries (**Target 8.1**). The World Bank (2019a), estimates that GDP growth averaged across all least developed countries has only met this target in four of the last 29 years (2005–2008). The average GDP per capita growth in least developed countries over the last decade is 2.3% per year, which although higher than OECD members' average growth (0.9% per year), is still well below the **SDG 8** ambitions (World Bank 2019b). So while there is an overall trend of increasing economic growth, further actions are needed to (i) increase the rate of growth, (ii) sustain that higher growth rate, and (iii) ensure that increased, sustained economic growth is sustainable by decoupling it from environmental degradation (**Target 8.4**). Here, we first explore the role of geoscientists in enabling green and inclusive economic growth and then consider how economic diversification and technological upgrading can help to increase and sustain economic growth.

8.2.1 Green and Inclusive Economic Growth

The European Commission (2011), defines the green economy as an '*economy that can secure growth and development, while at the same time improving human well-being, providing decent jobs, reducing inequalities, tackling poverty, and preserving the natural capital upon which we all depend*'. The emphasis is, therefore, not just on economic growth and environmental sustainability, but also on social development and equity (Avis 2018). The economic growth required to enable other SDGs cannot be promoted in isolation from social and environmental considerations, and should operate within planetary constraints to avoid irreversible and costly environmental damage, and be resilient to the effects of natural hazards (Avis 2018). The green economy focuses on resource efficiency and enhanced management of natural capital, underpinned by an acknowledgement that environmental degradation undermines long-term economic growth and threatens human development (European Commission 2018).

Rather than setting up a conflict between environmental conservation and economic growth (i.e., protecting the environment damages growth), this approach highlights that there are significant economic and development opportunities from environmentally-sensitive policy-making (European Commission 2018). For example, businesses will benefit from more efficient production practices. Decoupling economic growth from environmental degradation (**SDG 8.4**), and investing in environmental protection, can in itself drive economic growth while also achieving environmental objectives (European Commission 2018). Such decoupling requires key environmental variables to be stable or show

decreasing degradation while the economy continues to grow (Everett et al. 2010). For example, economic growth may continue to rise while carbon emissions fall, or water quality improves.

Decoupling environmental degradation from economic growth requires an understanding of the complex interactions between the natural environment (including the subsurface) and economic growth. This understanding can help to minimise environmental impacts of economic activities, reduce resource consumption (see Box 8.1), and '*avoid breaches in critical thresholds beyond which natural assets cannot be replaced and can no longer support the desired level of economic activity*' (Everett et al. 2010, p. 7). Understanding socio-environmental interactions requires enhanced characterisation of Earth resources, processes, and systems to inform environmental, human, and economic modelling and to explore interactions between these. New partnerships are, therefore, needed between those with environmental expertise (including geoscientists) and human, behavioural, and economic sciences.

Geoscientists may also be involved in innovations that combine economic opportunity and environmental and social responsibility. For example, improving the management and use of electronic waste (or e-waste) could be an economically important opportunity, with the World Economic Forum (2019), noting that the material value of e-waste is more than the GDP of many countries. Boliden, a mining and smelting company has been recycling scrap metal since the 1960s and has leveraged this expertise to become a global leader in capturing metals from e-waste.[1] Geoscience expertise can contribute to the transition to a circular economy, minimise the environmental impact of new technologies that themselves drive economic growth (e.g., mobile communications), and generate new economic opportunities. For further discussion and examples of resource efficiency, see **SDG 9** (sustainable industrialisation) and **SDG 12** (responsible consumption and production).

---

[1] https://www.boliden.com/sustainability/case-studies/largest-electronic-material-recycler-in-the-world/.

Box 8.1. The 10 Year Framework of Programmes on Sustainable
Consumption and Production (10YFP)

Adopted by Heads of States at the UN Conference on Sustainable Development (Rio + 20) in June 2012, the 10YFP is a global framework for action to enhance international cooperation and accelerate the shift towards sustainable consumption and production patterns, and resource efficiency, in both developed and developing countries, at national and regional levels. The 10FYP aims to support the decoupling of environmental degradation and resource use from economic growth, in turn, supporting enhanced productivity, poverty eradication, social development, and environmental sustainability. With capacity building and technology transfer at the heart of the 10FYP, it is hoped that innovation and cooperation will result in a major shift to sustainable consumption and production patterns. The 10FYP is written into the SDGs, with reference in **SDGs 8.4** and **12.1**.

An interim report on the 10FYP was published in 2014, by the UN Department of Economic and Social Affairs (UNDESA 2014). It highlighted work on six programmes

- Sustainable Public Procurement
- Consumer Information
- Sustainable Tourism
- Sustainable Lifestyles and Education
- Sustainable Building and Construction
- Sustainable Food Systems

The implementation of these programmes aims to support resource efficiency and decoupling of economic growth with environmental degradation. It demonstrates the need for us all, but higher income countries, in particular, to take actions to reduce per capita consumption and the environmental footprint of economic activities (Fig. 8.1). It also highlights that this is a shared responsibility of consumers, businesses, and governments.

8.2.2 Economic Diversification

Diversifying an economy and moving away from a reliance on a single income source towards multiple income sources can have a positive and stabilising effect, boosting employment and reducing vulnerability to economic, social, and environmental shocks. For example, local economies that are dependent on a small number of crops could be badly impacted if the region is affected by a particular pest or disease. National economies that are heavily dependent on one or a small number of mineral exports are particularly susceptible to commodity price fluctuations or falling demand, which can create economic instability (Hausmann and Rigobon 2003). In contrast, economic diversification can help to reduce vulnerability at local to global scales, and provides an opportunity to move towards products, markets, and jobs that produce less carbon emissions and are more climate resilient (UNFCCC 2016).

Natural resources (including minerals, metals, and hydrocarbons) provide many countries with an important source of income. Countries that are resource-rich are not necessarily *resource-dependent*, which depends on the extent of economic diversification. The International Council on Mining and Metals (ICMM) defines resource-dependence to be where a country has 20% or more of export earnings coming from natural resources, or where resource rents account for more than 10% of GDP (ICMM 2018a). Table 8.2 shows 53 countries that have been resource-dependent over the entire period of 1995–2015, accounting for 30% of the global population and 230 million people living on less than $1.90 a day (ICMM 2018a). For example, mineral rents in Suriname equated to 24% of GDP in 2016, with exports of metallic minerals, metals, and coal equating to 53.5% of export earnings (ICMM 2018b). The Extractives Industry Transparency Initiative (or EITI, see **SDG 16**) note that revenue from extractives makes up more than 80% of the total government revenue in Chad, Iraq and Timor-Leste (EITI 2019).

Fig. 8.1 Installation of solar panels as part of sustainable building practices. Image by skeeze from Pixabay

Table 8.2 Resource-dependent countries over the entire period of 1995–2015 (from ICMM 2018a)

	Hydrocarbons	Metals and Minerals	Both
Sub-Saharan Africa	Angola, Cameroon, Congo, Gabon, Equatorial Guinea, Libya, Nigeria	Botswana, Central African Republic, Democratic Republic of Congo, Ghana, Guinea, Mauritania, Namibia, Niger, South Africa, Togo, Zambia	
Asia	Brunei Darussalam, Indonesia, Turkmenistan	Mongolia	Australia, Papua New Guinea, Uzbekistan, Kazakhstan
Latin America and the Caribbean	Colombia, Ecuador, Trinidad and Tobago, Venezuela	Bolivia, Chile, Guyana, Jamaica, Peru, Suriname	
Middle East and North Africa, and Europe	Algeria, Azerbaijan, Egypt, Iran, Iraq, Kuwait, Norway, Oman, Qatar, Russia, Saudi Arabia, Syrian Arab Republic, United Arab Emirates, Yemen	Armenia, Georgia	Bahrain

Extractive sectors generally require large amounts of capital, but do not always strengthen the broader economy (World Bank 2015). 'Dutch Disease' describes declining activity in one or more sectors (e.g., manufacturing, agriculture) as it becomes less competitive due to an increase in the value of the national currency as a specific sector is heavily invested in and develops (e.g., natural resources). The term is named after the Netherlands which, following the discovery of major petroleum resources, experienced a decline in manufacturing as non-oil products became less competitive on the world market due to the strong Dutch Guilder. The declining sectors cannot compete internationally as a country fails to use appropriate monetary and fiscal policy to manage exchange rate fluctuations following rapid investment in natural resource development (Corden and Neary 1982).

Extractive sectors also do not generally generate large amounts of employment relative to their contribution to GDP (Fine et al. 2012; World Bank 2015). For example, they account for just 1% of Africa's workforce despite mining, oil, and gas contributing approximately 7.1% to Africa's GDP in 2015, through resource rents (Fine et al. 2012; Montt et al. 2018). There may be opportunities to create new jobs in industries related to resource extraction (e.g., processing minerals and manufacturing products) and sectors working to minimise the effects of resource extraction on the natural environment (e.g., water management agencies). Diversification will likely also require job creation in broader sectors, such as services, tourism, and research and development.

Investment in extractive sectors is still critical, however, if we are to address other sustainable development challenges, such as accessing the raw materials needed for renewable energy technologies (Nickless 2017). Governments can help to protect themselves from economic slumps from commodity price drops by using revenues from resource extraction to develop human capital and diversify their economy (World Bank 2012). The seven biggest Latin American mineral exporters were identified as significantly underperforming in terms of growth following the collapse in commodity prices in

mid-2008 (Camacho and Pérez-Quirós 2013). A country can shield itself to such price collapses through effective monetary policies (Roch 2017). Chile is considered to be resource-dependent, but it is also a highly diversified economy with a policy environment that has enabled the export of more than 2800 distinct products to more than 120 countries, including fruit, fish, wine, and chemicals (Fruman 2017).

8.2.3 Technological Upgrading

Research to develop new technologies and adapt existing technologies is critical to improving productivity and supporting economic growth. Where access to technologies is limited, there are opportunities for upgrading and technology transfer across different geographical settings, and from one discipline to another. Technologies are required both in the workplace and the home, recognising that for many workers the home is the workplace. Social or environmental factors may mean that the most appropriate technology in one context is not suitable in another context. Technological upgrading, therefore, requires (i) knowledge exchange, (ii) research and development, and (iii) access to economic capital to occur.

i. **Knowledge Exchange**. In many contexts, technology already exists to increase productivity and may be in place within the same or a different sector elsewhere. Improved knowledge exchange between sectors, and different national contexts, can help to build awareness of different technological possibilities. The occurrence of natural hazards, for example, can have a negative impact on economic growth and disrupt income generating activities thus reducing productivity. Many technologies are being used around the world to monitor volcanic activity, understand evolving meteorological conditions, and communicate with the affected population when an earthquake is detected. Not every country affected by natural hazards has access to or understanding of all the currently available technologies that may help to better charac-terise, reduce, or manage risk. It is often low-income countries that lack access to such technologies and need effective knowledge and technology transfer to upgrade and pro-tect economic assets and livelihoods. Sup-ported by the World Bank and the British Embassy in Guatemala, scientists and engi-neers from the Universities of Bristol and Birmingham (UK), have worked with Gua-temalan scientists at the National Institute for Seismology, Volcanology, Meteorology, and Hydrology (INSIVUMEH) to demonstrate how drones can be used to map Fuego vol-cano.[2] In a further example, UNESCO laun-ched the *International Platform on Earthquake Early Warning Systems* (IP-EEWS) initiative in 2015, to create a space that enhances collaboration and knowledge sharing within the scientific community and between scientists, decision- and policymak-ers in order to promote the development of early warning systems in earthquake-prone regions and countries and strengthen com-munities' preparedness and resilience against natural hazards. For more information on sharing information, experiences, best prac-tices, and policy relating to science, technol-ogy, and innovation, see the section on the UN *Technology Facilitation Mechanism* in **SDG 17**.

ii. **Research and Development**. While many technologies already exist that can help to increase economic productivity, in some cases research and development is needed to create new or advance existing tech-nologies. Gil et al. (2018) highlight *intelli-gent systems* to be an essential research agenda for the geosciences, noting that geoscience data is often uncertain, inter-mittent, sparse, multi-resolution, and multi-scale. They argue that approaches to understanding interactions between Earth processes and human activities may be limited due to the complexity of geoscience data. Machine learning, sensing, and robotics can help to enrich geoscience data and support the improved characterisation of links between Earth systems and human activities. For example, climate and Earth system models can be improved using new models that combine artificial intelligence and physical modelling (Reichstein et al. 2019), to support decision-making related to climate action (**SDG 13**).

iii. **Access to Economic Capital**. Upgrading technology also requires access to financial services, such as bank accounts and loans. Only 35% of adults in low-income countries have access to an account at a bank or other financial institution (UN 2018a). Innova-tions such as microfinance and mobile money agents are increasing access to and the ability to transfer capital, which can be used to fund technological upgrading in the workplace at home. Economic productivity is seriously hindered by a lack of access to clean, safe water, and associated health implications. In 2016, UNICEF estimated that women and girls spend 200 million hours every day collecting water, reducing the time available for education and income-generation (UNICEF 2016). Initia-tives such as WaterCredit are providing microfinance to improve householder water and toilet facilities (Water.Org 2019). By providing access to upgrade household technologies, microfinance is improving health and productivity.

Technological upgrading, therefore, requires multiple approaches, and their application at dif-ferent scales. While technology is fundamental, we should not ignore the important contribution that other forms of innovation make to increased pro-ductivity. This could be more research and development workers, novel partnerships, and improved methods to integrate western and traditional/indigenous scientific knowledge (see **SDG 9** and **SDG 17**). For example, Robbins (2018), cites the example of a researcher using traditional ecological knowledge from indigenous communities to gain a more holistic understanding

[2]https://www.bristol.ac.uk/cabot/news/2018/drone-expertise-volcanic-eruptions.html.

Fig. 8.2 Handheld XRF device. In this example the handheld XRF is used to evaluate the amount of metal contaminants in the soil. Photo by the US Department of Agriculture (Public Domain)

of the complex interactions between species. Knowledge exchange, research and development, and access to economic capital are critical for facilitating technological upgrading, but also contribute to a broader innovation that drives inclusive and sustainable economic growth.

Box 8.2 Handheld X-Ray Fluorescence Spectrometry Technology for Improved Productivity

Handheld X-ray fluorescence spectrometry (XRF) testing is now common in a variety of geoscience contexts as a field portable geochemical analytical tool (Fig. 8.2). If used effectively, it can provide geoscientists with live, geolocated data on the geochemical composition of rocks and soils.

This can be used to determine appropriate measures and types of fertiliser to use to enhance agriculture. It can also be used in mineral exploration to more effectively target zones with high-grade potential, reduce the environmental impact of sampling programmes and reduce the need to transport large volumes of samples to laboratories. This sampling method can boost productivity allowing exploration teams to design sample programmes to rapidly cover underexplored areas from a regional to local scale. The technology can also be applied in other contexts such as diamond drilling, mining, or investigating contaminated land (Young et al. 2016).

8.3 The Future of (Geo)Work

8.3.1 Employment Challenges and Opportunities

The creation of decent jobs—livelihoods that are stable, pay a fair wage, with safe working conditions—are central to **SDG 8** and critical to eradicating poverty, boosting economic growth, recovering from the impact of the COVID-19 pandemic, and ensuring the necessary tax revenues to improve services and infrastructure. The global labour force has grown by nearly 50% between 1990 and 2018, from 2.32 billion to 3.46 billion people (World Bank 2019c). In the world's least developed countries during the same timeframe, this growth was 112%, from 196 million to 416 million people (World Bank 2019c). The growth in the labour force is expected to continue, particularly in the Global South where populations are growing.

Creating jobs, however, is not enough to deliver **SDG 8**. UNDP (2019), estimates that in 2018, 700 million workers earned less than $3.20/day and still live in extreme or moderate poverty. Jobs must, therefore, pay enough to lift people out of poverty, have decent terms and conditions, and take place in safe and secure work environments. Globally, more than 60% of workers operate in the '*informal economy*'— generally not taxed or monitored by the government, with a lack of social protection, rights at work and decent working conditions (ILO 2018a). Informal employment can make it harder for workers to complain about poor treatment, access finance or support from official institutions, and escape extreme poverty.

The creation of jobs is influenced by some key trends shaping the future of work (see the *Further Reading* section of this chapter for more information). The first is the impact of artificial intelligence and robots on future employment. The Organisation for Economic Cooperation and Development (OECD 2019), suggests 14% of jobs in OECD countries are at high risk of automation, and a further 32% of jobs could be radically transformed due to automation. Some reports suggest that increases to productivity from automation will enable increases in workforce capacity, but acknowledge that reskilling of staff may be needed (Watson et al. 2019). New technologies have been introduced throughout history, improving productivity and creating new jobs (UNDESA 2017). The impacts of such technologies are uncertain, but policy and institutional responses (e.g., improved access to training and professional development for low-skilled workers), together with strategic foresight could help these technologies be used to leverage new and better opportunities for all (UNDESA 2017).

A second trend is the growth of the so-called gig economy, or "*labour market activities that are coordinated* via *digital platforms*" (Hunt et al. 2017), such as websites or smartphone applications. Customers use these platforms to request a service from an available worker, and the organisation operating the platform takes a fee or commission upon completion and payment (Hunt et al. 2017). This can provide flexibility for many workers, but often at the cost of workplace protections and benefits (e.g., sick pay). Estimates suggest that the 'gig' economy workforce in the UK doubled between 2016 and 2019, from 4.7 to 9.6% of workers (SSCU-HBS 2019). The extent of the gig workforce in the Global South is unclear (Hunt and Samman 2019). High levels of informal self-employment across Africa have been reported (Fine et al. 2012), with this much greater in scope than the gig workforce, but with similar challenges, such as the lack of social safety nets.

A third trend, particularly pertinent to geoscientists and geoscience-based sectors is a move towards sustainability, climate resilience, and the 'greening' of jobs. Strong links exist between environmental degradation, poverty and inequalities, with approximately 1.2 billion jobs around the world directly depending on ecosystem services (ILO 2018b; Montt et al. 2018). Failing to address environmental challenges will, therefore, impact current jobs and hinder the generation of new ones. This is discussed in detail in Sect. 8.3.2.

The creation of decent work, responding to the trends set out above, requires a strong enabling environment, including appropriate education, political stability, investment in innovation, and partnerships for development. Completing secondary and tertiary education (**SDG 4**) is important to increasing formal employment, with many of those in informal employment either having no education or only completing primary education (ILO 2018a). This education must be fit-for-purpose, with young people having access to appropriate high-quality education and vocational training that provide the skills to do the jobs available in the coming decades. Strategic job creation, aligned with appropriate education and training, can expedite the delivery of many SDGs, for example, by ensuring there is the technical and human capital required to design, build, and maintain infrastructures such as road, energy, or sewer networks. Job creation will also be supported by political stability (**SDG 16**), enhanced innovation spending (**SDG 9**), and technology transfer (**SDG 17**).

8.3.2 Job Creation and Environmental Sustainability

There is a strong push from UN frameworks (top-down) and both consumers and communities (bottom-up) to adopt policies that tackle ongoing environmental degradation and advance environmental sustainability, including decarbonisation and transitioning to a circular economy. This will affect the jobs created and lost in the future, with an overall net increase in jobs expected but this vary by region and sub-sector (ILO 2018b; Montt et al. 2018). Some sectors employing geoscientists will likely experience job demand growth, and others a decline in job demand. For example, AGI (2019) project that the total US geoscience workforce will increase by 6.2% between 2018 and 2028, but the oil and gas extraction industry is projected to decrease by 9.4%. It is of critical importance that geoscience educators and professional societies understand potential changes to future labour markets so that they can support geoscientists (particularly, but not limited to students) to have the skills and disciplinary knowledge required to gain employment.

The ILO (2018b) suggest decarbonisation will result in a net increase of jobs in the energy sector by 2030, with growth in the construction sector (6.5 million new jobs), mining of copper

ores and concentrates (1.2 million new jobs), and the production of electricity by hydropower (0.8 million new jobs). Other industries set to experience job demand growth as a percentage of the current workforce, include most types of electricity generation other than those using fossil fuels, including production by geothermal energy and wind (0.4% job demand growth), and production by nuclear energy (0.3% job demand growth). Spalek et al. (2013) project that the number of geothermal experts required in EU member states and associated countries would increase from 2500 in 2012 to 35,000 by 2030. The development of new energy infrastructure, including both energy generation and storage, sourcing of the raw materials to enable this, and ongoing environmental management of the waste generated by these growth industries will all require geoscientists.

The ILO (2018a) have also published estimates of how a transition to the circular economy will affect employment in different regions and sectors by 2030. This projection indicates net employment growth in the Americas and Europe, and net employment losses in Africa, the Middle East, Asia, and the Pacific (Montt et al. 2018), which underlines the importance of economic diversification to mitigate this potential impact on jobs and reduce employment losses. ILO (2018b) identify sectors benefitting from the transition to a circular economy to be those engaging in reprocessing of metals, repair and maintenance, and research and development. The same report identifies sectors suffering from the transition to a circular economy, with these almost entirely consisting of sectors linked to primary mining and hydrocarbon extraction.

Global ambitions are to transition to a more sustainable way of operating by 2030, but the impact on jobs will be felt before that. From the ILO (2018b) projections, and a broader look at the ambitions of the SDGs, sectors employing geoscientists will both grow and decline as we decarbonise and transition to a circular economy. Based on the targets within the SDGs, and our understanding of current progress in achieving these goals, it is likely that there will be an increased emphasis on environmental data collection, management, integration, and access to understand and manage complex environmental and social challenges (Gill et al. 2019). Improved subsurface mapping and environmental monitoring networks can support action to improve water and food security, decarbonise, reduce poverty, improve health, enable regeneration, and ensure resilient infrastructure. This is likely to mean increased demand for applied environmental geoscientists (including specialists in hydrogeology, engineering geology, geological hazard assessment, and contaminated land assessment), with experience in different geographic domains (e.g., coastal, marine, and urban environments), and the particular sustainable development challenges they contain.

In all contexts, there will be an increasing need for geoscientists who can integrate an understanding of sustainability (or social geoscience) into their professional practice (Stewart and Gill 2017). This will require a renewed emphasis on partnership building, training geoscientists to work effectively with varying disciplines (including the human and behavioural sciences) and sectors. These are skills that will also increase their employability prospects beyond traditional jobs for geoscientists. Geoscience graduates could help to meet the increasing demand for sustainable development specialists, given their ability to integrate diverse data, think across scales, apply lessons from the past to future thinking, and work in an interdisciplinary manner.

8.3.3 Job Creation and Sustainable (Geo-)Tourism

Sustainable tourism is expressed as a priority within the SDGs (Target 8.9), providing an opportunity to create new jobs that celebrate local culture and create new markets for local products. The UN World Tourism Organisation (UNWTO) defines sustainable tourism to be *"tourism that takes full account of its current and future economic, social and environmental impacts, addressing the needs of visitors, the industry, the environment and host communities"*

and emphasises the need to optimise the use of environmental resources, maintain essential ecological processes, and conserve natural heritage and biodiversity (UNEP/UNWTO 2005). As UNEP/UNWTO (2005) states, this requires

- *Investment in socio-economic benefits*: Ensuring a fair distribution of stable employment, income-earning opportunities and social services to the host communities (including indigenous communities), while contributing to poverty alleviation.
- *Respecting socio-cultural authenticity*: The host community have built and living cultural heritage and traditional values that should be conserved, contributing to intercultural understanding and tolerance.
- *Decarbonisation*. Increased use of renewable energy and the decarbonisation of transport to minimise the carbon emissions associated with tourism.
- *Resource management*: Effective management of the natural resources required by tourists (e.g., potable water) to ensure these are protected, used in a sustainable manner, and the immediate and long-term needs of local communities are met. Waste reduction and management is also needed, including reducing the use of plastic and minimising food waste.
- *Environmental management*: Effective management of geological and biological diversity in a given region (e.g., protecting landscapes, and the environmental integrity of water courses), which can help to draw people to a region.
- *Disaster preparedness and risk reduction*: Increased understanding of natural hazards in the region, and the steps needed to protect the lives and livelihoods of those living and working there, as well as visitors.
- *Promotion and monitoring of sustainable tourism practices*: Informed participation of all relevant stakeholders, consensus building, promotion of sustainable tourism practices, monitoring of the impacts and implementing corrections or preventive measures where needed.

Geotourism has been defined by the Arouca Declaration written by the Global Geoparks Network in November 2011, as tourism which '*sustains and enhances the identity of a territory, taking into consideration geology, environment, culture, aesthetics, heritage and the well-being of its residents*' (Arouca Declaration 2011). Geological tourism is just one of the multiple components of geotourism. Geotourism helps to conserve, disseminate, and cherish planet Earth. Geotourism enables visitors to grow in their understanding of Earth's 4.6 billion year history and to collectively explore shared futures for the Earth and humanity. UNESCO Global Geoparks (Box 8.3) play an important role in the development of geotourism.

Box 8.3 UNESCO Global Geoparks

UNESCO Global Geoparks are '*single, unified geographical areas where sites and landscapes of international geological significance are managed with a holistic concept of protection, education, and sustainable development*' (UNESCO 2017). Geoparks include geological heritage of international significance, and explore, develop, and celebrate the links between geology and the area's natural, cultural and intangible heritages. In 2019, there were 147 UNESCO Global Geoparks in 41 countries. Some examples include

Brazil: The Araripe UNESCO Global Geopark, holding large numbers of well-preserved fossils from the Lower Cretaceous, with a highly diverse paleobiology. The region has a distinct cultural identity, promoted through the Environmental Education and Interpretation Center of Araripe UNESCO Global Geopark. This centre provides an integrated understanding of the historical, cultural, socio-environmental, paleontological, and landscape aspects of sites with the Geopark.

Iceland: The Katla UNESCO Global Geopark includes volcanoes such as Katla and Eyjafjallajökull, large lava flows, waterfalls, and glaciers. The geography of this region resulted in the isolation of communities due to unbridged glacial rivers that were hard to cross, fostering innovation, entrepreneurship, and an understanding of geological processes.

Mexico: The Mixteca Alta, Oaxaca UNESCO Global Geopark is considered one of the most complex regions of Mexico from a geological perspective, with deposits from the Precambrian to the Cenozoic. Many sites within this Geopark are related to processes and landforms linked to land-use and farming by the Mixteca civilization, flourishing between the second century BC and fifteenth century AD.

Vietnam: The Dong Van Karst Plateau UNESCO Global Geopark, home to 17 ethnic groups, includes high mountains and deep canyons, with diverse geomorphological features and palaeontology. The diverse geology is complemented by a unique and rich cultural heritage.

UNESCO Global Geoparks can generate new job opportunities, new economic activities, and additional sources of income to local communities living within and in the surroundings of the Geopark, especially in rural regions. Information on all of the UNESCO Global Geoparks is available through the UNESCO website: www.unesco.org/geoparks (Fig. 8.3).

Geotourism can help to support economic growth and diversification. The financial benefits of UNESCO Global Geoparks were examined in a study by the UK National Commission for UNESCO (2013). While the cost of the

Fig. 8.3 Education activities at Mixteca Alta, Oaxaca UNESCO Global Geopark, Mexico. Field trip between local children, their parents and teachers, local authorities and guides in the Santa María Suchixtlán Geopark community ©Mixteca Alta, Oaxaca UNESCO Global Geopark, Mexico, Xóchitl Ramírez Miguel (used with permission)

Table 8.3 Economic impact of geotourism in Ireland, 2012–16 (€ million)

Revenue from	2012	2013	2014	2015	2016
Top Fee-Paying Sites	17.0	19.7	22.8	28.2	33.6
Top Free Sites	1.2	2.4	5.2	6.8	7.3
Hiking and Cross-Country Walking	149.8	189.8	225.1	286.3	329.8
Total Revenue	*167.9*	*211.9*	*243.1*	*321.3*	*370.7*
Gross Value Added	*108.6*	*137.0*	*163.6*	*207.7*	*239.6*

Source Indecon (2017)

UNESCO status for these Geoparks was estimated to be £330 k, the estimated annual financial benefit is estimated to be £19.17 m (UK National Commission for UNESCO 2013). This study was done prior to the official ratification by UNESCO of UNESCO Global Geoparks in November 2015, potentially increasing their value to the UK economy as a result of having greater recognition. A further study, commissioned by the Geological Survey of Ireland, showed that geotourism is a major contributor to the Irish economy (Table 8.3). Total revenues (expenditure by visitors) directly attributable to geotourism amount to over €370 million in 2016, with the sector contributing almost €240 million to the Irish economy GVA/GDP.

Geotourism can also crease new employment opportunities, including for geoscientists. UNESCO Global Geoparks employ geoscientists in roles such as Geopark managers, educators, facilitators, and scientists. Staff need a transferable skill set that allows them to communicate with local stakeholders including local authorities, business owners, citizens, visitors, and academics. This variety of skills is currently not adequately taught at the university level during a geoscience degree, and further professional development will likely be required. UNESCO Global Geoparks also create jobs (directly and indirectly) for other members of the local community. Many UNESCO Global Geoparks employ local inhabitants as guides or rangers. Locals have grown up in the region and often know the Geopark better than anyone else, with the enthusiasm and willingness to learn more about the geological, natural, and cultural heritage of the landscape and territory. In Mixteca Alta, Oaxaca UNESCO Global Geopark (Mexico), local inhabitants, some of them unable to write or read, have been trained to become Geopark guides, and provided with training on the local geology, geography, fauna and flora, history, and its many traditions. This provides additional opportunities to generate income. Guides are able to link their new geological knowledge with local indigenous knowledge and traditions. The Mixteca Alta, Oaxaca UNESCO Global Geopark has had a positive impact on what is a very poor region of Mexico. Many men have left the region to find jobs in the city or outside the country, resulting in a population of women, children, and elderly. The Geopark has helped to provide new opportunities for young people, and additional income-generating opportunities for the elderly and women living there.

Local citizens are also encouraged to expand or start new businesses connected to the Geopark, such as small hotels, outdoor sports services, restaurants serving local cuisine, and the sale of local crafts to visitors. Many UNESCO Global Geoparks support and encourage the formation of women's cooperatives. In the Qeshm Island UNESCO Global Geopark in Iran, a local women's cooperative runs the Star Valley visitor centre. They also display and sell their traditional handicrafts such as Golabaton, needlework portraying mostly flower patterns. This is a skill passed from one generation to another (Fig. 8.4).

The cooperative runs the local café, providing catering to one of the many Geopark guesthouses. The cooperative allows and empowers the women to actively contribute to the life of the community.

8.4 Safe and Secure Working Environments

The development of safe and secure working environments (Target 8.8) means creating spaces where injuries are prevented, and workplace violence and harassment of all kinds are eliminated. The International Labor Organisation (ILO) describes key standards that aim to ensure equity, non-discrimination, security, freedom, and dignity for all (ILO 2019a), with the SDG target aiming to ensure compliance in law and practice with these fundamental standards. Geoscientists have the right to expect a safe and secure work environment, and the responsibility to facilitate this in the diverse sectors and international contexts they operate in.

Safe and secure work environments are free from preventable fatalities and injuries. Geoscientists work in many dangerous situations, including mining (see Box 8.4), tunnelling, and oil and gas fields (both onshore and offshore). Geoscientists also conduct field research in extreme environments such as at high-altitudes, in active volcanic settings, and in Polar Regions. Effective planning and field safety preparations can help to reduce the likelihood of an accident and minimise the impacts of any accidents that do occur. Risk assessment and field safety is an essential part of the education of many geoscientists, however, there is no universal access to comprehensive courses and continued professional development to ensure all geoscientists can plan appropriately. Geologists should also be vigilant of practices that fail to take appropriate actions to mitigate the risks to their colleagues (e.g., companies not providing personal protective equipment), and report breaches in health and safety.

Fig. 8.4 Qeshm Island UNESCO Global Geopark, Iran, Golabaton, needlework portraying mostly flower patterns, is an art transferred from generation on generation in Qeshm Island UNESCO Global Geopark, Islamic Republic of Iran © Qeshm Island UNESCO Global Geopark, Iran/Asghar Besharati (used with permission)

Box 8.4 Safe and Secure Environments for Mineworkers

Despite considerable efforts in many countries, the rates of death, injury and disease amongst those working in the mining sector remain high (ILO 2019b). Many mineworkers work underground with limited access to natural light and ventilation, with others working close to cut-slopes with varying degrees of stability (Fig. 8.5). Problems include traumatic injury hazards, ergonomic hazards, psychosocial hazards, and exposure to extreme noise, heat, radon, solar ultraviolet, coal dust, crystalline silica, cyanide, mercury, and hydrofluoric acid (Donoghue 2004). Mining, therefore, remains one of the most hazardous occupations in the world. Accounting for 1% of the global workforce, mining is responsible for about 8% of fatal accidents at work.

Large-scale companies, responsible for larger mining operations, are generally compliant with national and international work laws and regulations. Further improving the work environment of large-scale formal mining is important, with the International Council on Mining and Metals (ICMM) and its members committed to reducing operational fatalities to zero. ICMM is an international organisation dedicated to a safe, fair and sustainable mining and metals industry, with 26 mining and metals companies and 35 regional and commodities associations as members. Amongst their members, ICMM reported

63 fatalities in 2016, 51 fatalities in 2017, and 50 fatalities in 2018 (ICMM 2019a). The number of recordable injuries in 2018, increased from 7,515 to 7,751.

Artisanal and small-scale mining (ASM) is also of great importance, expanding rapidly and often informally in many developing countries. ASM activities may be outside of legal and regulatory frameworks, with little monitoring and conditions not conforming to international labour standards (ILO 2019b). Small-scale mining employs large numbers of women and children, with accident rates estimated to be 6–7 times higher than in larger operations (ILO 2019b).

Safe and secure work environments are also free from modern slavery (including forced or compulsory labour). Approximately 20.9 million people around the world are still in forced labour, with more than half of these being women and girls (ILO 2019c). Modern slavery is a loss of personal freedom, resulting in diverse forms of exploitation from forced prostitution and forced marriage to forced labour and debt bondage. Forced labour is used in a variety of sectors, including mining and stone quarrying (Mendelsohn 1991; Upadhyaya 2004). The mining sector's high exposure to the risk of slavery is driven by large supply chains with suppliers having little incentive or ability to tackle exploitation (ICMM 2019b), as well as the overlap between sites of mineral extraction and active conflicts.

When mining takes place in active conflict zones, such as the Democratic Republic of Congo (DRC), instability can result in the use of forced labour. Demand for laptops and smartphones results in a corresponding demand for the natural resources integral to such consumer electronics (e.g., cobalt, gold, tantalum). In situations where regulation and governance are weak, the exploitation of natural resources can create profit for armed groups with human traffickers exploiting vulnerable communities. The Council on Foreign Relations (2018), note the use of both refugees and Congolese as forced labour in DRC mines, as well as women and children, being the victims of sexual slavery and the use of children as soldiers. A study in the south of Sierra Leone showed that in almost all

Fig. 8.5 Opencast coal mining in Siberia, Russia Image by Анатолий Стафичук from Pixabay

the mines visited, children were used as labour in breach of both the National Mining Act and international laws (Sheriff et al. 2018). Geoscientists share a responsibility with all of society for the eradication of modern slavery and must remain vigilant and report such practices in the sectors and geographic regions they operate in.

Safe and secure work environments do not tolerate harassment of any form. Harassment endangers the personal, profession, physical, and emotional well-being of individuals and their communities, and can have, especially injurious effects in disciplines with low diversity, such as geoscience. Studies show disturbing numbers of scientists who have experienced (sexual) harassment (e.g., Archie and Laursen 2013; Clancy et al. 2014), with many more cases reported in the media. In a 2010, member survey of the Earth Science Women's Network (ESWN), 51% of almost 500 respondents indicated to have experienced sexual harassment sometime during their career (Archie and Laursen 2013). Harassment is disproportionately targeted at groups that are already underrepresented in the geoscience community, such as women. In engineering, a bad workplace climate and culture resulted in 20% of women engineers leaving their field and thereby impacting the number of practicing women engineers (Fouad et al. 2012). In a 2014, survey of scientists conducting research in field settings (including social, life, and earth science disciplines), 71% of women respondents and 41% of men reported receiving inappropriate comments and 26% of women and 6% of men reported experiencing sexual assault while conducting field research (Clancy et al. 2014). Effective and enforced codes of conduct and grievance policies are often missing or lacking (Clancy et al. 2014). Harassment is further discussed in **SDG 5** (gender equality).

8.5 Conclusions

The natural environment is explicitly embedded into the targets and ambitions of **SDG 8**, recognising that sustainable economic growth requires resource efficiency and a decoupling from environmental degradation. This depends on effective environmental monitoring, geoscience research, and geoscience engagement in innovative technologies, practices, and policies. The breadth of geoscience engagement in **SDG 8** is, however, much broader than the environmental references made in **Target 8.4**. As articulated in this chapter, improving economic growth, diversification, and 'greening' depends on improved understanding of geological processes and resources (**Targets 8.1** to **8.4**). For a nation considering how to increase economic growth, sustain growth, build resilience into growth, and reduce the environmental impact of growth, it is imperative to understand natural capital (including both bio- and geodiversity) and how to manage this in a sustainable manner.

Targets 8.5 to **8.9** focus on full and productive employment and the provision of decent work for all. For all sectors, including those employing geoscientists, there is a need to ensure decent work conditions, provide safe and secure work environments, and remain vigilant to the use of modern slavery. Implementation of the SDGs will result in the creation and loss of jobs as major transitions are made around the world to end poverty, ensure inclusive growth and social development, and tackle environmental challenges. Geoscience professional and learned societies, geoscience unions, and bodies representing major sectors employing geoscientists should evaluate the implications of these transitions in order to support the geoscience community to adapt, mitigate potential challenges, and respond to increasing opportunities. For example, the focus on sustainable tourism in **Target 8.9** provides an opportunity for enhanced protection and celebration of our geological heritage. The UNESCO Global Geopark initiative provides one such opportunity for geoscience and geoscience's connection with culture to generate employment while simultaneously supporting environmental protection.

Delivering **SDG 8** requires an integrated approach, and recognition that its progress depends on engagement by diverse sectors (including geoscientists) and progress in other SDGs. Figure 8.6 illustrates some of the other

EXAMPLE INPUTS   DELIVERING SDG 8 FACILTATES

Fig. 8.6 Links between SDG 8 and other SDGs. The delivery of economic growth and decent work requires diverse inputs from improved infrastructure (**SDG 9**) to gender equality (**SDG 5**). In turn, progress in **SDG 8** delivers resources to invest in infrastructure, an improved natural environment, and enhanced social development. *SDG icons developed by the United Nations. The authors support the Sustainable Development Goals*

SDGs that act as enablers of **SDG 8** (including improved health, education, industrialisation, equalities and partnerships) or reinforce progress towards this goal. Figure 8.6 also illustrates the SDGs that depend on improved economic growth and decent work (amongst other factors) if they are to be achieved. This figure not only demonstrates the interconnectedness of the SDGs, through the lens of **SDG 8**, but also the critical role of geoscience research and practice to their ambitions and targets.

8.6 Key Learning Concepts

- Strong, sustainable and productive economies are secured through improved access to economic services, economic diversification, the creation of decent and fulfilling jobs, and the valuing and implementation of good practice regarding labour standards and workplace safety.
- Diversifying an economy can have a positive and stabilising effect on economic growth and employment, and reduce vulnerability to economic, social, and environmental shocks. The economies of many countries are dependent on natural resources. Governments can help to protect themselves from economic slumps from commodity price drops by using revenues from resource extraction to develop human capital and diversify their economy.
- Technological upgrading can help improve economic productivity. Improving access to appropriate technologies requires knowledge exchange, research and development, and access to financial services. This is complemented by broader innovation, including improved blending of western and indigenous

science and better integration of geoscience into policy.

- Sustainable development requires the creation of decent jobs, livelihoods that are stable, pay a fair wage, and have safe working conditions. Implementation of the SDGs will shape the future of work, and the types of jobs available to geoscientists. Geoscience educators and professional societies should seek to understand potential changes to labour markets so that they can support geoscientists to gain suitable employment. Growth areas are likely to link to applied environmental science and sustainability science.
- Geoheritage is an opportunity to boost sustainable tourism and build awareness of links between the natural environment, history, and culture. UNESCO Global Geoparks are playing an important role in the development of sustainable geotourism and help to create diverse livelihoods for those living in or near the geopark.
- The development of safe and secure working environments means creating spaces where injuries are prevented, and workplace violence and harassment of all kinds are eliminated. The field-based nature of many geoscience careers means access to the appropriate field safety training should be included in the education and ongoing professional development of all geoscientists. Geoscientists should be vigilant of poor practice that increases risks to colleagues and report health and safety breaches.
- The extractives industry has a high exposure to the risk of slavery, driven by complex supply chains and the demand for minerals found in active conflict zones where governance and regulation are weak.
- Effective and enforced codes of conduct relating to harassment and discrimination are also required to create decent work environments.

8.7 Educational Ideas

In this section, we provide examples of educational activities that connect geoscience, the material discussed in this chapter, and scenarios that may arise when applying geoscience (e.g., in policy, government, private sector international organisations, NGOs). Consider using these as the basis for presentations, group discussions, essays, or to encourage further reading.

- Chile and Zambia are two of the biggest producers of mined copper. Prepare a one-page briefing note that contrasts Chile and Zambia and outlines (i) their natural resource endowment, (ii) differences in their economic diversification, and (iii) lessons learned that would support a third country to responsibly govern their copper source.
- Reflect on the technologies that you use to assist your study of geoscience (e.g., hardware, software, laboratory equipment), and consider the productivity implications if these were not available. Research the global distribution of some of these technologies, and evaluate what steps may be needed to upgrade technology in one or more national contexts.
- Research the concept of '*technology readiness levels*', and consider how these relate to technology upgrading in key geoscience industries.
- Hold a debate in class about 'future employment prospects for geoscientists', exploring how demand for geoscientists may differ 10 or 30 years from now in terms of specific skills or themes of expertise. What actions could you take (e.g., online learning, module selection, extra reading) to help prepare you for a changing sector.
- Write a job description for a graduate 'Earth sustainability scientist' employed in a sector of your choice, thinking about what this job may involve and the skills and knowledge that may be requested.
- Prepare a comprehensive risk assessment and field safety plan for a geological mapping programme in a country of your choice. Consider the process you followed to prepare this plan, and design a simple training course that you could run with international colleagues to build capacity in safe work practice.

- Outline a geotouristic route (e.g., a hiking path or bike route) that highlights at least one geoheritage site, a site of natural and/or cultural value and another local attraction. Create a short information leaflet for visitors that explains what can be seen, making sure you use terms and wording that non-scientists understand.

Further Reading and Resources

European Commission (2018) The inclusive green economy in EU development cooperation. Tools and Methods Series Reference Document No 25. European Commission. https://publications.europa.eu/en/publication-detail/-/publication/a7a02150-01ad-11e9-adde-01aa75ed71a1/. Accessed 10 Sept 2019

Everett T, Ishwaran M, Ansaloni GP, Rubin A (2010) Economic growth and the environment. Defra Evidence and Analysis Series, UK Government

Montt G, Fraga F, Harsdorff M (2018) The future of work in a changing natural environment: climate change, degradation and sustainability. ILO Future of Work Research Paper Series. 48 p

Social Progress Index. https://www.socialprogress.org/resources?filter=2018

UN (2014) Interim progress report on the 10 year framework of programmes on sustainable consumption and production patterns (10YFP). https://sustainabledevelopment.un.org/content/documents/1444HLPF_10YFP2.pdf. Accessed 10 Sept 2019

UNDESA (2017) The impact of the technological revolution on labour markets and income distribution. UN Department of Economic and Social Affairs, Frontier Issues. https://www.un.org/development/desa/dpad/wp-content/uploads/sites/45/publication/2017_Aug_Frontier-Issues-1.pdf. Accessed 16 Sept 2019

References

AGI (2019) Geoscience Workforce Changes 2018-2028. www.americangeosciences.org/geoscience-currents/geoscience-workforce-changes-2018-2028. Accessed 27 Sept 2019

Archie T, Laursen S (2013) Summative report on the Earth Science Women's Network (ESWN) NSF Advance Paid Award (2009-2013), 149 p

Arouca Declaration (2011) Arouca Declaration. www.aroucageopark.pt/documents/78/Declaration_Arouca_EN.pdf. Accessed 17 July 2019

Avis W (2018) Inclusive and Green Growth in Developing Countries. K4D Helpdesk Report. https://assets.publishing.service.gov.uk/media/5af9702340f0b622dd7aa2c8/Inclusive_green_growth_in_developing_countries.pdf. Accessed 10 Sept 2019

BGS (2018) Battery Raw Materials. https://www.bgs.ac.uk/mineralsUK/statistics/rawMaterialsForALowCarbonFuture.html. Accessed 10 Sept

Camacho M, Pérez-Quirós G (2013) Commodity prices and the business cycle in Latin America: living and dying by commodities? CEPR Discussion Papers 9367, C.E.P.R. Discussion Papers

Clancy KBH, Nelson JN, Rutherford NJ, Hinde K (2014) Survey of academic field experiences (SAFE): trainees report on harassment and assault. PLoS ONE 9: e102172

Corden WM, Neary JP (1982) Booming sector and de-industrialisation in a small open economy. The Econ J 92 (December):825–48

Council on Foreign Relations (2018) Modern slavery—its root causes and the human toll. https://www.cfr.org/interactives/modern-slavery/. Accessed 17 July 2019

Donoghue AM (2004) Occupational health hazards in mining: an overview. Occup Med 54(5):283–289

EITI (2019) EITI Progress Report 2019. https://eiti.org/sites/default/files/documents/eiti_progress_report_2019_en.pdf. Accessed 22 July 2019

European Commission (2011) Rio + 20: towards green economy and better governance. Communication from the Commission to the European Parliament, the Council, the European Economic and Social Committee and the Committee of the Regions. https://eur-lex.europa.eu/legal-content/EN/TXT/?uri=CELEX:52011DC0363. Accessed 10 Sept 2019

European Commission (2018) The inclusive green economy in EU development cooperation. Tools and methods series rreference document No 25, European Commission. https://publications.europa.eu/en/publication-detail/-/publication/a7a02150-01ad-11e9-adde-01aa75ed71a1/. Accessed 10 Sept 2019

Everett T, Ishwaran M, Ansaloni GP, Rubin A (2010) Economic growth and the environment. Defra Evidence and Analysis Series, UK Government

Fine D, van Wamelen A, Lund S, Cabral A, Taoufiki M, Dörr N, Leke A, Roxburgh C, Schubert J, Cook P (2012) Africa at work: job creation and inclusive growth. www.mckinsey.com/featured-insights/middle-east-and-africa/africa-at-work. Accessed 16 Sept 2019

Fouad NA, Singh R, Fitzpatrick ME, Liu JP (2012) Stemming the tide: why women leave engineering. NSF Women Eng Rep 2012:64p

Fruman C (2017) Economic diversification: a priority for action, now more than ever. https://blogs.worldbank.org/psd/economic-diversification-priority-action-now-more-ever. Accessed 3 July 2019

Gil Y, Pierce SA, Babaie H, Banerjee A, Borne K, Bust G, Cheatham M, Ebert-Uphoff I, Gomes C,

Hill M, Horel J (2018) Intelligent systems for geosciences: an essential research agenda. Commun ACM 62(1):76–84

Gill JC, Mankelow J, Mills K (2019) The role of Earth and environmental science in addressing sustainable development priorities in Eastern Africa. Environ Dev 30:3–20

Hausmann R, Rigobon R (2003) An alternative interpretation of the 'Resource Curse': theory and policy implications. NBER Working Papers 9424, National Bureau of Economic Research, Inc

Hunt A, Samman E (2019) Gender and the gig economy. London: Overseas Development Institute. www.odi.org/sites/odi.org.uk/files/resource-documents/12586.pdf. Accessed 16 Sept 2019

Hunt A, Samman E, Mansour-Ille D (2017) Syrian women refugees: opportunity in the gig economy? London: Overseas Development Institute. www.odi.org/sites/odi.org.uk/files/resourcedocuments/11742.pdf. Accessed 16 Sept 2019

ICMM (2018a) Social Progress in Mining Dependent Countries. www.icmm.com/website/publications/pdfs/social-and-economic-development/180710_revised_spimdcs.pdf. Accessed 3 July 2019

ICMM (2018b) Role of Mining in National Economies. https://www.icmm.com/website/publications/pdfs/social-and-economic-development/181002_mci_4th-edition.pdf. Accessed 3 July 2019

ICMM (2019a) Benchmarking 2018 safety data: progress of ICMM members. www.icmm.com/safety-data-2018. Accessed 3 July 2019

ICMM (2019b) Tackling modern slavery in the mining supply chain. www.icmm.com/en-gb/case-studies/action-against-modern-slavery. Accessed 3 July 2019

ILO (2018a) Women and men in the informal economy: a statistical picture. International Labor Office, Geneva. www.ilo.org/wcmsp5/groups/public/—dgreports/—dcomm/documents/publication/wcms_626831.pdf. Accessed 3 July 2019

ILO (2018b) World Employment and Social Outlook 2018: greening with jobs. International Labor Office, Geneva. www.ilo.org/weso-greening/documents/WESO_Greening_EN_web2.pdf. Accessed 3 July 2019

ILO (2019a) Global Standards. International Labor Office, Geneva. www.ilo.org/global/standards/introduction-to-international-labour-standards/conventions-and-recommendations/lang–en/index.htm. Accessed 3 July 2019

ILO (2019b) The Mining Sector. International Labor Office, Geneva. www.ilo.org/global/industries-and-sectors/mining/lang–en/index.htm. Accessed 3 July 2019

ILO (2019c) Forced Labour. International Labor Office, Geneva. www.ilo.org/global/topics/dw4sd/themes/forced-labour/lang–en/index.htm. Accessed 3 July 2019

Indecon (2017) Sectoral Economic Review of Irish Geoscience Sector, 109 p. https://www.gsi.ie/en-ie/publications/Pages/An-Economic-Review-of-the-Irish-Geoscience-Sector.aspx. Accessed 25 Sept 2019

Mendelsohn O (1991) Life and struggles in the stone quarries of India: A case-study. J Commonw Comparat Polit 29(1):44–71

Montt G, Fraga F, Harsdorff M (2018) The future of work in a changing natural environment: Climate change, degradation and sustainability. ILO Future of Work Research Paper Series. 48 p.

Nickless E (2017) Delivering sustainable development goals: the need for a new international resource governance framework. Ann Geophys 60

OECD (2019) The Future of Work: Employment Outlook 2019. www.oecd-ilibrary.org/sites/9ee00155-en/index.html?itemId=/content/publication/9ee00155-en. Accessed 16 Sept 2019

Reichstein M, Camps-Valls G, Stevens B, Jung M, Denzler J, Carvalhais N, Prabhat (2019) Deep learning and process understanding for data-driven system science. Nature 566:195–204

Roch F (2017) The adjustment to commodity price shocks in Chile, Colombia, and Peru. IMF Working Paper No. 17/208

Robbins J (2018) Native knowledge: what ecologists are learning from Indigenous people. https://e360.yale.edu/features/native-knowledge-what-ecologists-are-learning-from-indigenous-people. Accessed 20 Sept 2019

Schrodt F, Bailey JJ, Kissling WD, Rijsdijk KF, Seijmonsbergen AC, Van Ree D, Hjort J, Lawley RS, Williams CN, Anderson MG, Beier P, Van Beukering P, Boyd DS, Brilha J, Carcavilla L, Dahlin KM, Gill JC, Gordon JE, Gray M, Grundy M, Hunter ML, Lawler JJ, Mongeganuzas M, Royse KR, Stewart I, Record S, Turner W, Zarnetske PL, Field R (2019) To Advance Sustainable Stewardship, We Must Document Not Only Biodiversity But Geodiversity. Proceedings Of The National Academy Of Sciences Of The United States Of America, 116, 16155–16158.

Sheriff I, Gogra AB, Koroma BM (2018) Investigation into the impacts of artisanal gold mining on the livelihood foundation of Baomahun community in southern Sierra Leone. Nat Resour 9(02):42

Spalek A, Schütz F, Bruhn D (2013) List of European universities offering training and education in the field of geothermal energy. Deliverable 5.5, Geo-Elec Project. https://geothermal.org/PDFs/Universities_offering_education_and_training%20in_geothermal_energy_in_Europe.pdf. Accessed 1 Oct 2019

SSCU-HBS (2019) Platform Work in the UK 2016-2019. Statistical Services and Consultancy Unit (SSCU), University of Hertfordshire and Hertfordshire Business School (HBS). www.feps-europe.eu/attachments/publications/platform%20work%20in%20the%20uk%202016-2019%20v3-converted.pdf. Accessed 16 Sept 2019

Stewart IS, Gill JC (2017) Social geology—integrating sustainability concepts into Earth sciences. Proc Geol Assoc 128(2):165–172

UK National Commission for UNESCO (2013). Wider Value of UNESCO to the UK (2012-2013), 70 p

UN (2018a) The Sustainable Development Goals Report 2018. https://unstats.un.org/sdgs/files/report/2018/

TheSustainableDevelopmentGoalsReport2018-EN.pdf. Accessed 3 July 2019

UN (2018b) Decent Work and Economic Growth: Why It Matters. www.un.org/sustainabledevelopment/wp-content/uploads/2018/09/Goal-8.pdf. Accessed 3 July 2019

UNDESA (2014) Interim Progress Report on The 10 Year Framework of Programmes on Sustainable Consumption and Production Patterns (10YFP). https://sustainabledevelopment.un.org/content/documents/1444HLPF_10YFP2.pdf. Accessed 10 Sept 2019

UNDESA (2017) The impact of the technological revolution on labour markets and income distribution. UN Department of Economic and Social Affairs, Frontier Issues. Available at: https://www.un.org/development/desa/dpad/wp-content/uploads/sites/45/publication/2017_Aug_Frontier-Issues-1.pdf (accessed 16 September 2019)

UNDP (2019) SDG 8—Overview. www.undp.org/content/undp/en/home/sustainable-development-goals/goal-8-decent-work-and-economic-growth.html. Accessed 3 July 2019

UNESCO (2017) UNESCO Global Geoparks. www.unesco.org/new/en/natural-sciences/environment/earth-sciences/unesco-global-geoparks/. Accessed 3 July 2019

UNFCCC (2016) The concept of economic diversification in the context of response measures. UNFCCC Technical Paper. https://unfccc.int/sites/default/files/resource/Technical%20paper_Economic%20diversification.pdf. Accessed 3 July 2019

UNEP/UNWTO (2005) Making Tourism More Sustainable - A Guide for Policy Makers. http://www.unep.fr/shared/publications/pdf/DTIx0592xPA-TourismPolicyEN.pdf. Accessed 16 Sept 2019.

UN (2014) Interim Progress Report on The 10 Year Framework of Programmes on Sustainable Consumption and Production Patterns (10YFP). Available at: https://sustainabledevelopment.un.org/content/documents/1444HLPF_10YFP2.pdf. Accessed 10 September 2019

UNICEF (2016) Collecting water is often a colossal waste of time for women and girls. www.unicef.org/press-releases/unicef-collecting-water-often-colossal-waste-time-women-and-girls. Accessed 3 July 2019

Upadhyaya K (2004) Bonded labour in South Asia: India, Nepal and Pakistan. In: The political economy of new slavery. Palgrave Macmillan, London, pp 118–136

Van Ree CCDF, Van Beukering PJH (2016) Geosystem services: a concept in support of sustainable development of the subsurface. Ecosyst Serv 20:30–36

Water.Org (2019) Water Credit Initiative. https://water.org/about-us/our-work/watercredit/. Accessed 3 July 2019

Watson J, Hatfield S, Wright D, Howard M, Witherick D, Coe L, Horton R (2019) Automation with Intelligence. Deloitte Insights. https://documents.deloitte.com/insights/Automationwithintelligence. Accessed 16 Sept 2019

World Bank (2012) Inclusive Green Growth: The Pathway to Sustainable Development, 192 p

World Bank (2015) World Bank Group Engagement in Resource-Rich Developing Countries: The Cases of the Plurinational State of Bolivia, Kazakhstan, Mongolia, and Zambia. Clustered Country Program Evaluation Synthesis Report. World Bank, Washington DC

World Bank (2019a) GDP Growth. https://data.worldbank.org/indicator/NY.GDP.MKTP.KD.ZG. Accessed 3 July 2019

World Bank (2019b) GDP Per Capita Growth. https://data.worldbank.org/indicator/NY.GDP.PCAP.KD.ZG. Accessed 3 July 2019

World Bank (2019c) Total Labour Force. https://data.worldbank.org/indicator/SL.TLF.TOTL.IN. Accessed 3 July 2019

World Bank Group (2017) The Global Findex Database. https://globalfindex.worldbank.org. Accessed 8 Oct 2019

World Economic Forum (2019) A New Circular Vision for Electronics: Time for a Global Reboot. http://www3.weforum.org/docs/WEF_A_New_Circular_Vision_for_Electronics.pdf. Accessed 8 Oct 2019

Young KE, Evans CA, Hodgesa KV, Bleacherc E, Graffbd TG (2016) A review of the handheld X-ray fluorescence spectrometer as a tool for field geologic investigations on Earth and in planetary surface exploration. www.sciencedirect.com/science/article/pii/S0883292716301226

Katrien An Heirman is an Earth scientist with expertise in climate and environmental change, geohazards and geoheritage, science communication and science policy. She graduated with Masters degrees from Ghent University (Belgium) and Royal Holloway, University of London (UK). She returned to Ghent University and gained her PhD and continued with several postdoctoral fellowships. Her multi-team, multi-disciplinary, international research has sent her to all corners of the globe. From 2015 to 2019, she worked as project officer at the International Geoscience and Geoparks Programme secretariat at the UNESCO Headquarters in Paris. In 2019, she joined the Geological Survey of the Netherlands as international policy advisor, building international development cooperations. In 2021, she moved to Rwanda to support the International Conference on the Great Lakes Region as a natural resource governance advisor for GIZ. This article/publication reflects Katrien's personal opinions.

Joel C. Gill is International Development Geoscientist at the *British Geological Survey*, and Founder/Executive Director of the not-for-profit organisation *Geology for Global Development*. Joel has a degree in Natural Sciences (Cambridge, UK), a Masters degree in Engineering Geology (Leeds, UK), and a PhD focused on multi-hazards and disaster risk reduction (King's College London, UK). For the last decade, Joel has worked at the interface of Earth science and international development, and plays a leading role internationally in championing the role of geoscience in delivering the UN Sustainable Development Goals. He has coordinated research, conferences, and workshops on geoscience and sustainable development in the UK, India, Tanzania, Kenya, South Africa, Zambia, and Guatemala. Joel regularly engages in international forums for science and sustainable development, leading an international delegation of Earth scientists to the United Nations in 2019. Joel has prizes from the London School of Economics and Political Science for his teaching related to disaster risk reduction, and Associate Fellowship of the Royal Commonwealth Society for his international development engagement. Joel is a Fellow of the Geological Society of London, and was elected to Council in 2019 and to the position of Secretary (Foreign and External Affairs) in 2020.

Sarah Caven Having started out in mineral exploration, Sarah now draws upon a diversity of global experience spanning private sector, government, social enterprise, and international development. A natural translator across scales and sectors, she contributes to artisanal and small scale mining programs through to regional prospectivity projects. In response to the resourcing future generations challenge, Sarah is passionate about enhancing collaboration, equity and business innovation in mining to unlock development opportunities. Sarah holds a master's in geology (University of Leicester), a Master of Business Administration, (University of British Columbia), and participated in Columbia University's executive education program, Extractive Industries and Sustainable Development.

Infrastructure, Industry, and Innovation

Joel C. Gill, Ranjan Kumar Dahal, and Martin Smith

J. C. Gill (✉)
British Geological Survey, Environmental Science Centre, Nicker Hill, Keyworth, Nottingham NG12 5GG, UK
e-mail: joel@gfgd.org; joell@bgs.ac.uk

J. C. Gill
Geology for Global Development, Loughborough, UK

R. K. Dahal
Central Department of Geology, Tribhuvan University, Kirtipur, Kathmandu, Nepal

M. Smith
British Geological Survey, The Lyell Centre, Research Avenue South, Edinburgh EH14 4AP, UK

Abstract

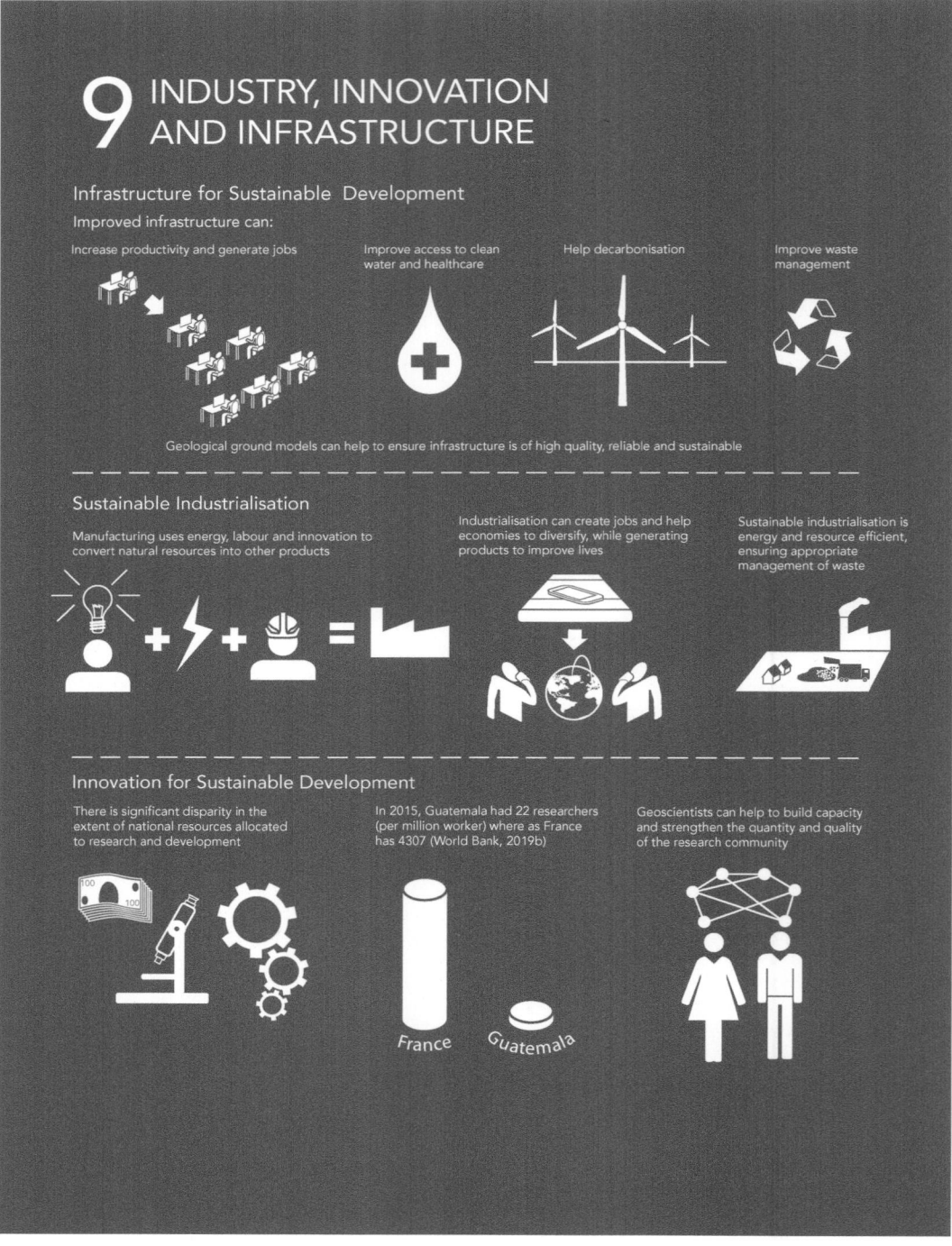

9.1 Introduction

Investment in energy, water and sanitation, telecommunications, transport, and waste management infrastructure is essential to efforts to improve economic growth and human development. Poor transport infrastructure adds 30–40% to the cost of goods traded amongst African countries, and current infrastructure constraints reduce economic productivity in the least developed countries of Africa by approximately 40% (Ayemba 2018). Improved infrastructure can increase productivity and generate jobs, supporting the ambitions of **SDG 8**. For example, if India were to increase investment in infrastructure by 1% of GDP, this could result in an estimated 3.4 million new jobs (EIU 2019). Reliable energy networks, resilient transport systems, and fast telecommunications are necessary for the expansion of industrialisation, the realisation of efficiency improvements, and the movement of goods to national and international markets.

Such industrialisation can create jobs and help economies to diversify (see **SDG 8**), as well as generating the products (e.g., solar panels, batteries, electric vehicles) required to decarbonise societies (see **SDG 7**), reduce food waste, and ensure well-resourced health and education facilities (see **SDGs 3** and **4**). For industrialisation to contribute to economic, social, and environmental sustainability, however, there needs to be close integration with research and development (or 'innovation'), including that by the geoscience community. Scientific research has enabled the development of technologies to increase the efficiency of industrial processes and reduce the environmental impact of industry. Innovation has resulted in diverse renewable technologies, water treatment facilities, and new materials that will underpin many future hubs of industry.

The specific targets of **SDG 9:** ***Build resilient infrastructure, promote inclusive and sustainable industrialisation and foster innovation*** are shown in Table 9.1, and focus on three themes (infrastructure, industrialisation, and innovation), but recognise the need for an integrated approach to all three.

The targets in Table 9.1, also emphasise a suite of cross-cutting themes, outlined below.

- *Resilience.* Quality infrastructure and industry should be resilient to external shocks, such as natural hazards or increases in raw material costs.
- *Equitable Access.* **SDG 9** emphasises the need to assist poorer and more vulnerable populations to better access and benefit from infrastructure, industry and innovation, recognising the '*leave no one behind*' theme that is within all the SDGs.
- *Resource Efficiency and Sustainability.* Both industry and infrastructure consume natural resources, and consideration must be given to make construction and manufacturing as energy and resource efficient as possible.
- *Knowledge Exchange between Countries.* Achieving **SDG 9** requires the free exchange of ideas, science, technologies, and innovation within and between countries. This includes effective management of transborder infrastructure, as well as international scientific collaborations.

These themes are not unique to **SDG 9**, but form a core part of the 2030 Agenda, and as such are characterised throughout this book. This chapter explores the targets in Table 9.1, and the cross-cutting themes above from the context of the geological sciences, highlighting the importance of involving geoscientists in diverse sectors (e.g., academia, the public sector, consultancy, and commercial practice), with different specialisms (e.g., engineering geology, hydrogeology, resource geology, and geohazards). Resilient infrastructure and industry facilities demand a comprehensive understanding of ground conditions, including the location of different geological materials and the geotechnical characteristics of these materials, as well as the dynamics of ground and surface water. Industry needs raw materials (e.g., industrial minerals, metals) and energy to manufacture products. Both infrastructure development and industrialisation require effective waste management,

Table 9.1 SDG 9 targets and means of implementation

Target	Description of Target (9.1–9.5) or Means of Implementation (9.A –9.C)
9.1	Develop quality, reliable, sustainable, and resilient infrastructure, including regional and transborder infrastructure, to support economic development and human well-being, with a focus on affordable and equitable access for all
9.2	Promote inclusive and sustainable industrialisation, and by 2030, significantly raise the industry's share of employment and gross domestic product, in line with national circumstances, and double its share in the least developed countries
9.3	Increase the access of small-scale industrial and other enterprises, in particular, in developing countries, to financial services, including affordable credit, and their integration into value chains and markets
9.4	By 2030, upgrade infrastructure and retrofit industries to make them sustainable, with increased resource-use efficiency and greater adoption of clean and environmentally sound technologies and industrial processes, with all countries taking action in accordance with their respective capabilities
9.5	Enhance scientific research, upgrade the technological capabilities of industrial sectors in all countries, in particular developing countries, including, by 2030, encouraging innovation and substantially increasing the number of research and development workers per 1 million people and public and private research and development spending
9.A	Facilitate sustainable and resilient infrastructure development in developing countries through enhanced financial, technological, and technical support to African countries, least developed countries, landlocked developing countries, and Small Island Developing States
9.B	Support domestic technology development, research, and innovation in developing countries, including by ensuring a conducive policy environment for, inter alia, industrial diversification, and value addition to commodities
9.C	Significantly increase access to information and communications technology and strive to provide universal and affordable access to the Internet in the least developed countries by 2020

protecting soils, water, and air from an array of contaminants.

Delivering **SDG 9** requires diverse inputs *beyond* geoscience. Resilient and sustainable infrastructure and industries will require economists, social scientists, ecologists, and engineers (i.e., economic, human, and environmental aspects), as well as the integration of geological understanding into planning. These additional factors are beyond the scope of this chapter, but we encourage the reader to explore an integrated perspective on infrastructure, industry and innovation through further reading. In this chapter, Sect. 9.2 focuses on infrastructure for sustainable development, setting out the key geological considerations to ensure infrastructure is reliable and resilient. Section 9.3 explores sustainable industrialisation, and Sect. 9.4 discusses the contribution of scientific research and development to sustainable development. We integrate these three themes in Sect. 9.5, focusing on the example of development corridors.

9.2 Infrastructure for Sustainable Development

9.2.1 Types and Benefits of Infrastructure

Infrastructure is a set of structures and facilities required for a society to function. This includes both hard infrastructure (e.g., roads, power stations, sewage treatment plants), and soft infrastructure (e.g., health facilities, financial systems, museums). Improved infrastructure results in economic, social and environmental dividends (EIU 2019).Infrastructure helps to create jobs and improve productivity. Infrastructure can support decarbonisation, reduce pollution, and increase resilience to natural hazards. Infrastructure can also improve quality of life, through better health and education. Improved infrastructure therefore directly and indirectly contributes to many of the SDGs, as outlined in

Table 9.2 Infrastructure and the SDGs

Infrastructure	Examples	Examples of direct and indirect and contribution to SDGs
Energy	Power stations	**SDG 7**. Energy infrastructure is needed to eliminate energy poverty, and ensure access to reliable, clean technologies
	Pipelines	
	Power lines	
Transport	Ports	**SDG 8**. Improved transport infrastructure can help products access new markets, supporting economic growth **SDG 13**. Reliable public transport can reduce the use of cars, and make more efficient use of renewable energy
	Airports	
	Roads	
	Train networks	
Water	Pipelines	**SDG 6**. Water infrastructure is needed to ensure universal access to safe water and sanitation **SDG 5**. Water infrastructure can improve gender equality by reducing the time spent collecting water **SDG 3**. Water infrastructure improves health and well-being, through a reduction in waterborne diseases
	Water Treatment Plants	
Solid waste	Landfill sites	**SDG 14**. Effective solid waste infrastructure is needed to reduce the amount of solid waste entering into and polluting the world's oceans **SDG 15**. Effective solid waste infrastructure is needed to reduce the amount of solid waste entering into and polluting terrestrial ecosystems
	Recycling plants	
Digital communications	Wireless broadband	**SDG 2**. Access to information via mobile applications and technologies can improve agricultural efficiency (e.g., weather and pest information) **SDG 4**. Access to ICT can improve education and enable greater freedom to pursue distance and online learning **SDG 8**. Access to broadband can improve connectivity and economic productivity **SDG 13**. Information about hazards (droughts and flooding, landslides)
	Mobile access	

Table 9.2. For example, ensuring universal access to clean water and safe sanitation (**SDG 6**) will require a suite of water and sanitation infrastructure in both rural and urban areas. Tackling climate change (**SDG 13**) will require a culture change away from the use of cars to instead using clean, efficient, mass transit mechanisms. Effective solid waste management infrastructure can help to reduce pollution and protect both terrestrial and ocean ecosystems (**SDGs 14** and **15**).

Improved information and communications technology (ICT, **Target 9.C**), benefits education (**SDG 4**), economic growth (**SDG 8**), and effective governance (**SDG 16**, for example, through the provision of online services). The rapid growth of distance learning, for example, through massive open online courses, is very exciting for continued professional development and lifelong learning. It offers access to teaching and educational resources from world-leading institutes. Courses currently on the *FutureLearn* website include.

- The Earth in My Pocket: an Introduction to Geology (The Open University)
- Data Science for Environmental Modelling and Renewables (University of Glasgow)
- Causes of Climate Change (University of Bergen)
- Exploring Possible Futures: Modelling in Environmental and Energy Economics (University of Basel)
- How to Survive on Earth: Energy Materials for a Sustainable Future (University of Wollongong)
- The Challenge of Global Water Security (Cardiff University)

These, and many other online courses, could enrich the understanding of geoscience and sustainable development both by geoscientists and by those shaping development policy. Making the most of these opportunities depends on reliable, high quality internet access, currently not available to many. While access to the internet in Africa has significantly increased since 2000, it is still far from universal. For example, 0.02% of the Ethiopian population had access in 2000 with this increasing to 18.62% by 2017; and 5.35% of the South African population had access in 2000 with this increasing to 56.17% by 2017 (ITU 2018). Planned investment in ICT, through submarine fibre cables and terrestrial networks, aims to further increase telecommunication and internet access. For example, the African Development Bank Group (AfDB) is supporting the Trans-Saharan fibre-optic backbone, increasing access to high-speed and affordable broadband, establishing and reinforcing ICT links between Niger, Chad, Algeria, Nigeria, Benin, and Burkina Faso (AFDB 2016).

When planning infrastructure to support sustainable development, it is helpful to think across different scales to ensure affordable and equitable access. In the context of transport, this may mean (i) improving regional access (e.g., investing in high quality, smaller roads to improve the connectivity of rural communities), (ii) investing in nationally important infrastructure (e.g., improving the efficiency of key ports, airports, and rail terminals), and (iii) international, transborder routes (e.g., working with neighbouring countries to develop a coherent plan for an integrated transportation network).

Many rural communities lack access to (are more than 2 km away from) an all-season road, and therefore connectivity to support economic and social development. World Bank (2016) examined six countries in sub-Saharan Africa and found that rural access (as defined above) varied: Ethiopia (22%), Kenya (56%), Mozambique (20%), Tanzania (25%), Uganda (53%), and Zambia (17%). This demonstrates a significant infrastructure gap, with a combined total of 148 million people in these six countries lacking access to roads (World Bank 2016). For many people living within 2 km of an all-season road, particularly those with disabilities, the elderly and other vulnerable groups, these distances will mean roads are still not accessible (Fig. 9.1).

For many countries, key national infrastructure includes power generation and transmission, transport hubs (e.g., airports), and the supply of potable water. The island of St Helena (population approximately 4500) in the Southern Atlantic is very remote. A monthly shipping service was the primary form of transport arriving into and leaving St Helen, running from Cape Town (South Africa). In 2017, an airport (Fig. 9.2), opened and with a weekly scheduled flight from Johannesburg (South Africa) to St Helena, via Windhoek (Namibia). This improved transport connection aims to improve tourism and economic development. In India (population approximately 1.3 billion people), there are 163 million people lacking access to safe water. India has a major desalination plant near Chennai, producing 36.5 million m^3 of water each year. This single piece of infrastructure is therefore essential to responding to increased demand for water.

In many contexts, infrastructure cuts across national borders. **Target 9.1** emphasises the need for regional and transborder infrastructure, with **Target 9.C** noting the need for support to be given to 'developing countries' to enable infrastructure development (see Table 9.1). Roads or railways may stretch from a landlocked site of production in one country to a major port in another country. For example, Kenya, is the largest exporting nation in East Africa, with the major port of Mombasa, handling approximately 30 million tonnes of cargo in 11 months (Akwiri 2019). The East Africa Railways Master Plan proposes connecting the ports of Mombasa and Lamu in Kenya, with other parts of Kenya, Uganda, Rwanda, Burundi, the Democratic Republic of Congo, South Sudan, and Ethiopia. The plan also proposes enhancing the connectivity of ports in Tanzania. This will allow a more efficient export of goods from these countries,

Fig. 9.1 Road construction in Tanzania. *Credit* Joel C. Gill

Fig. 9.2 St. Helena Airport. *Credit* Paul Tyson, CC BY 3.0, https://creativecommons.org/licenses/by/3.0/deed.en

reducing transport costs and times. It will also improve the efficiency of imports to landlocked countries from the same ports. China's Belt and Road Initiative (Box 9.3), includes land and maritime infrastructure corridors which cross more than 70 countries.

Infrastructure can, therefore, catalyse many aspects of sustainable development, but this requires action at multiple scales, with the engagement of diverse local and national stakeholders, often across national boundaries (see **SDG 16**).

9.2.2 Geoscience for Quality, Reliable, Resilient, and Sustainable Infrastructure

While the extent or distribution of infrastructure needs to increase, it must also be high quality, reliable, resilient, and sustainable. Infrastructure should conform to recognised quality standards, to ensure that it is safe and fit-for-purpose. It should take into account locally relevant environmental conditions (e.g., hydrology, seismic hazards) so that it is reliable and resilient—as well as make use of appropriate materials and designs, engineering methods, and safety mechanisms. To be sustainable, infrastructure designs should also embed social, economic, and cultural factors that may affect the type of infrastructure that is appropriate, and changing demographics that may affect how infrastructure is used. Sustainable infrastructure is emphasised in **Target 9.4,** with this requiring understanding of raw material (e.g., aggregates, sand, and water) sources and flows, discussed in detail in **SDG 12**, and waste management. Quality, reliable, resilient, and sustainable infrastructure, therefore, requires geoscientists (see Table 9.3), from a range of sectors, as well as engineers, architects, planners, and others.

Table 9.3 Geoscience for quality, reliable, resilient, and sustainable infrastructure

Infrastructure characteristics	Role of geoscientists (Examples)
Quality	Inform, adhere to, and monitor against internationally recognised standards for infrastructure design and development. For example, the European Standards include *Eurocode 7* for geotechnical design, setting out how to conduct a ground investigation and testing and design geotechnical structures
Reliable	Infrastructure should perform to an optimum for as long as possible. This requires an understanding of the subsurface and other environmental conditions. A tunnel closed regularly for repair work due to water ingress is not 'reliable'. In contrast, constructing a tunnel with a comprehensive understanding of groundwater dynamics in the vicinity, and appropriate mitigation measures put in place during construction is not likely to suffer the same problems
Resilient	Infrastructure should be able to withstand a set of external pressures likely to occur in a given region. Examples include soil creep, shrink-swell soils, scour, seismic shaking, and loading by volcanic ash. For example, landslides in Panama resulted in large amounts of sediment entering and closing down the water treatment plant of Panama City for almost a month. Infrastructure should also be climate resilient (IHA 2019) The initial role of a geoscientist is in understanding the geohazards that may occur in a given region, and their characteristics (e.g., spatial and temporal distribution, magnitude, potential impacts on built infrastructure). This includes desk-based reviews of existing literature, monitoring of instrumental data, and detailed site investigations and associated testing Infrastructure can itself change the natural landscape and environmental processes, to have negative impacts. For example, a hydropower dam may change sediment flux and the transport of nutrients into downstream ecosystems (Kummu and Varis 2007), loading or unloading during construction may increase the likelihood of landslides being triggered during a storm of an earthquake (Gill and Malamud 2017)
Sustainable	Consider the sourcing of geological materials for construction (e.g., aggregates, water), and any environmental implications of extraction. For example, if construction requires large amounts of water, abstracted from already stressed aquifers, this could increase the risk of saline intrusion Consider if geotechnical designs take into account changing infrastructure use (i.e., due to increasing populations, or a changing proportion of people commuting to work), ensuring it is fit-for-purpose in the future. For example, a bridge installed today may carry 200 cars/day, but by 2030, this could have increased to 2000 cars/day. The foundation and bridge construction should use guidance from socio-economic scientists to inform design

The development of a conceptual ground model is one way in which the characteristics outlined in Table 9.3 are integrated and communicated. Ground models (see Dearman and Fookes 1974; Fookes 1997; Brunsden 2002) integrate geological and geomorphological mapping, the results of detailed site investigation testing (conducted in accordance with recognised standards), and any existing literature relevant to the project (e.g., on resources, hazards, contamination) to capture and visualise our best understanding of the subsurface. Conceptual ground models inform decisions about the need for further invasive testing (e.g., to assess the spatial distribution of expansive soils), with these results then improving the next iteration of the ground model. Conceptual ground models then inform the overall project design, helping the early identification of

challenging ground conditions and ensuring that the budget and design reflects these conditions. This helps to create reliable and resilient infrastructure, and offer good value for money.

Two examples of ground models in the UK are shown in Figs. 9.3 and 9.4. Merritt et al. (2013) generated a three dimensional ground model of a site with an active landslide, integrating geophysical, geomorphological, and geotechnical investigations (Fig. 9.3). Linde-Arias et al. (2019) generate a ground model of the subsurface to inform the excavation of an open face cross passage on the underground Elizabeth Line in London, UK (Fig. 9.4). Both of these UK examples integrate diverse geological data to characterise the subsurface and inform geohazard management and infrastructure development. Linde-Arias et al. (2019) note that their ground

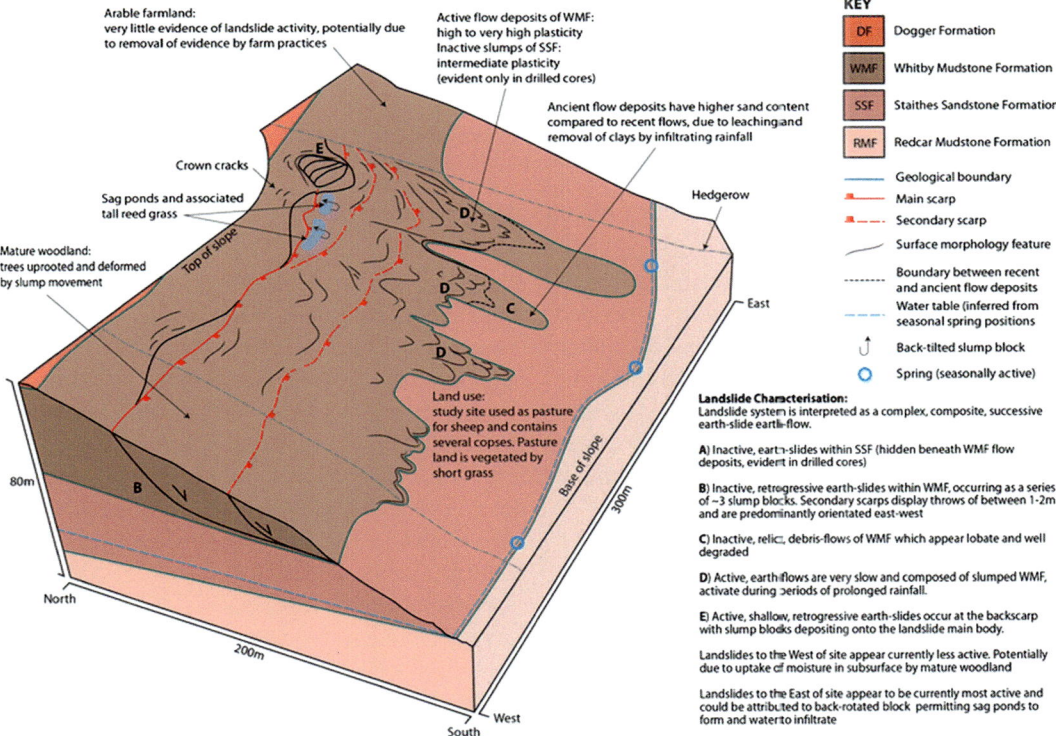

Fig. 9.3 3D ground model of the Hollin Hill study site based on geophysical, geomorphological and geotechnical investigations. Reprinted by permission from: Springer Nature, Landslides, 3D ground model development for an active landslide in Lias mudrocks using geophysical, remote sensing and geotechnical methods, Merritt, A.J., Chambers, J.E., Murphy, W., Wilkinson, P.B., West, L.J., Gunn, D.A., Meldrum, P.I., Kirkham, M. and Dixon, N., 2014. 3D ground model development for an active landslide in Lias mudrocks using geophysical, remote sensing and geotechnical methods. © Springer-Verlag Berlin Heidelberg 2013, 2013

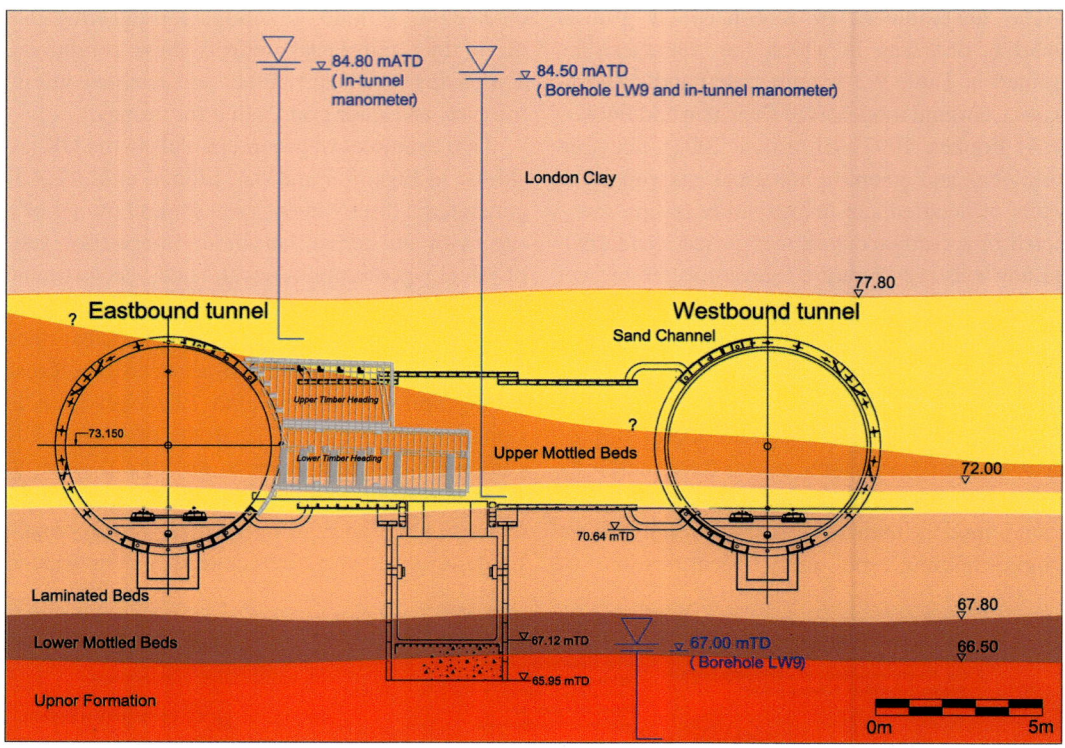

Fig. 9.4 2D ground model to inform the excavation of an open face cross passage on the underground Elizabeth Line, London, UK. Reprinted from Tunnelling and Underground Space Technology, 86, Linde-Arias et al., Development of a ground model, targeted ground investigation and risk mitigation for the excavation of an open face cross passage on the underground Elizabeth Line, London, 209–223, 2019, with permission from Elsevier

model resulted in changes to groundwater management during construction, which provided safer conditions for those working underground.

The development of ground models may be challenging in many global south contexts where pre-existing data to inform the ground model may be sparse, access to sites may be difficult (e.g., due to security concerns or the terrain), or due to the high costs of completing invasive testing. An initial conceptual ground model can be developed and informed by geological field mapping, descriptions of materials, and the development of cross-sections. The extent to which this model can be used to inform decision-making will depend on the purpose and may be restricted if it is solely based on inference from limited surface observations (Hearn and Massey 2009). For example, a lack of subsurface ground investigations in difficult, mountainous terrain in

Bhutan, made it challenging to use a ground model to understand deep-seated landslides (Hearn and Massey 2009). Many 'least developed countries' will face similar challenges. Where ground models are incomplete, their limitations not fully understood, and used to inform decisions about the siting of infrastructure this can result in costly delays, projects failing, or poor quality, unreliable infrastructure that is not resilient to external shocks.

National data repositories and geological surveys can help to address some of these challenges by collating and improving access to existing information, while reducing some duplication of costly drilling and testing. Depositing spatially referenced borehole records and materials, together with site investigation reports, into national data repositories can enhance capacity to inform the development of conceptual ground models, and advise on infrastructure development.

Box 9.1 Resilient Road Infrastructure in Nepal

Road construction is a key part of rural development in Nepal. This landlocked country, in the Himalayas, has 57,632 km of rural roads, with only 3.5% of these roads (2004 km) covered in tarmac as of 2016. Most of the roads in Nepal have not been constructed to international standards. Most rural roads are constructed through community participation, with little technical supervision. 74% of rural roads (as of 2016), are earthen roads and only serve people during the dry seasons when using an appropriately equipped vehicle. Roads and other transport have been identified as a major sector for investment to support development in Nepal. This requires public and future needs to be given careful consideration, and not just the needs or demands of donors.

Nepal has challenging geographic conditions that make road construction and maintenance difficult.

- Nepal is a tectonically active, multi-hazard region, with active faulting generating earthquakes. The $M_w = 7.8$ earthquake on 25 April 2015 resulted in 553 aftershocks with $M_w > 4$ in the following 45 days (Adhikari et al. 2015), together triggering snow avalanches and thousands of landslides, as well as increasing the probability of further landslides during monsoon rains (Bilham 2015; Collins and Jibson 2015). Triggered landslides blocked rivers and resulted in upstream flooding (Collins and Jibson 2015).

- Geological factors contribute to landslides and failures of cut slopes. Extreme rainfalls, loose soil deposits on slopes, rock discontinuities, and the overall structure of the rock mass all contribute to landslides, alongside the undercutting of slopes by humans.
- Extreme rainfall events are common in Nepal, with over 500 mm of rain falling in just 24 h, and 80% of annual rainfall falling in the three-month monsoon period.
- Climate change adds extra pressures, with infrastructure needing to cope with changing hydro-meteorological patterns including extreme heat, rain, snow, flooding, and storms.

This hazard landscape, together with the effects of climate change and difficult topography, resulting in a complex combination of threats to infrastructure, including roads and hydropower projects. These challenges apply to both smaller rural roads and large national highways (Fig. 9.5). In 2000, a large landslide blocked a major national road, carrying 5000 vehicles a day. This landslide blocked road access to Kathmandu for 11 days, resulting in an acute shortage of daily commodities. Prior to the landslide in 2000, the road was closed 18 separate times (for a total of 160 h). Geotechnical and engineering geological evaluations are essential to understand ground conditions and potential hazards, including landslides. Engineering geological mapping, discontinuity mapping in rock masses, and field and laboratory soil tests can help to understand ground conditions and obtain design parameters for engineered resilience.

Fig. 9.5 Low-cost mitigation measure in Krishnabhir Landslide at Prithi Highway connecting Kathmandu and Pokhara. The gabion retaining wall and associated bioengineering have helped to reduce the likelihood of slope disasters on this particular section of the landslide. *Credit* Ranjan Dahal

9.3 Sustainable Industrialisation

Industrialisation is a set of social and economic changes that result in manufacturing (the large-scale production of goods) becoming a primary economic activity of a country. Manufacturing takes natural resources and uses energy, labour, and innovation to convert these into more useful and higher-value products. For example, the manufacture of glass uses sand (SiO_2), limestone ($CaCO_3$), and soda ash (Na_2CO_3). Other companies may take products from one or more other manufacturing companies and integrate these to develop a new product, creating a complex supply chain. For example, a company making furniture may purchase glass, cut and treated wood, and machine processed metal (i.e., screws and

nails), which are then assembled to make display cabinets or tables. In the latter example, the manufacture of power tools and chemical wood preservation treatments also require separate industries and their own supply chains.

Inclusive and sustainable industrialisation is the focus of SDG Targets **9.2** and **9.3**. The United Nations (2016) have identified industrialisation to be an imperative for the African continent to help achieve economic and social development targets, including ending extreme poverty (see **SDG 1**). Many African economies are reliant on the production and export of raw commodities, and therefore, vulnerable to commodity price fluctuations or falling demand, and subsequent economic instability (Hausmann and Rigobon 2003). The importance of economic diversification for economic growth and social development is described in the previous chapter (**SDG 8**). A key aspect of that diversification is the transition from *producing and exporting* raw materials to processing raw materials and manufacturing goods for regional and global trade.

In advocating for more industrialisation, **SDG 9** is careful to note that this should be *sustainable*, with economic, human, and environmental considerations. The United Nations (2016), note a significant opportunity for Africa, to adopt alternative economic pathways to industrialisation, setting out a strategy for '*greener industrialisation*' that considers (i) increased energy efficiency, (ii) increased resource efficiency, and (iii) better waste management. These require an interdisciplinary suite of solutions, drawing in part on the skills and experiences of geoscientists. We discuss each of these themes below (recognising overlap and interactions), and briefly extend our discussion of resilient infrastructure in the previous section, to also consider the need for resilient industries. In each theme, there is an emphasis on technological innovations and more detailed understanding of physical processes. Enhanced research and development capabilities (the focus of Sect. 9.4), in all countries, are critical to ensuring industrialisation is facilitated and sustainable.

9.3.1 Industrialisation and Energy Efficiency

Industry accounted for approximately 37% of total global energy use in 2017 (IEA 2019). Many industries that extract raw materials (e.g., mining) and process raw materials into new products (e.g., manufacturing cements, iron and steel, glass, or chemicals) are energy-intensive (Jouhara et al. 2018). Growing industrialisation, particularly in emerging economies and the Global South will drive an increase in demand for energy in the next few decades meaning efficiency and greater use of renewable resources is essential to meet demand (**SDG 7**) and simultaneously tackle climate change (**SDG 13**).

Wind turbines, solar panels, hydropower, biogas, and geothermal energy sources (Fig. 9.6) will become increasingly important in industrial operations, reducing the use of fossil fuels and emission of greenhouse gases. For industries where lower temperatures are needed to enable processing (e.g., processes such as drying, evaporation, distillation, washing), geothermal offers an excellent alternative if the heat source and industrial operation are close (IRENA 2015). Geoscientists can support land-use planning by advising on where there may be the potential to use geothermal energy. For industries where higher temperatures are needed (e.g., the iron and steel sector), there are fewer methods capable of providing the sufficient temperatures (IRENA 2015), needed within the production process. Biomass (i.e., charcoal) is one possibility, however, production would need to be significantly scaled up to meet the technical substitution potential by 2030 (IRENA 2015). Coal and hydrocarbons are, therefore, likely to remain a key requirement for some forms of industrial development with carbon capture, storage, and use infrastructure required to minimise the environmental impact of this. Processing temperatures for other metals, however, are much lower (e.g., 140–280 °C for aluminium), potentially lending themselves to better integrate with renewable technologies (IRENA 2015).

Whether using renewable or non-renewable energy resources, greater energy efficiency is critical to sustainability. Energy efficiency requires the minimum amount of energy to be used to generate maximum productivity. This can be achieved through the design of efficient buildings and operations, and strategies to capture waste products (e.g., heat, CO_2) generated through industrial processes. Jouhara et al. (2018) review the different heat recovery technologies available for capturing waste heat, noting that this can be used to provide an additional energy source and reduce energy consumption. For example, Nordursalt in Iceland produce salt from seawater using hot wastewater from a nearby seaweed factory (otherwise discarded),

Fig. 9.6 Geothermal power generation in Iceland. Image by falco (from Pixabay)

and geothermal energy (Jóhannesson and Chatenay 2014), making this an energy-efficient manufacturing process.

The example of Nordursalt demonstrates how integrated development strategies and industrial partnerships can help to increase efficiency. Green industrialisation, including energy efficiency, requires thinking across and progress in multiple SDGs. For example, United Nations (2016), note that '*well-planned urban agglomeration can help ensure energy efficiency and facilitate resource efficiency in industrial production by enabling intra- and inter-industry interactions*' (p. 33). Many cities in the world's least developed countries have yet to be built. Sustainable and smart urban planning (**SDG 11**) can make it more cost-effective to implement greener technologies, and manage demand for energy resources, as well as enforce regulations that set minimum standards for energy efficiency in new buildings.

We refer the reader to **SDG 7** for a broader discussion of energy demand and challenges, **SDG 11** for a discussion of sustainable urbanisation, and **SDG 13** for discussion of climate change actions.

9.3.2 Industrialisation, Resource Efficiency, and Waste Management

In addition to energy efficiency, industrialisation also needs to consider efficiencies in the broader array of natural resources being used, reducing consumption and appropriately managing waste generated during manufacturing, from packaging, and the product itself (if not consumable) when it stops being useful. Industrialisation has the potential to produce large amounts of waste. Another aspect of resource efficiency is finding uses for this waste and ensuring waste is managed in an appropriate way to prevent air, water, and soil pollution. This aligns with at least four other SDG targets.

- **SDG 3.9**. *Substantially reduce the number of deaths and illnesses from hazardous chemicals and air, water, and soil pollution and contamination.*
- **SDG 8.4**. *Improve progressively, through 2030, global resource efficiency in consumption and production and endeavour to decouple economic growth from environmental degradation.*
- **SDG 12.4**. *By 2020, achieve the environmentally sound management of chemicals and all wastes throughout their life cycle, in accordance with agreed international frameworks, and significantly reduce their release to air, water, and soil in order to minimise their adverse impacts on human health and the environment.*
- **SDG 12.5**. *By 2030, substantially reduce waste generation through prevention, reduction, recycling, and reuse.*

Efficiency could be achieved by reducing the amount of a particular resource in a product or replacing a resource with an alternative and more sustainable material. These actions should not reduce the quality of a product, or reduce its lifetime, which could result in more resources being needed overall. Replacing Portland cement in concrete, for example, with blast furnace slag waste from the production of steel and fly ash waste from coal power generation can reduce carbon emissions associated with using Portland cement by an estimated 90%.[1] The product is also noted to have improved durability and increased fire resistance, thus contributing to the development of sustainable, resilient, and high quality infrastructure.

Slag waste left over from metal smelting or refining, is not only a useful addition to concrete. Research by the United States Geological Survey (USGS) has noted that the high calcium content

of slag can neutralise the acid from acid mine drainage or help slag absorb excess phosphates in the water when too much fertiliser is used (USGS 2017). Carbon8, a British company, is using a patented carbonation technology to support the circular economy by sequestering waste CO_2 gases and using this in their carbonation process to generate an improved aggregate source or treat contaminated soil and waste (Carbon8 2019). In the latter two examples, understanding of geological and geochemical processes is informing the development of innovative technologies to reduce waste.

Resource efficiency could also be achieved by increasing a product lifetime so that less raw materials are consumed over time. Modular technologies, such as Fairphone,[2] demonstrate resource efficiency principles with a focus on increasing how easy it is to repair a phone. The Fairphone is constructed using a set of modules that can be easily replaced rather than needing to replace the whole phone. Fairphone note:

> Consumer electronics are often viewed as semi-disposable objects, to be upgraded or discarded as soon as something better comes along. We're fighting against a market trend where the average phone is replaced every 18 months, creating a huge environmental impact. As technology advances rapidly, consumers are losing the ability to modify, repair, and truly understand how they can keep their devices longer.[3]

Given our understanding of the energy required to extract and process metals, and the waste generated, geoscientists should be the greatest advocates of approaches which minimise the extent of (unnecessary) mining that is required to replace products that could be easily repaired or a component replaced.

Resource efficiency could also include effective planning which repurposes waste and infrastructure after an initial industry closes. For example, mining generates infrastructure which can subsequently be used for large-scale fish farming in abandoned open pits and underground agriculture (mushroom farming) utilising abandoned underground tunnels and shafts

---

[1] https://www.wagner.com.au/main/what-we-do/earth-friendly-concrete/efc-home.

[2] https://www.fairphone.com/en/.

[3] https://www.fairphone.com/en/our-goals/design/.

(Engineering News 2002; Otchere et al. 2004; Centre for Development Support 2004). Some mine waste can also be repurposed. Waste from mining is used to make bricks and paving, or as fill in dams or subsided land (Zhengfu et al. 2010; CSIR 2019). Waste tips can also be reworked to extract new metals or more of the same metal, due to improved economic conditions or extraction techniques. Bell and Donnelly (2006) note examples of reworking of gold mining waste in South Africa, for additional gold, and reworking of galena mining waste in the UK, for barite and fluorspar, used in the offshore oil and steel industries, respectively.

We refer the reader to **SDG 12** for a broader discussion of responsible consumption and production, including waste management.

9.3.3 Resilient Industrialisation

In addition to *infrastructure* being resilient to external shocks (such as natural hazards), *industries* also need to be resilient to avoid negative economic, social, and environmental impacts. The priority actions within the *Sendai Framework for Disaster Risk Reduction* (UNISDR 2015), make widespread reference to businesses and the private sector. For example, Priority 1 (understanding disaster risk) emphasises the need to build the knowledge of the private sector in disaster risk and Priority 3 (investing in disaster risk reduction for resilience) explicitly notes the need to '*increase business resilience and protection of livelihoods and productive assets throughout the supply chains, ensure continuity of services and integrate disaster risk management into business models and practices*'.

Resilient industrialisation, therefore, needs to consider the resilience of.

- *Industrial Buildings* (e.g., the factory building). As with any infrastructure, industrial facilities are threatened by a range of external

pressures (e.g., seismic shaking, loading by volcanic ash, flooding, subsidence). Geoscientists help businesses to understand the spatial and temporal distribution of geohazards, and potential magnitudes and impacts of hazards. This information can inform the design of industrial facilities to help them withstand environmental shocks and avoid major technological disasters.

- *Industrial Operations* (e.g., the processing of chalcopyrite to produce copper). Commodity price fluctuations or increased demand may affect industrial operations through reduced availability of a given natural resource. This could be a raw material used within the product, or the energy required for processing. Resilient industrial operations may develop their own energy sources (e.g., through solar or geothermal) to help withstand price and demand fluctuations. Resilient operations also take into account changes to social license, recognising that operations are less likely to be stopped by public or government pressure if appropriate environmental regulations are adhered to.

- *Supply and Delivery Chains* (e.g., shipping copper to an industrial plant for further processing, and then shipping these to customers). For example, disruption to a transport network due to landslides triggered by heavy rain may hinder natural resources from reaching an industrial plant and cause them to cease operations temporarily. This could be financially costly and result in reputational damage if orders are delayed or cancelled. Access to and understanding of hazard maps could help businesses to plan alternative routes less likely to be affected by landslides during storms.

In all three cases, geological knowledge can inform the understanding of disaster risk, with geoscientists working alongside planners and logisticians to inform contingency plans.

9.4 Innovation for Sustainable Development

Science, research, and innovation are common themes throughout the SDGs, with multiple references within the SDG targets, particularly in the context of knowledge exchange (e.g., Targets **2.A**, **3.B**, **7.A**, **14.A**). **SDG 9** includes a specific target (**9.5**) focused on enhancing broad research and development (R&D) capabilities.

> Enhance scientific research, upgrade the technological capabilities of industrial sectors in all countries, in particular developing countries, including, by 2030, encouraging innovation and substantially increasing the number of research and development workers per 1 million people and public and private research and development spending (Target 9.5).

There is significant variation in the extent of national resources currently allocated to R&D. Figure 9.7 shows R&D expenditure as a percentage of GDP, from 2000 to 2015, for OECD members (solid line, top), the global average (dashed line, second from top), low and middle-income countries (dotted line, second from bottom), and Latin America and the Caribbean (dashed and dotted line, bottom). While OECD members spend an average of 2.4% of GDP on R&D, in low and middle-income countries, this is closer to 1.4% of GDP, and in Latin America and the Caribbean, it is approximately 0.78% of GDP. The largest increase between 2000 and 2015, however, was in low and middle-income countries growing from approximately 0.6% to 1.6% of GDP. Within this group, however, there is also significant variation. In 2015, Tajikistan spent 0.11%, Senegal spent 0.75%, and China spent 2.06% of GDP on R&D (World Bank 2019a).

The number of researchers per million workers, in a given country, also shows a large variation between Global North and Global South regions (World Bank 2019b). For example, North America (as of 2015) had 4313; European Union (as of 2016) had 3749; Latin America and the Caribbean (as of 2015) had 550; and South Asia (as of 2015) had 225 researchers per million people. In some countries, the number is

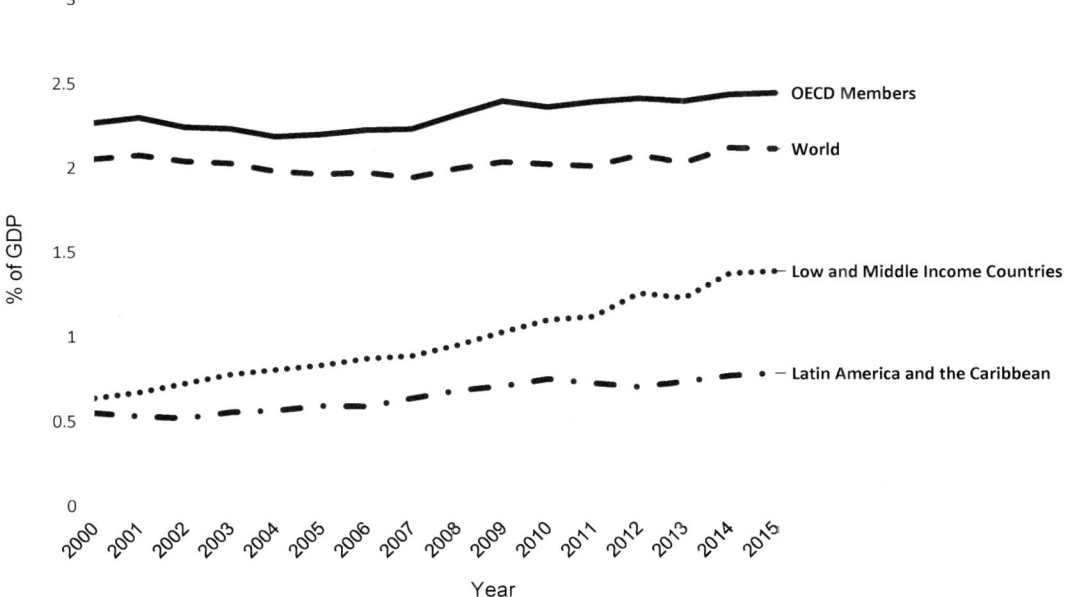

Fig. 9.7 Research and Development expenditure (% of GDP) from 2000 to 2015. Information for OECD members (solid line, top), the global average (dashed line, second from top), low and middle-income countries (dotted line, second from bottom), and Latin America and the Caribbean (dashed and dotted line, bottom). Data from World Bank (2019a)

substantially lower. For example, in Togo (as of 2017), there were 38 researchers, in Guatemala (as of 2015) there were 22, and in the Democratic Republic of Congo (as of 2015) there were 11 researchers in R&D, per million workers. In contrast, France (as of 2015) had 4307 researchers in R&D, per million workers, more than 391 times that of the Democratic Republic of Congo.

Meeting the ambitious demands of the SDGs will require an increase in all countries in the proportion of GDP spent on R&D and the number of researchers per million workers engaged in research. This requires political will and investment, demonstrating the importance of strong scientific institutions (see **SDG 16**) who often have greater advocacy power than individual geoscientists. At an individual level, it requires effective partnering and capacity building (see **SDG 17**) by the geoscience community. By working with existing researchers in countries such as the Democratic Republic of Congo or Guatemala, scientists help them to grow their own networks, sources of funding, and opportunities for dissemination. This can contribute to increased research income for scientists in the Global South to hire and train additional research staff, thereby strengthening the quantity and quality of the research community. Science graduates can be hired as research assistants, gaining experience and qualifications (e.g., graduate degrees), while being integrated into national and international science communities. Box 9.2 illustrates an example of international cooperation to strengthen science capacity, focused on managing and responding to volcanic hazards, which threaten infrastructure and industry in many regions.

Box 9.2. USGS Volcano Disaster Assistance Programme

The USGS Volcano Disaster Assistance Programme (VDAP[4]) builds international partnerships between the United States and nations affected by volcanic hazards and

disasters. The programme provides support during volcanic crises, trains those working on hazard monitoring and risk reduction in partner countries, and helps countries to establish and/or enhance capacity to monitor volcanoes, educate officials, and prepare for and manage volcanic events. Aspects of capacity building include.

- Travel grants, to support attendance at an annual 7-week training course in Hawai'i
- Sponsorship of workshops in the US and in partner countries on diverse topics such as seismology, volcanic gas emissions, lahar (mudflow) modelling, hazards mapping, volcano deformation, remote sensing, and monitoring of crater lakes.
- Mentoring to assist partners in writing and publishing their research.

VDAP partners include Argentina, Chile, Colombia, Costa Rica, Democratic Republic of Congo, Ecuador, El Salvador, Guatemala (Fig. 9.8), Indonesia, Mexico, Nicaragua, Papua New Guinea, Peru, and the Philippines.

Geoscientists should also look at the structural barriers hindering researchers in the Global South (e.g., the costs and visa barriers preventing attendance at major science conferences), as described in the chapter exploring **SDG 10**. Supporting increased innovation in the Global South can also require changes in collaboration practices—embedding the lessons articulated in **SDG 17** for effective and equitable partnerships. In 2015, a *Nature Geoscience* editorial (Vol. 8) reflected on how the geoscience community can strengthen geoscientific capacity in the Global South. This article showed the global distribution of *Nature Geoscience* author affiliations from January 2008 to May 2015, with no authors from many parts of the Global South (including much of sub-Saharan Africa). An overview of the same articles, however, shows that research

---

[4]https://volcanoes.usgs.gov/vdap/about.html.

Fig. 9.8 Fuego Volcano in Guatemala. The USGS volcano disaster assistance programme (Box 9.2) has included collaboration with the Instituto Nacional de Sismología, Vulcanología, Meterología e Hidrología (INSIVUMEH, National Institute for Seismology, Volcanology, Meteorology and Hydrology) in Guatemala. *Credit* Joel C Gill

was conducted in, or relevant to, some of these same geographical locations. Tanzania, for example, has no co-author affiliations and yet there are at least three studies published between 2008 and 2015, that likely included some form of collaboration with individuals in Tanzania.

There may be circumstances to explain why Tanzanian scientists supporting and contributing to data collection were not co-authors on the final publications. Research conducted by scientists in countries outside their own, however, often relies on the goodwill and support of in-country scientists and technicians (e.g., from universities, geological surveys). While scientists in the Global South are often included in the collection of data, they may not have the same opportunities to help analyse the data and contribute to the resulting publications. Addressing this lack of equity can help to enhance scientific research.

Strengthening scientific capacity, and increasing innovation for sustainable development, may involve changes in the Global South, but it also requires changes in broader scientific practice. For example, improving data sharing and encouraging co-authorship with scientists at host institutions based overseas. We all benefit when the international geoscience literature reflects the insights and intellect of a broader subset of the scientific community (Hewitson 2015).

9.5 Integration of 'Infrastructure, Industry, and Innovation' Through Development Corridors

Sustainable development requires resilient infrastructure, sustainable industrialisation, and enhanced innovation, but also coherent thinking across all three factors. This is exemplified by 'development corridors'—infrastructure networks that facilitate the movement of goods between sites of production (e.g., a copper mine, a gas field), processing zones, and national and international economic hubs (Enns 2018). Development corridors typically include transport (railways, roads, ports, airports) and energy infrastructure (pipelines, power generation, and transmission networks). Depending on the extent of integrated planning, they may also involve strategic investment in industry, innovation, and basic services to maximise the economic and social development opportunities arising from the resource extraction. Development corridors will likely be defined by multiple factors, linked to both the natural environmental and societal demand and constraints.

In sub-Saharan Africa, there are more than 30 development corridors at different stages of completion, with many of these cutting across national borders (Laurance et al. 2015). Figure 9.9 illustrates these corridors, and the extent to which they were completed (A), planned (F) or being upgraded (U) as of 2015. International cooperation at both political and technical levels is essential to the success of corridors, given their transboundary nature. This can be challenging, with the potential for conflict over the extraction and management of natural resources (including water), and political priorities changing when governments change.

A development corridor may solely focus on the transport of raw materials from the site of production and initial processing, to a transport hub for export. Alternatively, the corridor may involve multiple value addition industries that convert raw materials into a suite of manufactured goods, adding value and sending these to national and international markets. The latter is clearly preferable, resulting in more jobs, more economic activity, and more investment that has wider benefits. For example, industries will create jobs for local populations, generate greater tax revenues to invest in schools and hospitals, and drive investment in associated infrastructure (e.g., potable water, electricity, telecommunications) that benefits the wider community. Transport not only allows goods to flow from the site of production to economic hubs, but also makes it easier for people and their goods (e.g., handicrafts, agricultural products) and services (e.g., tailoring, car maintenance) to access bigger markets.

Geoscientists play a critical role in informing development corridors, and supporting long-term and integrated planning to achieve sustainable development. For example.

- *Understanding where future development corridors may occur, and how they may develop.* Given the importance of natural resources to many development corridors, understanding potential geological environments in which minerals or energy resources (including hydrocarbons) form, and their geographic distribution, can contribute to future planning and a proactive (vs. reactive) planning approach. Development corridors may actually follow geological features due to similar mineral and energy resource opportunities. Examples include trans-continental rift structures, large shear zones which host mineralisation, Archean cratons, or belts of granitic intrusions. Goodenough et al. (2016) outlined Europe's rare earth element resource potential, by examining environments of formation of alkaline igneous rocks and carbonatites, major hosts of many rare earth element deposits (Fig. 9.10). Such analyses can help inform the geographic locations where future extractive industries may be concentrated, and development corridors emerge. Gunn et al. (2018) reviewed the mineral potential of Liberia by (i) understanding the geology of Liberia, and (ii) contrasting this with similar geological environments in other countries, particularly in West Africa. This demonstrated that Liberia's geology is favourable for the occurrence of mineral and metal deposits including gold, iron ore, diamonds, base metals, and bauxite. Both examples demonstrate the importance of fundamental geological research, including geological mapping, as well as the need to connect this with socio-economic data (e.g., demographic projections) to inform sustainable development.

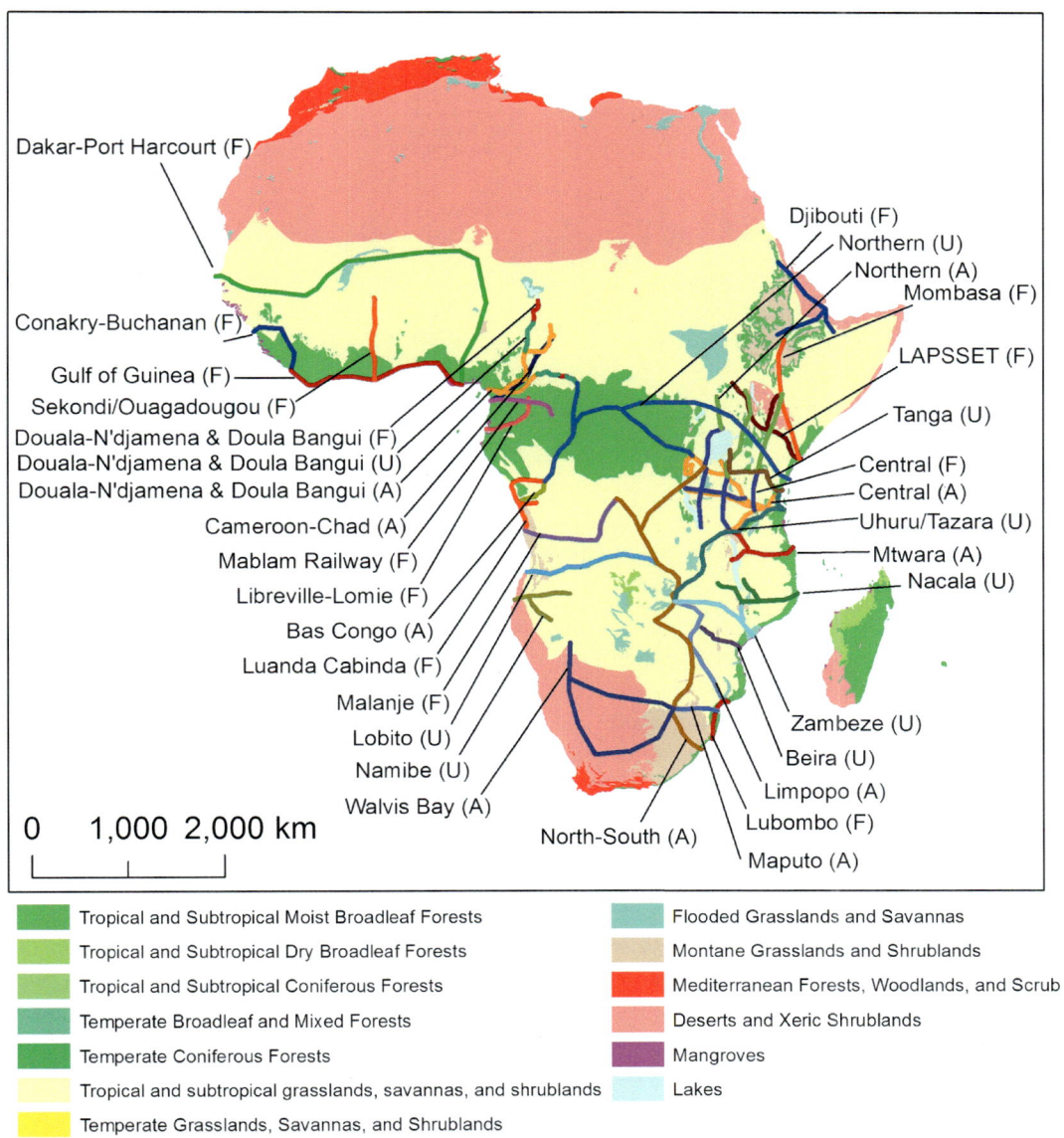

Fig. 9.9 Development Corridors in Sub-Saharan Africa. The status of each corridor at the time of original publication of this figure is indicated in parentheses (A, already active; F, planned for the future; U, upgrade planned or underway). Reprinted from Current Biology, 25 (24), William F. Laurance et al., Estimating the Environmental Costs of Africa's Massive "Development Corridors, 3202–3208. 2015, with permission from Elsevier

- *Informing analyses of resource flows to improve corridor sustainability.* The construction and sustainability of development corridors rely on an adequate flow of resources, including construction materials, water, energy, and food. Through this book, we describe the role of geoscientists in each of these (see **SDGs 12**, **6**, **7**, and **2**, respectively). For example, working together with water management professionals to estimate how much water a development corridor may be required to support both domestic and

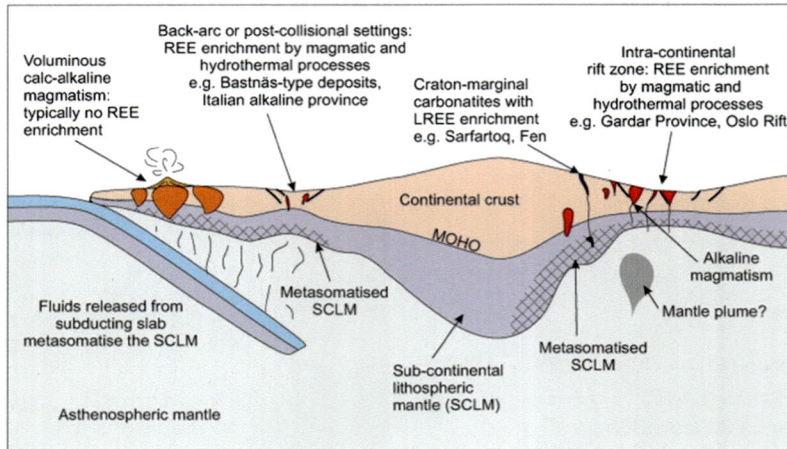

Fig. 9.10 Schematic diagram to illustrate the main environments of formation of alkaline igneous rocks and carbonatites, major hosts of many rare earth element deposits From Goodenough et al. (2016), https://doi.org/10.1016/j.oregeorev.2015.09.019, used under the CC BY 4.0 license (https://creativecommons.org/licenses/by/4.0/)

industrial use, how this may change over time, and how to meet this need through both ground and surface water supplies.

- *Characterising the subsurface to ensure resilient infrastructure and sustainable surface and subsurface land use planning.* Geological, hydrogeological, geophysical, and geotechnical data can integrate with environmental and socio-economic data to inform land use planning and infrastructure design (e.g., pipelines, roads, rail networks). The installing of sensors can facilitate real-time monitoring of ground conditions to assess how construction and/or infrastructure use is affected by or causing any change to the natural environment. This can inform ongoing construction, any mitigation steps required, and strategies to reduce environmental degradation.
- *Characterising coastal and near-shore environments.* Ports often form an important part of many development corridors, enabling the maritime export of raw materials and manufactured products. The dynamic nature of coastal zones means location-specific research on coastal and near-shore environmental processes will be needed to design resilient and quality marine infrastructure (Benveniste et al. 2019).

- *Supporting innovation through capacity building, knowledge exchange, and technology facilitation.* Development corridors can help to catalyse innovation, through investments in associated research and development (e.g., in approaches to improve the efficiency of metal extraction from ores), resulting in spin-off companies, attracting research institutes to set up campuses within the region, and the development of new environmental protection monitoring services (including in the public sector). Geoscientists involved in relevant research can work alongside these organisations to build capacity in data collection, management and analysis, share resources and disseminate ideas. Public–private research partnerships can help to explore innovative technologies while also encouraging these to be embedded within the industry. We discuss partnerships for development in detail in **SDG 17**.

Well-planned development corridors bring many socio-economic benefits, therefore, but they are not without problems. Development corridors may result in new environmental pressures on land used within the corridor, threatening carbon sinks or biodiversity (Laurance et al.

2015). Enns (2018) notes that development corridors can result in.

- *Additional competition for some struggling to earn a living.* For example, the inward movement of large amounts of seafood from ports to rural communities can threaten the livelihoods of those in rural communities reliant on fishing for an income.
- *The forced movement of populations.* Development corridors are land intensive and may involve the displacement of communities.
- *Migration of people towards corridors,* including to engage in illegal activities, to reside in informal settlements where there are no services, and away from vulnerable dependents in communities distant from the development corridor.

Development corridors may also create a dependency on a natural resource that is finite, subject to changes in commodity prices that may make it less economically viable to extract, or subject to changes in social acceptance that make it less appealing to extract. For example, a development corridor linked to coal may be subject to increasing pressure to cease extraction to meet international climate targets. Rapid closure of coal mines in the United Kingdom resulted in significant structural problems in coal communities, including higher levels of unemployment and incapacity benefit claims than other parts of the country (Johnstone and Hielscher 2017). Highly skilled, well paid jobs in industries were replaced by fewer jobs in services, paying less and with greater insecurity (Johnstone and Hielscher 2017). This illustrates the need for development corridors to integrate long-term planning into their management, including plans for economic diversification (away from a reliance on one natural resource), and investment in transferable skills that attract inward investment. We discuss economic diversification in more detail in **SDG 8**.

Baxter et al. (2017) propose five factors that are key to the success of development corridors and their aims to sustain economic development and support poverty reduction.

1. *Government support at the highest level,* of any corridor's development, with an intergovernmental agency established to oversee the implementation and ensure effective collaborations between national partners.
2. *Private sector involvement from the outset,* but recognising that success depends on engagement from other sectors also. Different actors will have different requirements, risks, and investment thresholds.
3. *Community engagement and capacity building,* throughout the development and operating of the corridor, involving civil society. Communication strategies should set out the benefits of development corridors, livelihood opportunities, and how vital ecosystems are being protected through effective planning and supervised implementation. Plans should include skills development for rural communities and those not directly near to the 'anchor project' (e.g., a mining operation).
4. *Access to geo-data and mapping.* The collection and integration of diverse spatial information to enable development corridors to proceed and allow all stakeholders in their planning. This includes topographical and socio-economic information, geological mapping, environmental data (e.g., vegetation, soils, water resources, and chemistry), and cadastral information.
5. *Good governance* Baxter et al. (2017). note that effective development corridors need (at least) two governance levels: the political and the technical, with both being transparent and accountable. Political governance should be multi-stakeholder, with representatives from government(s), donors, and the private sector. Technical governance should be independent of government.

Box 9.3. The Belt and Road Initiative: Transboundary Infrastructure Development

China's Belt and Road Initiative (BRI) aims to improve connectivity and cooperation between China and more than 70 countries, helping to boost trade and economic growth (Maliszewska and Van Der Mensbrugghe 2019), as well as Chinese visibility around the world. These objectives are achieved by multibillion dollar investments in overland and maritime infrastructure (e.g., roads, railways, and ports) linking China to Central Asia, South Asia, South East Asia, Europe, the Gulf Countries, and North Africa (Lall and Lebrand, 2019). Physical infrastructure will be complemented by policy, institutional and governance reforms to facilitate trade (Lall and Lebrand 2019).

Maritime corridors connect China with Indonesia, India, and Kenya, and through to Europe. Derudder et al. (2018) identify six BRI land corridors, encompassing more than 60 countries: (1) the China-Mongolia-Russia Economic Corridor, (2) the New Eurasian Land Bridge, (3) the China-Central Asia-Western Asia Corridor, (4) the China-Indochina Peninsula, (5) the China-Pakistan Economic Corridor, and (6) the Bangladesh-China-India-Myanmar Corridor.

Implications for Sustainable Development

The World Bank (2018), estimate improved and new infrastructure will reduce travel times along economic corridors by 12%, increase trade between 2.7–9.7%, increase income by up to 3.4%, and lift 7.6 million people from extreme poverty (World Bank 2018). BRI, therefore, has the potential to support economic and human development, although care is needed to ensure that these benefits are felt by a broad community of people, with a risk that some may get left behind (Lall and Lebrand 2019).

There are also implications for the natural environment, climate change, and broader societal resilience to environmental hazards. Roads modify the natural environment, changing hydrologic systems, erosion dynamics, and debris deposition (Losos et al. 2019). Cascading consequences can include increased risk of flooding, damage to aquatic ecosystems, as well as destabilisation of slopes and increased risk of landslides. Infrastructure construction requires significant volumes of natural resources (with concrete a major contributor to global greenhouse gas emissions), can irreparably damage bio- and geodiversity, and can generate large amounts of waste with potential environmental implications. For a detailed overview of potential environmental risks from BRI investments, see Losos et al. (2019).

Geoscientists can inform project design as set out in Sect. 9.2.2 including through geological hazard assessment and geomorphological mapping, characterisation of rock mass characteristics, and geotechnical parameters to inform infrastructure designs, analysis of raw material flows, and hydrogeological assessment to aid protection of groundwater supplies. This information can help to avoid environmental degradation. Losos et al. (2019) advocate for the incorporation of environmental assessment procedures into initial planning of entire BRI corridors. The scale of BRI also provides an exciting opportunity for innovation, the testing of new technologies, and the application of green infrastructure approaches.

The implementation of **SDG 9** through effective development corridors that integrate resilient infrastructure with diverse industries and innovation, has the potential to help achieve multiple other SDGs. Development corridors that embed the lessons set out in Baxter et al. (2017), can help to tackle poverty (**SDG 1**) and energy poverty (**SDG 7**), improve access to clean water and sanitation (**SDG 6**), create decent, high skilled and well paid jobs (**SDG 8**), and foster peace through transboundary dialogue (**SDG 16**). Geological data and analysis, collected through research, geological survey activities, and the private sector are fundamental to effective development corridors, and associated infrastructure and industrialisation, ensuring they have a positive economic, social, and environmental impact.

9.6 Key Learning Concepts

- Improved infrastructure can increase productivity and generate jobs, therefore, contributing to sustainable economic growth. Infrastructure can support human development, by improving access to essential services such as clean water and healthcare. Infrastructure can also improve our protection of the environment, by improving the efficiency of resource use, helping decarbonisation, and improving waste management.

- Conceptual ground models can help to ensure infrastructure is of high quality, reliable, resilient, and sustainable. Ground models integrate geological and geomorphological mapping, the results of detailed site investigation testing (conducted in accordance with recognised standards), and any existing literature relevant to the project (e.g., on resources, hazards, contamination) to capture and visualise our best understanding of the subsurface.

- Industrialisation is a set of social and economic changes that result in manufacturing (the large-scale production of goods) becoming a primary economic activity of a country. Greener industrialisation should increase energy efficiency, increase resource efficiency, and ensure better waste management. Resilient industrialisation requires buildings,

processes, and supply/delivery chains to be examined and strengthened.

- There is significant variation in the extent of national resources currently allocated to research and development, and the number of researchers (per million workers) in a given country. Through partnerships, geoscientists can help to build capacity and strengthen the quantity and quality of the research community. Equity needs to be at the heart of research partnerships, with opportunities for researchers in the Global South to build their own networks, access funding, and contribute to data analysis and publications.

- Development corridors are infrastructure networks that facilitate the movement of goods between sites of production, processing, and national and international economic hubs. To be effective, they need to take an integrated approach, bringing diverse stakeholders together, and underpinned by geological data and analysis. Geologists understanding of the formation and management of natural resources can inform the planning of future and current development corridors.

9.7 Educational Ideas

In this section, we provide examples of educational activities that connect geoscience, the material discussed in this chapter, and scenarios that may arise when applying geoscience (e.g., in policy, government, private sector international organisations, NGOs). Consider using these as the basis for presentations, group discussions, essays, or to encourage further reading.

- Select a major infrastructure project of national importance (e.g., road network, hydropower station, airport, nuclear power station). Consider the types and volumes of natural resources that are required to complete this project (e.g., aggregates, minerals, water).

Where could they be sourced to improve the sustainability of the project? What are the local and global factors that need to be considered in the design of the infrastructure to increase its resilience?

- Using World Bank data (https://data.worldbank.org/indicator/SP.POP.SCIE.RD.P6) explore the number of research and development workers in your country, and contrast this with 2–3 countries around the world. What differences do you notice, and what may be the implication on implementing the SDGs? What steps can geoscience organisations take to increase numbers of research and development workers? Consider this question for geoscience organisations based in (i) countries with low numbers of research and development workers, and (ii) countries with high numbers of research and development workers.
- What environmental data could inform decision making for those responsible for ensuring development corridors grow in a sustainable manner? To answer this we suggest you first explore what may sit on the development corridor (e.g., housing, transport infrastructure, sites of industry, water supplies, energy networks). This could then inform a broader conversation about the environmental data sets that would be necessary to ensure this development is high quality, reliable, resilient, and sustainable.

Further Reading and Resources

Adam Smith International (2015) Integrated Resource corridors initiative. Available at: www.adamsmithinternational.com/documents/resource-uploads/IRCI_Scoping_Report_Business_Plan.pdf. Accessed 29 July 2019

Conway G, Waage J, Delaney S (2010) Science and innovation for development. UK Collaborative on Development Sciences, London

Eurocode 7 (EN 1997) Available at: https://eurocodes.jrc.ec.europa.eu/showpage.php?id=137. Accessed 29 July 2019

Hearn GJ (Ed) (2011) Slope engineering for mountain roads. Geological Society of London. www.geolsoc.org.uk/SPE24

OECD (2018) Climate-resilient Infrastructure. Policy Perspectives. OECD Environmental Policy Paper No. 14 Available at: www.oecd.org/environment/cc/policy-perspectives-climate-resilient-infrastructure.pdf. Accessed 29 July 2019

Padmashree GS, Banji O (Eds) (2016) Sustainable industrialization in Africa—Toward a new development agenda. Palgrave Macmillan, London. https://doi.org/10.1007/978-1-137-56112-1

UNIDO (2019) Inclusive and Sustainable Industrialisation (including multiple regional guides). Available at: www.unido.org/inclusive-and-sustainable-industrial-development. Accessed 29 July 2019

References

Adhikari LB, Gautam UP, Koirala BP, Bhattarai M, Kandel T, Gupta RM, Timsina C, Maharjan N, Maharjan K, Dahal T, Hoste-Colomer R (2015) The aftershock sequence of the 2015 April 25 Gorkha-Nepal earthquake. Geophys Suppl Mon Notices R Astron Soc 203(3):2119–2124

AfDB (2016) AfDB commits €44 million to reinforce high-speed broadband access in Niger. Available at: www.afdb.org/en/news-and-events/afdb-commits-eur44-million-to-reinforce-high-speed-broadband-access-in-niger-16558. Accessed 5 Sept 2019

Akwiri (2019) Cargo handled by Kenya's Mombasa port up 6% in eleven months to May. Available at: https://af.reuters.com/article/drcNews/idAFL8N23O5AY. Accessed 23 July 2019

Ayemba D (2018) Infrastructure in Africa: Bridging the gap Available at: https://constructionreviewonline.com/2018/01/infrastructure-africa-bridging-gap/. Accessed 5 Sept 2019

Baxter J, Howard AC, Mills T, Rickard S, Macey S (2017) A bumpy road: maximising the value of a resource corridor. Extract Ind Soc 4(3):439–442

Bell FG, Donnelly LJ (2006) Mining and its impact on the environment. CRC Press.

Benveniste J, Cazenave A, Vignudelli S, Fenoglio-Marc L, Shah R, Almar R, Andersen O, Birol F, Bonnefond P, Bouffard J, Calafat F (2019) Requirements for a coastal hazards observing system. Frontiers Marine Sci 6:348

Bilham R (2015) Seismology: raising kathmandu. Nat Geosci 8(8):582

Brunsden D (2002) Geomorphological roulette for engineers and planners: some insights into an old game. Q J Eng Geol Hydrogeol 35(2):101–142

Carbon8 (2019) Carbon8 Technology https://c8s.co.uk/technology/. Accessed 22 August 2019

CSIR (2019) A CSIR perspective on South Africa's post-mining landscape. CSIR Report No.: stelgen16928, ISBN: 978–0–7988–5641–6, CSIR, Pretoria

Centre for Development Support (CDS) (2004) Proposals for the utilisation of redundant mine infrastructure for the benefit of local communities. CDS Research Report, LED and SMME Development 2004(1). University of the Free State (UFS), Bloemfontein

Collins BD, Jibson RW 2015 Assessment of existing and potential landslide hazards resulting from the April 25, 2015 Gorkha, Nepal earthquake sequence (No. 2015–1142). US Geological Survey

Dearman WR, Fookes PG (1974) Engineering geological mapping for civil engineering practice in the United Kingdom. Q J Eng Geol Hydrogeol 7(3):223–256

Derudder BJR, Liu X, Kunaka C (2018) Connectivity Along Overland Corridors of the Belt and Road Initiative (English). MTI discussion paper; no. 6. World Bank Group, Washington, D.C.

EIU (2019) The critical role of infrastructure for the Sustainable Development Goals. The Economist Intelligence Unit Limited. Available at: https://content. unops.org/publications/The-critical-role-of-infrastructure-for-the-SDGs_EN.pdf?mtime= 20190314130614. Accessed 23 July 2019.

Enns C (2018) Mobilizing research on Africa's development corridors. Geoforum 88:105–108

Engineering News (2002) Mushroom-mining project set for growth. Available at: https://www.engineering news.co.za/print-version/mushroommining-project-set-for-growth-2002-11-08. Accessed 1 October 2019.

Fookes PG (1997) Geology for engineers: the geological model, prediction and performance. Q J Eng GeolHydrogeol 30(4):293–424

Gill JC, Malamud BD (2017) Anthropogenic processes, natural hazards, and interactions in a multi-hazard framework. Earth Sci Rev 166:246–269

Goodenough KM, Schilling J, Jonsson E, Kalvig P, Charles N, Tuduri J, Deady EA, Sadeghi M, Schiellerup H, Müller A, Bertrand G (2016) Europe's rare earth element resource potential: An overview of REE metallogenetic provinces and their geodynamic setting. Ore Geol Rev 72:838–856

Gunn AG, Dorbor JK, Mankelow JM, Lusty PAJ, Deady EA, Shaw RA, Goodenough KM (2018) A review of the mineral potential of Liberia. Ore Geol Rev 101:413–431

Hausmann R, Rigobon R (2003) An alternative interpretation of the'resource curse: Theory and policy implications (No. w9424). National Bureau of Economic Research

Hearn GJ, Massey CI (2009) Engineering geology in the management of roadside slope failures: contributions to best practice from Bhutan and Ethiopia. Q J Eng GeolHydrogeol 42(4):511–528

Hewitson B (2015) To build capacity, build confidence. Nat Geosci 8(7):497

IEA (2019) Industry. Available at: https://www.iea.org/ tcep/industry/. Accessed 22 August 2019

ITU (2018) Country ICT Data (Percentage of Individuals using the Internet). Available at: https://www.itu.int/ en/ITU-D/Statistics/Pages/stat/default.aspx. Accessed 2 September 2019

IRENA (2015) A background paper to "Renewable Energy in Manufacturing", March 2015. IRENA, Abu Dhabi. Available at: https://irena.org/-/media/ Files/IRENA/Agency/Articles/2016/Nov/IRENA_ RE_Potential_for_Industry_BP_2015.pdf?la= en&hash=1214D8FDBD507297FC61073DACE78F8 F31927663. Accessed 22 August 2019

IHA (2019) Hydropower sector climate resilience guide. In: International hydropower association, p 63

Jóhannesson T, Chatenay C (2014) Industrial Applications of Geothermal Resources. Presented at "Short Course VI on Utilization of Low- and Medium-Enthalpy Geothermal Resources and Financial Aspects of Utilization", organized by UNU-GTP and LaGeo, in Santa Tecla, El Salvador. Available at: https://orkustofnun.is/gogn/unu-gtp-sc/UNU-GTP-SC-18-26.pdf. Accessed 22 August 2019

Johnstone P, Hielscher S (2017) Phasing out coal, sustaining coal communities? Living with technological decline in sustainability pathways. Extr Ind Soc 4(3):457–461. ISSN 2214–790X

Jouhara H, Khordehgah N, Almahmoud S, Delpech B, Chauhan A, Tassou SA (2018) Waste heat recovery technologies and applications. Thermal Sci Eng Progress 6:268–289

Kummu M, Varis O (2007) Sediment-related impacts due to upstream reservoir trapping, the Lower Mekong River. Geomorphology 85(3–4):275–293

Lall SV, Lebrand MSM (2019) Who wins, who loses? understanding the spatially differentiated effects of the belt and road initiative (English). In: Policy Research working paper; no. WPS 8806. World Bank Group, Washington, D.C.

Laurance WF, Sloan S, Weng L, Sayer JA (2015) Estimating the environmental costs of Africa's massive "development corridors." Curr Biol 25(24):3202–3208

Linde-Arias E, Lemmon M, Ares J (2019) Development of a ground model, targeted ground investigation and risk mitigation for the excavation of an open face cross passage on the underground Elizabeth Line, London. Tunn Undergr Space Technol 86:209–223

Losos EC, Pfaff A, Olander LP, Mason S, Morgan S (2019) Reducing environmental risks from belt and road initiative investments in transportation infrastructure. The World Bank

Maliszewska M, Van Der Mensbrugghe D (2019) The Belt and Road Initiative: Economic, Poverty and Environmental Impacts (English). Policy Research working paper; no. WPS 8814. World Bank Group, Washington, D.C.

Merritt AJ, Chambers JE, Murphy W, Wilkinson PB, West LJ, Gunn DA, Meldrum PI, Kirkham M, Dixon N (2014) 3D ground model development for an active landslide in Lias mudrocks using geophysical, remote sensing and geotechnical methods. Landslides 11(4):537–550

Nature Geoscience (2015) Globalize Geoscience—Editorial. Nat Geosci 8:491

Otchere FA, Veiga MM, Hinton JJ, Farias RA, Hamaguchi R (2004) Transforming open mining pits into fish farms: moving towards sustainability. Nat Resour Forum 28(3):216–223. Blackwell Publishing Ltd., Oxford, UK

United Nations (2016) Greening Africa's Industrialization. Available at: https://www.un.org/en/africa/osaa/pdf/pubs/2016era-uneca.pdf. Accessed 22 August 2019

UNOPS (2017) Evidence-Based Infrastructure: NIS-MOD-International

USGS (2017) Slag-What is it Good for? Available at: https://www.usgs.gov/news/slag-what-it-good. Accessed 22 August 2019

World Bank (2016) Measuring rural access: using new technologies (English). World Bank Group, Washington, D.C. Available at: https://documents.worldbank.org/curated/en/367391472117815229/Measuring-rural-access-using-new-technologies. Accessed 23 July 2019

World Bank (2019a) Research and development expenditure (% of GDP). Available at: https://data.worldbank.org/indicator/GB.XPD.RSDV.GD.ZS?end=2015&start=2000. Accessed 23 July 2019

World Bank (2019b) Researchers in RandD (per million people). https://data.worldbank.org/indicator/SP.POP.SCIE.RD.P6. Accessed 23 July 2019

Zhengfu BIAN, Inyang HI, Daniels JL, Frank OTTO, Struthers S (2010) Environmental issues from coal mining and their solutions. Mining Sci Technol (China) 20(2):215–223

Joel C. Gill is International Development Geoscientist at the *British Geological Survey*, and Founder/Executive Director of the not-for-profit organisation *Geology for Global Development*. Joel has a degree in Natural Sciences (Cambridge, UK), a Masters degree in Engineering Geology (Leeds, UK), and a Ph.D. focused on multi-hazards and disaster risk reduction (King's College London, UK). For the last decade, Joel has worked at the interface of Earth science and international development, and plays a leading role internationally in championing the role of geoscience in delivering the UN Sustainable Development Goals. He has coordinated research, conferences, and workshops on geoscience and sustainable development in the UK, India, Tanzania, Kenya, South Africa, Zambia, and Guatemala. Joel regularly engages in international forums for science and sustainable development, leading an international delegation of Earth scientists to the United Nations in 2019. Joel has prizes from the London School of Economics and Political Science for his teaching related to disaster risk reduction, and Associate Fellowship of the Royal Commonwealth Society for his international development engagement. Joel is a Fellow of the Geological Society of London, and was elected to Council in 2019 and to the position of Secretary (Foreign and External Affairs) in 2020.

Ranjan Kumar Dahal is an engineering geologist from Nepal, with a Ph.D. in Engineering and PostDoc in geohazards from Japan. He specialises in landslide, debris flow and earthquake risk. Ranjan is an Associate Professor in Central Department of Geology at Tribhuvan University, Nepal. He has published more than 70 technical/scientific papers in peer reviewed international and national journals, and has authored of two books and he has contributed in three books as co-author. He is serving Kagawa University and Ehime University of Japan as a visiting faculty member.

Martin Smith is a Science Director with the British Geological Survey and Principle Investigator for the BGS ODA Programme Geoscience for Sustainable Futures (2017–2021). He has a first degree in Geology (Aberdeen) and a PhD on tectonics (Aberystwyth, UK). A survey geologist by training Martin has spent a career studying geology both in the UK and across Africa and India. As Chief Geologist for Scotland and then for the UK he has worked closely with government and industry on numerous applied projects including in the UK on national crises, major infrastructure problems, decarbonisation research and urban geology and overseas for DFID-funded development projects in Kenya, Egypt and Central Asia. Martin is a Chartered Geologist and fellow of the Geological Society of London. He was awarded an MBE for services to geology in 2016.

Reduce Inequality Within and Amongst Countries

10

Melissa Moreano and Joel C. Gill

M. Moreano (✉)
Universidad Andina Simón Bolívar, Apartado
Postal: 17-12-569, N22-80 (Plaza Brasilia) Toledo,
Quito, Ecuador
e-mail: melissa.moreano@uasb.edu.ec

J. C. Gill
British Geological Survey, Nicker Hill, Keyworth
NG12 5GG, UK

J. C. Gill
Geology for Global Development, Loughborough,
UK

© Springer Nature Switzerland AG 2021
J. C. Gill and M. Smith (eds.), *Geosciences and the Sustainable Development Goals*,
Sustainable Development Goals Series, https://doi.org/10.1007/978-3-030-38815-7_10

Abstract

10 REDUCED INEQUALITIES

Inequality is multi-dimensional, including both economic and social factors

Unequal wealth accumulation can prevent those with less income from access to

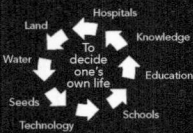

Less income and wealth may mean less power, which reinforces the challenge of accessing finance

Tackling inequality is therefore central to social justice

Inequality in time and space:

Inequality changes over time, increasing or decreasing depending on policies and practices. Since 1980, income inequality has increased in almost all regions

Inequality exists at a range of spatial scales, both within and between countries and groups of countries

Wealth may be extracted from poorer spaces (within or between countries) to support the economic development of richer spaces

Inequality, the environment, and sustainable development

Inequality drives people to live in places with greater exposure to harm (e.g., unstable slopes, or next to polluted water sources)

Reducing time in education or work

Those with a higher income generally consume and pollute more

Less equal societies contribute more to environmental degradation than more equal societies.

Environmental consequences of economic activities may be transferred from rich to poor communities.

than those with less income

Reducing Inequality
Actions to tackle inequality include:

Progressive taxation and pricing (including of natural resources)

Ensuring pro-poor economic growth

Redistribution of resources

Removal of discrimination from policy and practice

Geoscientists can support this through:

Considering how environmental research outputs are pro-poor

Knowledge redistribution within and between countries

Promoting equality in education and practice

10.1 Introduction

A primary aim of the Sustainable Development Goals (SDGs), as articulated in the 2030 Agenda for Sustainable Development, is to combat inequality within and amongst countries (UN 2015). In the context of **SDG 10—Reduce Inequality Within and Amongst Countries**— inequality is defined by the United Nations as *the gap between the richest and the poorest*, or between those people, groups of people or countries who have more and those who have less (Schorr 2018). As this indicates, inequality is a multi-dimensional phenomenon (Afonso et al. 2015; Costa et al. 2017), including aspects of both economic and social inequalities, with strong links between these.

Inequality is often reduced to a single dimension of **income**. Income inequality refers to an unequal distribution of wealth between the richest and the poorest. The UN Development Programme estimates that the richest 10% earn up to 40% of total global income, and the poorest 10% earn between 2 and 7% of total global income (UNDP 2019). If nothing changes, by 2050, the richest 0.1% of the population will have 26% of the world's wealth, the same as the global middle class (Alvaredo et al. 2018). This unequal accumulation of wealth can drive social inequality (alongside discriminatory legislation), preventing those with less income from accessing important **resources**, such as land, water, seeds, technology, schools, hospitals, knowledge, education, or the capacity to decide about one's own life. Costa et al. (2017, p. 6) note that.

> "Inequality is the distance between positions which individuals or groups of individuals assume in the context of a hierarchically organised access to *relevant social goods* (income, wealth, etc.) and **power resources** (rights, political participation, and positions)".

Income inequality reinforces power inequality and vice versa. People who have less income and wealth have less power, and less power may reinforce their lack of access to income and wealth (Boyce 2007; Therborn 2006, 2013), as well as other socially valuable goods like education, health services, and the possibility to make decisions about their lives (Schorr 2018). Tackling inequality is, therefore, central to social justice (Afonso et al. 2015; Costa et al. 2017). It underpins efforts to deliver all of the SDGs, including reducing poverty (**SDG 1**), tackling hunger (**SDG 2**), and improving access to health care (**SDG 3**), quality education (**SDG 4**) and water and sanitation (**SDG 6**), and gender equality (**SDG 5**). Lack of goods or access to services may not always be a result of scarcity, but of the unequal distribution of enough resources (Schorr 2018).

Table 10.1 shows seven targets (10.1–10.7), and three means of implementation (10.A–10.C), for **SDG 10**. These include ambitions to achieve and sustain income growth of the bottom 40% of the population at a rate higher than the national average, empower and promote the social, economic, and political inclusion of all, and encourage the direction of Official Development Assistance (ODA) to those countries with the greatest need (UN 2015).

SDG 10, Target 10.1 is primarily economic inequality, while **SDG 10, Target 10.2** captures the multi-dimensional nature of inequality, aiming to *'empower and promote the social, economic and political inclusion of all, irrespective of age, sex, disability, race, ethnicity, origin, religion or economic or other status"*. This target acknowledges that people may have different access to income, education, health or legal justice depending on whether they are poor or rich, women or men, asylum seekers or refugees, because they self-identify with or are identified with a gender different to the common male–female binary, or because of their age, place of birth, migratory condition, race or ethnicity. For example, in Latin America, in 2014, the levels of poverty of the Afro-descendant and indigenous population in the four countries for which information is available (Brazil, Uruguay, Ecuador, Peru) were one and a half to two times higher than those of the non-Afro-descendant or indigenous population (CEPAL 2016). In the United States, the incarceration of black and Latino population, who face higher levels of poverty, is far greater than the white population. In 2016, black males between 18 to 19 years old

Table 10.1 SDG 10 targets and means of implementation

Target	Description of Target (10.1–10.7) or Means of Implementation (10.A–10.C)
10.1	By 2030, progressively achieve and sustain income growth of the bottom 40 per cent of the population at a rate higher than the national average
10.2	By 2030, empower and promote the social, economic and political inclusion of all, irrespective of age, sex, disability, race, ethnicity, origin, religion or economic or other status
10.3	Ensure equal opportunity and reduce inequalities of outcome, including by eliminating discriminatory laws, policies, and practices and promoting appropriate legislation, policies, and action in this regard
10.4	Adopt policies, especially fiscal, wage, and social protection policies, and progressively achieve greater equality
10.5	Improve the regulation and monitoring of global financial markets and institutions and strengthen the implementation of such regulations
10.6	Ensure enhanced representation and voice for developing countries in decision-making in global international economic and financial institutions in order to deliver more effective, credible, accountable and legitimate institutions
10.7	Facilitate orderly, safe, regular and responsible migration and mobility of people, including through the implementation of planned and well-managed migration policies
10.A	Implement the principle of special and differential treatment for developing countries, in particular, least developed countries, in accordance with World Trade Organization agreements
10.B	Encourage official development assistance and financial flows, including foreign direct investment, to States where the need is greatest, in particular, least developed countries, African countries, small island developing States and landlocked developing countries, in accordance with their national plans and programmes
10.C	By 2030, reduce to less than 3 per cent the transaction costs of migrant remittances and eliminate remittance corridors with costs higher than 5 per cent

were 11.8 times more likely to be imprisoned than white males of the same age (Carson 2018). Black and Latino people make up approximately 32% of the population of the USA, but encompassed 56% of all incarcerated people in 2015 (NAACP 2015).

Factors resulting in economic and social exclusion do not impact on an individual or community in an additive manner, but their impacts may be much greater than the sum of their parts due to complex interactions. For example, a poor woman, of an ethnic minority, with minimal education could experience different types of inequality than an educated white woman. Complexities arising from different excluding factors need to be considered, when developing policies and practices to reduce inequality, so that our practices contribute to

reductions in inequality and do not reinforce them.

SDG 10 suggests that reducing inequality is key to promoting and achieving development that is sustainable and inclusive (Table 10.1). This chapter will explore these themes, introducing the concept of inequality, describing its multifaceted nature, and outlining the role of geoscientists in meeting the targets of SDG 10. We characterise inequality across space and time (Sect. 10.2), and describe links with poverty and sustainable development (Sect. 10.3), introducing terminology and examples that provide geoscientists with essential context to this global challenge. We proceed to outline approaches to tackling inequality (Sect. 10.4), and ways that geoscientists can support local to global efforts to advance this goal (Sect. 10.5).

10.2   Inequality in Space and Time

Inequality can exist at a range of spatial scales, due to the uneven distribution of particular sets of attributes (Ward 2009, 380), from villages, towns, and cities, through to countries, regions, and continents. Inequality occurs both between and within national boundaries, with actions to address both emphasised in **SDG 10**. For example.

- *Cities versus Hinterlands.* An uneven distribution of health and education services may exist, with poor access being a characteristic of poor rural regions or urban hinterlands and good access characteristic of rich, urban centres.
- *Global North versus Global South.* The distribution of technology may be uneven, resulting in a lack of access to energy in the Global South. The International Energy Agency estimates that just under 1 billion people still lack access to a reliable electricity supply (See **SDG 7**). The root causes of inequality between the Global North and Global South are analysed more in Sect. 10.3.
- *Urban Rich versus Urban Poor.* Wealth may be concentrated in the hands of a few in a given region. For example, in New York State in 2015, the average income of the top 1% was $2.2 million compared to an average income of the bottom 99% of $49,617, a ratio of 44.4 to 1 (Sommeiller and Price 2018).

Inequality also changes over time, increasing or decreasing depending on differences in policy and practice. In this section, we first examine the inequality between countries, with a focus on the Global North versus the Global South (Sect. 10.2.1). We proceed to examine inequality within a given country, demonstrating that inequality affects all nations to some degree even in those considered developed nations

(Sect. 10.2.2). We conclude by exploring how inequality has changed over time (Sect. 10.2.3).

10.2.1   Inequality Between Countries

It is estimated that 736 million people were living in extreme poverty in 2015, earning less than $1.90 a day, with almost half the world's population (3.4 billion people) living on less than $5.50 a day/$2000 a year (World Bank 2018). The concentration of this extreme poverty within some countries, described in **SDG 1** (No Poverty), shows the existence of inequality between countries.

10.2.2   Inequality Affects All Countries

Inequality affects both high-income and low-income countries. Figure 10.1 shows the share of national income accumulated by the richest 10% of the country's population. Almost all countries with data available show the richest 10% accumulating wealth far in excess of 10% of the income share. Another way to examine inequality within a population is the Gini coefficient or index, a commonly used measure of inequality that represents the income or wealth distribution of a population. Gini coefficients can vary from '0%', expressing perfect equality (i.e., everyone has the same income or wealth) to '100%', expressing total inequality (i.e., one person holds all the income or wealth, and everyone else receives or has nothing). These can also be presented on a scale from 0 to 1. There are some complexities with calculating Gini coefficients for those with negative incomes, or for small populations, but these are beyond the scope of this textbook. For readers interested in this theme, we include key references in the *Further Reading* section below. Alternative methods also exist, which may help to develop a more nuanced

Income share held by richest 10%, 2010
Percentage share of income or consumption accruing to the richest 10% of the population. In a country with 100
people, if you rank them by income the share of the top 10% corresponds to the sum of incomes of the top 10
people, as a proportion of total income in that country.

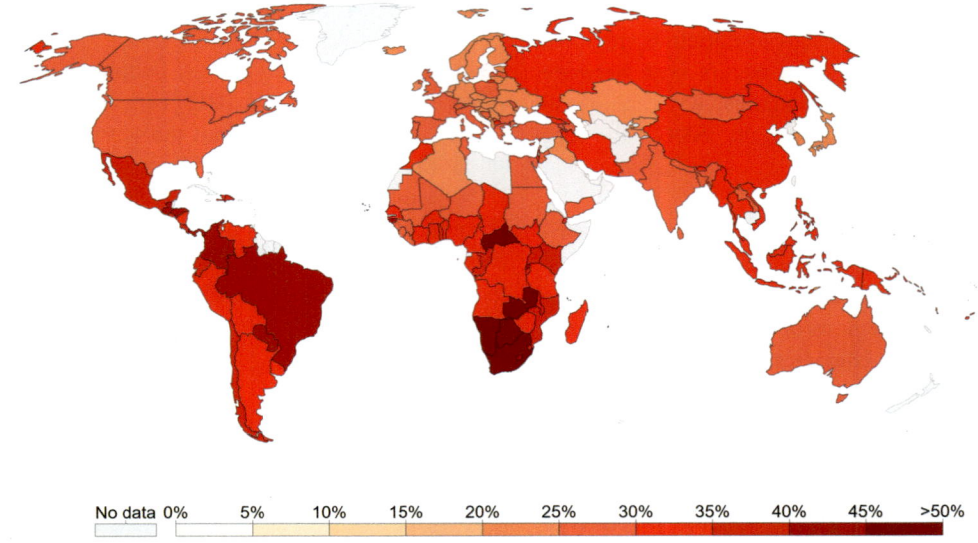

Source: World Bank   OurWorldInData.org/income-inequality/ • CC BY

Fig. 10.1 National income share accumulated by the top 10% of the population in 2010. Image from Roser and Ortiz-Ospina (2019), using data from the World Bank Poverty and Equity database (World Bank 2015). Reproduced under a CC BY license (https://creativecommons.org/licenses/by/4.0/)

understanding of income distributions (De Maio 2007).

Figure 10.2 shows within-country inequality, through a map of the Gini Index for each country where data was available, compiled using 2015, World Bank PovcalNet inequality data. This enables us to identify countries where income inequality is high (greens) and those where inequality is low (blues). While some limitations exist, in general, we note that there is higher inequality in Latin America and sub-Saharan Africa, and lower inequality in Europe, North Africa, and Southern Asia. While inequality may appear to be concentrated in low-income countries, there are some examples of high-income nations (e.g., the United States of America) with relatively high inequality. Whether considering the share of national income that top earners accumulate, or the Gini Coefficient, economic inequality affects all countries, although not to an equal degree.

10.2.3 Temporal Changes in Inequality, by Region

The World Inequality Report 2018, notes that income inequality (in terms of the percentage share of national income earned by the richest 10% of the population) has increased in nearly all countries and regions regardless of their state of development, albeit at different speeds (Alvaredo et al. 2018). When considering the Gini index, we see both declines and slight increases in inequality (Fig. 10.3). Differences between regions are much larger than the differences observed through this time period (Roser and Ortiz-Ospina 2016). For example, the Gini index values suggest that inequality is consistently and significantly greater in Latin America and the Caribbean than in other world regions, at any time between 1988 and 2013.

Economic inequality – Gini Index, 2015

Shown is the World Bank (PovcalNet) inequality data. This data includes both income and consumption measures and comparability across countries is therefore limited. A higher Gini index indicates higher inequality.

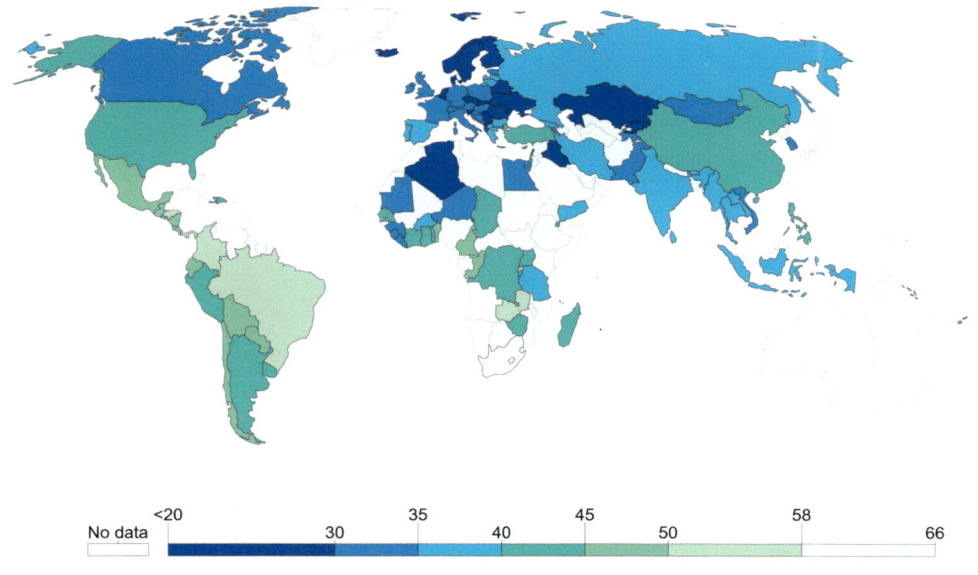

Source: World Bank   OurWorldInData.org/income-inequality/ • CC BY

Fig. 10.2 Economic Inequality—Gini Index, 2015. A map of the Gini Index for each country where data was available, compiled using World Bank PovcalNet inequality data. Image from Roser and Ortiz-Ospina (2019), using data from the World Bank Poverty and Equity database (World Bank 2015). Reproduced under a CC BY license (https://creativecommons.org/licenses/by/4.0/)

Fig. 10.3 Trends in the average economic inequality within countries, by world region (1988–2013). Lines show average within-country Gini index by region. A simple average is used without weighting countries by population. Industrialised countries are a subset of high-income countries. Image reused from World Bank (2016), using data from Milanović (2014) and PovcalNet (https://iresearch. worldbank.org/PovcalNet/). Reproduced under a CC BY 3.0 license (https:// creativecommons.org/ licenses/by/3.0/)

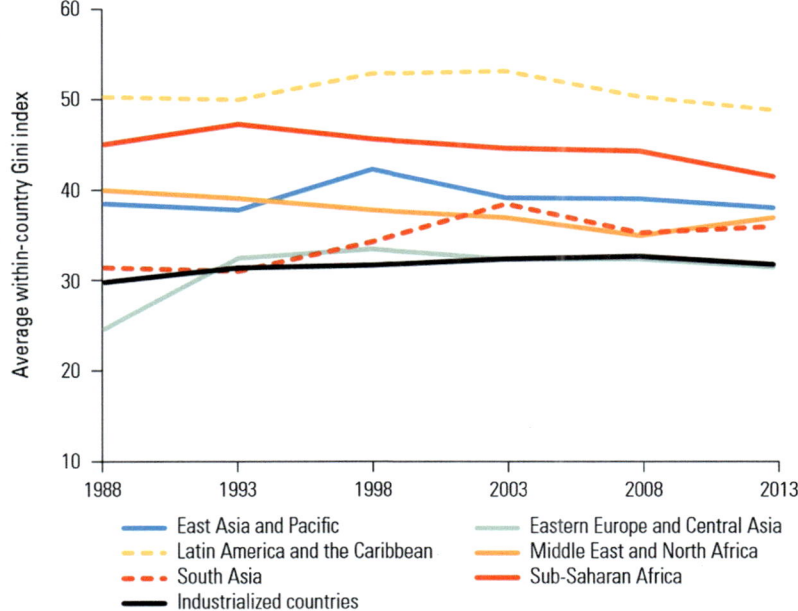

10.3 Inequality, the Environment, and Sustainable Development

The accumulation of material (e.g., money, infrastructure, technology, medicines) and symbolic (e.g., power, knowledge) resources in the hands of a minority can exacerbate inequality and poverty (Schorr 2018). The unequal distribution of wealth and resources drives, or originates, in the unequal distribution of power and progressive differentiation produced between people, groups of people and entire territories where people live. Those in power usually accumulate wealth and resources. Historically, those who accumulate power and wealth are usually white men, and those excluded are non-white, women, and minority groups (which could vary in different contexts). Schorr (2018) suggest that social inequalities exist because of power relationships that connect 'endowed actors and spaces' around the world. Such power relationships generate global inequality, and define wealthy and powerful centres and poorer and less powerful peripheries.

When analysing obstacles to economic development in the Global South since the middle of the twentieth century, some have suggested that wealth and poverty are interdependent (Alvaredo et al. 2018). Some countries are poor because they form part of a periphery, integrated at a disadvantaged position in market interactions, from which wealth is extracted to support the economic development of a high-income industrialised centre. Poor economic development exists because of structural inequalities, or unequal relationships between a centre and the periphery (Beigel 2010); the centre exists *because* the periphery exists, and both reproduce each other. This system is dynamic, with new parts of the world becoming centres as former centres become peripheries. This system is also prevalent at multiple scales, with centres and peripheries at various levels—within and across continents, nations, and cities and their hinterlands.

Box 10.1. Centres and Peripheries Relating to Environmental Resources

A poorer nation, or periphery, may hold significant mineral resources. With good governance and transparency, the country may benefit from internal investment and royalties derived from the extraction and sale of these raw materials (Fig. 10.4). However, they may lose out on greater benefits derived further down the value chain due to limited processing capacity (Fliess et al. 2017). Raw materials may be transported away from the poorer country (i.e., periphery), processed, and turned into valuable products for sale on the global market. Richer countries (i.e., centres), therefore, gain greater economic benefits from the raw materials.

Considering the origins of inequality in this way may help to design fairer and more effective development interventions that stop the reproduction of inequality. For example, instead of attributing land degradation and loss of soil fertility to poverty and lack of knowledge by subsistence farmers, we can recognise environmental degradation as both a consequence and cause of social marginalisation, or inequality. Maintaining and enhancing the integrity of the natural environment is critical to tackling inequality, and is threatened by persisting inequality.

How does environmental degradation trigger or exacerbate inequality? Degradation of the natural environment results in the destruction of habitats, livelihoods, and natural resources. For example, tailings (waste) produced from the mining of gold in South Africa, can result in uranium pollution entering shallow aquifers and adjacent streams (Winde and Sandham 2004; Winde 2013). This pollution of water resources can particularly impact poor, mainly black, South Africans living next to the mines where they work, who use the water for drinking,

Fig. 10.4 From raw material to high-value products. The value derived from iron ore (left, being mined in India) is much greater when processed and used to develop high-value products (see **SDG 12** for further discussion and examples of value addition). Image by Sarangib from Pixabay

agriculture, and bathing (Winde 2013). Low-levels of education and extreme poverty have limited the effectiveness of awareness raising campaigns (Winde 2013). In this example, inequality drives people to work and live in places with greater exposure to harm, and environmental degradation creates the potential for negative health impacts that can exacerbate inequality. Poor health, especially in contexts where access to health care and social security vary significantly, may reduce opportunities for employment or training. As environmental degradation of all types (e.g., climate change, air pollution, reduction in soil fertility, deforestation) persists, inequalities generated pose threats to social justice and sustainable development (Laurent 2014).

How does inequality result in environmental degradation? Those with higher levels of income generally consume and pollute more than those with less income (Nazrul Islam 2015), but less equal societies overall also contribute more to environmental degradation than those with higher levels of equality (Boyce 1994; Laurent 2014; Nazrul Islam 2015; Dorling 2017). Dorling (2017) notes that *"people in more equal rich countries consume less, produce less waste and emit less carbon, on average. Indeed, almost everything associated with the environment improves when economic equality is greater".* Contrast Japan and the USA, for example, both affluent nations, but with striking differences in both their inequality ratios (the income of the richest 10% of the population versus the income of the poorest 10%) and their ecological footprints. The inequality ratio in Japan is 4.5 and in the USA is 15.9, with the average per capita ecological footprint in Japan approximately half of that of the USA (Nazrul Islam 2015).

Laurent (2014) notes that inequality

- **Increases the need for environmentally harmful and socially unnecessary economic growth**. Decoupling economic growth from environmental degradation (e.g., production of waste, carbon emissions, natural resources consumption) is a significant challenge, and therefore, more economic growth (contrasted with a fairer distribution of national wealth) could exacerbate environmental challenges.
- **Increases the ecological irresponsibility of the richest, within each country and amongst nations**. Increasing inequality can result in the transfer of environmental damage of economic activities from the richest individuals, communities or countries, to the poorest. Consider the dumping of toxic electronic waste in the Global South, resulting in land, water, and air contamination, and significant health risks (Vidal 2016). With the aim of maximising profits, the material is sent overseas where recycling may be significantly cheaper, but safe working conditions and environmental protections are minimal (Vidal 2016).
- **Diminishes the social-ecological resilience of communities and societies and weakens their collective ability to adapt to accelerating environmental change**. Inequality can impact on the physical and mental health of communities, and therefore, increase their vulnerability to environmental shocks and change. Unequal power distribution hinders the possibility of poorer people to organise and decide how to confront the impacts of environmental change.
- **Hinders collective action aimed at preserving natural resources**. Laurent (2014) notes that inequality disrupts, demoralises and disorganises human communities, resulting in an inability to build political consensus on the legislation and actions needed to reduce environmental degradation. Inequality produces political polarisation, with environmental policy often becoming an issue of contention rather than collective action.
- **Reduces the political acceptability of environmental preoccupations and the ability to offset the potential socially regressive effects of environmental policies**. Environmental policies may be perceived to be (and actually) socially regressive, with a disproportionate amount of the costs being borne by the poor. Taxes on carbon or diesel, for example, could result in the poorest households paying a greater share of their income than richer households, making them politically unacceptable.

These relationships and examples, from Laurent (2014), highlight that effective strategies to tackle environmental degradation require inequality to be understood and reduced if they are to be effective and support efforts to ensure sustainable development. Inequality is a root cause of environmental degradation (and vice versa), and therefore, both must be addressed if we are to achieve **SDG 10** and other SDGs.

10.4 Approaches to Tackle Inequality

In this section, we discuss a range of approaches to address inequality, including **pro-poor economic growth, redistribution, progressive taxation and pricing**, and **removal of discrimination**.

The relationship between economic growth and reduction of inequality is complex. Where economic growth was once mandated to attain sustainable development (Osborne 2015), a greater complexity is now recognised with a need to understand available and required resources, and how to ensure equitable access to these. Economic growth does not automatically reduce inequality, but inequality will always negatively affect economic growth (Afonso et al. 2015). Economic growth that is pro-poor (i.e., increases the wealth of the poorest), however, ensures greater access to resources such as power and opportunities. Pro-poor economic growth ensures the benefits help to reduce inequality rather than maintaining the status-quo or increasing inequality. As noted previously, care needs to be taken that economic growth does not contribute to further ecological degradation, and hence to human suffering (Schorr 2018). **SDGs 8** and **12** explore strategies to decouple economic growth from environmental degradation.

Tackling inequality also requires the redistribution of economic and social resources between groups or places (Painter 2009; UNDP 2013), and to compensate those facing disadvantageous circumstances (Afonso et al. 2015). For example, some European welfare states redistribute wealth through diverse mechanisms, including public education and health systems, unemployment benefits, pensions, and universal minimum wages. These act together to help mitigate social inequality produced by wealth accumulation. The taxation system also plays an important role in helping to redistribute wealth. Progressive tax systems, where those who earn more pay more in tax, can help to tackle income and wealth inequality (Alvaredo et al. 2018).

Economic approaches can also be used to reduce inequality around access to natural resources, such as water. Progressive pricing, where the unit cost of water increases as more water is used, means that those consuming greater volumes (e.g., for industry, hospitality or wealthy households) pay more per unit than those using less (UNEP-DHI Partnership et al. 2017). This system can reduce the cost of water for the poorest in society, reduce overall consumption and waste, and increase equitable access to natural resources (UNEP-DHI Partnership et al. 2017; Zaied et al. 2017).

Box 10.2. Managing Water Resources in Tunisia

Increasing demand for water in Tunisia, one of the least water resources endowed countries in the Mediterranean basin (Benabdallah 2007), has resulted in the implementation of a progressive pricing policy (Zaied et al. 2017). Water is required for irrigation, the tourism sector (Fig. 10.5), industry and domestic use, with demand for all expected to grow. For example, the demand for water in the tourism sector is estimated to increase by 216% between 1996 and 2030 (Benabdallah 2007).

Agriculture, however, is the largest consumer of water, with approximately 84% of all demand as of 2007 (Benabdallah 2007). Reforms, including progressive pricing where the most expensive unit of water costs six times the least expensive unit, have, therefore, been put in place to reduce waste and ensure continued availability of water resources for domestic use (Benabdallah 2007; Zaied et al. 2017). Other reforms, such as guaranteeing free or very low-cost water for subsistence agriculture, while adopting progressive pricing for water used for agribusiness, could also be adopted.

Tackling inequality also requires the removal of discrimination from within legislation, ensuring all members of a community have the right to participate in politics, acquire political power, and access basic services and natural resources. For example, discrimination may prevent certain groups of people from having established legal rights over their land (Box 10.3).

Box 10.3. Inequality, Land Ownership and Agriculture in Latin America.

Ecuador is an agricultural country, with land unequally distributed amongst smallholder farmers and big landowners. Smallholder farmers provide food for national consumption in diversified plots, and big landowners produce food (typically banana, sugar cane, palm, and cacao) for international markets planted in monocultures that aggravate environmental pollution and soil degradation. Figure 10.6 shows that 712,035 low-income Ecuadorian families managing small farms together own 2,481,019 hectares of agrarian land, while 6,616 big landowners together own 3,593,496 hectares of land. Inequality is also prevalent in land distribution between men and women. This is observed in Ecuador, with Fig. 10.6, showing that women own 25.4% of the agricultural

Fig. 10.5 Use of water for tourism in Tunisia. Image by vk_photo from Pixabay

production units (UPA, in Spanish), and men owning the remaining 74.6% of units, despite the Food and Agricultural Organization (FAO 2011), reporting that 50–80% of the world's food supply being produced by women. A similar inequality is observed in other parts of Latin America. In Colombia, women manage 26% of agricultural holdings, men manage 61.4%, and a combination manages the remaining 12.6% (Bautista 2017).

10.5 Mobilising Geoscientists to Reduce Inequality

Having established the global and multi-dimensional nature of inequality, and the importance of tackling inequality if we are to ensure economic, social, and environmental sustainability (Schorr 2018), we now turn to the role of geoscientists. *How can we mobilise the knowledge and skills of geoscientists, and how can we embed the learning from this chapter into our structures and institutions, to help tackle inequality?* Geoscientists may need to make changes to the systems that determine access to our knowledge, education, and tools. This could be in terms of geoscientific data, publications arising from the analysis of that data, or the training courses that enable effective professional practice. Geoscientists also need to think about how to foster equality through inclusive work environments and the questions addressed in their research.

10.5.1 Improve Equitable Access to Scientific Data, Publications, and Meetings

Data repositories improve access, preservation and stability of scientific data. For example, the *National Geoscience Data Centre* (NGDC[1]) at

---

[1]https://www.bgs.ac.uk/services/ngdc/.

Fig. 10.6 Inequality in Ecuador. *Source* Observatory of Rural Change, OCARU, Ecuador (https://ocaru.org.ec/). Idea and design: Isabel Salcedo Quiroga and Jairo Erazo. (top) Land distribution in Ecuador between small-scale farmers and big landowners. (bottom) UPAs (agricultural production unit) distribution in Ecuador between women and men (used with permission)

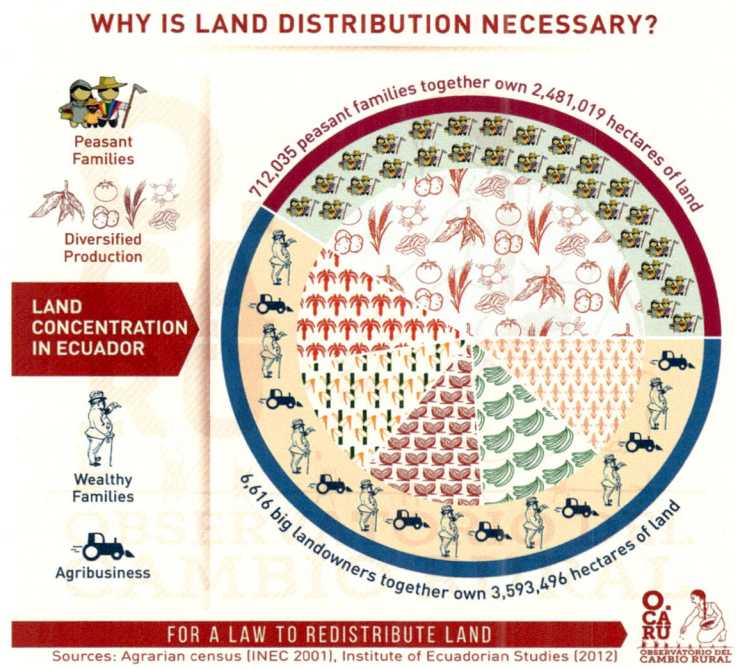

the British Geological Survey collects and preserves geoscientific data and information, making them available to a wide range of users and communities. Datasets are managed by the BGS for the long-term benefit of all. Data are available online and without charge where possible. Examples include real-time monitoring of environmental phenomena from sensors, GIS datasets, and elemental and stable isotope data. Free and open access to datasets may also include training for users on how to access and use available information.

Journal articles contain a rich source of scientific knowledge that can inform all aspects of the SDGs. For example, they may describe the history of seismic activity in a given region, the geotechnical properties of materials underlying a city, or the chemistry of groundwater in a nation lacking domestic access to safe water. If this information is not available to those making decisions (or scientists advising those making decisions) about disaster preparedness, urban infrastructure development, or water management, those decisions may not reflect the best available evidence.

The vision of the *International Network for the Availability of Scientific Publications* (INASP[2]) is that all people can access and contribute information, ideas and knowledge necessary to drive sustainable and equitable development. They work with partners and networks around the world to encourage the creation and production of information, to promote sustainable and equitable access to information, to foster collaboration and networking, and to strengthen local capacities to manage and use information and knowledge. Through INASP, the Geological Society of London have made access to their Lyell Collection available free of charge to approved institutions and NGOs in developing countries. The Lyell Collection includes more than 26,000 articles from the Society's journal titles, Special Publications and key book series, as well as journals published on behalf of other societies. This approach is in line with **SDG 10.A** which notes that 'special and

differential treatment' is appropriate for those in developing countries, especially least developed countries. A further challenge will be to make the information available in a range of languages other than English.

Equitable access to scientific knowledge can also be created if scientific meetings are more accessible and inclusive. Scientific conferences are a critical component of learning, networking, and dissemination for many research scientists, but these often have high registration fees and are based in expensive cities. The location of international conferences may also mean significant challenges in getting visas. For UK nationals to visit Uganda, a visa fee of $50 is required (paid on entry to Uganda) together with a valid passport. In contrast, for a Ugandan national to visit the United Kingdom, substantial paperwork is required, including a personal bank statement showing sufficient funds to cover the trip to the UK, invitation letters, biometric identifiers and a $70 fee that must be paid online (for other countries, larger fees apply). Finally, conferences may lack facilities for those with disabilities or children. In 2018, Langin (2018), explored the provisions for parents at scientific conferences in North America, and found that 94% provided a lactation room and 68% provided some form of childcare. The European Geosciences Union General Assembly also provides professional childcare facilities for those aged 3–11, free of charge.

10.5.2 Improve Equitable Access to Scientific Education and Skills

Training for geoscientists should encourage full participation for all, allowing field, laboratory, and classroom learning for those with diverse disabilities (e.g., visual or audio impairments, physical or mobility impairments, developmental, cognitive). Course leaders should give careful thought to how fieldwork courses (and other learning environments) can be designed to promote accessibility and inclusion. In a review by Atchison and Libarkin (2013), they note that inclusive instruction can lead to academic

---

[2]https://www.inasp.info/.

success for students with disabilities but also enhance the learning environment for all students. This is a really important point for educators to digest; embedding a commitment to inclusivity in the design of geoscience teaching can not only ensure we 'leave no one behind' but can also enrich understanding for the broader community of students.

The *International Association for Geoscience Diversity* (IAGD[3]) is a charitable organisation focused on creating access and inclusion for persons with disabilities in the Geosciences. They have resources on themes such as colour vision deficiency, visual impairment, physical (mobility) impairment, and deafness. *Diversity in Geoscience-UK* (a chapter of IAGD) has a broadened focus, going beyond disability to also cover wider aspects of diversity. The IAGD run a fully accessible field trip each year, with one of their aims being providing geoscience educators with an opportunity to learn how to accommodate students with disabilities in fieldwork.

10.5.3 Improve Knowledge Exchange Between and Within Countries

SDG 10.B emphasises the need for Official Development Assistance and financial flows to go to countries where the need is greatest (e.g., least developed countries, African countries, small island developing States, and landlocked developing countries), to support national plans and programmes. Science education and capacity building (at all levels, including both public understanding of science, citizen science, and enhanced research power) feature in many national and regional development strategies. For example, Gill et al. (2019), highlighted this to be a common theme of many development strategies in eastern Africa (e.g., East African Community Vision 2050, Southern African Development Community Regional Indicative Strategic Development Plan 2005–2020, Common Market for Eastern and Southern Africa

Medium Term Development Strategy 2016–2020, Kenya Vision 2030, Tanzania National Development Plan, Zambia Vision 2030).

There is an opportunity for geoscientists to engage in programmes that direct *Official Development Assistance* (ODA) to support the countries with the greatest need to meet these clearly expressed priorities. For example, the UK Global Challenges Research Fund uses some of the UK ODA commitment to support collaborative research that (a) builds capacity in partner countries, while (b) addressing clearly expressed development challenges. Key themes include equitable access to sustainable development, sustainable economies and societies, and human rights, good governance, and social justice.

The *UN Technology Facilitation Mechanism* is a set of processes aiming to facilitate collaboration and partnerships through the sharing of information, experiences, best practices, and policy advice. We discuss the UN Technology Facilitation Mechanism in detail in **SDG 17**.

10.5.4 Build Safe and Inclusive Educational and Professional Environments

SDG 10.2 advocates for the empowerment and social, economic, and political inclusion of all. This can be supported by educational and professional environments where all are safe, secure, and free from harassment, bullying, and discrimination. Tackling inequalities must be proactive though, recognising that the way in which we organise or structure events, for example, can also exacerbate inequalities. Geoscience events that include all-male panel discussions or keynote speakers do not help to foster an inclusive professional environment. Similar problems arise when women and those living in regions where research is conducted are excluded from scientific paper writing and the organisation of conference and dissemination meetings.

Contrast the international Resources for Future Generations conference (2018, Vancouver), with the proposed International Geological Congress

[3]https://theiagd.org/.

(2020, Delhi, although cancelled due to COVID-19). The former had 16 of 36 (44%) plenary speakers being female, and at the time of writing the latter had 1 of 14 (7%). International geoscience conferences that give little attention to the voices of early-career scientists, or scientists from a particular continent, may also exacerbate inequalities. Excluded groups and populations need to be better included in the structures of science, not out of tokenism, but because if we don't, we lose their vital ideas and insights. We discuss gender discrimination in-depth in **SDG 5** but recognise that our responsibilities extend beyond gender to also ensure inequalities linked to race, age, religion, or economic status are tackled. We discuss the importance of safe and secure work environments in **SDG 8**.

10.5.5 Integrate Geoscience Research with Issues of Intersectional Inequalities

While Sects. 10.5.1 to 10.5.4, have focused on structural reforms to tackle inequalities within the geoscience sector and inequalities in accessing geoscience knowledge, there are also ways in which we can deploy our skills to tackle the broader inequalities in the world. We have previously highlighted strong links between the natural environment and inequality, noting that environmental degradation can trigger or exacerbate inequality and that inequalities can result in environmental degradation (Sect. 10.3). Tackling inequality and understanding its impacts is, therefore, an interdisciplinary challenge that geoscience research can inform.

Inequality drives people to live in places with greater exposure to harm (Fig. 10.7). This may be unstable slopes, regions with enhanced levels of pollution, or places with poor soil fertility threatening food security. Geoscience research can.

1. Ensure our *focus* is on understanding the geology and environmental dynamics of least developed regions (helping to 'leave no one behind'), and not ignoring those places where data is harder to gather, access is more complicated, or there is a greater risk of the project not being successful. For example, fragile and conflict affected states may make partnership building, data collection, and dissemination of research challenging, resulting in these locations receiving disproportionately less *Official Development Assistance* to address environmental challenges,

Fig. 10.7 **Waste disposal sites can attract many living in poverty, exposing them to diverse harm**. Image by vkingxl from Pixabay

compared to more stable countries. For example, of the projects funded by the UK Global Challenges Research Fund, four have had Burundi as a country of focus, three the Central African Republic, and one in Chad—all of which are classified by the World Bank as being least developed countries. In contrast, 78 projects have Kenya as a country of focus, 61 have South Africa, 60 have India, and 24 have Brazil—all of which are lower or upper-middle income countries. Countries are likely to be selected based on many factors, including where existing partnerships exist and where language barriers can be crossed.

2. Help to *eliminate* some of the environmental degradations that trigger or exacerbate poverty, through supporting the improved management of waste materials. For example, the use of cyanide in gold mining can result in the risk of toxic chemicals polluting water, soil, and air. Geoscience research can help to ensure effective controls are in place to reduce the risk of environmental contamination. This could be through environmental monitoring or characterising the subsurface to inform the design of engineered structures (e.g., landfill).

3. Help to *reduce the level of harm communities are exposed to* (i.e., mitigation of negative impacts of living in a site), through monitoring of the environment and dialogue with communities. For example, geoscientists may provide advice and guidance on how to enhance soil fertility or water retention using locally available geological resources or ensure understanding of potential lahars and pyroclastic density currents is shared with those living in informal settlements around volcanoes, as well as more established, better resourced settlements.

4. Value *interdisciplinary* research opportunities, recognising that skilled interdisciplinary leaders can help to connect physical hazards, climate change, and natural resource insights into a broader analysis of current power inequalities and bring an understanding of pre-existing social conditions that could have contributed to existing inequalities. This can help to integrate research outputs to ensure that they are more than the sum of their parts.

5. Value *local knowledge* from local communities but also from local scientific communities. Make a real effort to transfer knowledge, technology, and skills to the point that local scientists can shape and complete their own research and interventions in the future, so to reduce dependence and epistemic colonialism (see **SDG 17** for a more detailed discussion of equitable and ethical partnerships).

10.6 Key Learning Concepts

- Inequality is the gap that exists between those people, groups of people, or countries who have more and those who have less. Inequality is multi-dimensional and includes both economic and social inequalities, with strong links between these. Income inequality reinforces social inequality and vice versa.

- Factors resulting in economic and social exclusion do not impact on an individual or community in an additive manner, but their impacts may be much greater than the sum of their parts due to complex interactions.

- Inequalities can exist at a range of spatial scales, both between and within national boundaries. Inequality also changes over time, increasing or decreasing depending on differences in policy and practice.

- Degradation of the natural environment results in the destruction of habitats, livelihoods, and natural resources, disproportionately affecting those who already suffer the effects of inequality. Inequality may drive people to live in places with greater exposure to harm, and environmental degradation creates the potential for negative health impacts that can exacerbate inequality.

- Those with higher levels of income generally consume and pollute more than those with less income, but less equal societies overall also contribute more to environmental degradation than those with higher levels of equality. Higher inequality can result in the environmental consequences of economic activities being transferred from the richest to the poorest communities.

- Tackling inequality requires pro-poor economic growth (i.e., growth that increases the wealth of the poorest), redistribution of resources, progressive taxation and pricing (e.g., of water resources), and the removal of discrimination.
- Geoscientists need to make changes to the systems that determine access to their knowledge, education, and tools. This includes placing data in open repositories, improving equitable access to scientific journals, meetings, and education, enhancing knowledge exchange between and within countries, building safe and inclusive educational and professional environments, and integrating geoscience research with issues of intersectional inequalities.

10.7 Educational Ideas

In this section, we provide examples of educational activities that connect geoscience, the material discussed in this chapter, and scenarios that may arise when applying geoscience (e.g., in policy, government, private sector international organisations, NGOs). Consider using these as the basis for presentations, group discussions, essays, or to encourage further reading.

- This chapter describes diverse types of inequalities and notes that these are 'intersectional'. What types of inequality may exist at your university, and how does the study of geoscience (a) reduce and (b) exacerbate these inequalities?
- In your national context, to what extent do you agree with the statement 'environmental degradation is a major driver of inequality'? Consider relevant examples and write a one-page summary for policymakers about how

actions to reduce environmental degradation could help reduce inequalities.

- Inequality drives people to live in places with greater exposure to harm (e.g., unstable slopes, regions with enhanced levels of pollution, places with poor soil fertility). Many of these regions are the focus of geoscience research. What steps can geoscientists take to involve those living in such regions in geoscience research, and increase access to the knowledge produced during this research? What challenges may you encounter as you follow these steps, and how may they be overcome? Ask those conducting relevant research in your department to reflect on how they have encountered inequalities while doing geoscience research, and any steps they have taken to reduce inequalities.
- Explore the Virtual Landscapes website (www.see.leeds.ac.uk/virtual-landscapes/), courtesy of the University of Leeds. What impact could resources such as this have on reducing inequalities? Research what other online tools exist to support field and laboratory geoscience skills.

Further Reading and Resources

Boyce JK (2007) Is inequality bad for the environment?. In: Equity and the environment. Emerald Group Publishing Limited, pp 267–288

Hamann M, Berry K, Chaigneau T, Curry T, Heilmayr R, Henriksson PJ, Hentati-Sundberg J, Jina A, Lindkvist E, Lopez-Maldonado Y, Nieminen E (2018) Inequality and the Biosphere. Annu Rev Environ Resour 43:61–83

Roberts JT (2001) Global inequality and climate change. Soc Nat Res 14(6):501–509

References

Afonso H, LaFleur M, Alarcón D (2015) Concepts of inequality development Issues No. 1

Alvaredo F, Chancel L, Piketty T, Saez E, Zucman G (eds) (2018) World inequality report 2018. Belknap Press, Cambridge, MA

Atchison CL, Libarkin JC (2013) Fostering accessibility in geoscience training programs. EOS Trans Am Geophys Union 94(44):400–400

Bautista R (2017) Informe 2017. Acceso a la tierra y territorio en Sudamérica. La Paz: Instituto para el Desarrollo Rural de Sudamérica

Beigel F (2010) Dependency Analysis: The creation of New social Theory in Latin America. In: Patel S (ed) The ISA Handbook of Diverse Sociological Traditions. SAGE Publications Ltd., London, pp 189–200

Benabdallah, S. (2007). The Water Resources and Water Management Regimes in Tunisia. Available online: https://www.nap.edu/read/11880/chapter/11#83. Accessed 4 Feb 2019

Boyce JK (1994) Inequality as a cause of environmental degradation. Ecol Econ 11:169–178

Boyce J (2007) Is inequality bad for the environment?, political economy research institute working paper series, No. 135, Amherst

Carson E (2018) Ann. Prisoners In 2016. Washington, DC: US Dept of Justice Bureau of Justice Statistics, January 2018, NCJ251149. Available Online: https://www.drugwarfacts.org/chapter/race_prison. Accessed 28 Aug 2019

CEPAL (2016) La matriz de la desigualdad social en América Latina. Naciones Unidas, Santiago de Chile

Costa S, Jelin E, Motta R (2017) Global Entangled Inequalities. Routledge, Conceptual Debates and Evidence from Latin America, London

De Maio FG (2007) Income inequality measures. J Epidemiol Community Health 61(10):849–852

Dorling D (2017) Is inequality bad for the environment? The Guardian. Available Online: https://www.theguardian.com/inequality/2017/jul/04/is-inequality-bad-for-the-environment. Accessed 4 Feb 2019

Fliess B, Idsardi E, Rossouw R (2017) Export controls and competitiveness in African mining and minerals processing industries, OECD Trade Policy Papers, No. 204, OECD Publishing, Paris. https://doi.org/10.1787/1fddd828-en

Food and Agricultural Organization FAO (2011) The role of women in agriculture. ESA Working Paper No. 11–02

Gunder Frank A (1969) The Development of Underdevelopment. In: Latin America: development or revolution (Monthly Review Press).

Laurent É (2014) Inequality as pollution, pollution as inequality - The social-ecological nexus. The Stanford Center on Poverty and Inequality Working Paper. Available online: https://inequality.stanford.edu/sites/default/files/media/_media/working_papers/laurent_inequality-pollution.pdf. Accessed 5 Feb 2019

Milanović B (2014) All the Ginis, 1950–2012. World Bank, Washington, DC. Available at: https://datacatalog.worldbank.org/dataset/all-ginis-dataset. Accessed 1 Oct 2019

National Association for the Advancement of Colored People (NAACP) (2015) Criminal Justice Fact Sheet. Available Online: https://www.naacp.org/criminal-justice-fact-sheet/. Accessed 27 Aug 2019

Nazrul Islam S (2015) Inequality and environmental sustainability. DESA Working Paper No. 145. ST/ESA/2015/DWP/145, p 30

Osborne T (2015) Tradeoffs in carbon commodification: A political ecology of common property forest governance. Geoforum 67:64–77

Painter J (2009) Redistribution. In: Gregory D, Johnston R, Pratt G, Watts M, Whatmore S (eds) The Dictionary of Human Geography. WileyBlackwell, Oxford, pp 625–626

Roser M, Ortiz-Ospina E (2019) Income Inequality. Available at: https://ourworldindata.org/income-inequality. Accessed 1 Oct 2019

Schorr B (2018) How Social Inequalities Affect Sustainable Development. Five Causal Mechanisms Underlying the Nexus, trAndeS Working Paper Series 1, Berlin: Lateinamerika-Institut, Freie Universität Berlin

Servicio Nacional de Calidad y Sanidad Vegetal y de Semillas, SENAVE (2017) Anuario Estadístico 2017

Sommeiller E, Price M (2018) The new gilded age: Income inequality in the U.S. by state, metropolitan area, and county. Economic Policy Institute, Washington DC, 66 p. Available online: https://www.epi.org/files/pdf/147963.pdf. Accessed 31 Jan 2019

Stewart IS, Gill JC (2017) Social geology—integrating sustainability concepts into Earth sciences. Proc Geol Assoc 128(2):165–172

Therborn G (2006) Inequalities of the world. Verso, London

Therborn G (2013) The killing fields of inequality. Polity, Cambridge

UNDP (2019) Goal 10: Reduced Inequalities. Available online: https://www.undp.org/content/undp/en/home/sustainable-development-goals/goal-10-reduced-inequalities.html. Accessed 31 Jan 2019

UNEP-DHI Partnership, UNEP-DTU, CTCN (2017) Progressive Pricing. Available online: https://www.ctc-n.org/resources/progressive-pricing. Accessed 4 Feb 2019

United Nations Development Programme (2013) Humanity Divided: confronting inequality in developing countries

United Nations (2015) Transforming Our World: The 2030 Agenda for Sustainable Development: United Nations, Geneva, p 35

Vidal J (2016) Toxic E-Waste Dumped in Poor Nations, Says United Nations. The Guardian. Available online: https://ourworld.unu.edu/en/toxic-e-waste-dumped-in-poor-nations-says-united-nations. Accessed 4 Feb 2019

Wallerstein I (2001) End of the World as We Know It: Social Science for the Twenty-First Century. University of Minnesota Press, Minneapolis

Ward K (2009) Inequality, spatial. In: Gregory D, Johnston R, Pratt G, Watts M, Whatmore S (eds) The dictionary of human geography. WileyBlackwell, Oxford, pp 380–381

Winde F (2013) Uranium pollution of water—a global perspective on the situation in South Africa. Vaal Triangle Occasional Papers: Inaugural lecture. Available online: https://repository.nwu.ac.za/bitstream/handle/10394/10274/Winde_F.pdf?sequence=1. Accessed 5 Feb 2019

Winde F, Sandham LA (2004) Uranium pollution of South African streams–An overview of the situation in gold mining areas of the Witwatersrand. Geo J 61 (2):131–149

World Bank (2015) World Bank Poverty and Equity database. Available at: https://data.worldbank.org/data-catalog/poverty-and-equity-database. Accessed 1 Oct 2019

World Bank (2016) Poverty and Shared Prosperity 2016: Taking on Inequality. World Bank, Washington, DC. https://doi.org/10.1596/978-1-4648-0958-3.License: CreativeCommonsAttributionCCBY3.0IGO

World Bank (2018) World Development Report 2018: Learning to Realize Education's Promise. World Bank, Washington, DC. https://doi.org/10.1596/978-1-4648-1096-1.License: CreativeCommonsAttributionCCBY3.0IGO

Zaied YB, Bouzgarrou H, Cheikh NB, Nguyen P (2017) The nonlinear progressive water pricing policy in Tunisia: equity and efficiency. Environ Econ 8(2):17–27. https://doi.org/10.21511/ee.08(2).2017.02

Joel C. Gill is International Development Geoscientist at the *British Geological Survey*, and Founder/Executive Director of the not-for-profit organisation *Geology for Global Development*. Joel has a degree in Natural Sciences (Cambridge, UK), a Masters degree in Engineering Geology (Leeds, UK), and a Ph.D. focused on multi-hazards and disaster risk reduction (King's College London, UK). For the last decade, Joel has worked at the interface of Earth science and international development, and plays a leading role internationally in championing the role of geoscience in delivering the UN Sustainable Development Goals. He has coordinated research, conferences, and workshops on geoscience and sustainable development in the UK, India, Tanzania, Kenya, South Africa, Zambia, and Guatemala. Joel regularly engages in international forums for science and sustainable development, leading an international delegation of Earth scientists to the United Nations in 2019. Joel has prizes from the London School of Economics and Political Science for his teaching related to disaster risk reduction, and Associate Fellowship of the Royal Commonwealth Society for his international development engagement. Joel is a Fellow of the Geological Society of London, and was elected to Council in 2019 and to the position of Secretary (Foreign and External Affairs) in 2020.

Melissa Moreano is Professor in the Department of Environment and Sustainability at Simón Bolívar Andean University in Quito, Ecuador. Melissa has a Ph.D. in human geography (King's College London), a BA in Biology and an M.Sc in Science and Technology Studies. Melissa has more than 15 years of experience in academic research, teaching and socio-environmental projects management. She is co-author of a chapter in the "*International Handbook of Political Ecology*" and co-editor of the book "*Political ecology in the middle of the world. Ecological struggles and reflections on nature in Ecuador*".

Sustainable Cities and Communities

11

Martin Smith and Stephanie Bricker

M. Smith (✉)
British Geological Survey, Lyell Centre, Edinburgh
EH14 4AP, UK
e-mail: msmi@bgs.ac.uk

S. Bricker
British Geological Survey, Environmental Science
Centre, Nicker Hill, Keyworth, Nottingham NG12
5GG, UK

© Springer Nature Switzerland AG 2021
J. C. Gill and M. Smith (eds.), *Geosciences and the Sustainable Development Goals*,
Sustainable Development Goals Series, https://doi.org/10.1007/978-3-030-38815-7_11

Abstract

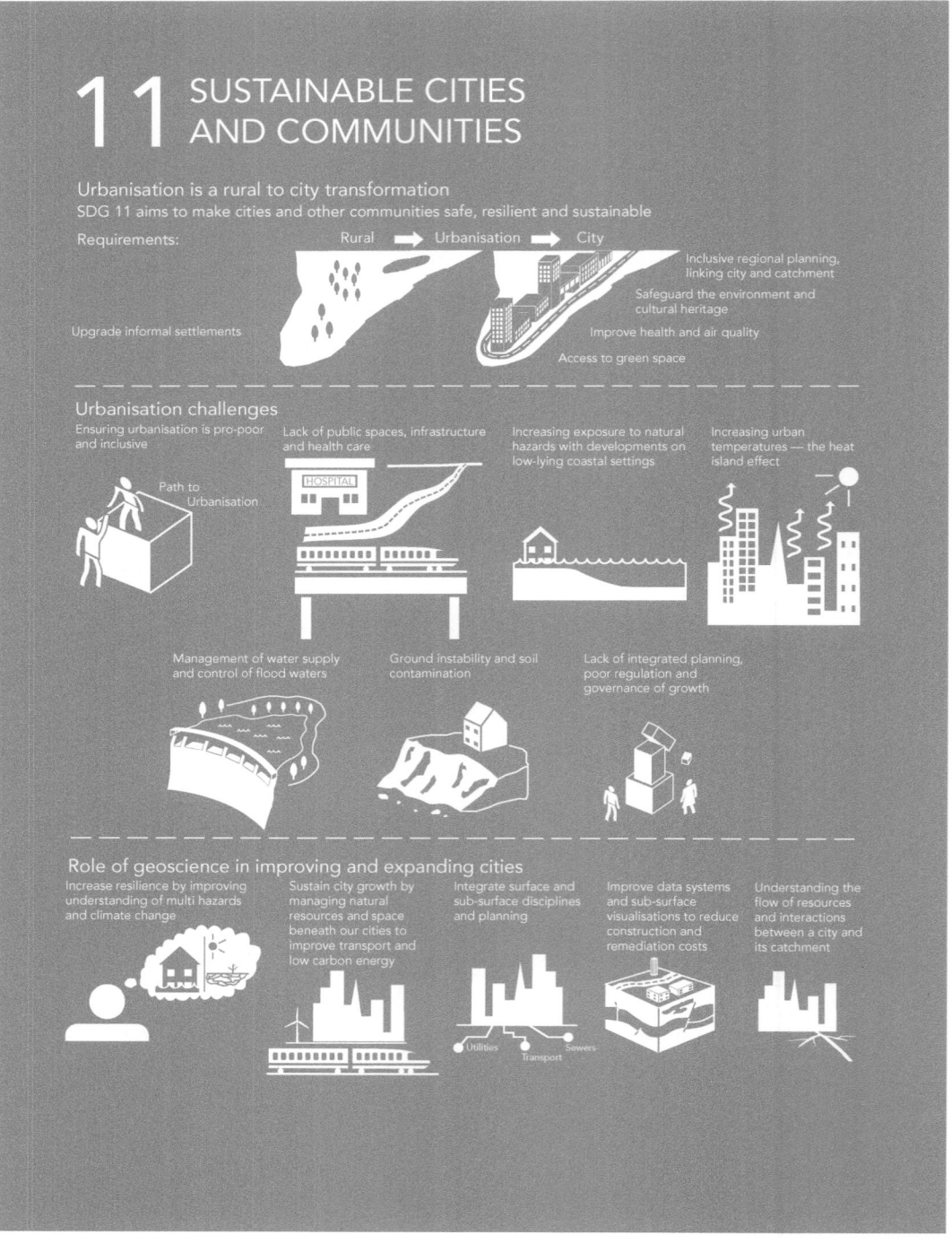

11.1 Introduction

Cities are an expression of modern life, the largest objects of human creation, a route to economic prosperity, and often regarded as complex organisms that link human and biophysical ecosystems (e.g., Wolman 1965; Graedel 1999; Van den Dobbelsteen et al. 2012; Bristow and Kennedy 2013). In 2018, they provided shelter, food, and employment to 3.5 billion people, 55% of humanity. World population projections estimate continued growth by at least two billion—equal to another China and India—in the next 33 years, and possibly by a further two billion by the end of the century (UN 2019a, b). So where will we all live?

By 2050, it is likely that we will have developed into an almost exclusively urban species with 70% of us living in cities, primarily as a result of rural to urban migration (IOM 2015), compounded by natural population growth. For example, the annual population increase in six major developing country cities—Dhaka, Karachi, Kinshasa, Lagos, Mumbai, and New Delhi—is greater than the entire population of Europe (UN-Habitat 2012). The metropolitan area of Kampala (Uganda, Fig. 11.1) has increased from 1.936 million people to 3.125 million people—an increase of 61% in just three years (United Nations 2018a).

The opposite problem of deurbanisation or shrinking cities (Biswas et al. 2018) is also part of the sustainability equation. Loss of manufacturing and falling population growth will impact on city services and facilities designed for greater volumes of people.

Whilst urbanisation[1] arguably constrains humanity's environmental impact to limited areas, which presently only cover approximately 3% of the land surface, they account for more than 70% of energy consumption and 75% of carbon emissions and are therefore major contributors to climate change. If by 2030

urbanisation is to grow to encompass up to 5 billion people (Figs. 11.2 and 11.3) and much of this expansion will take place in the Global South, where the pace of urbanisation is already overwhelming many cities, what will these cities look like? Sprawling, chaotic slums—with poor health from uncontrolled emissions and impoverished food and water—or clean, healthy environments with data smart efficient transport systems enabling the flow of people and resources?

The **Sustainable Development Goal (SDG) 11, Sustainable Cities and Communities**, is therefore a key global challenge engaging with a wide range of human needs and activities. It aims to, *Make cities and human settlements inclusive, safe, resilient, and sustainable* and includes 7 targets (11.1–11.7), 11 indicators, and 3 means of implementation (11.A, 11.B, and 11. C) as listed in Tables 11.1 and 11.2. The collective ambitions are to improve basic services and transport, upgrade informal (slum) settlements, develop inclusive and regional planning that acknowledges the links between city and rural, and peri-urban areas, safeguard the environment and the cultural and natural heritage, improve disaster risk management, and improve health, especially air quality with universal access to green space.

Urbanisation impacts and integrates with numerous other SDGs including **SDGs 1, 6, 7, 8, 9, 12, 15**, and **17**. Linkages have been considered by Misselwitz et al. (2015) who recognised that 10 of the 17 SDG goals are linked to **SDG 11** (5 explicitly; 9 implicitly) and 30% of the targets and 39% of the indicators are linked to **SDG 11**. Thus indicating the central integrating role urbanisation has in the SDG agenda. **SDG 11** also provides an overarching agenda for (i) the UN Habitat's National Urban Policy and the New Urban Agenda, (UN Habitat III Conference in 2016) which seek to offer national and local guidelines on the growth and development of cities through to 2036, and (ii) the UNISDR 2015 Sendai Framework for Disaster Risk Reduction (also discussed in **SDG 1** and **SDG 13**).

Cities are not only places to live, they are engines of the global economy, commerce, and

---

[1]*Urbanisation*, 'the process by which more and more people leave the countryside to live in cities' or 'proportion of people living in built environments such as towns and cities'.

Fig. 11.1 Urban Development in Kampala, Uganda © UKRI (used with permission)

Share of the population living in urban areas (projected to 2050), 2050

Share of the total population living in urban areas, with UN Urbanization projections to 2050. Urban areas are
defined based on national definitions which can vary by country.

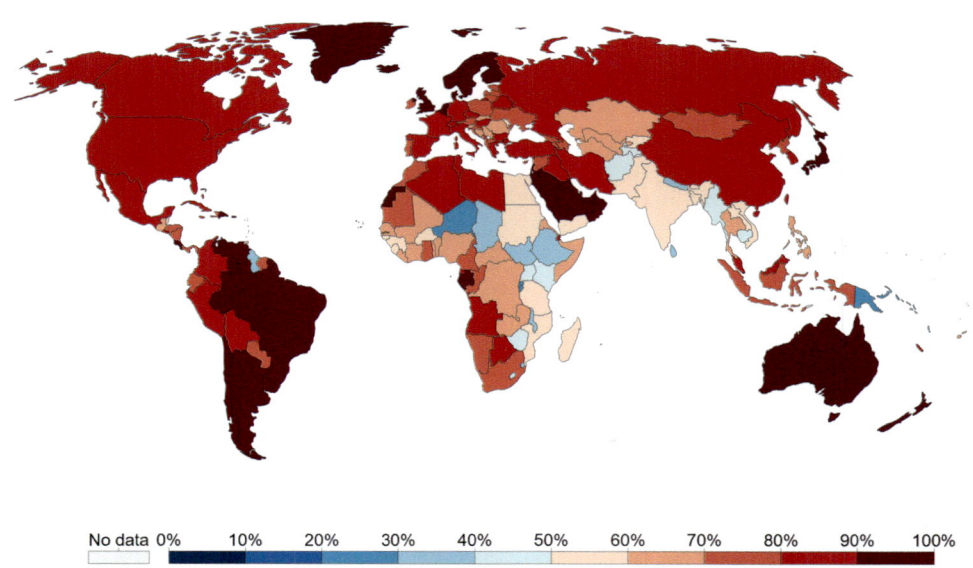

**Fig. 11.2 UN projections for the share of the popu-
lation living in urban areas by 2050**. Credit: Ritchie and
Roser (2019), reproduced under a CC BY licence (https://
creativecommons.org/licenses/by/4.0/). Note that even in
countries with less than 50% of the population living in
urban areas, there are still projected to be cities of millions
(United Nations 2014)

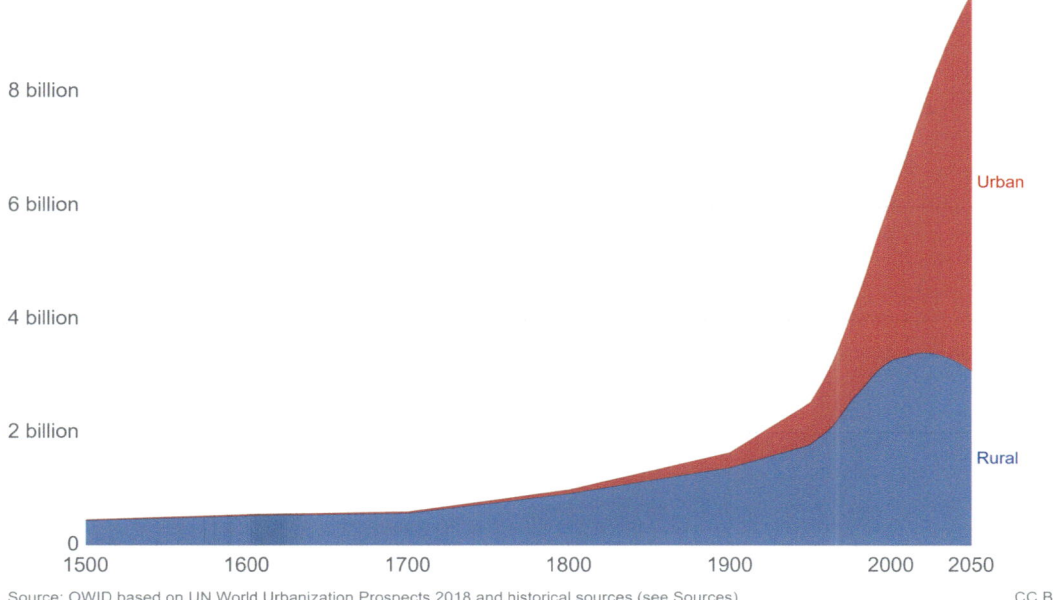

Fig. 11.3 Rural and urban population projections. Global urban and rural populations to 2016 (estimates), and UN projections to 2050, showing a significant increase in the urban population and a decline in the rural population. Credit: Ritchie and Roser (2019), reproduced under a CC-BY licence (https://creativecommons.org/licenses/by/4.0/)

centres of innovation and human cultural expression. Shorter trade links, economies of scales, and access to human capital and shared services within urban centres are providing the basis for economic growth (Ritchie and Roser 2019). There is also a strong link between the quality of life in cities and utilisation of resources, with many people migrating to cities in order to access more secure and stable basic utilities such as water and sanitation (WEF 2017). Urban health and sustainability therefore depend not only on their resilience (see Brand and Jax (2007) for definition) to the combined threats of natural hazards (e.g., rising temperatures and sea level) and anthropogenic change, but also on access to natural resources including water (**SDG 6**), energy (**SDG 7**), and green space and use of

the subsurface for communications, sharing information, waste, storage, and energy.

Thus, at all scales of development, urban resilience requires multi-stakeholder engagement (explored in **SDG 17**) to build consensus, inclusion, and sustainability with a holistic approach to planning and management of complex integrated systems and interdisciplinary science. For example, Dickson et al. (2012) present an urban risk assessment framework that combines hazard impact assessment with institutional and a socio-economic assessment. This is based on four principal building blocks: historical incidence of hazards, geospatial data, institutional mapping, and community participation—geoscience has a role to play in all of these (see Sect. 11.3).

Table 11.1 SDG 11 targets by 2030

Target	Description of target (11.1 to 11.7) or means of implementation (11.A to 11.C)
11.1	Ensure access for all to adequate, safe and affordable housing and basic services and upgrade slums
11.2	Provide access to safe, affordable, accessible and sustainable transport systems for all, improving road safety, notably by expanding public transport, with special attention to the needs of those in vulnerable situations, women, children, persons with disabilities and older persons
11.3	Enhance inclusive and sustainable urbanization and capacity for participatory, integrated and sustainable human settlement planning and management in all countries
11.4	Strengthen efforts to protect and safeguard the world's cultural and natural heritage
11.5	Significantly reduce the number of deaths and the number of people affected and substantially decrease the direct economic losses relative to global gross domestic product caused by disasters, including water-related disasters, with a focus on protecting the poor and people in vulnerable situations
11.6	Reduce the adverse per capita environmental impact of cities, including by paying special attention to air quality and municipal and other waste management
11.7	Provide universal access to safe, inclusive and accessible, green and public spaces, in particular for women and children, older persons and persons with disabilities
11.A	Support positive economic, social and environmental links between urban, peri-urban and rural areas by strengthening national and regional development planning
11.B	By 2020, substantially increase the number of cities and human settlements adopting and implementing integrated policies and plans towards inclusion, resource efficiency, mitigation and adaptation to climate change, resilience to disasters, and develop and implement, in line with the Sendai Framework for Disaster Risk Reduction 2015–2030, holistic disaster risk management at all levels
11.C	Support least developed countries, including through financial and technical assistance, in building sustainable and resilient buildings utilizing local materials

Table 11.2 SDG 11 indicators and geoscience relevance

Indicator	Geoscience link
11.1.1 Proportion of urban population living in slums, informal settlements or inadequate housing	Understanding the architecture and flow patterns of groundwater aquifers is key to sustainable water supply, sanitation and waste management. An understanding of exposure of urban populations to natural hazards and poor quality land
11.2.1 Proportion of population that has convenient access to public transport, by sex, age and persons with disabilities	Better knowledge of ground conditions and local ground hazards will inform routing options and improve affordability and efficient construction
11.3.1 The ratio of land consumption rate to population growth rate	Holistic planning for land use needs to involve the subsurface to mitigate sterilisation of resource potential and ensure effective use of underground space
11.3.2 Proportion of cities with a direct participation structure of civil society in urban planning and management that operate regularly and democratically	N/A
11.4.1 The total expenditure (public and private) per capita spent on the preservation, protection and conservation of all cultural and natural heritage	Ground conditions, ground motion, soil permeability and chemistry and groundwater flow and quality all impact upon preservation of heritage
11.5.1 The number of deaths, missing persons and directly affected persons attributed to disasters per 100,000 population	Disaster Risk reduction and understanding impacts of natural hazards both short and long term on city resilience

(continued)

Table 11.2 (continued)

Indicator	Geoscience link
11.5.2 The direct economic loss in relation to global GDP, damage to critical infrastructure and number of disruptions to basic services, attributed to disasters	As above and understanding how local geology can mitigate impacts of heat island effects—e.g., use of groundwater for cooling
11.6.1 The proportion of urban solid waste regularly collected and with adequate final discharge out of total urban solid waste generated, by cities	Knowledge of subsurface processes required for siting of waste disposal facilities and impact on water quality
11.6.2 The annual mean levels of fine particulate matter (e.g., PM2.5 and PM10) in cities (population weighted)	Understanding how different soil and geology types give rise to fine particulate matter under different urban environmental conditions
11.7.1 The average share of the built-up area of cities that is open space for public use for all, by sex, age and persons with disabilities	Understanding anthropogenic contamination and safe levels in soils and green spaces. Evaluation of the quality, connectivity and access to blue-green land cover and assessment of the multiple benefits of green space e.g., biodiversity net gain and flood risk reduction
11.7.2 The proportion of persons victim of physical or sexual harassment, by sex, age, disability status and place of occurrence, in the previous 12 months	N/A
11.A.1 The proportion of population living in cities that implement urban and regional development plans integrating population projections and resource needs, by size of city	Building geoscience into resource catchment studies and inclusion of urban subsurface assets and resource plans in urban and regional development plans
11.B.1 The number of countries that adopt and implement national disaster risk reduction strategies in line with the Sendai Framework for Disaster Risk Reduction (DRR) 2015–2030	Geoscience support for DRR—understanding hazards and mitigation strategies, improving communication and citizen engagement
11.B.2 The proportion of local governments that adopt and implement local disaster risk reduction strategies in line with national disaster risk reduction strategies	As above
11.C.1 The proportion of financial support to the least developed countries that is allocated to the construction and retrofitting of sustainable, resilient and resource-efficient buildings utilizing local materials	Geological input to construction materials, hazard evaluation and landscape interpretation (geomorphology) to assist with 'build back better'

In this chapter, we introduce the role of geoscience in contributing to the targets and indicators of **SDG 11** (Tables 11.1 and 11.2) and emphasise the importance of building an understanding of the sustainable management and supply of natural resources, risk and management of geological hazards (geohazards), and use of the subsurface beneath our cities in future planning and policy.

We begin with an assessment of the global context (Sect. 11.2) and then in Sect. 11.3 consider the role of geoscience in urban development and use a number of examples to explore specific SDG targets in detail and to highlight the importance of linking planning of the surface and subsurface; concluding statements are presented in Sect. 11.4.

11.2 Urbanisation: Global Context and Impact

The varying definitions of urban land produce conflicting statistics, but in general global analyses indicate that whilst 71% of the Earth's land surface area is defined as liveable and despite popular misconceptions (e.g., Easton 2018), only approximately 3% (around 3.5 million km^2) is urbanised[2] (CIESIN 2010).

---

[2]There is currently no consensus over the definition of an *urban area* and hence estimates of urbanised land and urban populations vary (Ritchie 2018).

The global spatial distribution of our urban centres is widely dispersed reflecting a range of geographical, economic, cultural, and political histories. Originally built close to natural resources (water, minerals) and geographical features for defence and the transport and trading of goods, for example, the ancient Silk Road, cities developed as commercial hubs located at key river crossings or sheltered coastal and estuary locations or along inland global trade routes.

Urbanisation is witnessed at a variety of scales. In terms of size, the largest cities are Tokyo, Shanghai, Mexico City, and Jakarta with populations >20 million followed by 33 megacities with populations >10 million, of which China has 15, India 6, and US, Brazil and Pakistan have 2 each. The number of megacities is projected to rise by 2030 to 43 to accommodate the growing population with 35% of the expansion occurring in just three countries, India, China, and Nigeria (UN 2018b). By contrast, amongst the smallest cities are the developing towns with populations of a few 10 s of thousands in the Small Island Developing States (SIDS), but whatever the scale, all face the same challenges (UN Habitat Report 2016).

Urbanisation is essentially a rural to city transformation driven by industrialisation and is clearly illustrated by comparing city maps and satellite imagery from the 1980s to the present day, see, for example, the Atlas of Urban Expansion.[3] As the urban population rapidly grows, it encourages the progression from low to middle incomes. As noted by the urbanist author Jane Jacobs, '*A metropolitan economy, if it is working well, is constantly transforming many poor people into middle-class people, many illiterates into skilled people, many greenhorns into competent citizens. Cities don't lure the middle class. They create it.*' (Jacobs 1961). Consequently, no country has developed without the growth of its cities and our global economic future is inextricably linked to urban growth (World Bank 2009). However, what is different today is that modern cities in the Global South

are developing rapidly and at a significantly larger scale than previously experienced during the industrial revolution in the developed Global North. The relationship between population growth, urban expansion, and resource use is also non-linear, meaning the impacts of urbanisation on the environment are increasing non-linearly (John et al. 2015). For example, the rate of demand for water has been twice the rate of population growth in recent decades (UN-WWAP 2015).

How is this broad trend in urbanisation away from developed to developing countries being expressed?

Characteristic of rapid urbanisation are high poverty rates and overcrowding in city centres with a lack of basic public services, infrastructure, and health care. Historically, this was just as true for Glasgow, Dublin, or Berlin as it is for Lagos, Nairobi, or Ho Chi Minh City today. The problem now is the pace of urbanisation, the volumes of people seeking to escape impoverished employment, poor to non-existent public services in rural areas, the increasing per capita demand on resources, and the lack of capacity to expand services in line with population growth. Kathmandu, for example, is currently able to supply only about one-third of the total water demanded by its one million plus residents, and experiences power cuts of up to 14 h per day in the dry season. Yet the city continues to grow by about 4% per year (CBS 2012).

Disturbingly, global urban land expansion forecasts (Schneider et al. 2009; Seto et al. 2011; Güneralp and Seto 2013) show that the fastest growth in urban land is occurring in the low-elevation coastal zones (LECZs, i.e., less than 10 m above sea level). Güneralp et al. (2015) predict that around the world, urban areas in the LECZ will increase by 230%, to 234,000 km^2, with the main growth around the west coast of Africa and China, and in the deltas of the Nile, Niger, Pearl, Red, Mekong, and Ganges–Brahmaputra rivers. As a result, by 2030 almost all urban land in SE Asia will be in the LECZ and vulnerable to high-frequency storm surge and flood events.

---

[3]http://www.atlasofurbanexpansion.org/.

Are the global urban initiatives making progress? The World Development Report noted in 2009 that over the preceding decade, most low- and middle-income countries had experienced absolute improvements on a range of basic welfare indicators including malnutrition, immunisation, and school participation in rural and urban areas (World Bank 2009). In 2014, pre-dating the SDGs the UN Habitat Programme established an agenda for a National Urban Policy (NUP) for its member states which was adopted by OECD (2017). Its stated aim is [a] *'coherent set of decisions derived through a deliberate, government-led process of coordinating and rallying various actors for a common vision and goal that will promote more transformative, productive, inclusive and resilient urban development for the long term'*. This was subsequently amended to the UN New Urban Agenda for Smart Cities in 2017 to, [the] *'… future we want includes cities of opportunities for all, with access to basic services, energy, housing, transportation and more'*.

In 2018, the UN progress report on **SDG 11** noted (from UN 2018b):

- Between 2000 and 2014, while the proportion of the global urban population living in slums dropped from 28.4 per cent to 22.8 per cent, the actual number of people living in slums increased from 807 million to 883 million,
- Based on data collected for 214 cities/municipalities, about three-quarters of municipal solid waste generated is collected,
- In 2016, 91% of the urban population worldwide were breathing air that did not meet the World Health Organization air quality guidelines value for particulate matter (PM), with >50% exposed to air pollution levels at least 2.5 times higher than the safety standard (PM 2.5). In 2016, an estimated 4.2 million people died as a result of high levels of ambient air pollution,
- From 1990 to 2013, almost 90% of deaths were attributed to internationally reported disasters that occurred in low- and middle-income countries. Reported damage to housing attributed to disasters shows a statistically significant rise from 1990 onwards.

At the time of writing, a progress tracker for **SDG 11** (Ritchie et al. 2018) contains little data or further feedback on progress.

In summary, with increasingly complex and changing policy and economic global environments, population migration; the development of new trade and capital flows between east and west, including the China Belt & Road initiative (HKTDC 2018) and the Arctic trade route (Lepczyk and Durkin 2018), then predicting and understanding how urbanisation will shape our planets future will continue to focus research effort. It will demand interdisciplinarity, research synthesisers, and harmonisation of links between different research disciplines.

11.3 The Contribution of Geoscience to Urbanisation

The science of urbanisation is complex and involves the interaction of natural, built and social systems, a system in which urban populations rely on ecosystem service provision to sustain life. The concept of 'urban metabolism' was originally conceived by Wolman (1965) as the inflow and outflow transactions required to sustain city functions, or more simply the supply and consumption of natural resources and disposal of waste by-products. The concept of 'urban metabolism' is important in that it acknowledges that urban populations disturb and modify natural processes and pathways and exert an influence beyond the city boundary, relying on the urban 'hinterland' or 'catchment' for natural resources. Cities' inputs include water, energy, food, materials, and nutrients, much of which is imported from the urban 'catchment'; In turn, cities produce outputs including solid waste, wastewater, and emissions that pollute the city and surrounding environment.

Whilst the connection between the city and its natural resource 'catchment' may seem obvious, more than 50 years since urban metabolism was introduced there is still a disconnect between the governance and planning of cities and the sustainable management of natural resources with the former managed at a city scale and the latter

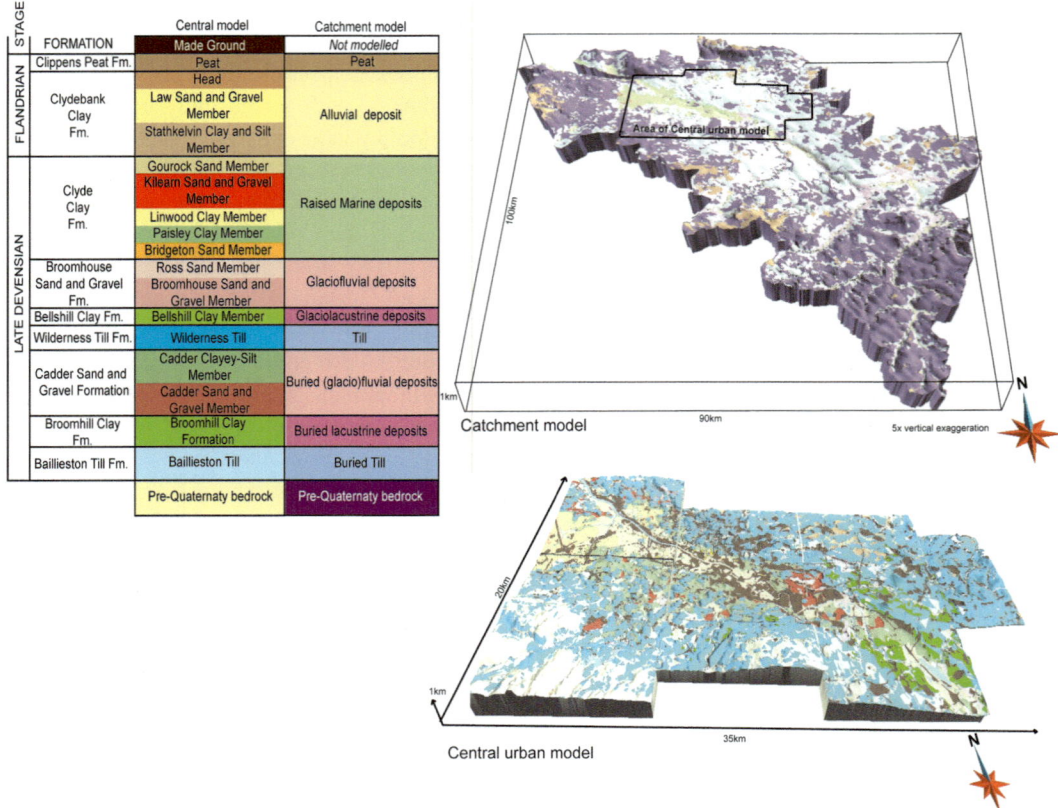

Fig. 11.4 The Clyde Catchment (top left) and Greater Glasgow (bottom left) superficial geology models. The area of the Clyde Catchment model which is covered by the higher resolution urban model is shown by the black box. The key to major stratigraphic units used in the superficial models is derived from Browne and McMillan (1989) for both the detailed central model and catchment models (right). Originally published in Kearsey et al. (2019) © UKRI (used with permission)

managed at the catchment scale. The failure of future city visions to embrace sustainability principles was evident in a recent review by John et al. (2015). In practical terms, this means that experts in natural resource evaluation and catchment management are often excluded from the design of new urban solutions. All cities are defined by a unique set of characteristics that inform their distinctiveness. Several attempts have been made to establish a classification system, or 'typology' to unify these characteristics for different city types to aid urban management and identify suitable interventions. For example, the Atkins Report on Future Proofing Cities (Godfrey and Savage 2012) recognises

different kinds of cities, i.e., disorganised, regulated, stable, historical, growing, shrinking, and newly designed. But few of these typologies include or value the physical geomorphology of the city catchment and its subsurface geology (Fig. 11.4). The geological typology of a city determines key opportunities and risks that the city is exposed to, including access to natural resources, susceptibility to natural hazards, and propensity for difficult ground conditions for the construction and maintenance of infrastructure and buildings.

Typically, the subsurface footprint of a modern city (including caverns, transport tunnels, building foundations, water wells, energy boreholes, etc.) involves construction costs that can run into £ billions. Failure to properly appraise the ground conditions before construction results in significant project overruns and overspend. In established cities, monetary costs continue to be incurred post construction. For example, declining groundwater levels affect existing utilities (sewers, telecoms, gas pipes), ground subsidence and chemical corrosion affect historical and cultural heritage (wooden building foundations), and archaeological sites. Thus urban geology and the subsurface at all levels are an important asset to the social and economic development of a city that needs to be recognised and integrated with surface policy and planning (Culshaw and Price 2011).

Historically, geological surveys and geoscience have ignored this urban dimension, and even in cities with underground resources (e.g., coal and ironstone mines) systematic surveys and assessment of borehole data from site investigations of the near surface was often overlooked. One of the key differences today is that cities now represent large and growing data economies where access to 3rd party digital data, improved technology, and a move away from maps to model simulations (e.g., Smith and Howard 2012; Schokker et al. 2017) underpin the increased flow of 3D data in the shallow subsurface to deliver an understanding of underground space and efficiencies in areas of water supply, energy, transport, and infrastructure.

Below we consider some examples that highlight the role geoscience plays to a number of key SDG target activities.

11.3.1 Cities Underground (Targets 11.2 and 11.3)

Since early times, caves have provided humans with shelter from elements and predators and the underground can be regarded as part of our core psyche (Hunt 2019). But in today's modern world it is an unknown landscape, and other than underground transport systems, largely 'out of sight—out of mind' and unforeseen. However, as cities in the developed world compete for space

and seek to limit urban sprawl, there is an increasing focus on the underground as an added dimension to our urban planning. Today, cities such as Montreal, Helsinki, St Petersburg, Singapore, and Shanghai are actively engaged in subsurface planning and construction of underground shopping malls, laboratories, and storage facilities.

Since the late 1990s, the British Geological Survey (BGS) has been developing an urban programme to deliver thematic applied geology maps, borehole and mining datasets, and 3D geological models for integrated sustainable urban planning of UK cities including Glasgow, Manchester, and London.[4] More recently, in 2013 this programme has been linked with ongoing studies in Hamburg and Paris to foster Sub-Urban, a European Cooperation in Science and Technology (COST) Action project (TU1206[5]). This action, for the first time, established a trans-European network of researchers and city authorities from an initial core of 7 partners (Box 11.1); it ultimately developed into 31 countries and 26 cities. The aim was to transform relationships between experts who develop urban subsurface geoscience knowledge, i.e., the National Geological Surveys, and those who can most benefit from it, i.e., urban decision makers, planners, consultants, and the wider research community.

Unlike 2D layer surface planning, underground spatial planning requires consideration of 3D volumes. The Sub-Urban project by looking at different city typologies and geological settings developed a community of practice for the growing field of urban geoscience and prepared an online toolbox[6] to aid research into the 3D/4D characterisation, prediction, and visualisation of the ground beneath cities. Similarly, in order to compliment Building Information Models (BIM) which are now used extensively above ground, an analogous concept GeoCim (Fig. 11.5) has been proposed as an approach that

---

[4]http://www.bgs.ac.uk/research/engineeringGeology/urbanGeoscience/home.html.

[5]https://www.cost.eu/actions/TU1206/#tabs|Name:overview.

[6]www.sub-urban.eu/toolbox.

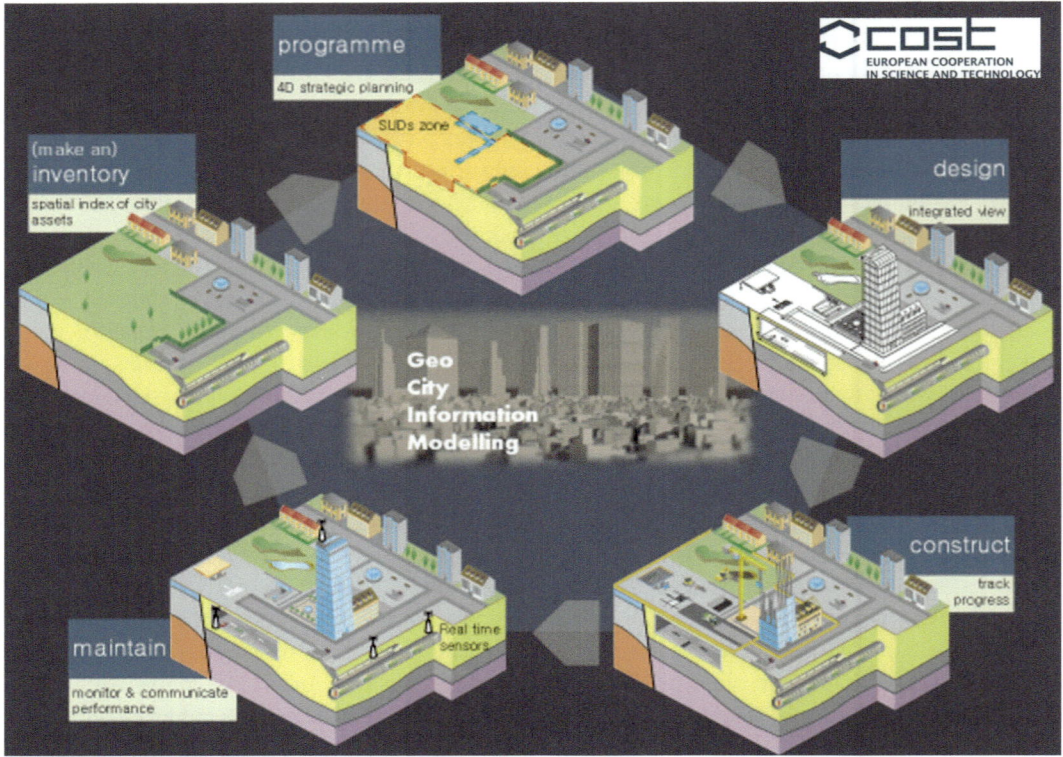

Fig. 11.5 GeoCIM life cycle diagram. From Mielby et al. (2017)

combines above and below ground city models, across site-to-city scales to identify subsurface opportunities, improving routing of transport systems and reducing uncertainty in ground conditions (Fig. 11.5).

Box 11.1 Sub-Urban: a Transport and Urban Theme COST Action project (TU 1206) 2013-2017

The Sub-Urban project drew together collective research capabilities across several European Geological Surveys (see image below) in 3D/4D characterisation, data systems, prediction, and visualisation of the subsurface to produce a series of good practice reports, and provide training and advice to city planners.[7]

Research partners and focus cities, include

- NGU: Geological Survey of Norway— Oslo
- GSN (TNO): Geological Survey of the Netherlands—Utrecht
- GLH/BSU: Ministry of Urban Development and Environment—Hamburg
- GEUS: Geological Survey of Denmark —Odense
- GSI: Geological Survey of Ireland— Dublin
- GSNI: Geological Survey of Northern Ireland—Belfast
- BGS: British Geological Survey— Glasgow

Across Europe city-scale 3D/4D models that draw on an extensive ground investigation (with tens to hundreds of thousands

---

[7]http://sub-urban.squarespace.com/#about.

of boreholes) and other data are being developed further by partners in the UK (Glasgow, London), Germany (Hamburg), and France (Paris). Model linkages are being used to look at the life cycle of urban planning and to enable prediction of groundwater, geothermal heat, Sustainable Urban Drainage Systems (see Sect. 11.3.2), and engineering properties. Combined with above-ground models (e.g., CityGML and BIMs), these provide valuable tools for holistic urban planning, identifying subsurface opportunities and saving costs by reducing uncertainty in ground conditions.

The BGS and Sub-Urban research has focused on post-industrial, redeveloping cities in the Global North with very low population growth rates and an established and strong regulatory environment and planning control. The issue of who owns the subsurface and how to regulate the competing uses of the underground space is a challenge, and applying this approach to megacities in the Global South with poorly understood geology, multiple hazards, resource issues, informal development, and variable regulatory controls remains to be tested.

11.3.2 Cities and Resources (Targets 11. 1 and 11.6)

Closely tied into the integrated planning approach discussed in Sect. 11.3.1 is an understanding of the resource opportunities beneath a city. These may include groundwater aquifers and the management of urban water, the presence of mineral and hydrocarbon deposits and past mining activities, geothermal energy, suitable rock formations for cavern storage, and soil type and quality for greening and open space.

SDG 6 addresses clean water and sanitation as a key service and is fundamental to many cities in the developing world that rely on groundwater from aquifers. With population increase and

climate change, it is forecast that the number of urban residents in perennial shortage of water will increase to approximately 160 million by 2050 (McDonald et al. 2011). As cities continue to grow, they are constrained by geology to go further beyond the city boundary or dig deeper to obtain water, this in turn increases the demand for energy (UN Water 2014, 2015). Therefore, in the Global South energy supply will have direct implications on availability as well as affordability of urban water in the rapidly growing cities.

Cities alter the topography and geomorphology of the landscape and human interventions including water abstraction, riverbank erosion and flood protection, canalisation, coastal barrages and impermeable concrete cover all affect river, groundwater and coastal systems and alter the natural hydrological and hydrogeological dynamics. From a groundwater perspective, the most significant impacts due to urbanisation are (i) direct groundwater abstraction, (ii) contamination of groundwater arising from polluting activities (industry, buried waste, and sewage), and (iii) the construction of low permeability infrastructure at the surface (buildings, roads, and car parks) and below ground (e.g., pipes, basements, and transport tunnels) which serve to reduce direct infiltration of water into the ground and alter shallow groundwater pathways (Bricker et al. 2017).

The potential of water-sensitive urban design (WSUD) or 'blue-green' city approaches to address issues of water resource consumption, water quality, urban drainage, and floodwater management as part of an integrated system are increasingly being realised through urban planning (e.g., Dorst et al. 2019; Li et al. 2019). This approach recognises the need for natural or 'non-structural' engineering such as rainwater harvesting, sustainable urban drainage systems (SuDS), riparian habitats, ponds, and wetlands, in addition to traditional engineered solutions. Where urban governance is strong and formal water management and urban planning policies exist, water-sensitive urban design can more readily be captured in urban master plans and implemented in a sustainable and measured way through the urban planning process. Where cities are developing in a more informal way, and where urbanisation is outpacing the implementation formal water infrastructure and the urban water system is seen as dysfunctional, it is more challenging to present a value case and implementation plan for natural or 'green' solutions. Though in the absence of formal urban policy, the role of community-led green-blue solutions implemented at a local neighbourhood level is acknowledged. Despite the growing desire for nature-based solutions to urban water management at both community- and government levels the challenge for functional 'blue-green' solutions is acute in the Global South where regions are increasingly exposed to unprecedented extreme rainfall and flooding events. Here options for sustainable drainage are more limited and the implementation of effective green-blue solutions at the catchment scale is needed. For example, as noted above by 2030 almost all of the urban land in SE Asia will be in the LECZ and vulnerable to high-frequency storm surge and flood events.

By mimicking natural surface water drainage patterns, SuDS aim to lower storm-water run-off rates and surface water flow into the formal drainage system, increase infiltration to the ground and water storage, and reduce the transport of pollutants to the water environment. SuDS are therefore designed on the principals of source control, infiltration, conveyance, retention, and detention, and include green roofs, infiltration systems, soakaways, permeable paving, swales, retention ponds, and storage tanks (Woods-Ballard et al. 2016). Each scheme must be designed in accordance with the site surface water run-off rates to ensure that the SuDS can cope with the volume of water generated during both frequent rainfall and extreme rainfall events (Fig. 11.6).

Whilst all SuDS interact with the urban water system, infiltration SuDS (infiltration systems, soakaways, permeable paving) promote direct infiltration of rainwater into the ground and therefore have the largest potential impact on groundwater. Issues such as ground permeability

and drainage potential, ground stability, and groundwater quality protection must be considered to negate any negative impacts associated with infiltration SuDS. For example, infiltration SuDS would be unsuitable in areas where groundwater levels are high or where there is already a risk of flooding, where infiltration of rainwater would mobilise contaminants, or where infiltration of rainwater would increase the risk of ground hazards such as unstable slopes (Dearden et al. 2013).

Box 11.2 Sustainable Urban Drainage Schemes in China

SuDS approaches are exemplified in Wuhan, China, where the concept of 'Sponge Cities', with permeable pavements, rain gardens, artificial ponds, riparian habitat restoration, and wetland creation, is promoted in favour of hard engineering solutions (Chunhui et al. 2019). Wuhan, known as the city of a hundred lakes (actually 166 lakes), is located at the confluence of the Yangtze and Hanjiang Rivers. With an annual average rainfall of 1150–1500 mm, it is prone to intense rainfall events and flooding particularly during the rainy season (June–July) (Wu et al., 2019). Rapid urbanisation had resulted in increased impermeable cover, an ageing and overwhelmed drainage network, and degradation and reclamation of the lakes. Compounded by climate change impacts, the city of Wuhan was exposed to a heightened flood risk, and in summer 2016 flooding across the city affected more than 750,000 people and resulted in economic losses of more than $3bn USD (Wu et al., 2019).

Following this flooding event, the authorities in Wuhan have introduced 'Sponge City' principles to integrate water-sensitive urban planning solutions. Working on a water source-control principle, the aim is to attenuate excessive rainfall through soil infiltration and retain it in

Fig. 11.6 Example of a retention sustainable drainage system installed in a new housing development in Derby, UK. Photo by author © UKRI (used with permission)

lakes and underground storage, only discharging it into the river once water levels there are low enough. For example, Wufengzha Wetland Park, originally built as a sluice for the river, now encompasses a series of permeable pavements, grassed infiltration areas, and rain gardens as a means to manage rainwater more effectively (Fig. 11.7).

11.3.3 Cities Living with Geohazards (Targets 11.5 and 11.B)

Future cities will need to be sustainable and disaster resilient and will require planning and emergency response strategies to mitigate and adapt to the impacts of climate change and address the goals of the Sendai Framework for Disaster Risk Reduction (as described in **SDGs 1** and **13**). A UNDESA report reveals that of the 1,146 cities

Fig. 11.7 Wuhan South lake. Image by Toehk. Licenced under CC BY-SA 2.0 (https://creativecommons.org/licenses/by-sa/2.0/)

with >0.5 million people, 59% are at high risk of exposure to at least one and often more than one type of natural hazards (United Nations 2019a, b). Geoscience is important to understand natural processes and geological controls on how landscape and subsurface will react to hazard-induced change. Below, we highlight two examples of how the geology of cities can input to natural and anthropogenic hazards.

Example 1 Cities with Multiple Hazards

As noted previously, the low-lying coastal cities of SE Asia include some of the most vulnerable urban conurbations. Here, seasonal climate change with increasing typhoon frequency, rainfall-induced landslides, heatwaves, and wildfires combine with earthquakes, tsunamis, and volcanic eruptions to create a complex hazard landscape. Monitoring of hazards, disaster risk management, and effective communication and public understanding are essential to preserving lives and sustaining economic development.

The World Risk Report (2019) ranks the Philippines as the 9th of 180 countries in terms of disaster risk. The capital, Manila, has a population of 1.78 million squeezed into an area of 15.4 km^2, and is currently the most densely populated city in the world, with more than 42,000 people per km^2 (Rith et al. 2019). The wider metropolitan area contains 16 cities with a further 12.8 million people. There are estimated to be 3.1 million homeless people in Manila, of which 0.6 million are children, who live in slums lacking adequate water, housing, sanitation, education, and health, and with deficient nutrition resulting in large numbers with stunted growth (Reuters 2018). Manila experiences numerous tropical cyclones per year, with 6–9 making landfall and often triggering landslides and flooding (GovPh 2019; NDRRMC 2019). Sea-level rise and the impact of far-field events (tsunamis and volcanic eruptions) add to the vulnerability of the population along the coastal strip.

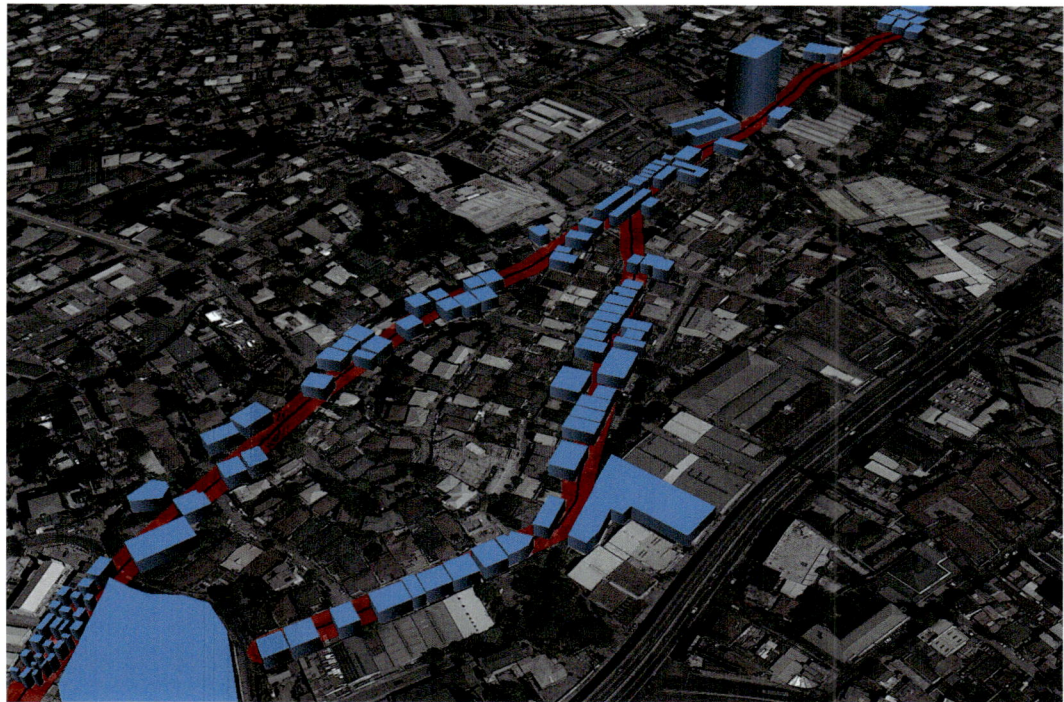

Fig. 11.8 Illustration of structures directly traversed by the Valley Fault System with 5-meter buffer zones along Pasig City (in Metro Manila, Philippines). Credit: Ervin Malicdem (Reproduced under a CC BY 4.0 licence, https://creativecommons.org/licenses/by/4.0/)

Geologically, Manila is built on a narrow isthmus between the coast and Laguna Lake and to the north, the greater metropolitan area is in part built on unstable slopes of volcaniclastic soils and alluvium that infills a valley extending down to the sea. The major vertical Marikina Fault System cuts across the region and presents a significant hazard (Nelson et al. 2000; Rimando and Knuepfer 2006), as illustrated in Fig. 11.8. Research into recent landslips and escarpment formation records reveals a history of movements associated with stress release along the tectonic plates (Nelson et al. 2000). Understanding the seismic risk associated with this fault, and the potential cascade of hazards (landslides, flooding, and liquefaction of sediments) that could result is needed to inform planning, development of mitigation strategies, and effective response. The Philippine Government is well organised, with detailed mapping at 1:10,000 scale of landslide deposits, and liquefaction and flood susceptibility maps. Thus in Manila, geology is actively used in planning and better engineering of foundations for resilient housing.

Example 2 Feeling the Heat

The heat island effect of urbanisation is well described (Gago et al. 2013; Yang and Santamouris 2018) and in many cities results from the escalating use of air conditioning and increased transport flows. These give rise to waste heat and air pollution which combined with climate change now represent a serious challenge to urban living conditions,[8] for example, the heatwave in India in 2015, where temperatures exceeded 45 °C in many parts of India, with an

---

[8]The Guardian newspaper has an interactive feature on 'Which cities are liveable without air conditioning—and for how much longer?' available at: https://www.theguardian.com/cities/ng-interactive/2018/aug/14/which-cities-are-liveable-without-air-conditioning-and-for-how-much-longer (accessed 28 October 2019).

estimated 2500 excess deaths (Wehner et al. 2016). The heatwave resulted in power outages due to increased electricity demand and impacted low-income urban neighbourhoods which lack open green spaces and are unsheltered from the heat trapped in the buildings during the night. Hotter nights are also affecting cities in Japan, leading to increased water demand and energy consumption (Fujibe 2011).

In 2018, 354 cities experienced average summer temperatures >35 °C and this is expected to rise to 970 by 2050 (UCCRN 2018). A report by the Asian Development Bank predicts a shift to a 'new climate regime' across SE Asia by the end of the century, when the coolest summer months would be warmer than the hottest summer months were in the 1950–80s (ADB 2017).

Extreme heat is injurious to health. As temperatures rise, asphalt heats up by 10–20 °C more than air, and so in a 50 °C city you can fry an egg on the pavement, animal paws blister, power grids are overwhelmed by air conditioning cooling demands (especially in India and Pakistan), and cultural events struggle to cope. To manage the heat during the annual Hajj in Makkah, attended by > 1.8 million pilgrims, the organisers now use retractable umbrellas for shade and air conditioning units weighing 25 tonnes to ventilate tents and spray mist in the streets. The World Cup in Qatar in 2022 and the 2021 Tokyo Olympics are the two major sporting events that will be impacted by this heat island effect. To address this challenge, an alternative is to move underground. There are many underground dwellings in Coober Pedy, for example, a community at the centre of Australia's opal mining industry (Gillies et al. 1981; Admiraal and Cornaro 2016), demonstrating that this is a sustainable proposition.

We need to cool our cities, and whilst the current focus in the developed world is on surface actions (e.g., developing garden cities, use of reflective paints, and heat adsorbing building materials) (Fig. 11.9), the opportunities in the subsurface remain to be investigated (e.g., using groundwater, using flooded former mine workings to cool circulating air or to create large water-filled caverns). These require an

Fig. 11.9 Green roof on Chicago City Hall. This approach can improve insulation, provide habitats for biodiversity, and lower urban air temperatures. Adapted from an image by TonyTheTiger (Reproduced under a CC-BY-SA 3.0 licence, https://creativecommons.org/licenses/by-sa/3.0/)

understanding of the underlying geology, fracture systems, local geothermal gradient, and heat and coolth demand before testing and inclusion in city energy planning.

11.3.4 Data Systems and Smart Cities (Targets 11A and 11B)

Cities are data economies and the flow and integration of temporal and spatial data, often in real time, is increasingly important to manage resilience. Delivered through smart city concepts, significant progress is being made to increase data discoverability, accessibility, and usefulness, to reduce siloed data management, in support of data-driven decision-making. This is particularly important in cities where support systems and services are often integrated.

The challenge of delivering effective data systems for cities should not be underestimated and requires significant investment, particularly in the Global South where data is often in analog format. Funding for IT infrastructure upgrades and software licences is often limited, organisational capability in digital data workflows is low, and government support for open-data initiatives is lacking due to security or political concerns. Under these circumstances, open-source software, cloud computing, and web services provide a viable proposition, not just for government authorities but also for private enterprises where the commercial market for data services exists. It is not uncommon under such circumstances for private companies to assume the role of data owner-provider, when by preference the government ought to be the custodian of the data.

From an environmental and geoscience perspective, cities operate at the intersection of natural-built-social systems. Data is needed, not just to be able to characterise natural environmental processes but also to understand the anthropogenic impacts on those processes. To understand the interactions with the built environment and how geoscience data and information can help support planning for sustainable urban growth, we must consider

- Data evidence-base needed to underpin urban planning and development.
- Spatial and temporal scales for which data is needed to characterise the complexity of the urban environment.
- Interoperability of the data with linked city data systems.
- Mechanisms through which decision makers can access and use the data.
- Ways in which geoscience data can be communicated effectively.

Cities generate large volumes of subsurface data (e.g., borehole logs, groundwater, geotechnical parameters, temperature, ground motion, and land quality) which, with modern technology, can be readily included and shared, combined with other data types, to deliver 2D, 3D, or 4D representation and modelled simulations that predict subsurface ground conditions and resource opportunities. Mandating, or incentivising, inclusion and sharing of near-surface (soil) and deeper subsurface data into the developing smart digital services will deliver more holistic urban planning and result in meaningful cost savings in remediation and underground construction, ultimately underpinning many of the SDG targets.

With open real-time surface sensor networks, the data economy of the urban environment is rapidly growing to monitor traffic people and goods, airflow, temperature, and pollution. We are developing cities as observatories and in the future, they will need to include dedicated underground sensor networks largely located in boreholes, on underground infrastructure and fibre-optic cables (Daley et al. 2013) to monitor the urban subsurface environment (built and natural).

To effectively monitor the condition of and impacts on our buried assets (e.g., utilities, transport tunnels, basements, and boreholes), sensors would need to be based on the chemistry of waves (sound, heat, and light), the physics of pressure from fluids and gases to detect movement and reactions and from gravity, and microseismicity and distributed acoustic sensing with fibre-optic cables to detect movement in response to stress. In the absence of dedicated buried sensor networks, we continue to rely on remote sensing using urban geophysics including, ground-penetrating radar, vibro-acoustics, and low-frequency electromagnetics (resistivity) to monitor buried assets (e.g., Metje et al. 2007; Miller 2013).

Monitoring networks, however, are expensive and should be targeted according to resource needs. They may be focused on groundwater, temperature, or ground motion. Monitoring networks help to understand the temporal and spatial changes in a city, and how the urban environment may respond to different scenarios. This allows for cost-effective and scientifically robust programmes for integrated regional and city planning.

11.4   Conclusions

It is no exaggeration to state that the volume of literature on cities is vast, reflecting the importance and complexity of our urban landscape and how it will respond to future change. Global progress towards achieving sustainable urban communities is slow and complicated by many factors. Transforming our understanding of the competing demands on resources to provide an equitable and healthy lifestyle for all city dwellers increasingly requires geosciences to play a fundamental role in underpinning planning and management of hazard risk and competing use of the ground at all depths beneath the cityscape. Too often, the subsurface is presented in a fragmented and incomprehensible way to urban planners and rarely considered in respect of climate change and hazard adaptation. Experience in developed countries indicates that this is often because

- Responsibilities for the subsurface layers are divided between various ministries.
- Surface and subsurface disciplines (e.g., hydrogeology, archaeology, geotechnics, and energy) rarely work together.

- Subsurface specialists (geologists) are not involved in urban planning, and so the ground beneath the city is only seen as a problem when things go wrong.

As our cities cannot be allowed to fail, geoscience must find ways of integrating, utilising, and communicating geology to help find solutions to building resilience to geohazards, sustainable management of resources, and limiting the impact of climate change that feed into policy and planning adaptation to future change. Within the Global South, this is challenging and requires a systems approach (Bai et al. 2018), exploring different ways of planning, building, and governing our cities that use interdisciplinary science combining social, built, and natural environment research.

11.5   Key Learning Concepts

- Urbanisation is a major feature in the humanisation of our planet, cities are fundamental to national and global economies, hubs of culture and innovation, and present particular development challenges and opportunities. By 2050, the majority of the global population will be urbanites living in cities built predominantly in low-elevation coastal zones.
- Cities are not isolated entities. They interact with their catchments stimulating migration, absorbing natural resources, and emitting waste. Ensuring sustainable urbanisation is therefore intrinsically linked to other SDGs, protecting biodiversity, meeting essential needs, and decoupling economic growth from environmental degradation.
- Many cities are vulnerable to multiple anthropogenic and natural hazards, and their cascading effects. Geoscientists' contributions to hazard and risk assessment can inform steps to reduce disaster risk and increase resilience. This aligns with the Sendai Framework for Disaster Risk Reduction.

- Geoscience has a role in many sectors providing information on natural resources (e.g., construction materials, and water resources), the nature of the subsurface (e.g., geological materials and dynamics), and providing data, interpretations, and visualisations to underpin integrated policy and efficient planning.

could be located in the subsurface, and which are better placed on the surface? What assumptions about the underlying geology of the city are you making in your analysis? For the infrastructure/services you have chosen to place in the subsurface, review what major cities have used the subsurface to develop these, and their geological environments.

11.6 Educational Ideas

In this section, we provide examples of educational activities that connect geoscience, the material discussed in this chapter, and scenarios that may arise when applying geoscience (e.g., in policy, government, private sector international organisations, NGOs). Consider using these as the basis for presentations, group discussions, essays, or to encourage further reading., and

- How may the surface expression of a city built on a bedrock limestone plateau differ to a city built on soft alluvial sand and gravel? The underlying geology influences, for example, the urban topography, how easy it is to build, the available natural resources such as groundwater and building materials, the density of the river drainage network, and migration of contaminants. What opportunities and challenges to sustainable urbanisation may each geological environment present?
- Many cities are exposed to multiple natural hazards, with potentially cascading effects in terms of (i) hazards triggering other hazards, and (ii) hazards triggering a cascade of impacts on infrastructure. For a case study city of your choice, identify the multiple hazards the city must plan for, and the potential relationships between these hazards (e.g., earthquakes may trigger landslides, storms may trigger flooding). If you identify potential relationships, what implications does this have for disaster risk reduction in the region?
- Consider the range of infrastructure and services located within a city. Which of these

Further Reading and Resources

Cities Alliance (2019). https://www.citiesalliance.org/

Culshaw MG, Reeves HJ, Jefferson I, Spink TW (2009) Engineering Geology for Tomorrow's. Cities. https://doi.org/10.1144/EGSP22 ISBN (print): 9781862392908 ISBN (electronic): 9781862393844. Geological Society of London

100 Resilient Cities (2019). https://www.100resilientcities.org/

Sterling R, Bobylev N (eds) (2016) Urban underground space: a growing imperative perspectives and current research in planning and design for underground space use. TUST 55: 1–342. ISSN: 0886-7798

UNDRR (2015) Sendai framework for disaster risk reduction. https://www.unisdr.org/we/coordinate/sendai-framework

UN Habitat (2019). https://new.unhabitat.org/

References

Admiraal H, Cornaro A (2016) Why underground space should be included in urban planning policy–and how this will enhance an urban underground future. Tunn Undergr Space Technol 55:214–220

Asian Development Bank (2017) A region at risk: the human dimension of climate change in Asia and the Pacific, 131pp. ISBN 978-92-9257-851-0. https://doi.org/10.22617/tcs178839-2

Bai X, Dawson RJ, Ürge-Vorsatz D, Delgado GC, Barau AS, Dhakal S, Dodman D, Leonardsen L, Masson-Delmotte V, Roberts DC, Schultz S (2018) Six research priorities for cities and climate change. Nat Comment 555:23–25. https://doi.org/10.1038/d41586-018-02409-z

Biswas AK, Tortajada C, Stavenhagen M (2018) In an urbanising world, shrinking cities are a forgotten problem. World Economic Forum agenda report. https://www.weforum.org/agenda/2018/03/managing-shrinking-cities-in-an-expanding-world. Accessed 27 Oct 2019

Brand FS, Jax K (2007) Focusing the meaning(s) of resilience: resilience as a descriptive concept and a boundary object. Ecol Soc 12(1):23. http://www.ecologyandsociety.org/vol12/iss1/art23/. Accessed 27 Oct 2019

Bricker SH, Banks VJ, Galik G, Tapete D, Jones R (2017) Accounting for groundwater in future city visions. Land Use Policy 69:618–630

Bristow DN, Kennedy CA (2013) Urban metabolism and the energy stored in cities. J Ind Ecol 17:656–667. https://doi.org/10.1111/jiec.12038

Chunhui L, Peng C, Chiang PC, Cai Y, Wang X, Yang Z (2019) Mechanisms and applications of green infrastructure practices for stormwater control: a review. J Hydrol 568:626–637. ISSN 0022-1694. https://doi.org/10.1016/j.jhydrol.2018.10.074

CIESIN (2010) Centre for International Earth Science Information Network (CIESIN), Columbia University, Gridded Population of the World (GPW), version 3 and Global Rural-Urban Mapping Project (GRUMP) Alpha Version. http://www.sedac.ciesin.columbia.edu/gpw. Accessed 1 Oct 2012

Culshaw MG, Price SJ (2011) The 2010 Hans Cloos lecture: The contribution of urban geology to the development, regeneration and conservation of cities. Bull Eng Geol Environ 70:333–376. https://doi.org/10.1007/s10064-011-0377-4

Daley TM, Freifeld BM, Ajo-Franklin J, Dou S, Pevzner R, Shulakova V, Kashikar S, Miller DE, Goetz J, Henninges J, Lueth S (2013) Field testing of fiber-optic distributed acoustic sensing (DAS) for subsurface seismic monitoring. Lead Edge 32(6):699–706

Dearden RA, Marchant A, Royse K (2013) Development of a suitability map for infiltration sustainable drainage systems (SuDS). Environ Earth Sci 70(6):2587–2602. https://doi.org/10.1007/s12665-013-2301-7

Dickson, E, Baker JL, Hoornweg D, Tiwari A (2012) Urban risk assessments: understanding disaster and climate risk in cities (English). Urban development series. World Bank report Washington, DC, 276pp

Dorst H, van der Jagt A, Raven R, Runhaar H (2019) Urban greening through nature-based solutions—key characteristics of an emerging concept. Sustain Cities Soc 49(2019):101620. https://doi.org/10.1016/j.scs.2019.101620

Easton M (2018) The illusion of a concrete Britain. https://www.bbc.co.uk/news/uk-42554635. Accessed 27 Oct 2019

Fujibe F (2011) Urban warming in Japanese cities and its relation to climate change monitoring. Int J Climatol 31:162–173. https://doi.org/10.1002/joc.2142

Gago EJ, Roldan J, Pacheco-torres R, Ordonez J (2013) The city and urban heat islands: a review of strategies to mitigate adverse effects. Renew Sustain Energy Rev 25:749–758. https://doi.org/10.1016/j.rser.2013.05.057

Gillies ADS, Mudd KE, Aughenbaugh NB (1981) Living conditions in underground houses in Coober Pedy Australia. In: Holthusen TL (ed) The potential of earth shelter and underground space, today's resource for tomorrow's space and energy viability, pp 163–177. https://doi.org/10.1016/B978-0-08-028050-9.50020-X

Godfrey N, Savage R (2012) Future proofing cities: risks and opportunities for inclusive growth in developing countries. Atkins, Epsom, p 188

GovPh (2019) Tropical cyclone information. http://bagong.pagasa.dost.gov.ph/climate/tropical-cyclone-information. Accessed 27 Oct 2019

Graedel TE (1999) Industrial ecology and the ecocity. The Bridge 29(4):10–14. http://www.complexcity.info/files/2011/10/CitiesOrganisms.pdf. Accessed 27 Oct 2019

Güneralp B, Seto KC (2013) Futures of global urban expansion: uncertainties and implications for biodiversity conservation. Environ Res Lett 8:014025

Güneralp B, Güneralp I, Liu Y (2015) Changing global patterns of urban exposure to flood and drought hazards. Glob Environ Chang 31:217–225

HKTDC Research—The Belt and Road Initiative (2018) Hong Kong Trade Development Council, May 2018. http://china-trade-research.hktdc.com/business-news/article/The-Belt-and-Road-Initiative/The-Belt-and-Road-Initiative/obor/en/1/1X000000/1X0A36B7.htm. Accessed 27 Oct 2019

Hunt W (2019) Underground a human history of the worlds beneath our feet. Simon & Schuster, 288pp

International Organization for Migration (IOM) (2015) World migration report 2015. Migrants and cities: new partnerships to manage mobility. International Organization for Migration. Geneva. ISBN 978-92-9068-709-2

Jacobs J (1961) The Life and death of Great American cities, 50th Anniversary edn. Modern Library, p 640. ISBN 970679644330

John B, Withycombe Keeler L, Wiek A, Lang DJ (2015) How much sustainability substance is in urban visions?— an analysis of visioning projects in urban planning cities, vol 48, November 2015, pp 86–98. https://doi.org/10.1016/j.cities.2015.06.001. ISSN 0264-2751

Kearsey TI, Whitbread K, Arkley SLB, Finlayson A, Monaghan AA, McLean WS, Terrington RL, Callaghan EA, Millward D, Campbell SDG (2019) Creation and delivery of a complex 3D geological survey for the Glasgow area and its application to urban geology. Earth Environ Sci Trans R Soc Edinb 108(2–3):123–140

Lepczyk A, Durkin A (2018) A breakthrough in Arctic trade routes. Trade vistas article. https://tradevistas.org/breakthrough-arctic-trade-routes/. Accessed 27 Oct 2019

Li C, Peng C, Chiang P-C, Cai Y, Wang X, Yang Z (2019) Mechanisms and applications of green infrastructure practices for stormwater control: a review. J Hydrol 568:626–637. https://doi.org/10.1016/j.jhydrol.2018.10.074

McDonald RI, Douglas I, Revenga C et al (2011) Global urban growth and the geography of water availability, quality, and delivery. AMBIO 40:437–446. https://doi.org/10.1007/s13280-011-0152-6

Metje N, Atkins PR, Brennan MJ et al (2007) Mapping the underworld—state of-the-art review. Tunn Undergr Space Technol 22:568–586. https://doi.org/10.1016/j.tust.2007.04.002

Mielby S, Eriksson I, Campbell SDG, de Beer J, Bonsor H, Le Guern C, van der Krogt R, Lawrence D, Ryzynrski G, Schokker J, Watson C (2017) TU1206 COST sub-urban WG2 report, 111pp. https://static1.squarespace.com/static/542bc753e4b0a87901dd6258/t/58aed9328419c2c2bcbb8aba/1487853904678/TU1206-WG2-001+Opening+up+the+subsurface+for+the+cities+of+tomorrow_Summary+Report.pdf. Accessed 27 Oct 2019

Miller R (2013) Introduction to this special section: urban geophysics. Lead Edge 32:233–369

Misselwitz P, Villanueva JS, Rowell A (2015) The urban dimension of the SDGs: implications for the new urban agenda. In: Paper presented at the cities alliance workshop in New York, 27 September 2015. Taken from cities alliance discussion paper no 3: sustainable development goals and habitat III: opportunities for a successful new urban agenda, 13–22pp

Multiple large earthquakes in the past 1500 years on a fault in metropolitan Manila, the Philippines. Bull Seismol Soc Am 90(1):73–85. https://doi.org/10.1785/0119990002

NDRRMC (2019) National Disaster Risk Reduction and Management Council. Available at: http://www.ndrrmc.gov.ph/. Accessed 27 Oct 2019

Nelson AR, Personius SF, Rimando RE, Punongbayan RS, Tungol N, Mirabueno H, Rasdas A (2000) Multiple large earthquakes in the past 1500 years on a fault in metropolitan Manila, the Philippines. Bull Seismol Soc Am 90(1):73–85

OECD (2017) National urban policy in OECD Countries. OECD Publishing, Paris, https://doi.org/10.1787/9789264271906-en. Accessed 27 Oct 2019

Reuters (2018) Manila's homeless story, 28 March 2018 by Rina Chandran. https://www.reuters.com/article/us-philippines-landrights-lawmaking/manilas-homeless-set-to-move-into-more-empty-homes-if-official-handover-delayed-idUSKBN1H41L7. Accessed 27 Oct 2019

Rimando RE, Knuepfer PL (2006) Neotectonics of the Marikina Valley fault system (MVFS) and tectonic framework of structures in northern and central Luzon, Philippines. Tectonophysics 415(1–4):17–38

Ritchie H (2018) How urban is the World? https://ourworldindata.org/how-urban-is-the-world. Accessed 27 Oct 2019

Ritchie H, Roser R (2019) Urbanization. https://ourworldindata.org/urbanization. Accessed 27 Oct 2019

Ritchie R, Mispy, Ortiz-Ospina (2018) Measuring progress towards the sustainable development goals. SDG-Tracker.org

Rith M, Fillone A, Biona JBM (2019) The impact of socioeconomic characteristics and land use patterns on household vehicle ownership and energy consumption in an urban area with insufficient public transport service–a case study of metro Manila. J Transp Geogr 79:102484

Schneider A, Friedl MA, Potere D (2009) A new map of global urban extent from MODIS satellite data. Environ. Res. Lett. 4:044003

Schokker J, Sandersen P, de Beer H, Eriksson I, Kallio H, Kearsey T, Pfleiderer S, Seither A (2017) 3D urban subsurface modelling and visualisation—a review of good practices and techniques to ensure optimal use of geological information in urban planning. TU1206 COST Sub-Urban WG2 report, p 100. https://static1.squarespace.com/static/542bc753e4b0a87901dd6258/t/58c021e7d482e99321b2a885/1488986699131/TU1206-WG2.3-004+3D+urban+Subsurface+Modelling+and+Visualisation.pdf. Accessed 27 Oct 2019

Seto KC, Fragkias M, Güneralp B, Reilly MK (2011) A meta-analysis of global urban land expansion. PLoS ONE 6(8):e23777

Smith M, Howard A (2012) The end of the map? Geoscientist online article. https://www.geolsoc.org.uk/Geoscientist/Archive/March-2012/The-end-of-the-map. Accessed 27 Oct 2019

UCCRN (2018) Impact 2050: the future of cities under climate change. Urban climate change research network, 59p

UN Habitat (2016). Habitat New Urban Agenda https://sustainabledevelopment.un.org/content/documents/17761NUAEnglish.pdf. Accessed 27 Oct 2019

UN Water (2014) The United Nations world water development report 2014: water and energy; facts and figures. https://unesdoc.unesco.org/ark:/48223/pf0000226961. Accessed 28 Oct 2019

UN Water (2015) The United Nations world water development report 2015: water for a sustainable world. https://www.unesco-ihe.org/sites/default/files/wwdr_2015.pdf. Accessed 27 Oct 2019

United Nations (2014) World urbanization prospects: the 2014 revision, highlights (ST/ESA/SER.A/352)

United Nations (2018a). World population prospects 2018. https://www.un.org/development/desa/publications/2018-revision-of-world-urbanization-prospects.html. Accessed 27 Oct 2019

United Nations (2018b) The Sustainable Development Goals Report. https://unstats.un.org/sdgs/report/2018. Accessed 27 Oct 2019

United Nations (2019a). World population projections. https://population.un.org/wpp/Graphs/Probabilistic/POP/TOT/900. Accessed 27 Oct 2019

United Nations (2019b). World population prospects 2019: Ten Key Findings. United Nations Department of Economic and Social Affairs. https://population.un.org/wpp/Publications/Files/WPP2019_10KeyFindings.pdf. Accessed 27 Oct 2019

Van den Dobbelsteen A, Keefe G, Tillie N, Rogeema R (2012) Cities as organisms. In: Roggema R (eds) Swarming landscapes. Advances in global change research, vol 48. Springer, Dordrecht, pp 195–206

Wehner M, Stone D, Krishnan H, AchutaRao K, Castillo F (2016) The deadly combination of heat and humidity in India and Pakistan in summer 2015. Bull Am Meteorol Soc 97(12):S81–S86

Wolman A (1965) The metabolism of cities. Sci Am 213 (1965):179–190

Woods-Ballard B, Wilson S, Udale-Clarke H, Illman S, Scott T, Ashley R, Kellagher R (2016) The SuDS manual. CIRIA C753 RP992. ISBN: 978-0-86017-760-9

World Bank (2009) World Development Report 2009: Reshaping Economic Geography. World Bank. https://openknowledge.worldbank.org/handle/10986/5991

World Economic Forum (WEF) (2017) Migration and its impact on cities. https://www.weforum.org/reports/migration-and-its-impact-on-cities. Accessed 27 Oct 2019

World Risk Report (2019) World risk report. https://reliefweb.int/sites/reliefweb.int/files/resources/WorldRiskReport-2019_Online_english.pdf. Accessed 27 Oct 2019

Wu H-L, Cheng WC, Shen SL et al (2019) Variation of hydro-environment during past four decades with underground sponge city planning to control flash floods in Wuhan, China: an overview (In Press). Undergr Sp. https://doi.org/10.1016/j.undsp.2019.01.003 Accessed 27 Oct 2019)

Yang J, Santamouris M (2018) Urban heat island and mitigation technologies in Asian and Australian cities —impact and mitigation. Urban Sci 2:1–6. https://doi.org/10.3390/urbansci2030074

Martin Smith is a Science Director with the British Geological Survey and Principle Investigator for the BGS ODA Programme Geoscience for Sustainable Futures (2017–2021). He has a first degree in Geology (Aberdeen) and a Ph.D. on tectonics (Aberystwyth, UK). A survey geologist by training Martin has spent a career studying geology both in the UK and across Africa and India. As Chief Geologist for Scotland and then for the UK he has worked closely with government and industry on numerous applied projects including in the UK on national crises, major infrastructure problems, decarbonisation research and urban geology and overseas for DFID-funded development projects in Kenya, Egypt and Central Asia. Martin is a Chartered Geologist and fellow of the Geological Society of London. He was awarded an MBE for services to geology in 2016.

Stephanie Bricker leads the Urban Geoscience research programme at the British Geological Survey. She is a chartered hydrogeologist with 15 years' experience across research and regulatory fields. Her research portfolio covers the role of geology to support sustainable urban development in the UK and overseas, including urban groundwater systems, the development of integrated city ground models and urban subsurface planning. She is interested in the physical interactions between natural and human systems in cities and supports sustainable use of subsurface through planning and policy reform. She is a founder-member of Think Deep UK and Chair of the EuroGeoSurveys Urban Expert Group.

Ensure Sustainable Consumption and Production Patterns

Joseph Mankelow, Martin Nyakinye, and Evi Petavratzi

J. Mankelow (✉) · E. Petavratzi
British Geological Survey, Environmental Science
Centre, Nicker Hill, Keyworth, Nottingham NG12
5GG, UK
e-mail: jmank@bgs.ac.uk

M. Nyakinye
Directorate of Geological Surveys, Ministry of
Petroleum and Mining, Machakos Road, Nairobi,
Kenya

© Springer Nature Switzerland AG 2021
J. C. Gill and M. Smith (eds.), *Geosciences and the Sustainable Development Goals*,
Sustainable Development Goals Series, https://doi.org/10.1007/978-3-030-38815-7_12

Abstract

12 RESPONSIBLE CONSUMPTION AND PRODUCTION

Since 1900, annual extraction of natural resources has increased greatly:

Construction materials (×34)   Ores and minerals (×27)   Fossil fuels (×12)   Biomass (×3.6)

Demand for resources will increase as populations grow, move to cities, and become wealthier.

SDG 12 aims to:

Reduce the links between resource use, environmental degradation, and economic growth

Understand the resource nexus to improve resource management

Move towards a circular economy

Geoscientists can support SDG 12 by:

Help maximise value and efficiency of mineral resource use

Reduce environmental impacts of mineral development

Integrating resource management into urban development

Contribute to global actions to decarbonise

12.1 Introduction

Global demand for natural resources has increased greatly in recent decades as countries seek to develop their economies and enhance the standard of living of growing local populations. During the twentieth century, the annual extraction of construction materials grew by a factor of 34, ores and minerals by a factor of 27, fossil fuels by a factor of 12, and biomass by a factor of 3.6 (UNEP 2014a). Global material use has tripled over the past four decades (1977–2017), with annual global extraction of materials growing from 30 billion tonnes in 1977 to 92 billion tonnes in 2017 (Fig. 12.1). Material extraction per capita (Fig. 12.2) increased from 7 to 10 tonnes between 1970 and 2010 indicating improvements in the material standard of living in many parts of the world; however, large gaps in material standard of living exist between North America and Europe and all other world regions and, in particular, Africa (UNEP 2016).

The global population is predicted to grow from an estimated 7.7 billion in 2019 to 8.5 billion in 2030 and 9.7 billion by 2050 (UN 2019a). It will also become increasingly urban, rising from 55% of the global population in 2018 to 68% in 2050 (UN 2018a). The bulk of that growth will take place in Africa and Asia. This growth in the global population will at the same time be accompanied by a significant increase in global middle classes—from 1.8 billion in 2009 to 4.9 billion in 2030 (Pezzini 2012). While the bulk of growth in middle classes will be in Asia (Pezzini 2012), there will also be significant growth in Africa (which has already tripled over the last 30 years) to 1.1 billion (42% of the continent's population) by 2060 (Deloitte 2014).

Global population growth accompanied by the expected rise in the middle classes means that demand for an improved quality of life will drive a need to access goods and services, increasing pressure on the use of natural resources. If current resource consumption patterns were to continue (Figs. 12.1 and 12.2), it is estimated that global material use of metals, non-metallic minerals, fossil fuels, and biomass would reach between 167 billion (OECD 2018) and 190 billion (UN 2019b) tonnes per year by 2060 of which non-metallic minerals such as construction aggregates (e.g., crushed rock, sand, and gravel) will represent more than half of the total raw material use (OECD 2018).

The aim of **SDG 12**[1] (Table 12.1)—**Ensure Sustainable Consumption and Production Patterns**—is *to achieve equitable development while at the same time ensuring sustainable management of resources*. It has been recognised that delivering sustainable consumption and production patterns requires coordinated action in order to reduce unsustainable resource use, to minimise waste, and to improve the management of hazardous substances. As a result, the United Nation's 10-Year Framework of Programmes on Sustainable Consumption and Production Patterns[2] has been incorporated into **SDG 12**. This is a global framework for action to enhance international cooperation and accelerate the shift towards sustainable consumption and production patterns in both developed and developing countries.

12.2 Global Challenges and Progress

The traditional global material footprint (the total amount of raw materials extracted to meet final consumption demands) as shown in Figs. 12.1 and 12.2 currently supports unequal standards of living. In 2017, the average person in North America required about 30 tonnes of raw materials to support their standard of living; this compares with 20.6 tonnes per capita material footprint in Europe, 11.4 tonnes in the Asia Pacific region, 10.2 tonnes in Latin America and the Caribbean, 9.6 tonnes in West Asia and less than 3 tonnes for a person living in Africa (IRP 2017). If sustainable development is to be

---

[1]A full listing of all the SDG targets and their indicators in MS Excel format can be accessed via https://unstats.un.org/sdgs/indicators/Global%20Indicator%20Framework%20after%202019%20refinement.English.xlsx.

[2]https://sustainabledevelopment.un.org/index.php?page=view&type=400&nr=1444&menu=35.

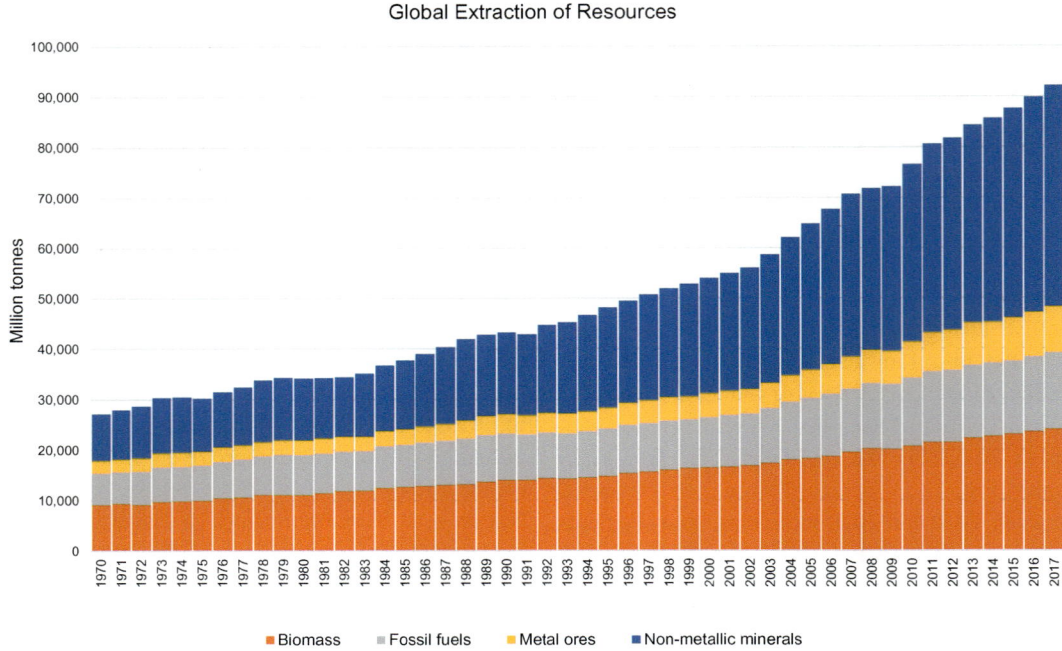

Fig. 12.1 Global resource consumption 1970–2017. Data from IRP (2019)

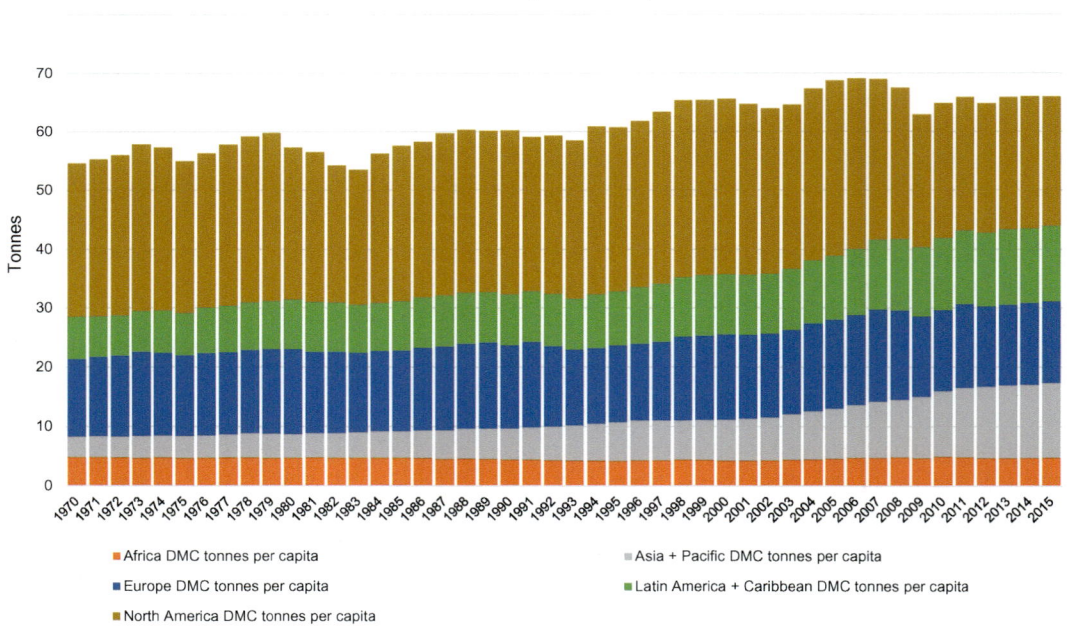

Fig. 12.2 Material consumption. Data from IRP (2019)

achieved while at the same time ensuring equitable development and delivery of **SDG 12**, there is a need to

- Reduce the intrinsic link between (to decouple) escalating natural resource use and environmental degradation from economic development.
- Recognise the critical interlinkages between different types of resources themselves by employing a systems approach and taking these into account, help to avoid burden shifting—the resource nexus.
- Move away from material supply and use within the traditional linear economy (take–make–use–dispose) to that of a circular economy (which maximises resource efficiency, innovation across the whole product life cycle, value addition, reuse, and recycling while minimising the generation of waste and negative environmental and social impacts).
- Develop and employ fit-for-purpose methodologies to monitor progress towards achieving the **SDG 12** targets.

All of the above are intrinsically linked with each other and are able to contribute to ensuring sustainable consumption and production patterns. Geoscience has an integral role in delivering them.

12.2.1 Decoupling Natural Resource Use from Economic Development

If sustainable consumption and production patterns are to be achieved while at the same time meeting the material requirements to ensure human well-being within the growing global population, there is a need for all countries (high as well as low and middle income) to break the link (to 'decouple') escalating resource use and environmental degradation from economic growth (UNEP 2011, 2014). Decoupling means reducing the amount of natural resources such as water, biomass, and minerals used to produce economic growth. While the sustainable use of natural resources and materials is the focus of **SDG 12**, decoupling resource use and environmental degradation from economic development is also targeted in **SDG 8.4**.

Enhancing resource efficiency means achieving the same (or greater) production of goods and services (economic) output with fewer inputs and delinking economic development from environmental deterioration. UNEP (2011) and Hennicke et al. (2014) differentiate between two types of decoupling as applied to sustainable development:

- **Resource decoupling**—it means reducing the use of (primary) natural resources per unit of economic activity.
- **Impact decoupling**—it means raising economic output while reducing negative environmental impacts that arise from the extraction of natural resources (e.g., groundwater pollution due to mining or agriculture), production (e.g., land degradation, wastes, and emissions), use of commodities (e.g., CO_2 emissions from transportation), and in the post-consumption phase (e.g., wastes and emissions).

While there are trends to decouple resource use and economic growth in resource-intensive mature economies, this is less the case for low- and middle-income countries (Angrick et al. 2014). However, by implementing policies which reflect the need for decoupling and by utilising different and emerging technologies, it is possible for such countries to grow their economies without following the same levels of historic resource use as those which occurred in mature economies around the world (UNEP 2014; IRP 2017). While global resource efficiency grew by around 27% between 1980 and 2009, it rose by 98% in India and 118% in China (Hennicke et al. 2014). Barriers related to, for example, technological innovation, resource-efficient infrastructure, and poor policy implementation currently disadvantage investments in resource productivity (see **SDG 9** also). The countries which are able to overcome such barriers will be able to lead the next wave of

Table 12.1 SDG 12 targets and means of implementation

Target	Description of target (12.1–12.8) or means of implementation (12.A–12.C)
12.1	Implement the 10-Year Framework of Programmes (10YFP) on Sustainable Consumption and Production Patterns, all countries taking action, with developed countries taking the lead, taking into account the development and capabilities of developing countries
12.2	By 2030, achieve the sustainable management and efficient use of natural resources
12.3	By 2030, halve per capita global food waste at the retail and consumer levels and reduce food losses along production and supply chains, including post-harvest losses
12.4	By 2020, achieve the environmentally sound management of chemicals and all wastes throughout their life cycle, in accordance with agreed international frameworks, and significantly reduce their release to air, water and soil in order to minimise their adverse impacts on human health and the environment
12.5	By 2030, substantially reduce waste generation through prevention, reduction, recycling and reuse
12.6	Encourage companies, especially large and transnational companies, to adopt sustainable practices and to integrate sustainability information into their reporting cycle
12.7	Promote public procurement practices that are sustainable, in accordance with national policies and priorities
12.8	By 2030, ensure that people everywhere have the relevant information and awareness for sustainable development and lifestyles in harmony with nature
12.A	Provision of support to developing countries to strengthen their scientific and technological capacity to move towards more sustainable patterns of consumption and production
12.B	Develop and implement tools to monitor sustainable development impacts for sustainable tourism that creates jobs and promotes local culture and products
12.C	Rationalising inefficient fossil-fuel subsidies that encourage wasteful consumption by removing market distortions, in accordance with national circumstances, including by restructuring taxation and phasing out those harmful subsidies, where they exist, to reflect their environmental impacts, taking fully into account the specific needs and conditions of developing countries and minimising the possible adverse impacts on their development in a manner that protects the poor and the affected communities

development. It is also easier to design for resource efficiency at the start of a process or project, rather than when a specific route is already in place thus potentially offering developing countries an advantage over more mature economies. In order to both monitor and assist in the delivery of 'decoupling', the UNEP incorporated it within the mandate of the International Resource Panel (IRP) which was established in 2007. The IRP aims to improve the evidence base for monitoring and policymaking, in particular, through systems-based assessment of the resource-related challenges and opportunities supporting the transition towards sustainable development (IRP 2017).

12.2.2 The Resource Nexus

The critical and complex interlinkages between different resources have received increased recognition in recent years. As a result, the concept of the resource nexus has increasingly been adopted to facilitate an integrated approach to the assessment of the resource life cycle. Commonly adopted is the water–energy–food nexus (see Ferroukhi et al. 2015; D'Odorico et al. 2018). Others, however, have adopted a four- (see Ringler et al. 2013) or five-node nexus (Fig. 12.3) which incorporates the essential systems of water (**SDGs 6** and **15**), energy (**SDG 7**), food (**SDG 2**), land (**SDG 15**), and material resources proposing that it better captures the realities and complexities of the human–environment system and related goals as specified in the SDGs (see Bleischwitz et al. 2018).

In order to meet material demand from the present and future generations, strategic and holistic thinking about the potential factors that may affect supply and demand for resources is paramount (de Ridder et al. 2014; WEF 2014; UNEP 2015a; Wakeford et al. 2016). At present, sustainability assessments and governance frameworks aim to address issues around individual resources (e.g., raw materials or water) without taking into account potential interdependencies between them. In other words, a 'singular thinking approach' is favoured at the moment rather than a holistic or systemic one (Giampietro 2018). Nexus governance offers an opportunity to be both adaptive and innovative (Marx 2015). Bleischwitz et al. (2018) note that if the SDGs are implemented in ways that overlook the critical interlinkages between different resources, all of the SDGs (not just **SDG 12**) may well risk a further acceleration of natural resource demand and degradation, ensuring numerous knock-on effects on individuals, communities, businesses, and societies—and the ecosystems on which all depend. Table 12.2 lists some of the direct and indirect impacts related to the water–energy–food nexus.

To be effective in overcoming such impacts, nexus-style solutions need to be adopted at the policy and planning levels. However, this will require a significant change in institutional thinking and working. Single-sector, top-down, and compartmentalised approaches are insufficient in tackling the challenges surrounding sustainable utilisation of water, energy, food, and other natural resources. There is a need to move away from exploring impacts in isolation and move towards a systemic approach. All too often agricultural policies (e.g., those linked to **SDG 2**), for instance, continue to be drafted in isolation of water policies (**SDG 6**) and vice versa while institutions with higher level objectives in common (such as food economic growth or socio-economic transformation) fail to cooperate, and instead compete for resources, both financial

and natural (Riddell Associates 2015). For example, 29 national and county departments and agencies have responsibility for water–energy–food-related functions in a single county of Kenya (Thuo et al. 2017). The level of institutional coordination required to overcome such a silo approach is often significant. Several different stakeholder groups should be considered, including their needs and requirements when following a systemic approach (Fig. 12.4).

To date, while the resource nexus offers a promising conceptual approach, the use of nexus methods to systematically evaluate resource interlinkages or support the development of socially and politically relevant resource policies has been limited (Albrecht et al. 2018).

12.2.3 Transitioning to a Circular Economy

The global system of production and consumption has historically been predominantly linear whereby the focus has been on ensuring the supply of materials to meet demand. It has been facilitated by a century of declining commodity prices. If decoupling economic growth and the future well-being of the global population from the use of natural resources is to be achieved, there is a need to change from the historic linear economic approach (take–make–use–dispose) to that of a circular economy. A circular economy relies on sustainably sourced natural resources, and products that are designed for repair, reuse, remanufacture, and recycling (Fig. 12.5) (see Lee et al. 2012; Hennicke et al. 2014). Within a circular economy, the environmental and social consequences of primary resource extraction and processing continue to be minimised, while maximum value is extracted from resources (and their derived products) thus keeping them in use for as long as possible and minimising the disposal of materials as waste. While some aspects of natural resource use (such as forestry and agriculture) are 'restorative and regenerative' by design (Ellen MacArthur Foundation 2017), for other natural resources, in particular minerals, importance within the circular economy is placed

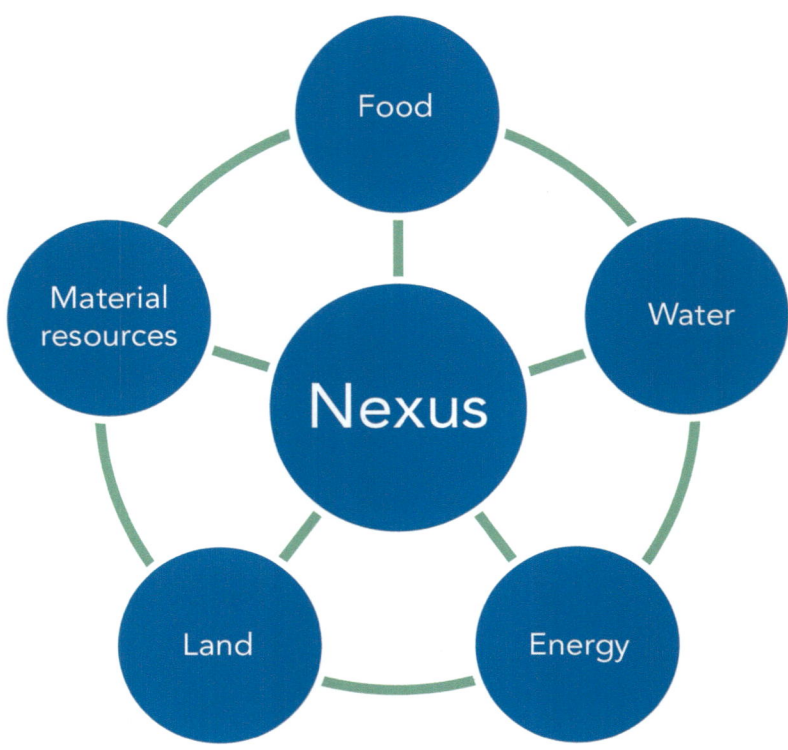

Fig. 12.3 The resource nexus

Table 12.2 Impacts of risks related to the water–energy–food nexus (non-exhaustive) (WEF 2011)

Impacts	Direct impacts	Indirect impacts
Impact on governments	• Stagnation in economic development • Political unrest • Cost of emergency food relief • Significantly reduced agricultural yields • Threats to energy security	• Increased social costs linked to employment and income loss as agriculture is negatively effected • National security risks/conflicts over natural resources
Impact on society/populations	• Increased levels of hunger and poverty • Increased environmental degradation • Severe food and water shortages • Social unrest • Food price spikes	• Migration pressures • Irreparably damaged water sources • Loss of livelihoods
Impact on business	• Export constraints • Increased resource prices • Commodity price volatility as shortages ripple through global markets • Energy and water restrictions	• Lost investment opportunities

Fig. 12.4 Stakeholder groups often involved in the nexus approach

on maximising length of use followed by reuse and recycling to enhance sustainability. Increased efficiency across the entire life cycle of resource use means more effective extraction and production, sustainable and smarter consumption as well as prevention and minimisation of negative environmental impacts (Hislop and Hill 2011; Preston 2012; UNEP 2012; IRP 2017).

In order to monitor the development of the circular economy, it is essential to quantify and understand the amount of materials flowing in and out of the economy, how they are used in society, and their level of circularity (Bloodworth 2013). Quantification of materials flowing into and out of the economy along with the stocks of materials being used in the economy is undertaken using Material Flow Analysis (see Brunner and Rechberger 2017; Nuss et al. 2017; EIPRM 2018; Allesch et al. 2018).

The circular economy has been receiving attention not only by countries with developed economies such as Japan, the European Union, and China who have all instituted high-level policy agendas (see UNEP 2016) but also by lower- and middle-income countries who increasingly look to enhance existing or adopt new circular economy approaches as a means for achieving sustainable economic growth (see Republic of Rwanda 2015; Gower and Schroeder 2016; Soezer 2016; Preston and Lehne 2017). For developing countries, circular economy policies (combined with urban planning that enables the beneficial exchange of materials and energy across different industry and infrastructure sectors in cities) are found to yield economic gains, natural resource conservation, greenhouse gas mitigation, and air-pollution reductions (IRP 2017).

As noted by Schroeder et al. (2018), while adopting circular economy practices will help to achieve **SDG 12**, they also contribute directly to achieving **SDG 6**, **SDG 7**, **SDG 8**, **SDG 9**, and

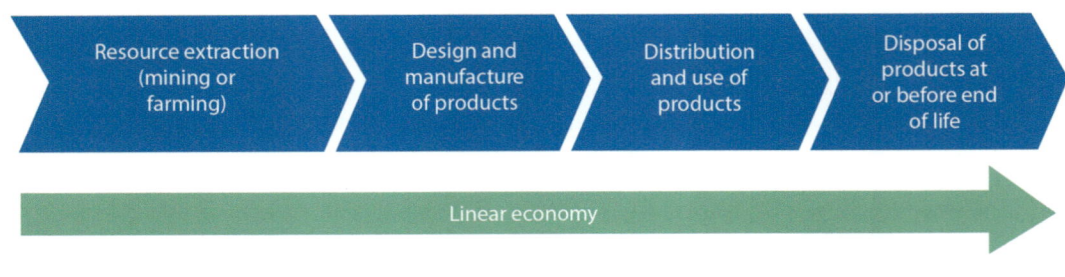

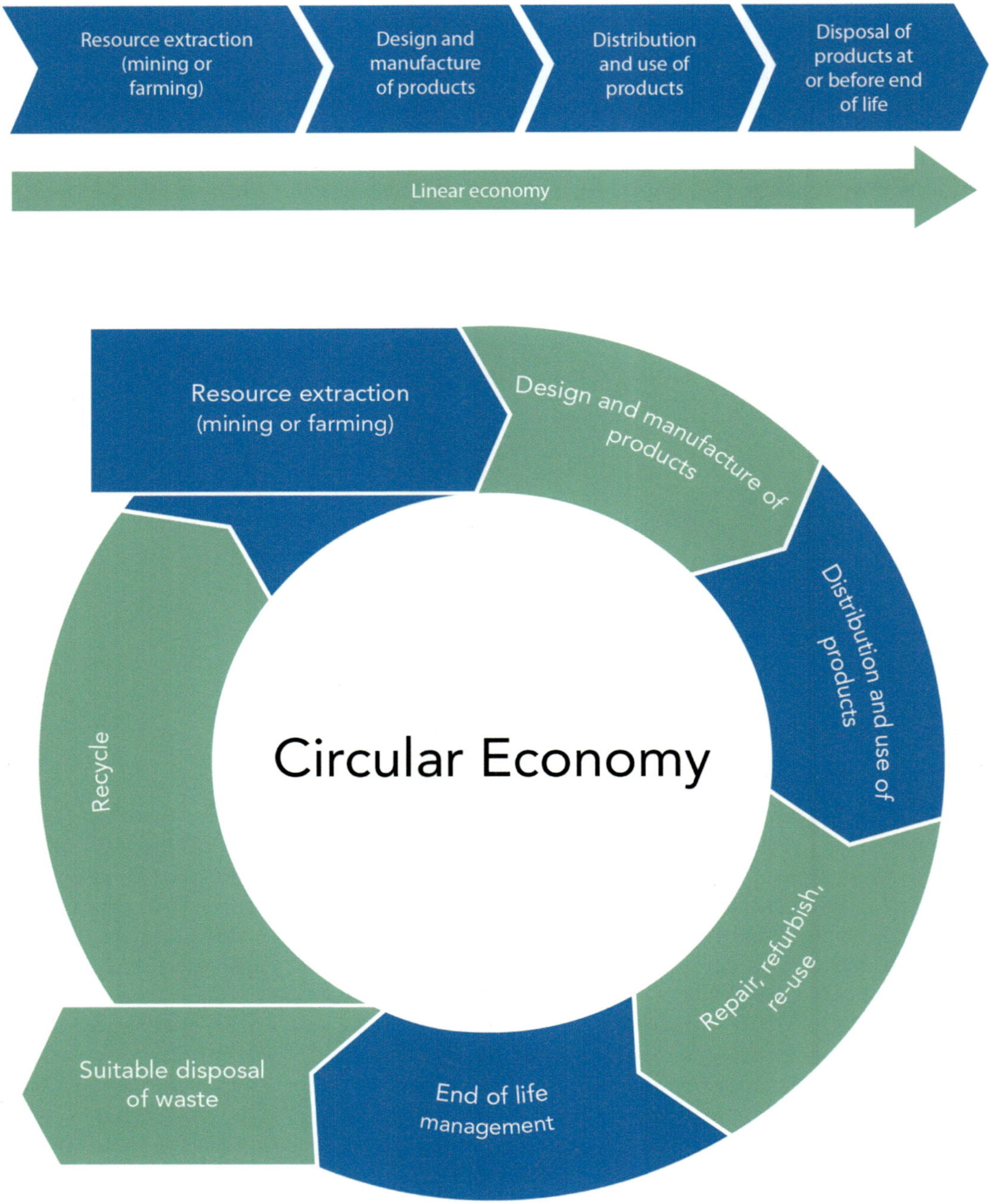

Fig. 12.5 The linear (top) versus circular (bottom) economy

SDG 15 and indirectly to many of the others. However, like all sustainable development initiatives, increasingly adopting a circular economy is likely to have trade-offs with other SDG targets. For example, while transitioning to circular economic approaches clearly delivers benefits in terms of maximising reuse and recycling of materials thus reducing demand for primary raw materials, any approaches or strategies implemented as part of the circular economy may only partially address barriers to economic and industrial development and they also should not be assumed to be optimal from a social or environmental perspective (Preston and Lehne 2017).

Whilst the drive towards achieving a circular economy is viewed as one of the key solutions to significantly decreasing demand for natural resources, it has to fit within the context of understanding the length of time materials stay in use (lifetime) within an economy along with the increasing global population and its material demand requirements. When the demand for a commodity increases over time, recycling alone cannot meet the higher demand—even if all products were collected and recycled with 100% efficiency at the end of their life. For example, global copper consumption in 1970 was about 8 million tonnes; by 2010, this had increased to 23 million tonnes (BGS 2018). If all the copper incorporated into products in 1970 were

recovered at the end of their life in 2010, there would still be a supply shortfall (the recycling gap) of 15 million tonnes which could be filled only by primary production (Fig. 12.6). As long as consumption increases, the need for primary extraction of minerals to meet global demand will continue (Bloodworth et al. 2017; Wellmer et al. 2019). However, a long-term future situation where consumption begins to level off and secondary resources progressively displace mined material can be envisaged (Bloodworth et al. 2019).

Like all global industries, companies operating in the mining sector have an opportunity to contribute towards achieving not only **SDG 12** but also all the SDGs. Common opportunities for the mining sector to contribute positively to **SDG 12** include enhancing material stewardship, minimising waste, and incorporating life cycle thinking into operations (WEF 2016).

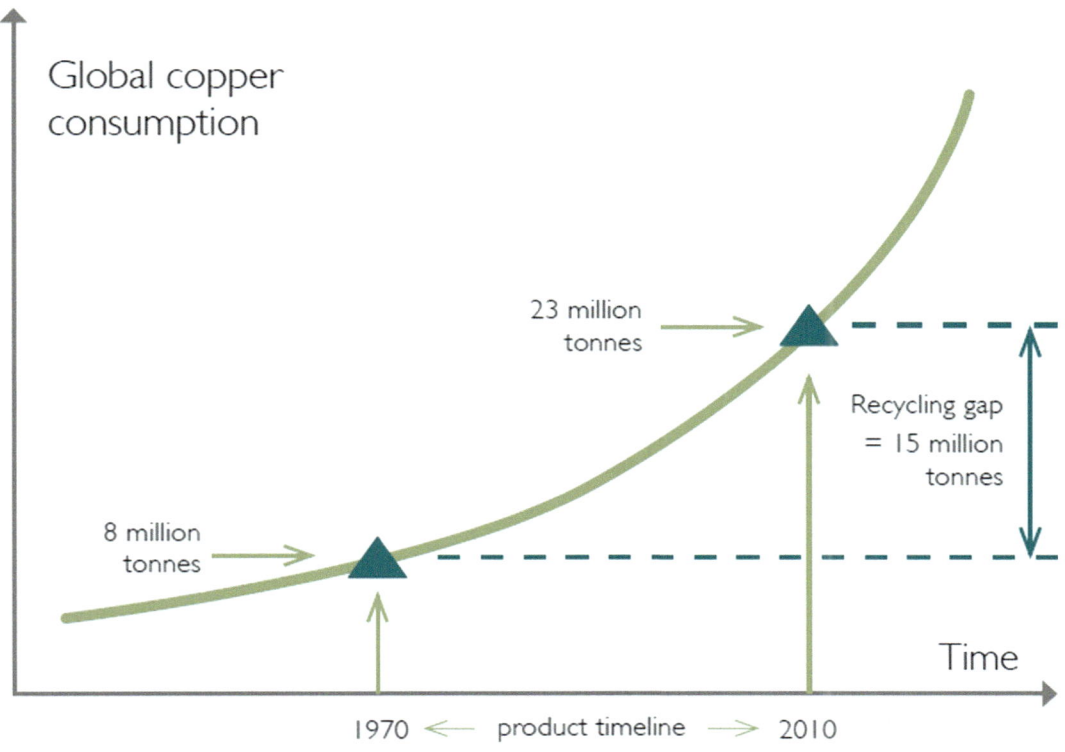

Fig. 12.6 Global copper consumption **and the recycling gap**. After Graedel et al. (2014) © British Geological Survey

12.2.4 Progress with SDG Targets

As with all the SDGs, the UN monitors progress towards achieving the **SDG 12** targets and reports this on an annual basis in the Sustainable Development Goals Report. While demonstrable progress is recognised as being made in some critical areas of the SDGs (e.g., a decline in extreme poverty, enhanced efforts to combat climate change, development of national policies to respond to the challenges of urbanisation, and positive engagement with the SDGs), many areas require urgent collective attention and more ambitious responses are required to achieve the 2030 SDG targets (UN 2019b). Specifically in relation to the progress of **SDG 12** (UN 2019b):

- The global material footprint (Fig. 12.1) continues to both rapidly grow and outpace population and economic growth (estimated to be between 167 billion and 190 billion tonnes by 2060 without concerted political action). Therefore, the efficiency with which natural resources are used to support economic growth remains unchanged at the global level and negative trends continue to be seen. There are no visible signs yet of decoupling economic growth and natural resource use at the global level. Good progress being made in sub-Saharan Africa, Central and Southern Asia, and Oceania (excluding Australia and New Zealand), mostly as a result of increases in GDP, is being offset by increased primary raw material consumption in other regions.
- There continues to be a significant discrepancy between the per-capita material footprint in high-income countries when compared with upper-middle-income countries and low-income countries.
- The material footprint for fossil fuels is more than four times higher for developed than developing countries. Decoupling the use of fossil fuels from economic growth is considered key to achieving sustainable consumption and production (UN 2018b).
- Construction of new infrastructure in emerging and transitioning economies (a pattern that many developing countries are likely to follow) and the outsourcing of the material- and energy-intensive stages of production from high-income countries to less resource-efficient countries have resulted in a significant increase in domestic material consumption in Eastern and Southeastern Asia (10 billion tonnes more or about two-thirds of the increase at the global level in 2017 than in 2010).
- Well-designed national policy frameworks and associated instruments (regulatory, voluntary, economic) remain necessary to enable the fundamental shift required towards sustainable consumption and production patterns. In 2018, 71 countries and the European Union reported a total of 303 such policies and instruments continuing the overarching positive trend in the formulation of such policies/instruments seen since 2002. The primary focus in such policies and instruments is the economic benefits offered by more sustainable consumption and production patterns with the social benefits being largely overlooked. While the formulation of relevant policies assists in the sustainable management of natural resources, the implementation of these to deliver tangible positive changes in sustainable consumption and production remains limited (UN 2018b).

To assist in implementing the 10-Year Framework of Programmes on Sustainable Consumption and Production Patterns, the UN has established the One Planet Network.[3] The Network provides a platform to bring together the global community for sustainable development, and its 2018–2022 strategy outlines how it will become a leading implementation, monitoring, and evaluation mechanism for **SDG 12** (One Planet Network 2018). In addition to the One Planet Network, the International Resources Panel exists to build and share the knowledge needed to improve our use of resources worldwide and so also acts as an enabler for monitoring progress towards achieving **SDG 12**.

Table 12.3 presents a summary of the latest progress towards achieving the **SDG 12** targets.

---

[3]https://www.oneplanetnetwork.org/.

Table 12.3 High-level summary of progress towards achieving **SDG 12**. *Sources* UNEP (2015b, 2017a), UN (2018b, 2019b)

SDG 12 target	Summary of progress
12.1 Implement the 10-Year Framework of Programmes on Sustainable Consumption and Production Patterns (10YFP)	An increase in the number of national sustainable consumption and production policies compiled since 2002 demonstrates overarching positive trends. However, implementation of these to foster tangible changes in impact remain limited and only a few are mandated to coordinate policy implementation across ministries
12.2 Achieve the sustainable management and efficient use of natural resources	A small change in global Domestic Material Consumption has been recorded between 2010 and 2015 (from 1.2 kg per dollar of GDP to 1.1 kg). Therefore, fewer raw materials are required to produce a unit of output. However, DMC per capita and in absolute terms has continued to grow from 2000 to 2017 with consequences for global resource depletion and environmental impacts
12.3 Halve per capita global food waste at the retail and consumer levels and reduce food losses along production and supply chains, including post-harvest losses	A range of interventions designed to tackle food loss and waste (e.g., research into the causes and identification of solutions, target-setting, policy formulation, legislation, and education/awareness raising campaigns) are being implemented
12.4 Achieve the environmentally sound management of chemicals and all wastes throughout their life cycle, in accordance with agreed international frameworks, and significantly reduce their release to air, water and soil in order to minimise their adverse impacts on human health and the environment	Mixed progress to global compliance rates for protocols and conventions relating to environmentally sound management of chemicals and wastes. Reporting of information by parties signed to the Montreal Protocol and the Basel, Rotterdam and Stockholm Conventions varies with an average of 70%. The Minamata Convention has been signed by 128 countries[a]
12.5 Substantially reduce waste generation through prevention, reduction, recycling and reuse	Information on total recycling rates is sparse and variable, with data being better at a city level. While recycling rates are highest in the high-income countries, some low- and lower-middle-income countries do collect quite reasonable percentages of their total municipal solid waste for recycling (20–40%). There is some evidence that recycling rates are lower in some of the more developed, upper-middle-income countries. The collection and management of waste electrical and electronic equipment (WEEE) heavily depends on the legislation in each country
12.6 Encourage companies, especially large and transnational companies, to adopt sustainable practices and to integrate sustainability information into their reporting cycle	Sustainability reporting has been gaining momentum, driven by new private sector partnerships to achieve the SDGs along with growing interest from companies (especially large companies), regulators, investors and other stakeholders. 93% of the world's 250 largest companies (in terms of revenue) are now reporting on sustainability

(continued)

Table 12.3 (continued)

SDG 12 target	Summary of progress
12.7 Promote public procurement practices that are sustainable, in accordance with national policies and priorities	Driving sustainability along value chains through public procurement processes that consider social, economic and environmental factors has gained increasing recognition. It is being progressively embraced by national and local bodies. The relevance of sustainable public procurement as a strategic tool to drive sustainability and transform markets is no longer questioned. However, the need to ensure that it is better integrated into broader sustainable consumption and production policies continues
12.8 Ensure that people everywhere have the relevant information and awareness for sustainable development and lifestyles in harmony with nature	Numerous on-going public engagement and awareness raising activities. Includes the UN's One Planet Network

[a] Montreal Protocol on Substances that Deplete the Ozone Layer, Basel Convention on the Control of Transboundary Movements of Hazardous Wastes, Rotterdam Convention on the Prior Informed Consent Procedure for Certain. Hazardous Chemicals and Pesticides in International Trade, Stockholm Convention on Persistent Organic Pollutants, and Minamata Convention on Mercury.

Despite recognising some broad progress made towards achieving **SDG 12**, a recent survey (which 454 sustainability experts completed) rated progress on **SDG 12** (as well as six other of the SDGs) as poor (GlobeScan-Sustainability 2019). However, respondents to the survey were also of the view that Climate Action (**SDG 13**) and Ensuring Sustainable Consumption and Production (**SDG 12**) are the most critically urgent, and noted that these two SDGs receive most attention within their organisation.

The UN (2018a, b) reports that serious concern exists in relation to the lack of an adequate monitoring framework for many of the targets under **SDG 12**. No internationally established methodologies or standards exist currently for monitoring 10 of the 13 indicators for the targets (but methodologies are being developed or tested). As an example, in its latest report on monitoring of progress towards the SDGs for the Asia and Pacific region, the UN Economic and Social Commission for Asia and the Pacific indicated that only 10% of the required data to assess progress towards achieving **SDG 12** exists (ESCAP 2019).

In order to facilitate the monitoring of progress towards achieving **SDG 12**, there is a need for better data collection and more consistent data collection. However, the qualitative rather than the quantitative nature of several of the **SDG 12** targets makes it difficult for countries to measure their achievements (Chan et al. 2018). Adopting methodologies that incorporate social science approaches for qualitative assessment may assist in overcoming some of the problems. Limited resources along with limited technical capacity and fragmented institutional systems mean that many countries also have difficulty developing monitoring processes and collecting the required data (Steinbach et al. 2016). Steinbach et al. (2016) note that substantive efforts in institutional and technical capacity development as well as financial resources are required to effectively monitor changes in consumption and production patterns, indicating that where they do exist, it is important for current international reporting systems to converge in order to assist the process.

12.3 Geoscience and SDG 12

As with all the SDGs, if the ambitions of **SDG 12** are to be met, geoscience will have an important role to play. Chosen because they utilise some of the concepts described above, the following case studies provide just some examples of where an understanding of geoscience can make a positive contribution towards achieving **SDG 12**.

12.3.1 Resource Decoupling in a Green Economy Strategy for Kenya

In Kenya, the Ministry of Devolution and Planning is mandated to coordinate the management, implementation, monitoring, and reporting of the SDGs. Within the Ministry, the SDG Coordination Department has been established with the responsibility for ensuring appropriate monitoring and evaluation of all SDG activities, and the country has reported good progress on several of the SDGs (see Government of Kenya 2017).

Like many low- and middle-income countries, Kenya's economy is heavily reliant on natural resources. For 2014, the UN estimated that 42% of Kenya's GDP and 70% of overall employment were derived from natural resource-related sectors (UNEP 2014b), many of which are acutely vulnerable to climate change and variability. In recent years, viable deposits of oil, natural gas, coal, and other minerals have been discovered. However, the prospective benefits to the economy and enhanced national energy security from the exploitation of these minerals need to be weighed against the risks of negative environmental impacts. To mitigate these impacts, the Government of Kenya is to embrace sustainable consumption and production approaches in order to transition to a green economy.

In 2016, the government published its Green Economy Strategy and Implementation Plan, 2016–2030 (Government of Kenya 2016). The strategy outlines activities to achieve a low carbon, resource-efficient, equitable, and inclusive socio-economic transformation in the country. The strategy aims to facilitate Kenya attaining a higher economic growth rate (consistent with the country's development blueprint, Vision 2030) but which firmly embeds the principles of sustainable development as the country's growth continues. The strategy is designed to guide Kenya's transition to a sustainable path in the following five key areas: sustainable infrastructure development; building resilience; sustainable natural resources management; resource efficiency; and social inclusion and sustainable livelihood.

In order to optimise the contribution of the agriculture, forestry, water, fisheries, wildlife, land use, and extractive industry sectors to the Kenyan economy, there is an emphasis on the need for decoupling economic development from sustainable natural resource management and the conservation of Kenya's natural resources. There is recognition that Kenya's natural resources are under intense pressure from global and local drivers such as population increase, over-extraction of natural resources, poaching of wildlife, urbanisation, changing consumption patterns among the population, climate change, and the use of chemicals. Intrinsically linked to achieving such decoupling is the promotion of resource efficiency. Within this area, the strategy recognises that the challenge for Kenya is to develop its resource efficiency agenda, reduce the environmental impact of production and consumption while addressing the policy and technical challenges of waste management. The country aims to optimise the contribution of Kenya's natural resources to the economy, industrialisation, and livelihoods by recognising the interlinkages of the economy–environment nexus.

Kenya sees the delivery of the green growth path outlined in the green economy strategy as a way to ensure faster growth, a cleaner environment, and higher economic productivity and is highlighted as one of the means by which the country will meet the **SDG 12** targets.

12.3.2 Maximising Value and Efficiency of Mineral Resource Use in Kenya

The Government of Kenya, through the Ministry of Petroleum and Mining, State Department for Mining, is in the process of establishing four Value Addition Centres (VAC) located in different mineral-rich regions of the country. Through this initiative, the government aims to maximise efficiency in mineral resource use by adding value to mineral products at the source while enhancing the economic growth and well-being of the local communities.

Fig. 12.7 The Voi Gem Centre. Credit: Martin Nyakinye (DGS)

Included among the VACs is the Voi Gem Centre (Fig. 12.7). This is located within Voi town in Taita-Taveta County in the south-east of the country. The area has been the main source of Kenyan gemstone production since independence. The Voi Gem Centre has been operational since 2017 and offers several value addition and support services to the gemstone mining industry. The Voi Gem Centre is expected to increase the value of gemstones at the source, reduce waste, and facilitate sustainable mining and trading of gemstones. It will achieve this by providing the facilities necessary to maximise the value of gems extracted locally. Facilities at the centre include

- An administration block offering all the essential services in a single location.
- An accredited gemology laboratory for the identification, verification, and valuation of gems.
- A lapidary providing gem cutting, faceting, and polishing.
- A jewelry-making section.
- A security safe for the secured custody of gemstones.
- The provision of export permits.
- An exhibition hall for gemstone auction and trading.
- Gemstone dealers' booths.
- A bank and a restaurant.

The other VACs are still at the feasibility stage with a tender award for construction expected by the end of 2019 and full operation by 2022. There is the Kakamega gold refinery which is to be located at the Old Rosterman mine in Kakamega County. This will include a demonstration of the mine facility for safe and sustainable mining as well as providing mercury-free gold recovery services to the artisanal and small-scale gold mining (ASGM) communities. Others include the Vihiga Granite Slab Processing Centre to be located at Emuhaya in Vihiga County and the Soapstone Value Addition Centre to be located in Kisii county. Prior to construction, all of these VACs will be subjected to Kenya's Environmental and Social Impact Assessment (ESIA) process.

12.3.3 Reducing Mercury Emissions into the Environment and Use in Artisanal and Small-Scale Gold Mining (ASGM) in Kenya

The effects of mercury on human health are well documented (UNEP 2019 and Bell et al. 2017) yet globally, the use of mercury by artisanal and small-scale gold mining (ASGM) shows no sign of declining (UNEP 2013, 2017b). In the Kenyan context, estimates of mercury use in ASGM mining ranging 1.3–2 units of mercury per unit of gold produced (Government of Kenya unpublished report 2016). One particular study (Ogola et al. 2002) found mercury use by ASGM miners in a single county (Migori) in the west of the country ranged 60–80 kilogrammes per month. Since 2002, the area has experienced an acceleration in the growth of ASGM activity. Barreto et al. (2018) estimate that the engagement by locals in ASGM in parts of Migori and Siaya Counties could be up to 70% while the figure could be as high as 100% in parts of other counties in the west of Kenya, such as Kakamega and Vihiga. Although clear, credible data is scant; mercury use among the ASGM miners is now estimated at multiple times the figures of 2002. This points to a serious and increasing mercury use problem that needs to urgently be addressed.

Mining operations involve deep unstable shafts (Fig. 12.8), manual extraction, crushing of ore by hand followed by milling, or alternatively by panning gold directly from rivers in the region (Fig. 12.9). Mercury is mixed with the resulting heavy mineral ore concentrates forming a mercury–gold amalgam. The amalgam is then heated, vaporising the mercury to obtain the gold.

To minimise the environmental pollution and health impacts caused by the use of mercury, the government of Kenya is implementing a strategy to reduce the anthropogenic emissions of mercury by artisanal and small-scale mining activities. This will contribute to **SDG 3** (good health and well-being) and **SDG 15** (life on land). In 2018, the Ministry of Environment and Forestry issued a 'Request for Proposals' for consultants to develop a national overview of the ASGM sector in Kenya, and for the development of a national action plan for the ASGM sector in Kenya. The preliminary results from the studies (McKay 2019; ISSITET 2019) are already providing a framework for intervention by the Kenyan government to steer the ASGM sector towards greener mining. Key recommendations proposed by the studies to be implemented include

Fig. 12.8 Manual winching of ore from an ASGM gold mine. Credit: Martin Nyakinye (DGS)

Fig. 12.9 ASGM miners panning for gold. Credit: Martin Nyakinye (DGS)

- Formalisation of the ASGM sector in order to provide easier entry points for interventions.
- Enhancing capacity awareness, education, and capacity building among the ASGM practitioners about the dangers of mercury use and existing alternatives to mercury.
- Lower barriers to access for miners to finance in order to increase uptake of greener, safer technologies in their activities.
- Strengthening the human capacity and resources available to relevant enforcement agencies in order to ensure the adherence to greener production pathways.

The development of Kenya's *National Action Plan to Reduce and, Where Feasible, Eliminate Mercury Use in Artisanal and Small-Scale Gold Mining* (UNEP 2017c) will provide concrete actions and clear framework to reduce ASGM mercury emissions in Kenya, and thus contribute to the global effort to reduce overall levels of anthropogenic mercury emissions into the atmosphere and ecosystems. The end result is expected to be a reduction in the negative environmental

and human health impacts of ASGM activity in Kenya.

12.3.4 Estimating Demand for Minerals Required for Future Construction in Hanoi

Between 1990 and 2016, the percentage of Vietnam's population living in urban areas increased from about 20% to approximately 35% (GSOV 2018). Such an increase in urban population is resulting in an expansion in the size of cities in the country much reflected elsewhere in the world. For Hanoi, one such city in Vietnam, current and future population trends have led to the formulation of ambitious plans for additional urban development (Iwata 2007; Leducq and Scarwell 2018) as detailed in the Hanoi Masterplan to 2030 (Perkins Eastman 2011).

Planned urban growth requires the construction of houses, schools, hospitals, roads, and other infrastructure. The Hanoi Masterplan

2030–2050 includes a major new road network, new rail links, an expanded city core, five satellite urban areas, and three eco-townships. This development all requires an input of raw materials, particularly aggregates in the form of crushed rock and sand. Without a steady supply of these construction minerals, the expansion of Hanoi cannot be delivered.

By undertaking a top-down mineral supply–demand mass balance, Bide et al. (2018) were able to estimate future demand for crushed rock and construction sand required to meet the future growth of Hanoi. The top-down approach (Fig. 12.10) estimates future supply based solely on trends in mineral production, trade in minerals, and population growth. By calculating the current apparent consumption of minerals and then using projected population growth figures, estimates were made for future demand for construction minerals. Such a top-down approach is 'simplistic' in comparison to a detailed material flow analysis (MFA) (see Bide et al. 2018). However, the methodology does lend itself to situations where data (in particular, at the relevant region or better still individual city level) is sparse or non-existent. Such is the case for Hanoi where detailed data for production and trade in mineral commodities are either not collected or are not publicly available on either a city or regional level.

Following the methodology depicted in Fig. 12.10, the apparent consumption of crushed rock and construction sand was calculated for Hanoi for 2007–2016 (Fig. 12.11). The effects of a slowdown in the construction industry can be seen from around 2010 to 2012. Changes in policy regarding international exports of sand have also had an effect. However, despite the impact of these external forces the general trend has been one of increased consumption for crushed rock and consistent consumption for sand.

Having calculated apparent consumption over a period of time, the data were then combined with forecasts of future population growth for the city to estimate likely future demand for crushed rock (Fig. 12.12) and construction sand (Fig. 12.13) to 2030. The vertical lines are added to illustrate the potential discrepancies in the main trend line that may be observed in the future. The forecasts are based on projections using a compound annual growth rate calculation over a 10-year period.

Results show that considerable increases in demand for both crushed rock, and construction sand are to be expected as the population of Hanoi increases and homes and infrastructure are constructed to accommodate this growth. For crushed rock, there is a 2.5-fold increase in demand in 2030 (86 million tonnes), over 2016 levels (33 million tonnes) and there is a twofold increase for construction sand with 14 million tonnes required in 2030 compared with 6 million tonnes in 2016. Within a circular economy, some of this demand would be met by secondary and recycled aggregates. However, secondary and recycled aggregates are generally only suitable for lower grade applications (such as for building foundations). The degree to which aggregates are recycled in Hanoi is currently unknown. If the increased demand for the minerals is to be met while at the same time minimising negative environmental and social impacts, planned new extraction capacity is likely to be required.

Ideally, consideration of such mineral supply requirements should be considered when plans for urban development such as the Hanoi Masterplan are compiled. If due consideration is not given to future demand, issues can arise, such as high price volatility, increased informal mining with the associated negative environmental and social impacts, slow economic growth, and hindered development.

12.3.5 The Resources Impact of Decarbonising Economies

Over the last decade as the effort for the transition to a low-carbon future has increased, a variety of technologies have been developed (currently being scaled up) which rely on the availability of numerous materials, including the so-called critical metals. The growth in electrification in transport (Fig. 12.14), energy storage, and renewable energy generation is leading to increased demand for minor metals that are often

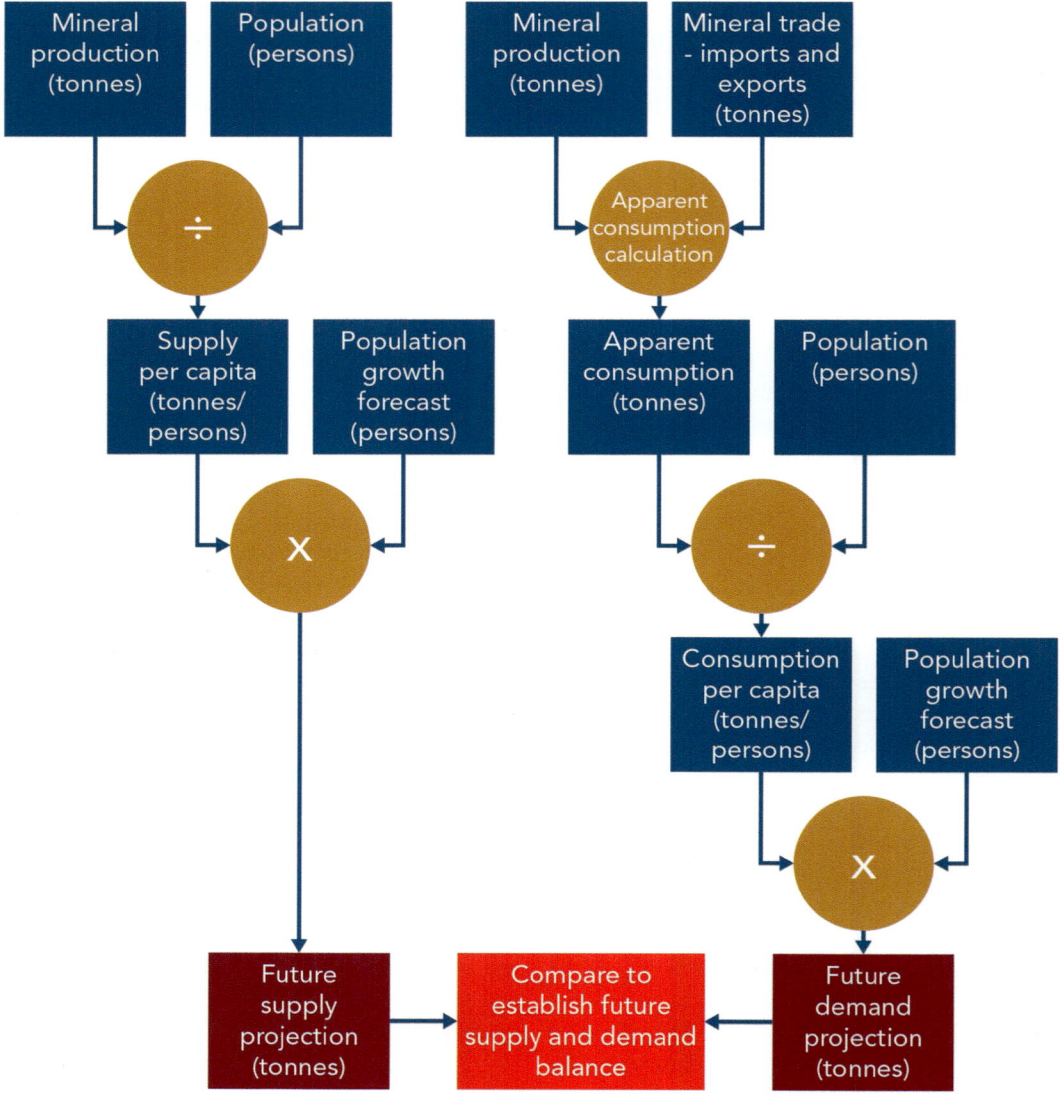

Fig. 12.10 Schematic diagram explaining the steps involved in the top-down future mineral supply and demand calculation. Adapted from Bide et al. (2018)

deemed critical, due to their associated supply constraints. The demand for many of these minor metals is expected to rise significantly in the coming years. For example, for cobalt and lithium (both used in batteries), the International Energy Agency projects that cobalt production will have to triple and lithium production will have to increase fivefold by 2030 in order to satisfy anticipated growth in the number of electric vehicles (International Energy Agency 2019). For other metals, such as tellurium (used in photovoltaics) and neodymium (used in wind turbines), similar upward growth is expected. Developed scenarios are suggesting an increase in production from a few hundred tonnes in 2017 to over a couple of thousand tonnes in 2030 and for neodymium a doubling in production by the same year (Watari et al. 2018).

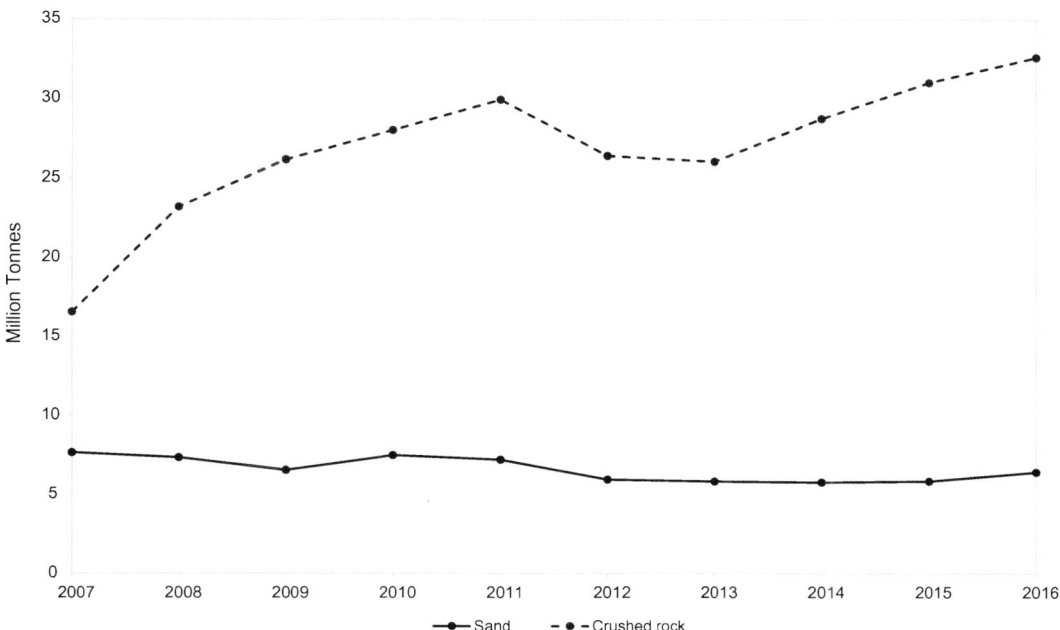

Fig. 12.11 Material consumption **for Hanoi.** Consumption of crushed rock (dashed line) and construction sand (solid line) for Hanoi, 2007–2016. Credit: Bide et al. (2018)

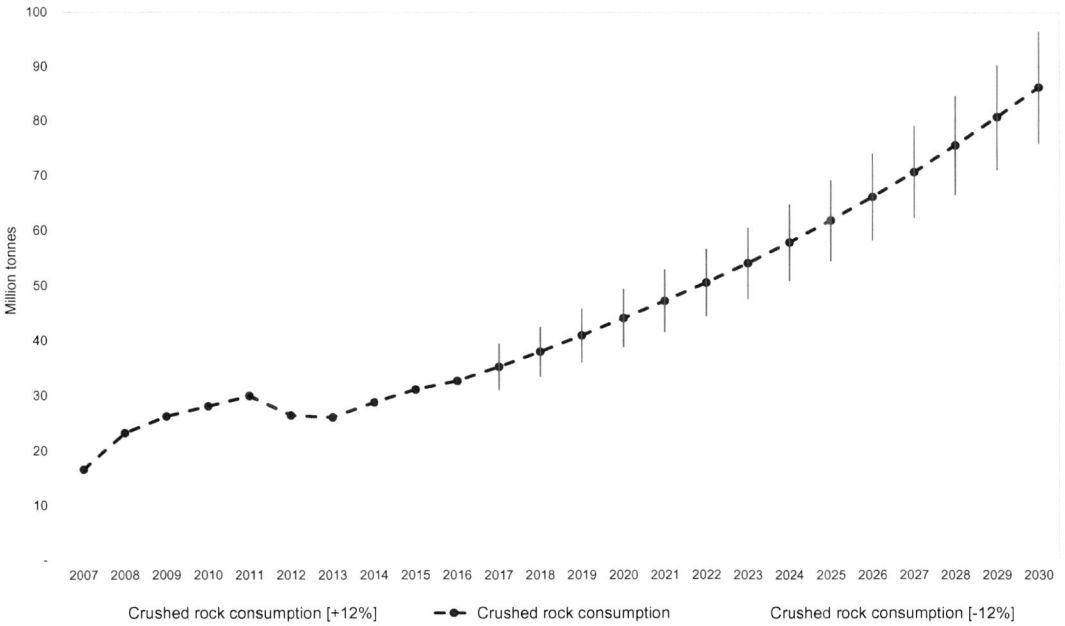

Fig. 12.12 Hanoi consumption **of crushed rock (2007–2016) and estimated future consumption (2017–2030).** Estimated via material flow analysis, with 12% (one standard deviation) error bars on projected years. Credit: Bide et al. (2018)

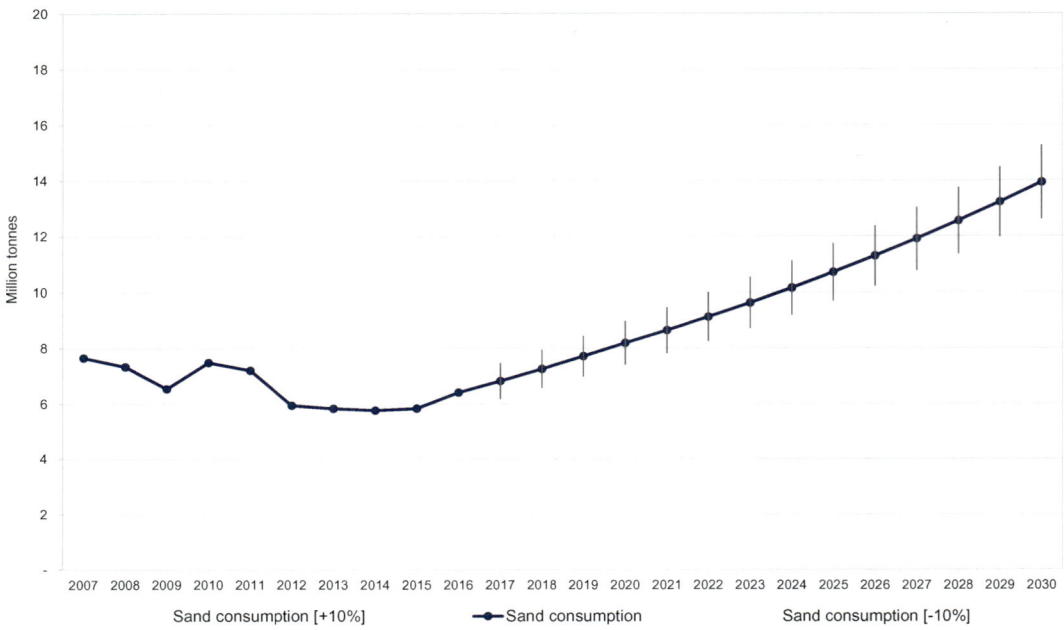

Fig. 12.13 Hanoi consumption **of construction sand (2007–2016) and estimated future consumption (2017–2030).** Estimated via material flow analysis, with 10% (one standard deviation) error bars on projected years. Credit: Bide et al. (2018)

Fig. 12.14 The introduction of electric buses is helping to reduce emissions and pollution. © Andrew Bloodworth (BGS), used with permission

Such an increase in demand for these minerals will have an impact on global consumption and production patterns. The markets of industrial metals such as iron, aluminium, copper, and others will also grow as they comprise essential parts of decarbonisation technologies. If the availability of metals is constrained, then the upscaling of technologies essential for decarbonising our economies will slow down, with corresponding negative impacts on climate change mitigation. The supply risk is particularly high for minor metals that are often produced as by-products of industrial metals. They are not mined on their own, they are dependent financially on the recovery of other metals, and their market dynamics are influenced by the market of the major metals they are connected with.

For many of the metals needed in decarbonisation technologies, their extraction and production are linked with environmental consequences and they are carbon intensive. Between seven to eight per cent of the global energy production is consumed for extracting and refining metals (Wellmer et al. 2019). Lower ore grades of extraction and continuous demand growth for raw materials have led to huge amounts of waste rock being produced. If decarbonisation is the aim, then it is important that the extraction and production of minor and critical metals to be consumed by green technologies are responsibly sourced and follow the principles of sustainability. Successful interventions will have to consider the whole material cycle and should be based on the systemic understanding of the interaction between the economy, environment, and society.

12.4 Conclusions

Achieving many of the SDGs will ultimately depend on the responsible stewardship of the Earth's finite natural resources. Natural resource management is directly tied to at least 12 of the 17 Sustainable Development Goals (SDGs) and therefore achieving the targets of **SDG 12** will help deliver the other goals (IRP 2017). Despite showing some positive progress in several targets of **SDG 12** in recent years, overall the need

to significantly modify current trends to ensure sustainable consumption and production patterns remains if the goal is to be achieved by 2030.

Demand for raw materials from emerging and transitioning economies as a result of their growing populations and the drive to enhance the quality of life will continue to grow into the future. Achieving the **SDG 12** targets will require a fundamental shift in how natural resources are managed and used. The concepts of resource efficiency, resource productivity, decoupling, resource nexus, and circular economy have entered mainstream science and policy development, and having gained traction they are now increasingly being implemented into practices. All have a role to play in supporting this shift. The need to improve natural resource management continues as does the requirement for agreeing and implementing mechanisms of data collection needed to monitor progress towards achieving **SDG 12**.

Geoscience with its inherent links to the understanding and management of natural resources has and will continue to have an important role in helping to achieve sustainable consumption and production patterns. With an understanding of the links between natural resource (or raw material) supply and demand, geoscientists can contribute by

- *Assisting in the development of relevant policies and ensuring that decision makers are informed of the likely material demand impact of emerging policies.* The Green Economy Strategy and Implementation Plan for Kenya as summarised in this chapter is just one example where governments around the world are increasingly adopting the concepts outlined in this chapter in order to better manage and utilise their natural resources as their economies develop. Likewise, undertaking analyses such as the supply–demand mass balance assessment for Hanoi to better understand likely future raw material needs as cities around the world grow in size is important if better development plans for such cities are to be implemented.
- *Participating in initiatives such as the establishment of value addition centres and*

mercury reduction strategies. The examples from Kenya show how this engagement can assist those working in the informal/small-scale mining sectors to maximise the value from the minerals they extract while at the same time reduce the use of chemicals harmful to the environment and human health. This supports **SDG 3** (good health and well-being), **SDG 8** (decent work and economic growth), and **SDG 15** (life on land).

- *Alleviating some of the supply risks associated with minor and critical metals essential for low-carbon technologies.* For many of these metals, our understanding of where they are found, how we can explore for them, and how we can enhance existing geological data to promote new discoveries is limited, as their historic use was small or non-existent. Competitive geoscience data into deposit formation, new exploration techniques, and geological mapping will be essential to ensure risk-free supply chains for low-carbon technologies supporting **SDG 7** (clean and affordable energy) and **SDG 13** (climate action).

The last point in the list above can be considered as being along the lines of the 'traditional' role of a geoscientist—understanding the formation, location, and viability (environmental, social, and economic) of extracting and utilising the natural resources required for manufacturing the goods we need or producing the food we eat. The other examples indicate the much wider contribution geoscience, however, can make towards achieving the targets of **SDG 12**.

12.5   Key Learning Concepts

- Global demand for natural resources (e.g., construction materials, ores and minerals, fossil fuels, and biomass) is rapidly increasing as countries seek to develop their economies and enhance the standard of living of growing local populations, many of whom are getting wealthier.
- Responding to this increased demand, **SDG 12** aims to help the world transition to sustainable

consumption and production patterns, so as to achieve equitable socio-economic development while also protecting resources for future generations, minimising waste, and reducing pollution of the natural environment (i.e., decoupling development from environmental degradation). This requires appropriate policies, technological innovation, and the development of resource-efficient infrastructure.

- The resource nexus recognises that there are critical and complex interlinkages between different resources (e.g., water, energy, food, land, and material resources). Sustainability requires the consideration of what may affect the supply and demand of each resource, recognising these interlinkages. Institutions should work together to develop coherent and comprehensive policies for resource management, recognising that a diverse ecosystem of stakeholders can inform planning and practice.
- A circular economy relies on sustainably sourced natural resources, and the use of products that are designed for repair, reuse, remanufacture, and recycling (so as to reduce the disposal of materials). Increases in efficiency across the entire life cycle of resource use means more effective extraction and production, sustainable and smarter consumption, and preventing or reducing negative environmental impacts.
- Geoscientists are part of the complex ecosystem of stakeholders that can deliver **SDG 12**, given their comprehensive knowledge of the formation, location, and viability (economic, social, and environmental) of extracting and using natural resources. Geological data and expertise can inform policymaking and implementation of measures to decouple resource extraction from environmental damage, and decarbonise societies.

12.6   Educational Resources

In this section, we provide examples of educational activities that connect geoscience, the material discussed in this chapter, and scenarios that may arise when applying geoscience (e.g., in policy, government, private sector international organisations, and NGOs). Consider using these as the basis for presentations, group discussions, essays, or to encourage further reading:

- Consider the social and environmental impacts associated with the development of a standard smartphone. Start by determining what minerals they contain, and consider where these may have originated and the energy and water resources required to extract and process these. How well does a standard smartphone fit on a model of the 'circular economy'? What changes would you recommend to make smartphones adhere to ambitions of SDG 12?
- Walk around a local urban environment, close to your institution or place of learning. What natural resources that you can observe or deduce were required to develop the built environment you see? Reflecting on your local and national geological setting, where do you think of these resources came from? What do you think this urban environment will look like in 20 years, and what are the implications on the types and quantities of materials used?
- Earth Overshoot Day (www.overshootday.org/) is focused on the exhaustion of biological resources that the Earth can regenerate in a year, showing excessive consumption. Review the solutions offered (www.overshootday.org/solutions/), and in small groups choose one of these, consider what contribution geoscientists can make, and present this to the class. Calculate your own environmental footprint (https://www.footprintcalculator.org/) and think what steps you personally can take to reduce consumption.

Acknowledgements The authors kindly acknowledge the contribution of Tom Bide and Teresa Brown of the British Geological Survey (BGS) to the Hanoi Material Flow Analysis. Martin Smith (also of the BGS) is thanked for reading and commenting on this chapter. The compilation of this chapter was supported by BGS NC-ODA grant NE/R000069/1 entitled Geoscience for Sustainable Futures and has been delivered via the BGS eastern Africa ODA Research Platform. Joseph Mankelow and Evi Petavratzi publish with the permission of the Director of the BGS. Martin Nyakinye publishes with the permission of the Director of the DGS.

Recommended Reading and Websites

One Planet Network. The UN's hub for **SDG 12** with the aim of being a platform to bring together the global community helping to make this happen. www.oneplanetnetwork.org

Further information on **SDG 12** from the UN. https://www.un.org/sustainabledevelopment/sustainable-consumption-production/

Food and Agricultural Organisation of the United Nations. http://www.fao.org/sustainable-development-goals/goals/goal-12/en/

Information on monitoring of progress of **SDG 12**. https://sustainabledevelopment.un.org/**SDG 12**, https://sdg-tracker.org/sustainable-consumption-production and https://unstats.un.org/sdgs/metadata/

The International Resources Panel. www.resourcepanel.org

African Circular Economy Network. https://www.acen.africa/

Extractives Hub. https://www.extractiveshub.org/main/default/

The Responsible Minerals Initiative. http://www.responsiblemineralsinitiative.org/

The Elsevier Journal specifically for Sustainable Consumption and Production. https://www.journals.elsevier.com/sustainable-production-and-consumption

Taylor & Francis Sustainable Development Goals Online. A curated library to support the United Nations' SDGs. https://www.taylorfrancis.com/sdgo/goal/ResponsibleConsumptionAndProduction/all

References

Albrecht TR, Crootof A, Scott CA (2018) The water-energy-food nexus: a systematic review of methods for nexus assessment. Environ Res Lett 13. https://doi.org/10.1088/1748-9326/aaa9c6

Allesch A, Cao Z, Cullen J, Heldal T, Liu G, Lundhaug M, Petavratzi E, Rechberger H, Simoni M, Müller D (2018) Developing the data base a framework for monitoring the physical economy. CEC4Europe Factbook. Circular Economy Coalition for Europe, 8p. https://www.cec4europe.eu/publications/

Angrick M, Burger A, Lehmann H (eds) (2014) Factor X: policy, strategies and instruments for a sustainable resource use. Springer Science and Business Media, Dordrecht, 311p. https://doi.org/10.1007/978-94-007-5706-6

Barreto L, Schein P, Hinton J, Hruschka F (2018) Economic contributions of artisanal and small-scale mining in Kenya: gold and gemstones, 105p. https://assets.publishing.service.gov.uk/media/5a392bb8e5274a79051c9d7c/Kenya_case_study.pdf

Bell L, Evers D, Johnson S, Regan K, DiGangi J, Federico J, Samanek J (2017) Mercury in women of child-bearing age in 25 countries. IPEN and Biodiversity Research Institute, 70p. https://ipen.org/sites/default/files/documents/updateNov14_mercury-women-report-v1_6.pdf

Bide TP, Brown TJ, Petavratzi E, Mankelow JM (2018) Vietnam – Hanoi city material flows. British Geological Survey Report, OR/18/068, 49p. http://nora.nerc.ac.uk/id/eprint/522143/1/OR18068.pdf

Bleischwitz R, Hoff H, Spataru C, van der Voet E, VanDeveer SD (2018) The resource nexus: preface and introduction to the Routledge Handbook. In: Bleischwitz R, Hoff H, Spataru C, van der Voet E, VanDeveer SD (eds) 2018. Routledge handbook of the resource nexus. Routledge, Oxford, pp 3–14

Bloodworth A (2013) Resources: track flows to manage technology-metal supply. Nature News Nature Publishing Group December 2013, pp 19–20

Bloodworth AJ, Gunn AG, Petavratzi E (2017) Mining and resource recovery. In: From waste to resource productivity: evidence and case studies, 95–108. Report of the Government Chief Scientific Adviser. Government Office for Science and Department for Environment, Food and Rural Affairs, 45p. https://www.gov.uk/government/uploads/system/uploads/attachment_data/file/667480/from-waste-to-resource-productivity-evidence-case-studies.pdf

Bloodworth A, Gunn AG, Petavratzi E (2019) Metals and decarbonisation: a geological perspective. British Geological Survey Science Briefing Paper, 8p. https://www.bgs.ac.uk/downloads/start.cfm?id=3559

Brunner PH, Rechberger H (2017) Handbook of material flow analysis for environmental, resource and waste engineers. CRC Press, Taylor & Francis Group

British Geological Survey (2018) World mineral statistics database. https://www.bgs.ac.uk/mineralsuk/statistics/wms.cfc?method=searchWMS

Chan S, Weitz N, Persson Å, Trimmer C (2018) **SDG 12**: responsible consumption and production - a review of research needs, 25p. https://www.sei.org/wp-content/uploads/2018/11/sdg-12-responsible-consumption-and-production-review-of-research-needs.pdf

Deloitte (2014) The Deloitte consumer review. Africa: a 21st century view. Deloitte, 32p. https://www2.deloitte.com/content/dam/Deloitte/ng/Documents/consumer-business/the-deloitte-consumer-review-africa-a-21st-century-view.pdf

D'Odorico P, Davis KF, Rosa L, Carr JA, Chiarelli D, Dell'Angelo J et al (2018) The global food-energy-water nexus. Rev Geophys 56. https://doi.org/10.1029/2017RG000591

EIPRM (2018) Raw materials scorecard. European Innovation Partnership on Raw Materials, 120 p. https://doi.org/10.2873/08258

Ellen MacArthur Foundation (2017) What is the circular economy? Concept. https://www.ellenmacarthurfoundation.org/circular-economy/concept

ESCAP (2019) Asia and the Pacific SDG progress report, 2019. Economic and Social Commission for Asia and the Pacific. United Nations, 73p. https://www.unescap.org/sites/default/files/publications/ESCAP_Asia_and_the_Pacific_SDG_Progress_Report_2019.pdf

Ferroukhi R, Nagpal D, Lopez-Peña A, Hodges T, Mohtar RH, Daher B, Mohtar S, Keulertz M (2015) Renewable energy in the water, energy & food nexus. International Renewable Energy Agency, 124p. https://www.irena.org/documentdownloads/publications/irena_water_energy_food_nexus_2015.pdf

Giampietro M (2018) Perception and representation of the resource nexus at the interface between society and the natural environment. Sustainability 10:2545. https://doi.org/10.3390/su10072545

GlobalScan-Sustainability (2019) Evaluating progress on the SDGs, 29p. https://globescan.com/wp-content/uploads/2019/03/GlobeScan-SustainAbility-Survey-Evaluating-Progress-Towards-the-Sustainable-Development-Goals-March2019.pdf

Gower R, Schroeder P (2016) Virtuous circle: how the circular economy can create jobs and save lives in low and middle-income countries, Tearfund, London. http://www.tearfund.org/ ∼ /media/files/tilz/circular_economy/2016-tearfundvirtuous-circle.pdf

Government of Kenya (2016) Green economy strategy and implementation plan - Kenya, 2016–2030, 46pp. http://www.environment.go.ke/wp-content/uploads/2018/08/GESIP_Final23032017.pdf

Government of Kenya (2017) Implementation of the Agenda 2030 for sustainable development in Kenya. Ministry of Devolution and Planning, 76p. https://sustainabledevelopment.un.org/content/documents/15689Kenya.pdf

Graedel TE, Gunn G, Espinoza LT (2014) Metal resources, use and criticality. In: Gunn G (ed) Critical metals handbook. AGU and Wiley

GSOV (2018) Statistics Database. General Statistics Office of Vietnam. http://www.gso.gov.vn/Default_en.aspx?tabid=766

Hennicke P, Khosla A, Dewan C, Nagrath K, Niazi Z, O'Brien M, Thakur MS, Henning W (2014) Decoupling economic growth from resource consumption a transformation strategy with manifold socio-economic benefits for India and Germany. Deutsche Gesellschaft für Internationale Zusammenarbeit (GIZ) GmbH, Berlin, 33p. https://www.giz.de/de/downloads/giz2014-en-IGEG_2_decoupling-econimic-growth.pdf

Hislop H, Hill J (2011) Reinventing the wheel: a circular economy for resource security. Green Alliance, 52p. https://www.green-alliance.org.uk/resources/Reinventing%20the%20wheel.pdf

IEA (2019) Material efficiency in clean energy transition. International Energy Agency, 162p. https://www.iea.org/publications/reports/MaterialEfficiencyinCleanEnergyTransitions/

IRP (2017) Assessing global resource use: a systems approach to resource efficiency and pollution reduction. In: Bringezu S, Ramaswami A, Schandl H, O'Brien M, Pelton R, Acquatella J, Ayuk E, Chiu A, Flanegin R, Fry J, Giljum S, Hashimoto S, Hellweg S, Hosking K, Hu Y, Lenzen M, Lieber M, Lutter S, Miatto A, Singh Nagpure A, Obersteiner M, van Oers L, Pfister S, Pichler P, Russell A, Spini L, Tanikawa H, van der Voet E, Weisz H, West J, Wiijkman A, Zhu B, Zivy R (eds) A report of the international resource panel. United Nations Environment Programme. Nairobi, Kenya. 104 p. https://www.resourcepanel.org/reports/assessing-global-resource-use

IRP (2019) Global material flows database. International Resource Panel. https://www.resourcepanel.org/global-material-flows-database

ISSITET Consulting Ltd (2019) Developing a national action plan for reducing mercury use in the artisanal and small-scale gold mining sector in Kenya: national overview of education, information and technology. Report submitted to Ministry of Environment and Forestry. Unpublished

Iwata S (2007) The comprehensive urban development programme in Hanoi capital city of the Socialist Republic of Vietnam (HAIDEP). Final report, JICA, Hanoi

Leducq D, Scarwell H-J (2018) The new Hanoi: opportunities and challenges for future urban development. Cities 72:70–81

Lee B, Preston F, Kooroshy J, Bailey R, Lahn G (2012) Resources futures. Chatham House (The Royal Institute of International Affairs), London, 212p. https://www.chathamhouse.org/sites/default/files/public/Research/Energy%2C%20Environment%20and%20Development/1212r_resourcesfutures.pdf

Marx S (2015) Governing the nexus for sustainability. Chang Adapt Socio-Ecol Syst 2:79–81. https://www.degruyter.com/downloadpdf/j/cass.2015.2.issue-1/cass-2015-0008/cass-2015-0008.pdf

McKay & Company Advocates (2019) Draft legal and policy report on artisanal and small-scale gold mining (ASGM) sector in Kenya. Report submitted to Ministry of Environment and Forestry. Unpublished

Nuss P, Blengini GA, Haas W, Mayer A, Nita V, Pennington D (2017) Development of a Sankey diagram of material flows in the EU economy based on Eurostat data, EUR 28811 EN, Publications Office of the European Union, Luxembourg, 45p. https://doi.org/10.2760/362116

OECD (2018) Global material resources outlook to 2060 - economic drivers and environmental consequences. OECD Publishing, 214p. https://doi.org/10.1787/9789264307452-en

Ogola JS, Mitullah WV, Omulo MA (2002) Impact of gold mining on the environment and human health: a case study in the Migori Gold Belt, Kenya. Environ Geochem Health 24:141. https://doi.org/10.1023/A:1014207832471

One Planet Network (2018) One plan for one planet - 5 year strategy 2018–2022. United Nations, 12p. https://www.oneplanetnetwork.org/sites/default/files/strategy_one_planet.pdf

Perkins Eastman (2011) Hanoi capital master plan to 2030. http://www.perkinseastman.com/project_3407114_hanoi_capital_master_plan_to_2030

Pezzini M (2012) An emerging middle class. OECD Observer. http://oecdobserver.org/news/fullstory.php/aid/3681/An_emerging_middle_class.html

Preston, F (2012) A global redesign? Shaping the circular economy. Briefing paper. Chatham House, 20p. https://www.chathamhouse.org/sites/default/files/public/Research/Energy%2C%20Environment%20and%20Development/bp0312_preston.pdf

Preston F, Lehne J (2017) A wider circle? The circular economy in developing countries. Chatham House Briefing, 24p. https://www.chathamhouse.org/sites/default/files/publications/research/2017-12-05-circular-economy-preston-lehne-final.pdf

Republic of Rwanda (2015) National e-waste management policy for Rwanda, 13p. http://www.minirena.gov.rw/fileadmin/Environment_Subsector/Laws__Policies_and_Programmes/Laws/E-waste_policy_-_FINAL.pdf

Riddell Associates (2015) Nexus trade-offs and strategies for addressing the water, agriculture and energy security nexus in Africa. Riddell Associates Ltd. on behalf of the International Water Association, the Infrastructure Consortium for Africa and the International Union for Conservation Nature, 135p. https://iwa-network.org/wp-content/uploads/2016/03/1450107971-Nexus-Trade-off-and-Strategies_-ICA-Report_-Dec-2015_2.pdf

de Ridder M, van Duijne F, de Jong S, Jones J, van Luit E, Bekkers F, Auping W, Gehem W (2014) The global resource Nexus - impact on sustainable security of supply of agri-food imports for the Netherlands. The Hague Centre for Strategic Studies and TNO, 125p. https://hcss.nl/sites/default/files/files/reports/Report_19_Nexus.pdf

Ringler C, Bhaduri A, Lawford R (2013) The nexus across water, energy, land and food (WELF): potential for improved resource use efficiency? Curr Opin Environ Sustain 5(6):617–624. https://doi.org/10.1016/j.cosust.2013.11.002

Schroeder P, Anggraeni K, Weber U (2018) The relevance of circular economy practices to the sustainable development goals. J Ind Ecol. https://doi.org/10.1111/jiec.12732

Soezer A (2016) Nationally appropriate mitigation action on a circular economy solid waste management approach for urban areas in Kenya. Ministry of Environment and Natural Resources of Kenya and UNDP Low Emission Capacity Building (LECB) Programme, 108p. https://www.undp.org/content/undp/en/home/librarypage/environment-energy/mdg-carbon/NAMAs/nama-on-circular-economy-solid-waste-management-approach-for-urb.html

Steinbach N, Palm V, Constantino S, Pizzaro R (2016) Monitoring the shift to sustainable consumption and production patterns in the context of the SDGs. Statistics Sweden, Chilean Ministry of Environment and United Nations, 84p. https://sustainable development.un.org/content/documents/2298SCP%20 monitoring.pdf

Thuo S, Schreiner B, Byakika S (2017) The water-energy-food nexus and poverty eradication in Kenya. Pegasys Institute Policy Brief 2/17, 10p. https://cdkn.org/wp-content/uploads/2015/06/PI-%E2%80%93-Policy-Brief-002.pdf

UN (2018a) 2018 revision of world urbanization prospects – key facts. https://population.un.org/wup/Publications/Files/WUP2018-KeyFacts.pdf, https://population.un.org/wup/

UN (2018b) HPLF review of SDGs implementation: **SDG 12** - ensure sustainable consumption and production patterns. United Nations High-Level Political Forum on Sustainable Development, 7p. https://sustainabledevelopment.un.org/content/documents/196532018backgroundnotes**SDG12**.pdf

UN (2019a) World population prospects: highlights. United Nations, Department of Economic and Social Affairs, Population Division (ST/ESA/SER.A/423), 39p. https://population.un.org/wpp/Publications/Files/WPP2019_Highlights.pdf

UN (2019a) World population prospects: highlights. United Nations, Department of Economic and Social Affairs, Population Division (ST/ESA/SER.A/423), 39p. https://population.un.org/wpp/Publications/Files/WPP2019_Highlights.pdf

UN (2019b) The Sustainable Development Goals report. United Nations, 64p. https://unstats.un.org/sdgs/report/2019/The-Sustainable-Development-Goals-Report-2019.pdf

UNEP (2011) Decoupling natural resource use and environmental impacts from economic growth. In: Fischer-Kowalski M, Swilling M, von Weizsäcker EU, Ren Y, Moriguchi Y, Crane W, Krausmann F, Eisenmenger N, Giljum S, Hennicke P, Romero Lankao P, Siriban Manalang A, Sewerin S (eds) A report of the Working Group on Decoupling to the International Resource Panel. United Nations Environment Panel, 176p. https://www.resourcepanel.org/file/400/download?token=E0TEjf3z

UNEP (2013) Global mercury assessment 2013: sources, emissions, releases and environmental transport. United Nations Environment Program, 44p. http://wedocs.unep.org/bitstream/handle/20.500.11822/7984/-Global%20Mercury%20Assessment-201367.pdf?sequence=3&isAllowed=y

UNEP (2014a) Decoupling 2: technologies, opportunities and policy options. In: von Weizsäcker EU, de Larderel J, Hargroves K, Hudson C, Smith M, Rodrigues M (eds) A report of the Working Group on Decoupling to the International Resource Panel. United Nations Environment Panel, 158p. https://www.resourcepanel.org/sites/default/files/documents/document/media/-decoupling_2_technologies_opportunities_and_policy_options-2014irp_decoupling_2_report-1.pdf

UNEP (United Nations Environment Programme) (2014b) Green Economy Assessment Report: Kenya. https://www.greengrowthknowledge.org/research/green-economy-assessment-report-kenya

UNEP (2015a) Policy coherence of the Sustainable Development Goals - a natural resource perspective. United Nations Environment Programme, 50p. https://wedocs.unep.org/bitstream/handle/20.500.11822/9720/-Policy_Coherence_of_the_Sustainable_Development_Goals_A_Natural_Resource_Perspective-2015Policy_Coherence_of_the_Sustainable_Development_Goals_-_A_N.pdf?sequence=3&isAllowed=y

UNEP (2015b) Global waste management outlook. United Nations Environment Programme, 346p. https://wedocs.unep.org/bitstream/handle/20.500.11822/9672/-Global_Waste_Management_Outlook-2015Global_Waste_Management_Outlook.pdf.pdf?sequence=3&amp%3BisAllowed=

UNEP (2016) Global material flows and resource productivity. In: Schandl H, Fischer-Kowalski M, West J, Giljum J, Dittrich M, Eisenmenger N, Geschke A, Lieber M, Wieland HP, Schaffartzik A, Krausmann F, Gierlinger S, Hosking K, Lenzen M, Tanikawa H, Miatto A, Fishman T (eds) An assessment study of the UNEP International Resource Panel. United Nations Environment Panel, 200p. https://www.resourcepanel.org/file/423/download?token=Av9xJsGS

UNEP (2017a) Global review of sustainable procurement - 2017. United Nations Environment Panel, 111p. https://www.oneplanetnetwork.org/sites/default/files/globalreview_web_final.pdf

UNEP (2017b) Global mercury supply, trade and demand. United Nations Environment Program, 83p. https://wedocs.unep.org/bitstream/handle/20.500.11822/21725/global_mercury.pdf?sequence=1&isAllowed=y

UNEP (2017c) Developing a national action plan to reduce and, where feasible, eliminate mercury use in artisanal and small-scale gold mining - guidance document. United Nations Environment Program, 94p. https://wedocs.unep.org/handle/20.500.11822/25473

UNEP (2019) Global mercury assessment 2018. United Nations Environment Program, 59p. https://wedocs.uep.org/handle/20.500.11822/27579

Wakeford JJ, Mentz-Lagrange S, Kelly C (2016) Managing the energy-food-water-nexus in developing countries: case studies of transition governance. Quantum Global Research Lab Working Paper 2016/01, 33p. https://www.researchgate.net/publication/315700591_Managing_the_Energy-food-water_Nexus_in_Developing_Countries_Case_Studies_of_Transition_Governance

Watari T, McLellan BC, Ogata S, Tezuka T (2018) Analysis of potential for critical metal resource constraints in the International Energy Agency's long-term low-carbon energy emission. Minerals 8:156. https://doi.org/10.3390/min8040156

WEF (2011) Global risks 2011, 6th edn. World Economic Forum, 56p. http://www3.weforum.org/docs/WEF_Global_Risks_Report_2011.pdf

WEF (2014) The water-energy nexus: strategic considerations for energy policy-makers. World Economic Forum, 10p. http://www3.weforum.org/docs/GAC/2014/WEF_GAC_EnergySecurity_WaterEnergyNexus_Paper_2014.pdf

WEF (2016) Mapping mining to the sustainable development goals: an atlas. World Economic Forum, 10p. https://www.undp.org/content/dam/undp/library/Sustainable%20Development/Extractives/Mapping_Mining_SDGs_An_Atlas_Executive_Summary_FINAL.pdf

Wellmer FW, Buchholz P, Gutzmer J, Hagelüken C, Herzig P, Littke R, Thauer RK (2019) Raw materials for future energy supply. Springer International Publishing AG, Switzerland

Martin Nyakinye is a Chief Superintending Geologist at the Kenyan Directorate of Geological Surveys, within the Ministry of Petroleum and Mining. He is also Deputy Director of the Geological Data Management Division, overseeing the development of the Kenya National Geological Data Centre. Martin has a BSc in Geology from the University of Nairobi, and a Masters degree specialising in Mineral Exploration, from the Institute of Geo-information Science and Earth Observation, at the University of Twente (The Netherlands). He has interests in the application of geospatial technologies in mineral exploration. Martin is registered geologist, and member of the Geological Society of Kenya.

Joseph Mankelow is a Principal Geologist at the British Geological Survey (BGS) with over 20 years' experience. He is responsible for delivery of high quality outputs from applied research on issues related primarily to understanding mineral resource potential and using this research to inform associated policy development. A focus of his work has been enhancing access to and understanding of the minerals data of countries to encourage inward investment. For the past three years, Joseph has also been responsible for managing delivery of the BGS's integrated resource management research in eastern Africa delivered via its Official Development Assistance Programme—Geoscience for Sustainable Futures.

Evi Petavratzi is a Senior Mineral Commodity Geologist at the British Geological Survey (BGS), with over 13 years of working experience in industry, academia and the public sector. At BGS, her research is in the field of security of supply and the circular economy, which in recent years, has focused on the move towards a low carbon future. She is involved in several multidisciplinary projects in the areas of battery raw materials and critical metals, material flow analysis, circular economy for the transport sector and advancing the knowledge base for raw materials. She holds a Ph.D. in Mining Engineering, is a member of the Institute of Materials, Minerals and Mining (IOM3) and a Chartered member with the Chartered Institute of Waste Management and. In the past, she has worked in the areas of resource efficiency, sustainability and industrial ecology.

Climate Action

Joy Jacqueline Pereira, T. F. Ng, and
Julian Hunt

J. J. Pereira (✉)
Southeast Asia Disaster Prevention Research
Initiative (SEADPRI-UKM), Universiti Kebangsaan
Malaysia, 43600 Bangi, Malaysia
e-mail: joy@ukm.edu.my

T. F. Ng
Department of Geology, University of Malaya,
50603 Kuala Lumpur, Malaysia

J. Hunt
Trinity College, Cambridge CB2 1TQ, UK

University College London, London, UK

© Springer Nature Switzerland AG 2021
J. C. Gill and M. Smith (eds.), *Geosciences and the Sustainable Development Goals*,
Sustainable Development Goals Series, https://doi.org/10.1007/978-3-030-38815-7_13

Abstract

13 CLIMATE ACTION

SDG 13 aims to:

Strengthen resilience and reduce the number of deaths attributed to disasters

Build the capacity of all countries to adapt to climate change

Build climate change into national policies

 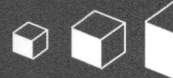 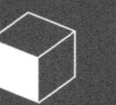 

This requires:

Reductions in greenhouse gas emmissions

Inclusion of climate change science and strategies into education systems

Adoption of disaster risk reduction policies and operational plans to adapt to adverse impacts

Budgetary commitment and mobilisation of resources to develop sub-surface carbon storage

Climate change challenges

Communicating the science of how natural processes and anthropogenic activities contribute to climate change

Encouraging greater uptake of mitigation and adaptive strategies, leaving no one behind

Adoption of the precautionary principle, limiting global warming <1.5°C above pre-industrial levels

Achieving a balance between economic development and carbon emissions

Improving implementation and monitoring of the Sendai Framework for Disaster Risk Reduction

Role of geoscience in mitigating and adapting to climate change

Provide sub-surface geological models that support mitigation strategies (e.g., for energy and carbon storage) and low carbon energy resources (e.g., geothermal)

Understand climate influenced hazards, and the roles of natural processes and human activities

Understand multi-hazard cascades and their short and long term impacts

Provide scientific advice to inform policy, urban planning, and protection of cultural heritage

Input to education on earth systems and raising awareness at local community level

  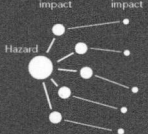  

13.1 Introduction

Climate in a narrow sense refers to weather, in a wider sense it refers to the state of the climate system that is statistically described over a period ranging from months to thousands or millions of years. The World Meteorological Organization (WMO) generally averages the mean and variability of climate parameters such as temperature, precipitation, and wind over 30 years. The Intergovernmental Panel on Climate Change (IPCC) defines climate variability as differences in the mean and other statistics of climate parameters on all spatial and temporal scales beyond that of individual weather events. Climate change refers to changes in the state of the climate that is depicted by differences in the mean and other statistics of climate parameters that persist over an extended period, typically decades or longer (IPCC 2014).

An extreme weather event is a rare event at a particular place and time of year where the value of the weather variable is above (or below) a threshold value near the upper (or lower) ends of the range of observed values of the variable. The characteristics of extreme weather vary from place to place. A persistent pattern of extreme weather over a season can be classified as an extreme climate event (e.g., drought or intense rainfall). Both extreme weather events and extreme climate events could be collectively referred to as 'climate extremes'.

Natural processes and human activity cause climate change. Changes in solar cycles and volcanic eruptions are examples of natural phenomena that affect the global climate. However, since the industrial revolution economic development, population growth and land-use change have augmented natural change by contributing unprecedented levels of greenhouse gases (GHGs) including carbon dioxide, methane, and nitrous oxide. There is growing evidence that human influence has contributed substantially to surface warming of continents (Bindoff et al. 2013), affected the global water cycle, caused glaciers to retreat as well as increased surface melting of the Greenland ice cap, loss of Arctic sea ice, and raised the upper oceanic heat content and global mean sea levels (IPCC 2014). Recent findings indicate that human activities have caused around 1 °C of global warming since pre-industrial times (IPCC 2018).

The impacts of climate change have been observed on natural and human systems on all continents and oceans. These include alteration of hydrological systems that affect the availability and quality of water resources, variations in geographic ranges of seasonal activities, animal and bird migration patterns, abundances and interactions of terrestrial, freshwater, and marine species, fluctuations in crop yield as well as ocean acidification.

The prognosis for climate change over the twenty-first century is not very favourable under all assessed emission scenarios (Fig. 13.1), even without the addition of GHGs due to natural sources. Surface temperature is expected to rise, more frequent and longer lasting heatwaves are anticipated, extreme precipitation events will be more intense and frequent, and the ocean will continue to warm and acidify while global mean sea level is projected to rise albeit unevenly across regions. Past emissions have already committed the forthcoming climate to warming conditions. The gravity of the situation is worse at 2 °C compared to 1.5 °C. Compared to 1.5 °C, global warming of 2 °C is projected to result in more extreme weather, higher impact on biodiversity and species, lesser productivity of maize, rice, and wheat, 50% more of the global population exposed to water shortages and several hundred million more people exposed to climate-related risk and susceptible to poverty by 2050.

Climate change will amplify existing climate-related risks and create new risks for natural and human systems in all stages of development. There are two complementary approaches to reducing and managing the risks of climate change. These are climate change mitigation and adaptation. Mitigation refers to human actions to reduce the source or enhance the sinks of GHGs in order to restrict future climate change. Substantial emission reduction is required over the next decade including the removal of carbon

Global greenhouse gas emissions scenarios

Potential future emissions pathways of global greenhouse gas emissions (measured in gigatonnes of carbon dioxide equivalents) in the case of no climate policies, current implemented policies, national pledges within the Paris Agreement, and 2°C and 1.5°C consistent pathways. High, median and low pathways represent ranges for a given scenario. Temperature figures represent the estimated average global temperature increase from pre-industrial, by 2100.

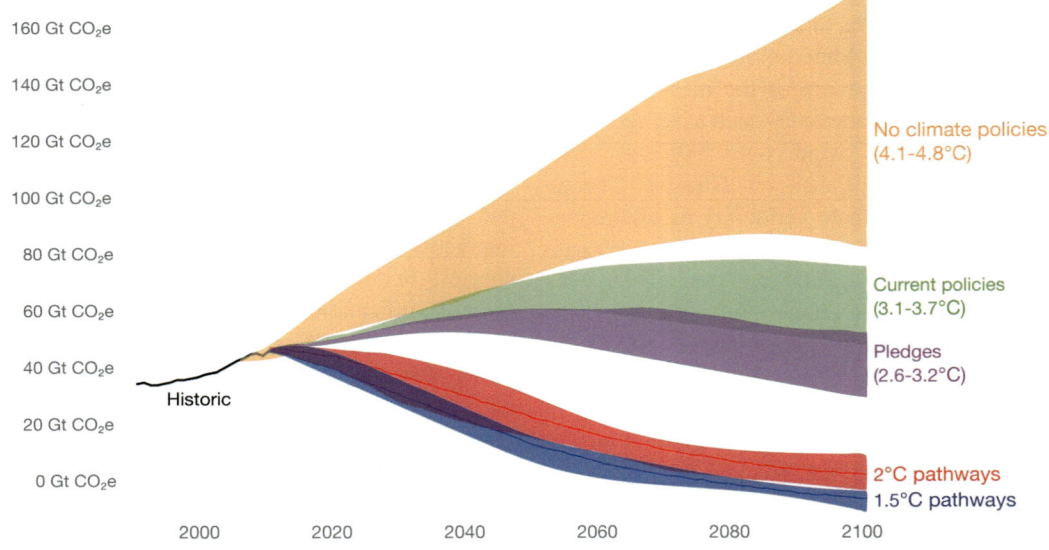

Based on data from the Climate Action Tracker (CAT).
The data visualization is available at OurWorldinData.org. There you find research and more visualizations on this topic.   Licensed under CC-BY-SA by the authors Hannah Ritchie and Max Roser.

Fig. 13.1 Global greenhouse gas emission scenarios. All scenarios result in some degree of warming. Current policies are likely to result in warming of more than 3 °C, double the warming desired in the Paris Agreement. Credit: Ritchie and Roser (2019). Reproduced under a CC-BY-SA licence (https://creativecommons.org/licenses/by-sa/2.0/)

dioxide from the atmosphere, to limit global warming to 1.5 °C and reduce climate risks (IPCC 2018). Adaptation is the process of adjustment to actual or expected climate and its effects. Adaption is required to respond to committed warming due to past emissions. However, the increased extent of climate change limits the potential for adaptation. With increased mitigation, there is a better opportunity for effective adaption and reduced costs.

Sustainable Development Goal (SDG) 13, Climate Action, is therefore a key global challenge for managing climate change with effective policies, investment, and technologies as well as behavioural and lifestyle choices. The goal is to *'Take urgent action to combat climate change*

and its impacts' with three targets (13.1–13.3) and two means of implementation (13.A and 13. B), as listed in Table 13.1.

The collective ambition of **SDG 13** is to strengthen resilience and adaptive capacity to climate-related hazards and disasters, integrate climate change measures into policies, strategies, and planning, as well as improve education, awareness-raising and capacity building by meeting the commitments of the United Nations Framework Convention on Climate Change (UNFCCC), the primary platform for negotiating global action on climate change.

Climate actions are linked to numerous SDGs including **SDG 1** (end poverty), **SDG 2** (zero hunger), **SDG 3** (good health and well-being),

Table 13.1 SDG 13 targets and means of implementation

Target	Description of target *(13.1 to 13.3)* or means of implementation *(13.A to 13.B)*
13.1	Strengthen resilience and adaptive capacity to climate-related hazards and natural disasters in all countries
13.2	Integrate climate change measures into national policies, strategies and planning
13.3	Improve education, awareness-raising and human and institutional capacity on climate change mitigation, adaptation, impact reduction and early warning
13.A	Implement the commitment undertaken by developed-country parties to the United Nations Framework Convention on Climate Change to a goal of mobilizing jointly $100 billion annually by 2020 from all sources to address the needs of developing countries in the context of meaningful mitigation actions and transparency on implementation and fully operationalize the Green Climate Fund through its capitalization as soon as possible
13.B	Promote mechanisms for raising capacity for effective climate change-related planning and management in least developed countries and small island developing States, including focusing on women, youth and local and marginalized communities

SDG 7 (energy), **SDG 11** (sustainable cities), **SDG 14** (life below water), and **SDG 15** (life on land). Using the SDGs as an analytical framework, the IPCC underscored that the potential synergies for climate mitigation actions that limit global warming to 1.5 °C far outweigh the negative outcomes in various sustainable development dimensions (IPCC 2018).

Science is closely connected to the implementation of climate actions in building the resilience of the poor, ensuring sustainable practices in the agriculture, health, and energy sectors as well as conserving oceans and coastal ecosystems (ICSU 2017; IPCC 2018; Pereira et al. 2019). The contribution of geoscience towards these objectives is set out in Table 13.2, which also shows the indicators used to monitor progress towards **SDG 13**. Examples of geoscience being relevant to **SDG 13** include modelling the susceptibility of multiple hazards to inform disaster risk reduction and understanding where ground conditions are appropriate for carbon and energy storage. Climate extremes are expected to be unprecedented as the climate changes. The risk of disasters will be determined by the exposure of assets and the vulnerability of society. For example, the impact of a tropical cyclone depends on where it makes landfall. Likewise, the impact of a heatwave will depend on the vulnerability of the population. The cumulative impacts of disasters can affect the livelihood options and resources of a society as well as their capacity to prepare for and respond to future climate extremes. This situation calls for enhanced synergies between climate change adaptation (CCA) and disaster risk reduction (DRR).

This chapter explores these contributions, setting out how geoscientists can contribute to the targets of **SDG 13**. We emphasise the importance of common framing for climate and disaster risks over different time frames and spatial settings as well as knowledge of the subsurface to contribute to the multidisciplinary solution space of climate action. We begin with an overview of global progress in tackling climate change (Sect. 13.2). We then illustrate the role of geoscience in climate change adaptation and mitigation (Sect. 13.3) as well as examples of actions on resource mobilisation and capacity building (Sect. 13.4). In the conclusion (Sect. 13.5), we highlight the important ways in which geoscience knowledge plays a critical role in limiting global warming to 1.5 °C and addressing future climate risks, to achieve **SDG 13** for the global community.

13.2 Progress in Tackling Climate Change

Scientific work on climate change goes back as early as the fifteenth century and the industrial revolution served as an impetus for investigating atmospheric carbon dioxide and surface

Table 13.2 SDG 13 indicators by 2030 and geoscience relevance

SDG 13 indicator	Relevance to geoscience
13.1.1 Number of deaths, missing persons and directly affected persons attributed to disasters per 100,000 population 13.1.2 Number of countries that adopt and implement national disaster risk reduction strategies in line with the Sendai Framework for Disaster Risk Reduction 2015-2030 13.1.3 Proportion of local governments that adopt and implement local disaster risk reduction strategies in line with national disaster risk reduction strategies	Susceptibility modelling of multiple hazards such as landslides, flash floods, coastal hazards, etc. for developing local disaster risk reduction strategies
13.2.1 Number of countries that have communicated the establishment or operationalization of an integrated policy/strategy/plan which increases their ability to adapt to the adverse impacts of climate change, and foster climate resilience and low greenhouse gas emissions development in a manner that does not threaten food production (including a national adaptation plan, nationally determined contribution, national communication, biennial update report or other)	Enhance forecasting of climate-related hazards through susceptibility modelling and improve knowledge of ground conditions for storage of carbon and energy
13.3.1 Number of countries that have integrated mitigation, adaptation, impact reduction and early warning into primary, secondary and tertiary curricula 13.3.2 Number of countries that have communicated the strengthening of institutional, systemic and individual capacity-building to implement adaptation, mitigation and technology transfer, and development actions	Improve communication on relevance of geoscience for integrating climate change mitigation, adaptation and disaster risk reduction as well as better geoscience curricula for water supply and sanitation, ground conditions, land use planning, subsurface development, siting of critical infrastructure, multi-hazard early warning; heat island effect, etc.
13.a.1 Mobilized amount of United States dollars per year between 2020 and 2025 accountable towards the $100 billion commitment	Mobilisation of resources on a bilateral basis from the Global North to the Global South through national geological organisations for subsurface evaluation, carbon capture and storage, etc.
13.b.1 Number of least developed countries and small island developing States that are receiving specialized support, and amount of support, including finance, technology and capacity-building, for mechanisms for raising capacities for effective climate change-related planning and management, including focusing on women, youth and local and marginalized communities	Reinforcing existing regional geoscience networks such as the CCOP in East Asia and SPC in the Pacific Islands for raising capacity in least developed countries and small island developing States, covering aspects of geoscience for integrated adaptation, disaster risk reduction and mitigation

temperature (Koh et al. 2013). Progress in the science domain gained traction in the 1970s and culminated in the establishment of the Intergovernmental Panel on Climate Change[1] in 1988 by the World Meteorological Organization (WMO[2]) and the United Nations Environment Programme (UNEP[3]). The initial task for the IPCC as outlined in UN General Assembly Resolution 43/53 of 6 December 1988 was to prepare a comprehensive review and recommendations with respect to the state of knowledge of the science of climate change, the social and economic impact of climate change, and possible response strategies and elements for inclusion in a possible future international convention on climate (IPCC 2018).

The First Assessment Report of the IPCC in 1990 led to the creation of the United Nations Framework Convention on Climate Change

---

[1]https://www.ipcc.ch/.

[2]https://public.wmo.int/en.

[3]https://www.unenvironment.org/.

(UNFCCC[4]), the key international treaty to reduce global warming and cope with the consequences of climate change. Since then, the IPCC has been conducting periodic assessments on the scientific basis of of risk of human-induced climate change, its potential impacts, and options for adaptation and mitigation. The IPCC plays a critical role in linking the science and policy domains for climate change, where findings from this platform serve as the basis for climate change negotiations at the UNFCCC (Gao et al. 2017). The risk framing approach to climate change introduced by the IPCC provides a conceptual basis for the integration of climate change and disaster risk reduction over a range of time frames and spatial settings (Fig. 13.2).

Notwithstanding this, there is a fundamental difference in the use of the term 'climate change' in the science and policy domains with respect to its attribution. The science perspective as represented by the IPCC ascribes climate change to natural variability or external forcings that are both natural and due to human activity. In the policy domain, the UNFCCC restricts the definition of climate change to changes that are directly or indirectly attributed to human activity and that is in addition to natural climate variability over comparable time periods. Further clarification has been provided to improve the science policy discourse by mainstreaming terms such as detection and attribution (IPCC 2014).

Detection is the process of demonstrating a statistical change in climate or a system without providing a reason for that change. *Attribution* is the process of evaluating the range of causes for a change or event to ascertain relative contributions, with the provision of statistical confidence. Scientific communication on climate change is expected to be more nuanced with an increase in studies on the impacts of climate change that take into account aspects of detection and attribution.

The UNFCCC was ratified in 1994 and serves as the primary platform for enacting mechanisms to stabilise greenhouse gas concentrations to prevent dangerous human interference with the climate system. The stabilisation is to be achieved within a duration to allow ecosystems to adapt naturally, maintain food production, and enable sustainable development. Industrialised nations spearhead emission reduction taking into account values of 'equity' , 'common but differentiated responsibilities and respective capacities', and the 'precautionary principle'. Figures 13.2 and 13.3 show the annual and cumulative total CO_2 emissions (by world region), respectively, from 1751 to 2017. It is evidence from these that while China is currently the largest emitter of CO_2, European Union states and the USA have made by far the largest contributions to CO_2 emissions over time. The UNFCCC has sought to enhance global engagement in climate actions through various means including the Kyoto Protocol (1997), Cancún Agreements (2010), Durban Platform for Enhanced Action (2011), and most recently the Paris Agreement (2015) (Ha and Teng 2013; Gallo et al. 2018; Kuriyama and Abe 2018).

Progress in tackling climate change has accelerated in the policy domain through the Paris Agreement[5]. The Paris Agreement is a legal framework of the UNFCCC in which the Global North and Global South share the burden of reducing the emission of GHGs to manage the risks of climate change. A political target has been set to hold the increase in the global average temperature well below 2 °C compared to pre-industrial levels, and if possible pursue a warming limit of 1.5 °C. It is economically feasible to achieve this target through stringent mitigation efforts whilst enabling effective adaptation measures to cope with the climate impacts despite the political challenges (Yu and Zhu 2015; Leemans and Vellinga 2017; Cooper 2018; Travis et al. 2018).

The robust review of the global literature by the IPCC has confirmed that many impacts will be less potent by limiting the global warming to 1.5 °C compared to 2 °C. Limiting the global limit to 1.5 °C is possible but requires deep emission cuts, deployment of a range of

---

[4]https://unfccc.int/.

[5]https://unfccc.int/process-and-meetings/the-paris-agreement/the-paris-agreement.

Annual total CO₂ emissions, by world region

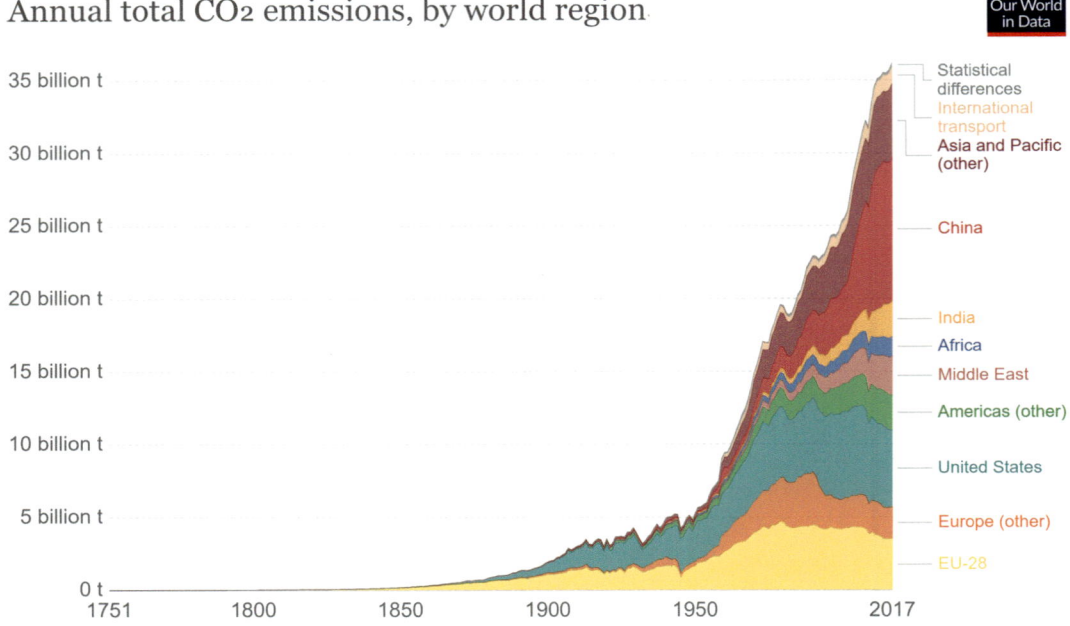

Source: Carbon Dioxide Information Analysis Center (CDIAC); Global Carbon Project (GCP)
Note: "Statistical differences" notes the discrepancy between estimated global emissions and the sum of all national and international transport emissions.
OurWorldInData.org/co2-and-other-greenhouse-gas-emissions • CC BY

Fig. 13.2 Annual total CO₂ emissions, by world region. Credit: Ritchie and Roser (2019). Reproduced under a CC-BY licence (https://creativecommons.org/licenses/by/4.0/)

technologies, behavioural changes, and increased investment in low carbon options, at an unprecedented scale (IPCC 2018). Four plausible pathways with no or limited overshoot of 1.5 °C have been proposed, including a scenario with lesser use of technology through afforestation to one where emission reduction is mainly achieved through technological means such as carbon capture and storage. Early action is expected to be cheaper and would lead to better outcomes as well as reduce the need for adaptation.

Global progress on **SDG 13** and implementation of the Sendai Framework is tracked using the Sendai Monitor,[6] released in March 2018. The Sendai Monitor is an online tool where official information is uploaded by Governments based on a set of indicators that were negotiated in 2016 (United Nations 2016). Only nine of the 195 countries have validated their data while 108 have not started the process as of June 2019.

Reports from the remaining countries are still in progress. Nearly half of the nations in the world have adopted and implemented national disaster risk reduction strategies in line with the Sendai Framework while data is not available for the remaining countries. Information is lacking for all other indicators where some have not been identified. The situation is expected to improve after the means for implementing the Paris Agreement, currently being negotiated by Governments, are established.

13.3 The Contribution of Geoscience to Climate Action

Geoscientists have made major contributions in multidisciplinary settings to enhance understanding of climate change in the science domain. Geoscience knowledge drives models

[6]https://sendaimonitor.unisdr.org/.

Cumulative CO₂ emissions by world region

Cumulative carbon dioxide (CO₂) emissions by region from the year 1751 onwards. Emissions are based on territorial emissions (production-based) and do not account for emissions embedded in trade.

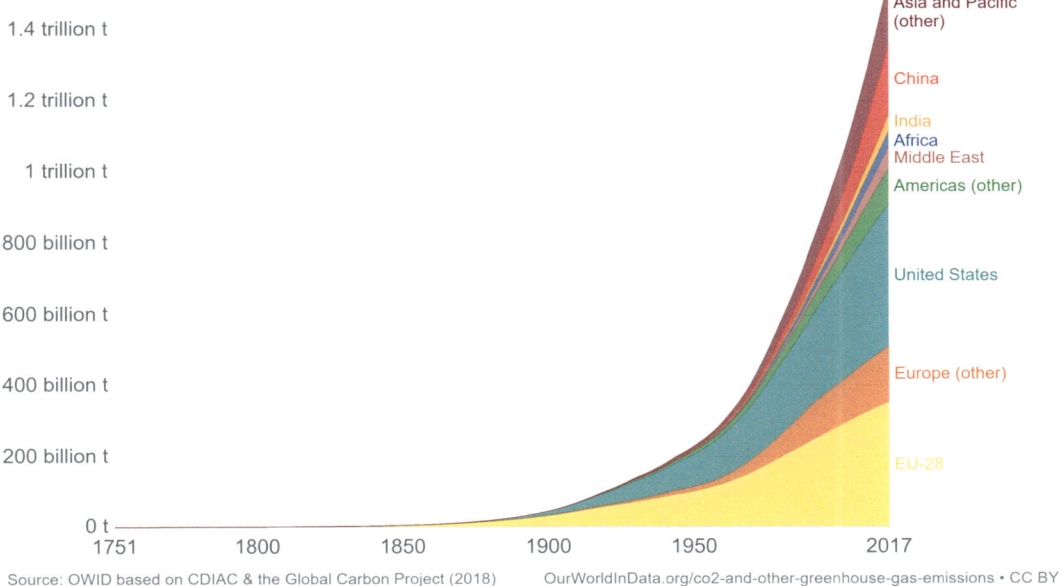

Fig. 13.3 Cumulative CO₂ emissions, by world region. Credit: Ritchie and Roser (2019). Reproduced under a CC-BY licence (https://creativecommons.org/licenses/by/4.0/)

that provide basic information for climate change adaptation. Subsurface geoscience information is also critical for key technologies that support climate change mitigation. There is significant potential for geoscientists to contribute further in the scientific discourse, for example, on the attribution of climate change to determine the causes of climate-driven geohazards. Geoscience inputs in this aspect could provide insights to delineate the contribution of natural and anthropogenic causes and serve as the basis for developing climate change policies that are equitable.

The geoscience community has also played a significant role in the policy domain, specifically in advancing progress in climate change mitigation. For example, geologists in the UK, Norway, and Canada played a critical role in the adoption of carbon capture and storage (CCS) as a climate change mitigation option under the Clean Development Mechanism, a cooperative instrument of the Kyoto Protocol (Lovell 2009) and in

developing sound regulatory advice to underpin safe storage and enhance public confidence in CCS. The potential deployment of CCS technology to the Global South will help to balance energy resource development with emission reduction.

13.3.1 Climate Change Adaptation (Target 13.1)

The 2015 Paris Agreement calls for measures to strengthen climate change adaptation and this is important to address impacts due to historical emissions, which have already committed the future climate to warming conditions. Even if the climate eventually equilibrates at 1.5 °C or less above pre-industrial levels, anticipatory adaptation planning is required to eliminate the risk of large damages and adaptation costs in exposed and vulnerable areas (Travis et al. 2018).

In line with **Target 13.1** of **SDG 13**, the Paris Agreement and the Sendai Framework for Disaster Risk Reduction mutually support development goals by strengthening resilience and adaptive capacity to climate-related hazards and disasters, particularly within national systems. The Sendai Framework on Disaster Risk Reduction emphasises climate change as a driver of hazards (**Box 13.1**). The transformation of scientific knowledge on climate, systematic observation, and early warning into tools, products, and services that support decision-making at the local level is critical for this purpose (Dolman et al. 2016; Giuliani et al. 2017; Forino et al. 2017). Greater involvement of local governments in implementing disaster risk reduction measures is expected to bring positive outcomes in reducing the number of deaths and injuries due to climate extremes and change.

Box 13.1. The Sendai Framework and Climate Change

The UNISDR (now UNDRR) Sendai Framework on Disaster Risk Reduction has seven global targets covering mortality, affected people, economic loss, damage to critical infrastructure, national and local disaster risk reduction strategies, international cooperation, and early warning including risk information and assessments. Four priority areas have been identified covering

- Understanding disaster risk.
- Strengthening disaster risk governance to manage disaster risk.
- Investing in disaster risk reduction for resilience.
- Enhancing disaster preparedness for effective response, and to 'Build Back Better' in recovery, rehabilitation, and reconstruction.

For more information see the relevant sections of SDG 1, ending poverty.

The Sendai Framework considers climate change and variability as a significant impediment to sustainable development. Climate is recognised as an underlying driver of increasing disaster risk both in terms of severity and increased frequency and intensity. In addition to large catastrophic disasters, the Sendai Framework equally applies to the risk of small- to large-scale, frequent and infrequent, sudden and slow-onset events, caused by natural or anthropogenic hazards. New risks and a steady rise in disaster losses are expected in the short, medium, and long terms, especially at the local level.

The Sendai Framework promotes partnerships and multi-hazard management of disaster risk in development, at all levels and across all sectors. There is a wide range of complementary approaches to deal with current disaster risks and future risks due to climate change within national systems, which constitute 'no-regrets' options (IPCC 2012; Forino et al. 2017). The use of science is advocated, for example, in comprehensive surveys on multi-hazard disaster risks, development of regional disaster risk assessments and climate change scenarios. Geoscientists should foster partnerships with climate scientists and other specialists to advance knowledge on multi-hazard risks.

Geological hazards have been widely investigated and significant contributions have been made to reduce the risk of catastrophic disasters (Marriner et al. 2010). Notwithstanding this, national-level investigation on the susceptibility of climate-influenced hazards using the wealth of information available from national geoscience agencies is not widespread (Cigna et al. 2018). Geoscience information has also not been mainstreamed into the policy domain with respect to climate change adaptation, to benefit society through cross-sectoral planning. This is reflected by the limited geoscience inputs and participation of national geoscience agencies in preparing progress reports such as the National Communications and Biennial Update Reports to the UNFCCC. National Adaptation Programmes are conducting climate risk assessments in designated conservation sites, which draw on geoscience information (Wignall et al. 2018). However, this is not a common practice, particularly in the Global South.

Earth processes and society are connected in multiple ways and geoscience inputs provide invaluable insights to understanding risk, exposure, and vulnerability. Whilst knowledge and technology is progressing in many areas, the geoscience community has to enhance the effort to contribute to the multidisciplinary solution space to meet the challenges expressed in the Paris Agreement and the Sendai Framework (Rogelj and Knutti 2016; Pereira 2018). In this context, a key challenge is to forecast and lessen the impact of natural hazards as the climate changes, particularly in the Global South. In Asia and Africa, lack of data, poor understanding of interactions between geology, climate change, and land-use change coupled with weak institutions and capacity have caused much damage and destruction to infrastructure (Hearn 2016; Broeckx et al. 2018; Maes et al. 2018).

Susceptibility modelling is advancing in the evaluation of climate-related hazards such as landslides, floods, erosion, and subsidence (e.g., Cigna et al. 2018; Reichenbach et al. 2018; González-Arquerosa et al. 2018; Hosseinalizadeh et al. 2019). Susceptibility modelling enables spatial demarcation of areas where a hazard event could occur, depending on contributing surficial features, geological conditions, and processes that vary depending on the hazard. Modelling of hazards at the global, regional, and national scale is not sufficient to provide specific adaptation measures at the local level, as indicated by the experience of landslides (Fig. 13.4), glacial lake outburst flooding, and coastal hazard assessments (Radosavljevic et al. 2016; Allen et al. 2018; Broeckx et al. 2018).

Local-level studies offer the best options for monitoring and early warning adaptation measures. The British Geological Survey (BGS) has

Fig. 13.4 Landslide (Cusco, Peru). Credit: Galeria del Ministerio de Defensa del Perú. Reproduced under a CC-BY 2.0 licence (https://creativecommons.org/licenses/by/2.0/)

assessed the entire range of nationally available datasets to delineate areas of the United Kingdom that are susceptible to hazards including landslides, flooding, and subsidence, targeting World Heritage Sites in the UK (Cigna et al. 2018). This approach can be applied to target other areas where there may be a risk to infrastructure such as dams, transport routes, and coastal power stations, among others. Areas at risk can be subject to further detailed investigation using conventional engineering geology methods (see **SDG 9**, covering resilient infrastructure).

Landslide susceptibility modelling is progressing to incorporate the effects of climate and environmental changes at different spatial and temporal scales (Gariano and Guzzetti 2016; Reichenbach et al. 2018). Machine learning algorithms are advancing to use a small number of samples for landslide susceptibility modelling with periodic updates to take into account climate change (Huang and Zhao 2018). Susceptibility modelling based on terrain morphology, geology, soils, and land cover has been found to be cost effective, applicable at large or small scales, complementary to hydrological models and suitable for land-use decision-making (van Westen et al. 2008; Perucca and Angilieri 2011). For example, high-resolution terrain mapping is taking into account the identification of and linkage of landslides and erosional processes as a response to tectonic activity and climate change (Geach et al. 2017).

Machine learning models have also been found to be effective in delineating areas susceptible to internal erosion or piping processes, which contribute to loss of agricultural productive capacity, land degradation, and increased sediment yields (Hosseinalizadeh et al. 2019). The monitoring of areas susceptible to landslides prior to the occurrence of wildfire has been identified as an adaptation option to help reduce the effects of erosion (Peterson and Halofsky 2018). Similarly, the monitoring of periglacial degradation of bedrock and moraine has been identified as a key option for early warning of

debris flow in high mountain regions as the temperature increases (Wei et al. 2018).

Flood and flash flood risk assessment and solutions to flood and flash flooding have also benefitted from the susceptibility approach. Intensive data requirements and access to expert knowledge for standard engineering flood models are a challenge to governments of the Global South (Cunha et al. 2017; Teng et al. 2017; Komi et al. 2017). Flash floods and water shortages have also been reported to occur in the same areas at different seasons. Geoscience knowledge is significant in this respect, particularly to promote sustainable urban drainage systems, for storage for excess water in underground reservoirs and engineered structures to ensure consistent water supply (Stephenson 2018; Nguyen et al. 2019).

Coastal hazards such as storms and floods as well as slow-onset sea-level rise, inundation, and erosion are expected to impact communities that live in susceptible coastlines with high exposure. Recent geoscience findings indicate that a small rise of 0.5 m in sea level is expected to double the frequency and the intensity of tsunami-induced flooding of the coasts of Macau due to earthquakes along the Manila Trench (Li et al. 2018). Sea-level rise also threatens coastal aquifers and exposes infrastructure such as waste disposal sites that could emerge as future pollution sources (Jamaludin et al. 2016; Yahaya et al. 2016; Stephenson 2018). Geomorphological features are an important factor in the development of decision-support tools that deal with coastal disaster risk reduction and multi-hazard risk of social–ecological systems (Fischer 2018; Ferreira et al. 2018; Hagenlocher et al. 2018). Geoscientists have the capacity to advance disaster preparedness in a variety of sea-level scenarios, to build the resilience of coastal communities.

There is great potential for geosciences to progress susceptibility modelling for multi-hazards at the local level under a variety of climate settings, in collaboration with experts from diverse disciplines.

The Asian Network on Climate Science and Technology (ANCST[7]) facilitates the advancement of science, technology, and innovation through multi-sector and multidisciplinary partnerships, to support the implementation of the Sendai Framework on Disaster Risk Reduction and the Paris Agreement.

ANCST has been instrumental in bringing together geoscience, climate, and atmospheric experts from Malaysia and the UK, to jointly develop the project on 'Disaster Resilient Kuala Lumpur', summarised in Fig. 13.5. In this project, selected meteorological and hazard models including susceptibility approaches are being adapted for tropical circumstances and integrated onto a common multi-hazard platform for the City Hall of Kuala Lumpur (DBKL) to improve forecasting.

Improved forecasting capacity for flash floods, landslides (Fig. 13.6), sinkholes, strong winds, urban heat, and air pollution at the city and neighbourhood scales is expected to contribute greatly to enhance disaster resilience as the climate changes in tropical terrain.

13.3.2 Climate Change Mitigation (Target 13.2)

The Paris Agreement set the global goal of limiting warming to below 2 °C above pre-industrial levels while 'pursuing efforts to limit the increase to 1.5 °C' (UNFCCC 2016). Geoscience contributes to this goal by supporting the transition to a low-carbon energy regime to mitigate greenhouse gas emissions and address future energy security. Renewable energy resources and technologies such as geothermal, wind, solar power, hydropower, tidal wave, and biomass as well as their applications and services to buildings, industry, electricity, and transport utilise geoscience information. Geoscience knowledge, integrated into policies, strategies, and planning at various levels, has great potential to support integrated tools to facilitate the transition to a low-carbon energy regime, adapt to the adverse climate change impacts and foster disaster resilience, and achieve **Target 13.2** of **SDG 13** (Lovell 2009; Barrie and Conway 2014; Martens and Kühn 2015; Kühn et al. 2016).

Wind, solar power, hydropower, tidal wave, biomass, and other renewable forms of energy generation are dependent on weather and climate. Modelling and measurement for resource assessment and site selection for these energy sources draw on geoscience information. For example, the placement of renewable energy facilities may extend to complex terrain and offshore regions that are difficult to model. More effort is required to combine geoscience information with climate and other data sources to enable a multidisciplinary and dynamic analysis of the suitability of renewable energy facilities (see also the chapter exploring **SDG 7**, affordable and clean energy).

Carbon capture and storage (CCS) is increasingly accepted as a viable, feasible, and safe technology for climate change mitigation. However, CCS alone cannot be expected to support the goal of maintaining global temperatures below 2 °C, particularly in the absence of effective policy drivers. The CCS technology essentially involves the separation of carbon dioxide from a source and subsequently storing the carbon for long-term isolation from the atmosphere (Metz et al. 2005). The process involves three stages: capture, transport, and storage. The CO_2 is collected from a static emitter such as a power plant, compressed and then routed to a storage site through pipelines. Storage sites are essentially geological formations with suitable porosity and permeability including oil and gas reservoirs, deep coal seams, saline aquifers, and salt caverns (Fig. 13.7).

The deployment of CCS has been shown to be geologically viable, safe, effective and its costs are expected to decrease (Szulczewski et al. 2012; Cook 2017). However, there are concerns regarding pressure increase and saltwater

[7]http://ancst.org.

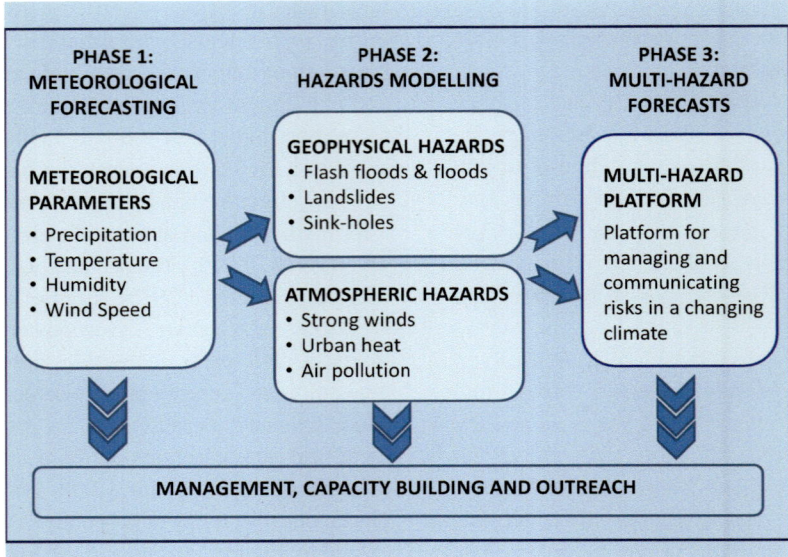

Fig. 13.5 Established in 2013 with seed-funding from the Cambridge Malaysia Education Development Trust Fund and Malaysia Commonwealth Studies Centre, ANCST under the coordination of Universiti Kebangsaan Malaysia's Southeast Asia Disaster Prevention Research Initiative (SEADPRI-UKM) was instrumental in bringing together a multidisciplinary team of scientists from meteorological, geological and atmospheric backgrounds for the Disaster Resilient Kuala Lumpur project supported by the Newton Ungku Omar Fund, a joint initiative of the Governments of UK and Malaysia. Credit: Authors' Own

displacement in deep aquifers. Pressure increase could lead to the disintegration of cap rocks or reactivation of faults and subsequently cause leakage of carbon dioxide. Saltwater displacement may contaminate drinking water reservoirs in shallow groundwater systems above the storage complex, if they are connected. The effort to increase the understanding of sequestration mechanisms and technology is continuously ongoing (Charalampidou et al. 2017; Kühn et al. 2017; Renforth and Henderson 2017).

Geoscientists can contribute to limiting negative emissions by providing safe storage capacity to meet the temperature target of the Paris Agreement. Advances in integrating CCS technology with other types of energy production (e.g., biomass) and energy storage require significant geoscience knowledge (Martens and Kühn 2015). Another emerging geoscience sequestration technology is coupled carbonate weathering (CCW), from carbonate mineral weathering in combination with aquatic photosynthesis on the continents, which may help to offset atmospheric CO_2 at a global scale (Liu et al. 2018).

Geothermal energy is increasing in use for both heating and cooling (Lund and Boyd 2016), requiring more enhanced knowledge on subsurface conditions. This includes information on natural and induced fractures as well as permeability characteristics to better predict mechanical and flow response of heat and ensure safe and economical energy supply from shallow and deep geothermal resources (Kühn et al. 2016). A range of modelling is applied in all phases of geothermal exploitation, from the prediction of geothermal potentials to the optimisation of borehole locations as well as in improving the efficiency of existing geothermal facilities (Bocka et al. 2013; Hong et al. 2017). A thorough understanding of the subsurface is critical for managing geothermal systems effectively, particularly in urban areas, to avoid overexploitation and conflicts with other subsurface use.

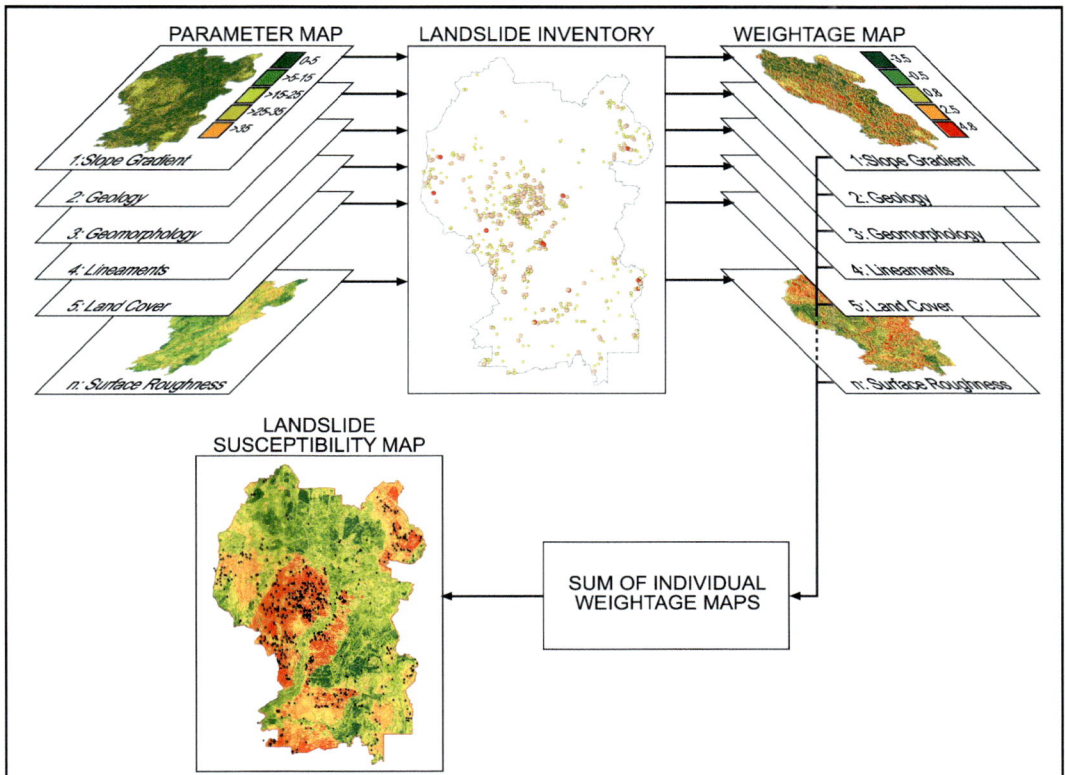

Fig. 13.6 *Landslide susceptibility modelling* in the Disaster Resilient Kuala Lumpur project *used the statistical approach, where* parameters that influence the hazard were correlated with the inventory to obtain the weightage that is used to derive the level of susceptibility at the city scale, which is subsequently validated. Credit: Authors' Own

Long-term utilisation of geothermal systems requires numerical simulation and three-dimensional models. These research fields should be further expanded in the geosciences.

Wind and solar are becoming increasingly important energy sources. However, energy production from these facilities is intermittent and alternate sources are required to compensate for fluctuating power generation.

Geological formations offer a great potential to store energy over various timescales in the form of subsurface storage of heat, subsurface hydrogen storage, and compressed air energy storage (Kabuth et al. 2017). Storage of energy is primarily in salt caverns (Ozarslan 2012; Bauer et al. 2013; Bauer 2016). Compressed air energy storage is most promising in wind farms, converting electricity into mechanical energy in the form of highly pressurised air, which is then stored in the subsurface. The pressurised air is then used to generate electricity through wind turbines, which is integrated into the grid during peak loads. The use of porous geological formations is currently under investigation to expand the deployment of compressed air energy storage (Wang and Bauer 2017). This is expected to advance the expansion of wind and solar energy as porous geological formations are more widely available and can offer even larger storage capacities. The use of geological formations to integrate energy storage and carbon storage is also being explored to close the entire carbon cycle (Martens and Kühn 2015).

Offshore, the marine environment offers much potential for renewable energy and carbon storage (see also **SDG 14**). The development of renewable energy is most advanced for wind while wave and tide energy sources are expanding

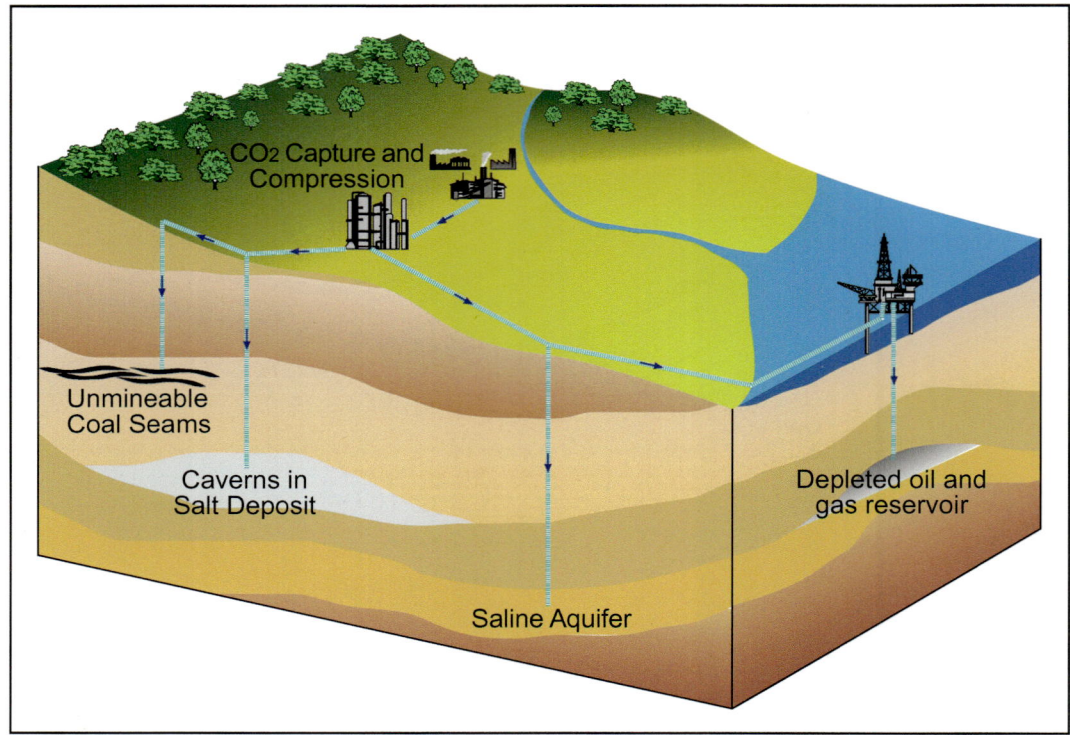

Fig. 13.7 There are various ways in which geological formations can sequester and store carbon dioxide as well as stockpile energy, and this requires substantial geoscience knowledge of subsurface conditions and processes. Credit: Authors' Own

rapidly (Barrie and Conway 2014). Much of the advancement in the marine sector is challenged by a range of geoscience issues (Barrie et al. 2014). Geological characteristics and physical environment parameters need to be properly assessed to facilitate the safe deployment of marine renewable energy and support offshore carbon storage. Geological information is critical for understanding geotechnical conditions on which the energy system will be anchored. The routing of cables is also dependent upon the geological and physical conditions of the seabed, where the understanding of ocean phenomena such as subaqueous landslides needs to be improved (Mengerink et al. 2014; Reichenbach et al. 2018). The assessment criteria for site suitability and potential assessment for carbon storage is under development, drawing on geological characteristics and their potential for leakage. This new feature of continental shelf

research is expected to advance strategies for carbon storage in offshore sedimentary basins.

13.3.3 Education and Awareness Raising (Target 13.3)

Education, awareness raising, and strengthening of institutional capacity are key to achieve **Target 13.3** of **SDG 13** (Fig. 13.8). However, educational curriculum reforms in institutions of higher learning have had limited success in ensuring that all students are exposed to climate science, climate change, disaster risk reduction, and sustainability issues (Hess and Collins 2018; Brundiers 2018; Nakano and Shaw 2018). Scientists tend to converse with their peers and require new skill sets to communicate with the media, policymakers, and other stakeholders to influence public discussion about climate change.

Fig. 13.8 Students in Ladakh express their perspectives on climate change through paintings and artwork © Geology for Global Development (used with permission)

This contributes to a poor understanding of the role of geoscience in climate change and disaster risk in many policy and planning institutions. There is also the need to develop a good narrative for the geoscience story of climate change (Filho et al. 2018; Harris 2017; Reis and Ballinger 2018).

A wide range of education, awareness-raising, capacity building, and policy engagement orientations is required, including more inclusive and transformative social learning approaches, to effectively support the Paris Agreement (Macintyre et al. 2018). The geoscience community needs to strengthen linkages with multiple disciplines including the social sciences, forge strategic partnerships, and participate actively in science-policy platforms. The rich tradition of geo-conservation and emerging capacity for providing web-based 'smart geo-services' can be leveraged upon for this purpose. Enhanced communication on the relevance of geoscience knowledge in integrating climate change adaptation and disaster risk reduction will also support sustainable development. These aspects should be explicitly integrated into geoscience education, training, and continued professional development (Stewart and Gill 2017; Bolden et al. 2018; Zhang et al. 2018).

Geoscience institutions have played a leading role to support education, awareness raising, and capacity building to mitigate greenhouse gas emission and meet future demands for renewable energy supply. The Sleipner storage site in the North Sea is the world's first and longest operating CCS demonstration project, which is being successfully monitored by using geoscience knowledge to raise awareness and provide assurance to policymakers, investors, and the public on the safety of the technology. Awareness raising of CCS is also emphasised in Canada, where geoscience expertise was utilised at the Weyburn CO_2-Enhanced Oil Recovery project to refute claims of leakage (Jones et al. 2011; Sacuta et al. 2017). Another major CCS demonstration project is in Australia. The demonstration site is the result of two decades of work focused solely on carbon capture and geological formations, initially drawing on the expertise of geologists, geophysicists, geochemists, and hydrogeologists (Cook 2017).

The number of specialised courses on geothermal energy is limited worldwide (Zarrouk 2017). Raising awareness of geothermal energy in Germany uses a geoethical approach where a generic underground laboratory has been established (Meller et al. 2017), and transparent communication is encouraged. This is tangible science that can serve to enhance mutual understanding of stakeholder groups and increase public awareness to facilitate responsible exploitation of geothermal energy. Capacity building of geoscientists in climate mitigation technology is advanced by the European Geosciences Union, which periodically brings together geoscientists from all over Europe and the rest of the world to discuss future challenges during the General Assembly. The issue of renewable energy and carbon storage is of explicit concern at this platform where advances in multidisciplinary approaches and future research needs are highlighted, providing insights on the role of geoscience (Kühn et al. 2016).

13.3.4 Resource Mobilisation and Capacity Building (Targets 13.A and 13.B)

Resource mobilisation in the context of meaningful mitigation actions (**Target 13.A**) as well as capacity building in the least developed countries and small island developing States (**Target 13.B**) are important means of implementation for **SDG 13**. The Global South faces many challenges in maintaining economic growth while increasing energy efficiency and shifting from carbon to renewable energy to reduce GHG emissions (Liobikienė and Butkus 2018), with links to many other SDGs. Such challenges include capacity limitations and the availability of and access to technology and financial resources. Economic development is a necessity for the Least Developed Countries (LDCs) as they manage the risks of climate change. Issues such as extensive poverty (**SDG 1**), widespread unemployment (**SDG 8**), poor access to clean water (**SDG 6**), rural electrification (**SDG 7**), deforestation as well as dryland and desert expansion (**SDG 15**) are critical for LDCs (Teklu 2018). Small island developing states (SIDS) are already experiencing the impacts of climate change (see **SDG 14**), particularly in the tourism and fisheries sectors

(Nurse et al. 2014). Existing vulnerabilities and weak adaptive capacities need to be urgently addressed to ensure the sustainability of SIDS in a changing climate (Robinson 2018).

Resource mobilisation and capacity building related to geoscience have had a long history of creating enabling conditions for contributing to economic development and poverty eradication in the mineral, energy, and construction sectors of the Global South. Enhanced capacity in geoscience knowledge has also played an important role in ensuring the well-being of society by the provision of information on groundwater resources, disaster risks, and environmental pollution as well as food security and human health. The practice of resource mobilisation for enhancing geoscience capacity is now supporting climate change mitigation actions in the Global South. Climate change adaptation is of particular focus in LDCs and SIDs. In both these cases, the importance of ensuring that the recipient country has the capacity to absorb and sustain the technology being deployed by the donor cannot be overemphasised.

Resources have been mobilised for enhancing geoscience capacity in carbon capture and storage (CCS) from the British Geological Survey and Geoscience Australia to emerging economies such as China and India (Feitz et al. 2017; BGS 2018). Regional bodies such as the Coordinating Committee for Geoscience Programmes in East and Southeast Asia (CCOP) and the Geosciences Division of the Pacific Community (SPC) serve as important one-stop-networks for strengthening geoscience knowledge in CCS and renewable energy in the Global South. The CCOP convenes training and capacity building in CCS on a routine basis for national geoscience organisations in the region (CCOP 2018). The SPC facilitates opportunities for exploration of geothermal energy and ocean thermal energy conversion to advance sustainable development in the region (Petterson and Tawake 2016).

Groundwater has a significant relationship with the water, health, food, and energy nexus in the context of climate change (Jakeman et al. 2016). Countries in the Pacific islands are frequently exposed to climate extremes, and sea-level rise is an imminent threat to groundwater resources, as characterised in the chapter exploring **SDG 14**. Many LDCs and SIDs need enhanced geoscience capacity for groundwater resource management and climate extremes. National geoscience institutions are addressing this need through the mobilisation of resources. For example, rural Africa has benefited from work conducted by BGS to support adaptation and build resilience to climate change (BGS 2011). Among the products of this project is the aquifer resilience map for Africa, which draws on existing information on geology and hydrogeology (see **SDG 6**). Geoscience Australia is developing tools that encompass desktop software designed for local governments and communities to provide insights into the likely impacts of future risks, so that appropriate responses can be taken for planning, preparing, and responding to disasters.

13.4 Summary and Conclusions

All countries are exposed to increased risks due to climate change. Climate change adaptation and mitigation are two complementary approaches for reducing and managing the risks of climate change. Adaption is required to respond to committed warming due to past emissions but its potential is limited for high levels of warming. Climate change mitigation by substantially reduced emissions over the next few decades is critical to reduce future climate risks. The Paris Agreement sets an ambitious target to maintain global average temperatures well below 2 °C and if possible, limit warming to 1.5 °C above pre-industrial levels, whilst supporting adaption efforts in the least developed countries and small island developing states. The complementary goals of the Paris Agreement and Sendai Framework on Disaster Risk Reduction have strengthened policy coherence under **SDG 13** on urgent actions for combating climate change and its impacts (Djalante 2019; Mizutori 2019).

Geoscience is steadily increasing its contribution to the multidisciplinary solution space that addresses the challenge of climate change.

Susceptibility modelling of hazards such as landslides, floods, erosion, and subsidence offers invaluable insights for understanding risk, exposure, and vulnerability to predict and lessen the impact of natural hazards as the climate changes. Nationally available geoscience datasets must be leveraged to develop local-level monitoring and early warning adaptation measures, under a variety of climate settings. Geoscience also fundamentally supports emission reduction and the transition to a low-carbon energy regime. This is done primarily through carbon capture and storage (CCS) and the development of geothermal energy which are viable, feasible, and safe options to mitigate carbon emissions.

Geoscience knowledge has been integrated into policies, strategies, and planning at various levels. At the global level, carbon capture and storage has been successfully promoted at the UNFCCC (Lovell 2009). But increasingly geoscience data and understanding need to be mainstreamed into the policy domain with respect to climate change adaptation, to benefit society through cross-sectoral planning. This calls for a major transformation in the geoscience community with respect to education, awareness raising, capacity building, and policy engagement. The geoscience community needs to strengthen linkages with multiple disciplines including the social sciences, forge strategic partnerships, and participate actively in science-policy platforms. Such aspects should be explicitly integrated into geoscience education, training, and continued professional development. The long-term benefits would include enhanced resource mobilisation and strengthening of geoscience capacity for climate change mitigation actions in the Global South, particularly for the least developed countries and small island developing states.

13.5 Key Learning Concepts

- Climate change refers to changes in the state of the climate that persists over an extended period, typically decades or longer, which is caused by natural processes and human activity. Natural processes that affect the global climate are solar cycles and volcanic eruptions. Human activities have contributed to natural change by contributing unprecedented levels of greenhouse gases (GHGs) including carbon dioxide, methane, and nitrous oxide.

- Human activities have caused around 1 °C of global warming since pre-industrial times, contributing to surface warming of continents, affecting the global water cycle, causing glaciers to retreat as well as increasing surface melting of the Greenland ice cap, loss of Arctic sea ice, raising upper oceanic heat content and global mean sea levels.
- Past emissions have already committed the future climate to warming conditions. The gravity of the situation is worse at 2 °C compared to 1.5 °C. More extreme weather, higher impact on biodiversity and species, lesser productivity of maize, rice, and wheat, 50% more of the global population exposed to water shortages and several hundred million more people will be exposed to climate-related risk and susceptible to poverty by 2050 should global warming increase to 2 °C compared to 1.5 °C.
- All countries are exposed to increased risks due to climate change. Climate change mitigation and adaptation are two complementary approaches for reducing and managing the risks of climate change. Climate change mitigation is critical for substantially reducing emissions over the next few decades and reduce future climate risks. Climate change adaption is required to respond to committed warming due to past emissions but its potential is limited for high levels of warming.
- Geoscientists support emission reduction and the transition to a low-carbon energy regime. This is done primarily through carbon capture and storage (CCS), development of geothermal energy, and subsurface energy storage by providing options that are viable, feasible, and safe. Geoscience knowledge is recognised at the global policy level for carbon capture and storage.
- The risk of disasters is determined by the extent of a hazard, exposure of assets, and

vulnerability of society. Susceptibility modelling of hazards such as landslides, floods, erosion, and subsidence offers invaluable insights for understanding risk, exposure, and vulnerability to predict and lessen the impact of natural hazards as the climate changes. Nationally available geoscience datasets can be leveraged to develop local-level monitoring and early warning measures under a variety of climate settings, requiring enhanced synergies between climate change adaptation (CCA) and disaster risk reduction (DRR).

- Geoscience knowledge needs to be mainstreamed into the policy domain with respect to climate change solutions that benefit society. A major transformation is required in geoscience education, training, and continued professional development with respect to awareness, capacity building, policy engagement, strategic linkages, and transdisciplinary networking for climate change actions.

13.6 Educational Ideas

In this section, we provide examples of educational activities that connect geoscience, the material discussed in this chapter, and scenarios that may arise when applying geoscience (e.g., in policy, government, private sector international organisations, and NGOs). Consider using these as the basis for presentations, group discussions, essays, or to encourage further reading.

- How may climate changes affect the frequency and magnitude of natural and environmental hazards? Prepare a matrix with characteristics of climate change (e.g., rising temperatures andsea-level rise) on one axis, and diverse natural hazards relevant to your region (e.g., landslides, flooding, and subsidence) on the other axis. For each cell, consider if there is an effect of the climate change characteristic on the natural hazard, and if so, describe it. How can the contents of your matrix inform steps taken to reduce disaster risk?

- What are the geological characteristics associated with good locations for carbon capture and storage? Review a geological map of your region (use the OneGeology Portal[8] or any available paper/digital maps) to determine if there are potential geological units that may be suitable.
- Communicating geoscience to public audiences is a valuable skill. Reflecting on the theme 'what does geological history teach us about climate change today', design a public engagement activity that helps *children* understand key lessons from the geological record for climate action.
- Research the four plausible pathways proposed with no or limited overshoot of 1.5 °C warming (mentioned in Sect. 13.2). What are the geoscience contributions to each? Divide into four groups, with each giving a summary of different pathways, the role of geoscientists, and any assumptions made when assuming this pathway would have no or limited overshoot of 1.5 °C warming. Debate the merits of each pathway as a class, and vote on which you would choose to pursue.

Further Resources

Geological Society of London (2019) The role of geological science in the decarbonisation of power production, heat, transport and industry. https://www.geolsoc.org.uk/Lovell19

IPCC (2019) The intergovernmental panel on climate change. https://www.ipcc.ch/

Stephenson MH, Ringrose P, Geiger S, Bridden M, Schofield D (2019) Geoscience and decarbonization: current status and future directions. Pet Geosci petgeo2019-084

UNISDR (2019) Sendai framework for disaster risk reduction. https://www.unisdr.org/we/coordinate/sendai-framework

Kelman I (2015) Climate change and the Sendai framework for disaster risk reduction. Int J Disaster Risk Sci 6(2):117–127

References

Allen SK, Ballesteros-Canovas J, Randhawa SS, Singha AK, Huggel C, Stoffel M (2018) Translating the concept of climate risk into an assessment framework to inform adaptation planning: Insights from a pilot study of flood risk in Himachal Pradesh, Northern India. Environ Sci Policy 87:1–10

Barrie J, Todd B, Heap A, Greene HG, Cotterill C, Stewart H, Pearce B (2014) Geoscience and habitat mapping for marine renewable energy—introduction to the special issue. Cont Shelf Res 83:1–2

Barrie JV, Conway KW (2014) Seabed characterization for the development of marine renewable energy on the Pacific margin of Canada. Cont Shelf Res 83:45–52

Bauer S (2016) Energy storage in the geological subsurface: dimensioning, risk analysis and spatial planning: the ANGUS + project. Environ Earth Sci 76(1). https://doi.org/10.1007/s12665-016-6319-5

Bauer S, Beyer C, Dethlefsen F, Dietrich P, Duttmann R, Ebert M, Feeser V, Görke U, Köber R, Kolditz O, Rabbel W, Schanz T, Schäfer D, Würdemann H, Dahmke A (2013) Impacts of the use of the geological subsurface for energy storage: an investigation concept. Environ Earth Sci 70(8):3935–3943

BGS (2011) Project Information Sheet: groundwater resilience to climate change in Africa. www.bgs.ac.uk/GWResilience. Accessed 25 Aug 2018

BGS (2018) Geoscience for sustainable futures: the 'Official Development Assistance' (ODA) Programme of the British Geological Survey. https://www.bgs.ac.uk/research/international/oda/home.html. Accessed 25 Aug 2018

Bindoff NL, Stott PA, AchutaRao KM, Allen MR, Gillett N, Gutzler D, Hansingo K, Hegerl G, Hu Y, Jain S, Mokhov II, Overland J, Perlwitz J, Sebbari R, Zhang X (2013) IPCC, 2013: climate change 2013: the physical science basis. Contribution of working group I to the fifth assessment report of the intergovernmental panel on climate change (Stocker TF, Qin D, Plattner G-K, Tignor M, Allen SK, Boschung J, Nauels A, Xia Y, Bex V, Midgley PM (eds)). Cambridge University Press, Cambridge, United Kingdom and New York, NY, USA, 1535pp

Bocka M, Scheck-Wenderotha M, GeoEn Group (2013) Research on utilization of geo-energy. Energy Proc 40:249–255

Bolden I, Seroy S, Roberts E, Schmeisser L, Koehn J, Rilometo C, Odango E, Barros C, Sachs J, Klinger T (2018) Climate-related community knowledge networks as a tool to increase learning in the context of environmental change. Clim Risk Manag. https://doi.org/10.1016/j.crm.2018.04.004

Broeckx J, Vanmaercke M, Duchateaua R, Poesen J (2018) A data-based landslide susceptibility map of Africa. Earth Sci Rev 185:102–121

Brundiers K (2018) Educating for post-disaster sustainability efforts. Int J Disaster Risk Reduct Mar Policy 27:406–414

---

[8]http://portal.onegeology.org/OnegeologyGlobal/.

CCOP (2018) Annual report 2017. Coordinating committee for geoscience programmes in East and Southeast Asia (CCOP) Technical Secretariat: Bangkok, 60p

Charalampidou E, Garcia S, Buckman J, Cordoba P, Lewis H, Maroto-Valer M (2017) Impact of CO_2-induced geochemical reactions on the mechanical integrity of carbonate rocks. Energy Proc 114:3150–3156

Cigna F, Tapete D, Lee K (2018) Geological hazards in the UNESCO World Heritage sites of the UK: from the global to the local scale perspective. Earth Sci Rev 176:166–194

Cook P (2017) CCS research development and deployment in a clean energy future: lessons from Australia over the past two decades. Engineering 3(4):477–484

Cooper M (2018) Governing the global climate commons: The political economy of state and local action, after the U.S. flip-flop on the Paris Agreement. Energy Policy 118:440–454

Cunha NS, Magalh MR, Domingos T, Abreu MM, Küpfer C (2017) The land morphology approach to flood risk mapping: an application to Portugal. J Environ Manag 193:172–187

Djalante R (2019) Key assessments from the IPCC special report on global warming of 1.5 °C and the implications for the Sendai framework for disaster risk reduction. Prog Disaster Sci 1. Article 100001

Dolman A, Belward A, Briggs S, Dowell M, Eggleston S, Hill K, Richter C, Simmons A (2016) A post-Paris look at climate observations. Nat Geosci 9(9):646

Feitz AJ, Zhang J, Zhang X, Gurney J (2017) China Australia geological storage of CO^2 (CAGS): summary of CAGS2 and introducing CAGS3. Energy Proc 114:5897–5904

Ferreira O, Viavattene C, Jiménez J, Bolle A, das Neves L, Plomaritis T, McCall R, van Dongeren A (2018) Storm-induced risk assessment: evaluation of two tools at the regional and hotspot scale. Coast Eng 134:241–253

Filho L, Morgan E, Godoy E, Azeiteiro U, Bacelar-Nicolau P, Veiga Ávila L, Mac-Lean C, Hugé J (2018) Implementing climate change research at universities: barriers, potential and actions. J Clean Prod 170:269–277

Fischer A (2018) Pathways of adaptation to external stressors in coastal natural-resource-dependent communities: implications for climate change. World Dev 108:235–248

Forino G, von Meding J, Brewer G, van Niekerk D (2017) Climate change adaptation and disaster risk reduction integration: strategies, policies, and plans in three Australian local governments. Int J Disaster Risk Reduct 24:100–108

Gallo C, Faccilongo N, La Sala P (2018) Clustering analysis of environmental emissions: a study on Kyoto protocol's impact on member countries. J Clean Prod 172:3685–3703

Gao Y, Gao X, Zhang X (2017) The 2 °C global temperature target and the evolution of the long-term goal of addressing climate change—from the United Nations framework convention on climate change to the Paris agreement. Engineering 3(2):272–278

Gariano S, Guzzetti F (2016) Landslides in a changing climate. Earth Sci Rev 162:227–252

Geach M, Stokes M, Hart A (2017) The application of geomorphic indices in terrain analysis for ground engineering practice. Eng Geol 217:122–140

Giuliani G, Nativi S, Obregon A, Beniston M, Lehmann A (2017) Spatially enabling the global framework for climate services: reviewing geospatial solutions to efficiently share and integrate climate data and information. Clim Serv 8:44–58

González-Arquerosa ML, Mendozab ME, Boccob G, Castillo BS (2018) Flood susceptibility in rural settlements in remote zones: The case of a mountainous basin in the Sierra-Costa region of Michoacán, Mexico. J Environ Manag 223:685–693

Ha Y, Teng F (2013) Midway toward the 2 degree target: adequacy and fairness of the Cancún pledges. Appl Energy 112:856–865

Hagenlocher M, Renaud F, Haas S, Sebesvari Z (2018) Vulnerability and risk of deltaic social-ecological systems exposed to multiple hazards. Sci Total Environ 631–632:71–80

Harris D (2017) Telling the story of climate change: geologic imagination, praxis, and policy. Energy Res Soc Sci 31:179–183

Hearn G (2016) Managing road transport in a world of changing climate and land use. Proc Inst Civ Eng—Munic Eng 169(3):146–159

Hess D, Collins B (2018) Climate change and higher education: Assessing factors that affect curriculum requirements. J Clean Prod 170:1451–1458

Hong T, Jeong K, Chaec M, Changyoon (2017) Framework for the analysis of the potential of ground source heat pump system in elementary school facility. Energy Proc 105:1051–1057

Hosseinalizadeh M, Kariminejad N, Rahmati O, Keesstra S, Alinejad M, Behbahani AM (2019) How can statistical and artificial intelligence approaches predict piping erosion susceptibility? Sci Total Environ 646:1554–1566

Huang Y, Zhao L (2018) Review on landslide susceptibility mapping using support vector machines. CATENA 165:520–529

ICSU, International Council for Science (2017) A guide to SDG interactions: from science to implementation (Griggs DJ, Nilsson M, Stevance A, McCollum D (eds)). International Council for Science, Paris, 238pp

IPCC (2012) Managing the risks of extreme events and disasters to advance climate change adaptation. A special report of working groups I and II of the intergovernmental panel on climate change (Field CB, Barros V, Stocker TF, Qin D, Dokken DJ, Ebi KL, Mastrandrea MD, Mach KJ, Plattner G-K, Allen SK, Tignor M, Midgley PM (eds)). Cambridge University Press, Cambridge and New York, 582pp

IPCC (2014) Climate change 2014: synthesis report. In: Contribution of working groups I, II and III to the fifth assessment report of the intergovernmental panel on climate change. IPCC, Geneva, 151pp

IPCC (2018) Summary for Policymakers. In: Masson-Delmotte V, Zhai P, Pörtner H-O, Roberts D, Skea J, Shukla PR, Pirani A, Moufouma-Okia W, Péan C, Pidcock R, Connors S, Matthews JBR, Chen Y, Zhou X, Gomis MI, Lonnoy E, Maycock T, Tignor M, Waterfield T (eds) Global warming of 1.5°C. An IPCC special report on the impacts of global warming of 1.5°C above pre-industrial levels and related global greenhouse gas emission pathways, in the context of strengthening the global response to the threat of climate change, sustainable development, and efforts to eradicate poverty. World Meteorological Organization: Geneva, Switzerland, 32pp

Jakeman AJ, Barreteau O, Hunt RJ, Rinaudo JD, Ross A (eds) (2016) Integrated groundwater management: concepts, approaches and challenges. Springer Open, New York, p 756

Jamaludin U, Yaakob J, Suratman S, Pereira J (2016) Threats faced by groundwater: a preliminary study in Kuala Selangor. Bull Geol Soc Malay 62:61–68

Jones DG, Beaubien SE, Barkwith AKAP, Barlow TS, Bellomo T, Braimant G, Gal F, Graziani S, Joublin F, Lister TR, Lombardi S (2011) Near surface gas monitoring at the Weyburn unit in 2011. British Geological Survey Commissioned Report, CR/12/014, 91pp

Kabuth A, Dahmke A, Beyer C, Bilke L, Dethlefsen F, Dietrich P, Duttmann R, Ebert M, Feeser V, Görke U, Köber R, Rabbel W, Schanz T, Schäfer D, Würdemann H, Bauer S (2017) Energy storage in the geological subsurface: dimensioning, risk analysis and spatial planning: the ANGUS + project. Environ Earth Sci 76:23

Koh F, Pereira J, Aziz S (2013) Platforms of climate change: an evolutionary perspective and lessons for Malaysia. Sains Malays 42(8):1027–1040

Komi K, Neal J, Trigg MA, Diekkrüger B (2017) Modelling of flood hazard extent in data sparse areas: a case study of the Oti River basin, West Africa. J Hydrol: Reg Stud 10:122–132

Kühn M, Kempka T, De Lucia M, Scheck-Wenderoth M (2017) Dissolved CO_2 storage in geological formations with low pressure, low risk and large capacities. Energy Proc 114:4722–4727

Kühn M, Ask M, Juhlin C, Bruckman VJ, Kempka T, Martens S (2016) Interdisciplinary approaches in resource and energy research to tackle the challenges of the future. Energy Proc 97:1–6

Kuriyama A, Abe N (2018) Ex-post assessment of the Kyoto Protocol—quantification of CO 2 mitigation impact in both Annex B and non-Annex B countries-. Appl Energy 220:286–295

Leemans R, Vellinga P (2017) The scientific motivation of the internationally agreed 'well below 2 °C' climate protection target: a historical perspective. Curr Opin Environ Sustain 26–27:134–142

Li L, Switzer A, Wang Y, Chan C, Qiu Q, Weiss R (2018) A modest 0.5-m rise in sea level will double the tsunami hazard in Macau. Sci Adv 4(8):1180

Liobikienė G, Butkus M (2018) The challenges and opportunities of climate change policy under different stages of economic development. Sci Total Environ 642:999–1007

Liu Z, Macpherson G, Groves C, Martin J, Yuan D, Zeng S (2018) Large and active CO 2 uptake by coupled carbonate weathering. Earth Sci Rev 182:42–49

Lovell B (2009) Challenged by carbon: the oil industry and climate change. Cambridge University Press, Cambridge, p 135

Lund J, Boyd T (2016) Direct utilization of geothermal energy 2015 worldwide review. Geothermics 60:66–93

Macintyre T, Lotz-Sisitka H, Wals A, Vogel C, Tassone V (2018) Towards transformative social learning on the path to 1.5 degrees. Curr Opin Environ Sustain 31:80–87

Maes J, Parra C, Mertens K, Bwambale B, Jacobs L, Poesen J, Dewitte O, Vranken L, de Hontheim A, Kabaseke C, Kervyn M (2018) Questioning network governance for disaster risk management: Lessons learnt from landslide risk management in Uganda. Environ Sci Policy 85:163–171

Marriner N, Morhange C, Skrimshire S (2010) Geoscience meets the four horsemen? Tracking the rise of neocatastrophism. Glob Planet Chang 74:43–48

Martens S, Kühn M (2015) Geological Underground will contribute significantly to the implementation of the energy policy towards renewables in Germany. Energy Proc 76:59–66

Meller C, Schill E, Bremer J, Kolditz O, Bleicher A, Benighaus C, Chavot P, Gross M, Pellizzone A, Renn O, Schilling F, Kohl T (2017) Acceptability of geothermal installations: a geoethical concept for GeoLaB. Geothermics 73:133–145

Mengerink K, Van Dover C, Ardron J, Baker M, Escobar-Briones E, Gjerde K, Koslow J, Ramirez-Llodra E, Lara-Lopez A, Squires D, Sutton T, Sweetman A, Levin L (2014) A call for deep-ocean stewardship. Science 344(6185):696–698

Metz B, Davidson O, de Coninck H, Loos M, Meyer L (2005) IPCC special report on carbon dioxide capture and storage. United States, Washington

Mizutori M (2019) From risk to resilience: Pathways for sustainable development. Prog Disaster Sci 2. Article 100011

Nakano G, Shaw R (2018) Education governance and the role of science and technology. Science and technology in disaster risk reduction in Asia. In: Rajib Shaw R, Shiwaku K, Izumi T (eds) Science and technology in disaster risk reduction in Asia: potentials and challenges. Academic, London, pp 175–196

Nguyen TT, Ngo HH, Guo W, Wang XC, Ren N, Li G, Ding J, Liang H (2019) Implementation of a specific urban water management—Sponge City. Sci Total Environ 652:147–162

Nurse L, McLean R, Agard J, Briguglio L, Duvat-Magnan V, Pelesikoti N, Tompkins E, Webb A, Barros V, Field C, Dokken D, Mastrandrea M, Mach K, Bilir T, Chatterjee M, Ebi K, Estrada Y, Genova R, Girma B, Kissel E, Levy A, MacCracken S, Mastrandrea P, White L (2014) Small islands. In: Climate change 2014: impacts, adaptation, and vulnerability. Part B: regional aspects. contribution of working group II to the fifth assessment report of the intergovernmental panel on climate change. Cambridge University Press, Cambridge, pp 1613–1654

Ozarslan A (2012) Large-scale hydrogen energy storage in salt caverns. Int J Hydrogen Energy 37(19):14265–14277

Pereira JJ (2018) Science and technology to enhance disaster resilience in a changing climate. In: Rajib Shaw R, Shiwaku K, Izumi T (eds) Science and technology in disaster risk reduction in Asia: potentials and challenges. Academic, London, pp 31–38

Pereira JJ, Muhamad N, Lim CS, Aziz S, Hunt J (2019) Making cities disaster resilient in a changing climate: the case of Kuala Lumpur, Malaysia. IRDR working paper series, 17pp. https://doi.org/10.24948/2019.05

Perucca LP, Angilieri YE (2011) Morphometric characterization of del Molle Basin applied to the evaluation of flash floods hazard, Iglesia Department, San Juan, Argentina. Quat Int 233:81–86

Peterson D, Halofsky J (2018) Adapting to the effects of climate change on natural resources in the Blue Mountains, USA. Clim Serv 10:63–71

Petterson MG, Tawake AK (2016) Toward inclusive development of the Pacific region using geoscience. Geol Soc Am Spec Pap 520:459–478

Radosavljevic B, Lantuit H, Pollard W, Overduin Couture N, Sachs T, Fritz M (2016) Erosion and flooding—threats to coastal infrastructure in the Arctic: a case study from Herschel Island, Yukon Territory, Canada. Estuaries Coasts 39:900–915

Reichenbach P, Rossia M, Malamud BD, Mihir M, Guzzettia F (2018) A review of statistically-based landslide susceptibility models. Earth Sci Rev 180:60–91

Reis J, Ballinger R (2018) Creating a climate for learning-experiences of educating existing and future decision-makers about climate change. Marine Policy. https://doi.org/10.1016/j.marpol.2018.07.007

Renforth P, Henderson G (2017) Assessing ocean alkalinity for carbon sequestration. Rev Geophys 55 (3):636–674

Robinson S (2018) Climate change adaptation in small island developing states: Insights and lessons from a meta-paradigmatic study. Environ Sci Policy 85:172–181

Rogelj J, Knutti R (2016) Geosciences after Paris. Nat Geosci 9(3):187

Sacuta N, Daly D, Botnen B, Worth K (2017) Communicating about the geological storage of carbon dioxide—comparing public outreach for CO_2 EOR and saline storage projects. Energy Proc 114:7245–7259

Stephenson M (2018) Geological aspects energy and climate change. Elsevier, London, pp 123–146

Stewart I, Gill J (2017) Social geology—integrating sustainability concepts into Earth sciences. Proc Geol Assoc 128(2):165–172

Szulczewski M, MacMinn C, Herzog H, Juanes R (2012) Lifetime of carbon capture and storage as a climate-change mitigation technology. Proc Natl Acad Sci 109 (14):5185–5189

Teklu T (2018) Should Ethiopia and least developed countries exit from the Paris climate accord?—geopolitical, development, and energy policy perspectives. Energy Policy 120:402–417

Teng J, Jakeman AJ, Vaze J, Croke BFW, Dutta D, Kim S (2017) Flood inundation modelling: a review of methods, recent advances and uncertainty analysis. Environ Model Softw 90:201–216

Travis W, Smith J, Yohe G (2018) Moving toward 1.5°C of warming: implications for climate adaptation strategies. Curr Opin Environ Sustain 31:146–152

UNFCCC (2016) Adoption of the Paris agreement. www.unfccc.int/resource/docs/2015/cop21/eng/10a01.pdf. Accessed 27 Oct 2019

United Nations (2016) Report of the open-ended intergovernmental expert working group on indicators and terminology relating to disaster risk reduction. New York: United Nations General Assembly (A/71/644). http://www.preventionweb.net/drr-framework/open-ended-working-group

van Westen CJ, Castellanos E, Kuriakose SL (2008) Spatial data for landslide susceptibility, hazard, and vulnerability assessment: an overview. Eng Geol 102:112–131

Wang B, Bauer S (2017) Compressed air energy storage in porous formations: a feasibility and deliverability study. Pet Geosci 23(3):306–314

Wei R, Zeng Q, Davies T, Yuan G, Wang K, Xue X, Yin Q (2018) Geohazard cascade and mechanism of large debris flows in Tianmo gully, SE Tibetan Plateau and implications to hazard monitoring. Eng Geol 233:172–182

Wignall R, Gordon J, Brazier V, MacFadyen C, Everett N (2018) A climate change risk-based assessment for nationally and internationally important geoheritage sites in Scotland including all Earth science features in Sites of Special Scientific Interest (SSSI). Scottish Natural Heritage Research Report No. 1014

Yahaya N, Lim C, Taha M, Pereira J (2016) Exposure of municipal solid waste disposal sites to climate related geohazards: case study of Selangor. Bull Geol Soc Malays 62:53–59

Yu H, Zhu S (2015) Toward Paris: China and climate change negotiations. Adv Clim Chang Res 6(1):56–66

Zarrouk S (2017) Postgraduate geothermal energy education worldwide and the New Zealand experience. Geothermics 70:173–180

Zhang X, Chen N, Chen Z, Wu L, Li X, Zhang L, Di L, Gong J, Li D (2018) Geospatial sensor web: a cyber-physical infrastructure for geoscience research and application. Earth Sci Rev 185:684–703

T. F. Ng is registered with the Board of Geologists Malaysia and has expertise in engineering geology, structural geology and geographical information systems. He has served as Head of the Department of Geology and is former Editor-in-Chief of the Bulletin of the Geological Society of Malaysia. He had led several research projects on integrated risk assessment of geohazards, characterisation of geotechnical engineering properties of peat, and evaluation of climatic hazards for local level adaptation response.

Joy Jacqueline Pereira is a Principal Research Fellow and Professor at Universiti Kebangsaan Malaysia's Southeast Asia Disaster Prevention Research Initiative and Fellow of the Academy of Sciences Malaysia. She is Vice-Chair of the Intergovernmental Panel on Climate Change (IPCC) Working Group 2 on Impacts, Adaptation and Vulnerability and served as a Review Editor of the Special Report on Global Warming of 1.5 °C. Joy is a Board Member of the UN Office for Disaster Risk Reduction (UNDRR) Asia Pacific Science, Technology and Academic Group and co-founder of the Asian Network on Climate Science and Technology.

Julian Hunt Baron Hunt of Chesterton, is a Senior Fellow at Trinity College and Honorary Professor at Department of Applied Mathematics and Theoretical Physics (DAMTP) at the University of Cambridge. Lord Hunt is also Professor of Climate Modelling in the Department of Space and Climate Physics and Department of Earth Sciences at University College London. He is an internationally-renowned scientist and a Fellow of the Royal Society, with over 150 peer-reviewed publications. Lord Hunt is a former Chief Executive of the UK Met Office and co-founder of the Asian Network on Climate Science and Technology.

Conserve and Sustainably Use the Oceans, Seas, and Marine Resources

14

Michael G. Petterson, Hyeon-Ju Kim, and Joel C. Gill

M. G. Petterson (✉)
School of Science, Auckland University of
Technology, St Pauls Street, Auckland
New Zealand
e-mail: michael.petterson@aut.ac.nz

H.-J. Kim
Seawater Energy Plant Research Center, Marine
Renewable Energy Research Division,
Korea Research Institute of Ships and Ocean
Engineering, 32, 1312 Beon-gil Yuseong-daero
Yuseong-gu, Daejeon 305-343, South Korea

J. C. Gill
British Geological Survey, Nicker Hill,
Keyworth, Nottingham NG12 5GG, UK

Geology for Global Development,
Loughborough, UK

© Springer Nature Switzerland AG 2021
J. C. Gill and M. Smith (eds.), *Geosciences and the Sustainable Development Goals*,
Sustainable Development Goals Series, https://doi.org/10.1007/978-3-030-38815-7_14

Abstract

14 LIFE BELOW WATER

Healthy Oceans:

Host biodiversity, which helps to reduce pollution

Absorb carbon dioxide

Provide ecosystem to services which support the global economy

Provide livelihoods to billions of people living in coastal communities

Provide an essential source of food and nutrition

   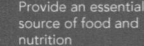  

Protecting Marine Environments:

Geoscientists can help to assess, trace, and reduce marine pollution and its impacts

Climate change is contributing to sea-level rise, resulting in flooding, erosion, and groundwater contamination

Some CO_2 released by human activity is dissolved into the oceans, rivers and lakes, resulting in acidification

Marine resources provide potential economic and development benefits

Marine parks help to protect important biological and geological diversity

Geoscientists can advise on the benefits and risks of extracting seabed mineral deposits

Geoscientists can help leverage the renewable energy potential of ocean resources

Geoscientists inform coastal infrastructure development, critical for trade and tourism

Improving ocean health and supporting development requires:

Increased scientific knowledge and research capacity

Science instituitions and programmes that facilitate learning across countries and disciplines

Increased sharing of marine technologies and data.

  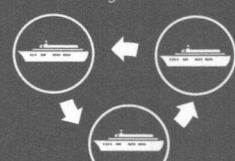

14.1 Introduction

Covering 71% of Earth's surface, and making up 97% of the water on Earth, the sheer scale and size of oceans on Earth underline their importance to the evolution of our planet. Oceans probably formed on Earth as soon as its surface was cool enough for liquid water to exist. In geological terms, modern oceans are 'born' and subducted within 200–300 million year lifecycles. The world's oceans, seas and coastal areas are critical to sustainable development, helping to advance social and economic development (Halpern et al. 2012). For example:

- Oceans are complex ecosystems hosting significant biodiversity, with careful management needed to reduce pollution of all kinds. This biodiversity is not surprising given the geological and geomorphological diversity shown in ocean seascapes. This includes trenches with depths of almost 11 km, around 80,000 km of mid-ocean ridges, mountains, seamounts, ocean islands, and ocean plateaus, island arc systems, and voluminous abyssal plains some 4 km beneath sea level.
- Oceans play a critical role in climate, absorbing heat and carbon dioxide from the atmosphere. Approximately one-third of the carbon dioxide released into the atmosphere by human activity has been dissolved into oceans (Sabine et al. 2004), contributing to the problem of ocean acidification.
- Coastal and marine environments (Fig. 14.1) add approximately US$2.5 trillion to the global economy each year, when considering tangible outputs including fishing, shipping traffic, and carbon absorption (Hoegh-Guldberg 2015), estimates that the overall ecosystem services provided by coastal and marine resources contribute US$28 trillion to the global economy each year (United Nations 2016).
- Oceans provide livelihoods and opportunities for billions in coastal communities. The UN estimates that over three billion people depend on marine and coastal resources for their livelihoods, including fishing and tourism (United Nations 2019a, b).
- Oceans provide food, with total fish production in 2016 (including both inland and marine sources) of 171 million tonnes, with 88% of this for direct human consumption (FAO 2018). 63% of this 171 million tonnes are from marine sources.
- Access to oceans is needed to extract a range of energy and mineral resources including oil, gas, gas hydrates, aggregates and construction materials, and seabed minerals (including rare earth elements, gold, copper, and silver).

Delivering **SDG 14** is therefore necessary if we are to deliver many other **SDGs**, including reducing the vulnerability of the world's poorest communities (**SDG 1.5**), reducing hunger through access to nutritious proteins (**SDGs 2.1** and **2.3**), promoting sustained, inclusive and sustainable economic growth and employment (**SDG 8**), and strengthening resilience to climate-related hazards (**SDG 13.1**). Protecting oceans from degradation is therefore of primary importance, as is the responsible harvesting of living and non-living resources. Sea-level rise threatens entire coastal communities, and even countries, increasing their vulnerability to flooding (Wahl 2017). Ocean acidification, due to increased absorption of carbon dioxide in seawater, is making it harder for ocean species (e.g., oysters, and corals) to develop carbonate shells and structures (NOAA 2019a), and therefore poses a threat to both jobs and food security. Warmer oceans are also associated with reductions in some fish stocks. Sea-surface temperatures have risen by 0.7 °C over the last 100–130 years, with this projected to increase to 1.2–3.2 °C by 2100 due to greenhouse gas emissions. This may change the geographic regions where aquaculture of some species is viable, and lead to a decline in cold-water fish species.

Table 14.1 outlines seven targets and three means of implementation relating to **SDG 14**. These include ambitions to address marine pollution, restore marine environments, and improve the management of ocean resources. This chapter

Fig. 14.1 Indian Ocean (Maldives). Key economic sectors in the Maldives rely on access to coastal and marine environments, including fisheries, tourism, shipping, and boat building. Image by David Mark from Pixabay

Table 14.1 SDG 14 targets and means of implementation

Target	Description of target (14.1 to 14.7) or means of implementation (14.A to 14.C)
14.1	By 2025, prevent and significantly reduce marine pollution of all kinds, in particular from land-based activities, including marine debris and nutrient pollution
14.2	By 2020, sustainably manage and protect marine and coastal ecosystems to avoid significant adverse impacts, including by strengthening their resilience, and take action for their restoration in order to achieve healthy and productive oceans
14.3	Minimize and address the impacts of ocean acidification, including through enhanced scientific cooperation at all levels
14.4	By 2020, effectively regulate harvesting and end overfishing, illegal, unreported and unregulated fishing and destructive fishing practices and implement science-based management plans, in order to restore fish stocks in the shortest time feasible, at least to levels that can produce maximum sustainable yield as determined by their biological characteristics
14.5	By 2020, conserve at least 10 per cent of coastal and marine areas, consistent with national and international law and based on the best available scientific information
14.6	By 2020, prohibit certain forms of fisheries subsidies, which contribute to overcapacity and overfishing, eliminate subsidies that contribute to illegal, unreported and unregulated fishing and refrain from introducing new such subsidies, recognizing that appropriate and effective special and differential treatment for developing and least developed countries should be an integral part of the World Trade Organization fisheries subsidies negotiation
14.7	By 2030, increase the economic benefits to Small Island developing States and least developed countries from the sustainable use of marine resources, including through sustainable management of fisheries, aquaculture and tourism
14.A	Increase scientific knowledge, develop research capacity and transfer marine technology, taking into account the Intergovernmental Oceanographic Commission Criteria and Guidelines on the Transfer of Marine Technology, in order to improve ocean health and to enhance the contribution of marine biodiversity to the development of developing countries, in particular small island developing States and least developed countries
14.B	Provide access for small-scale artisanal fishers to marine resources and markets
14.C	Enhance the conservation and sustainable use of oceans and their resources by implementing international law as reflected in UNCLOS, which provides the legal framework for the conservation and sustainable use of oceans and their resources, as recalled in paragraph 158 of The Future We Want

examines how geoscientists can help to deliver these targets, with an emphasis on *Small-Island Developing States* (commonly abbreviated to SIDS) in the Pacific. SIDS have significant and challenging development issues (e.g., resilience to climate change and natural hazards) and opportunities (e.g., renewable energy and sustainable tourism). They give us a valuable perspective on the health of the oceans and coastal regions, with opportunities for geoscientists to contribute to ocean conservation and restoration. The lessons we present here, however, have broader relevance. Coastal areas in countries such as the United Kingdom are associated with deprivation due to a decline in industries such as tourism (House of Lords 2019), and in other contexts, easy access to the ocean is a key driver of urban development with associated pollution challenges. The future of the oceans therefore has relevance for communities across the globe.

In this chapter, we first describe diverse coastal communities, but with a particular emphasis on Small Island Developing States (or SIDS, Table 14.2), their distribution, development challenges, and the geological processes that determine their existence (Sect. 14.2). We proceed to explore the role of geoscientists in protecting such marine environments (Sect. 14.3), highlighting sources of pollution with a geological origin and geoscience activities to identify, monitor, and reduce potential pollution to marine environments. We then explore the diverse and valuable ocean resources that drive sustainable development (Sect. 14.4). The final section synthesises key lessons to increase scientific knowledge, develop research capacity, and transfer marine technology for ocean management (Sect. 14.5).

14.2 Coastal Environments, Small Island Developing States, and Sustainable Development

Coastal communities take many forms, with both general and location-specific sustainability challenges. Communities include rural areas with low population density, agriculture and small-scale fishing, and urban areas (including megacities) with high population density, busy ports, and heavy industry. Coastal communities may form a small part of one nation (e.g., approximately 3 million out of 60 million people (5%) live on the coast of the United Kingdom), or the entire population of a nation. Small Island Developing States, or SIDS, consist of one or more small islands, or island groups, forming archipelago countries or overseas territories. For example, the Solomon Islands include more than 900 islands, and Vanuatu consists of at least 82 islands. While each individual island typically has a small land area, the total land and marine territory attributed to any individual nation can be large.

The Pacific islands region, with 22 countries and territories, covers an area approximately the same size as Africa, around 300 million km². Papua New Guinea is the only 'Pacific Island' to have a large land area, being around the same size

Table 14.2 Small Island developing states (both countries and territories)

Region	Country or territory
Caribbean	Anguilla, Antigua and Barbuda, Aruba, Bahamas, Barbados, Belize, British Virgin Islands, Cuba, Dominica, Dominican Republic, Grenada, Guyana, Haiti, Jamaica, Montserrat, Netherlands Antilles, Puerto Rico, Saint Kitts and Nevis, Saint Lucia, Saint Vincent and the Grenadines, Suriname, Trinidad and Tobago, and the United States Virgin Islands
Pacific	American Samoa, Cook Islands, Federated States of Micronesia, Fiji, French Polynesia, Guam, Kiribati, Marshall Islands, Nauru, New Caledonia, Niue, Northern Mariana Islands, Palau, Papua New Guinea, Pitcairn, Samoa, Solomon Islands, Timor-Leste, Tonga, Tuvalu, Vanuatu, Wallis & Futuna
Atlantic, Indian, Mediterranean Oceans, and the South China Sea	Bahrain, Cape Verde, Comoros, Guinea-Bissau, Maldives, Mauritius, São Tomé and Príncipe, Seychelles, Singapore

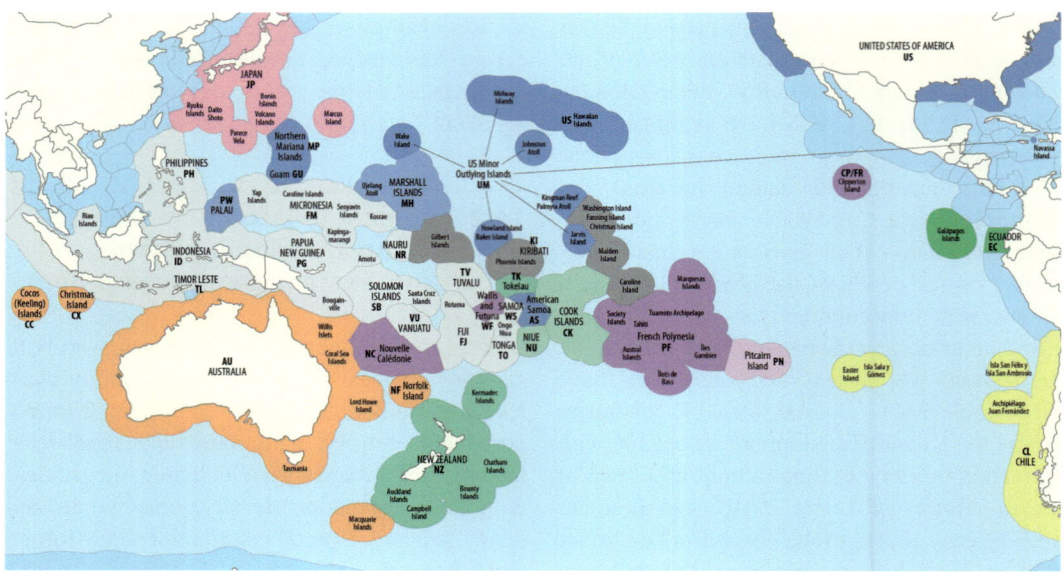

Fig. 14.2 Exclusive Economic Zones in the Pacific Ocean. *Credit* Maximilian Dörrbecker, CC BY-SA 2.5 (https://creativecommons.org/licenses/by-sa/2.5/)

as France. The remaining countries have a cumulative land area of approximately 85,000 km², or the same size as Austria, within an ocean area close to the size of Africa. They are also, therefore, *large ocean states*, some with Exclusive Economic Zones (areas of sea where a state has special rights regarding marine resources, illustrated in Fig. 14.2) of enormous dimensions. A classic example is Cook Islands with around 200 km² of land and 2 million km² of the ocean (Petterson and Tawake 2016, 2018). Papua New Guinea has a population of approximately 8.5 million people, with other Pacific SIDS having populations between approximately 900,000 (Fiji) and 10,000 (Tuvalu).

In geological terms, SIDS include ocean structures such as island arcs, atolls, raised or obducted parts of the oceanic crust that are currently situated above sea level, and rifted pieces of continental crust. Oceanic island arcs form from the subduction of oceanic crust beneath other oceanic crust, whilst seamounts and ocean plateaus form from mantle plumes. SIDS are given special attention in development discussions because of a set of shared characteristics including small land area, remote geographic locations,

limited economic development opportunities, and enhanced vulnerability to disaster and risk. SIDS also create some unique opportunities for sustainable development. Issues of relevance include

- *Energy*. A lack of reliable and sufficient power has constrained development within many SIDS. Many SIDS have a large dependency on (mostly imported) oil for a relatively small amount of electricity generation. In some Pacific island small states (e.g., Papua New Guinea, Vanuatu, and Solomon Islands), large proportions of the population do not have access to electricity, with the exception of local diesel generators or solar energy. Access to solar-powered batteries is rapidly increasing, and this is mitigating energy-poverty for some. Increasing use of renewable technologies, including new ocean energy technologies, could bring greatly improved, reliable, and increased output to provide a renewed basis for future economic and social development. Cheaper and more reliable energy production could help to achieve additional SDGs including improving food and water security, and enhanced employment opportunities.

- *Water and Sanitation.* Many SIDS, particularly small atoll-SIDS, can rely on fragile freshwater lenses that sit on top of the denser saline ocean, rainwater harvesting, and expensive desalination that requires large amounts of energy. While the mortality rate attributed to unsafe water, unsafe sanitation, and lack of hygiene (per 100,000 population in 2016, with data from World Bank, 2019a) for Pacific island small states (5.2) is significantly below that of the Least Developed Countries (34.3) and even the world average (11.8), it is still more than 17 times higher than the European Union (0.3). Atoll small islands may experience extended times of drought, which bring severe water stress. Tuvalu, for example, experienced a La Niña-influenced drought, lasting five months in 2011. This required water to be brought by ship from Australia, New Zealand, and elsewhere.

- *Food Security.* In the Caribbean, 1 in 5 people is undernourished, compared to less than 1 in 20 people in neighbouring Latin America (United Nations 2015). The prevalence of undernourishment (% of the population in 2017, with data from World Bank 2019b) for Pacific island small states (5.6%) is significantly below that of the Least Developed Countries (23%) and even the world average (11%), but more than twice that of the European Union and North America (2.5%). Some islands within SIDS, and particularly within atolls have limited soil and land resources. Their soils are sandy, lack organic material, and are saline, reducing the potential for diverse agriculture.

- *Economic Growth and Livelihoods.* SIDS are typically small, remote, have a narrow resource and export base, and have a high exposure to external economic shocks. Given their large exclusive economic zones, there may be new opportunities and livelihoods through seabed mining. This would need to be managed in such a way so as to minimise environmental damage and maximise wealth return for the national good. The creation of marine parks and reserves, with geoscience as an integral part of their design to take account of issues such as the geology of the ocean floor, and the range of ocean landscapes could also provide economic, social and environmental benefits.

- *Environmental Protection.* SIDS have a high exposure to global environmental challenges including climate change and natural hazards. SIDS crystallise environmental change issues in an acute manner, often capturing global attention because of their susceptibility to climate change impacts, including rising sea levels. The small size of many SIDS necessarily leads to high percentages of island populations and infrastructure being impacted by a single disaster event. Cyclone Pam in 2015, for example, had an economic impact in Vanuatu equivalent to 64% of the value of the national economy (ILO 2015). Many SIDS are atolls, which are low lying islands, only a metre or several metres above sea level. They are largely formed of sand and broken coral limestone clasts of varying sizes, that are deposited, and cemented, upon a slowly subsiding ocean island volcano. In plan view they are circular to sub-circular in shape reflecting the geomorphology of the uppermost subsiding volcanic edifice. Atolls enclose an inner shallow lagoon on one side but drop off steeply to open ocean on their outer side. Atoll islands tend to be long and narrow and can form a number of widespread discrete archipelago units within one country or territory. A sea-level rise of a few metres over a short space of time could lead to the inundation of these islands, threatening their existence and entire (unique) cultures. Even relatively small sea-level rises increase the impact of the largest spring tides (so-called 'King Tides') on atoll islands. Pacific atolls have a combined population of approximately 200,000 people. The Maldives, the world's most populous atoll country in the Indian Ocean, has a population of approximately 436,000. When considering a global population of 7.6 billion, atoll peoples thus represent distinct cultural centres.

We return to some of these themes later in this chapter, exploring how geoscience understanding

Fig. 14.3 A typical atoll: Gilbert islands, Kiribati. Note the annular island geometry, the low-lying nature of the islands and the thin nature of most islands (© Michael G. Petterson)

and engagement can provide solutions to challenges and maximise the benefits of potential opportunities.

Box 14.1. History of Settlement in Pacific Atolls

For around 4000 years, Micronesia (the atoll islands of the Pacific, Nauru, Marshall Islands, Federated States of Micronesia, Kiribati, and Tokelau) and Polynesia (Samoa, Tonga, Tuvalu, and Cook Islands) have been populated (Irwin 1998; Dickinson 2009). Humans have survived and colonised a widely dispersed series of sand- and coral-dominated islands and islets across millions of square kilometres of theocean. For the great majority of this history, humans have existed on fragile freshwater lenses that form as a lower density layer, sitting on top of ocean water, within atolls, supplemented by rainwater capture. They have eaten a diet largely comprising seafood, added to from sparse land agricultural produce such as coconuts, breadfruit, and slow-growing, salt-resistant taro. Materials for shelter and sailing were entirely supplied from local bush materials.

Perhaps more than any other people, atoll people are ocean people. The history of the settlement of the many scattered islands and archipelagos is a testament to the close oceanic affinity. Countless ocean-going

journeys into the unknown involving thousands of indigenous peoples from areas such as Taiwan, China, Malaysia, and India resulted in the Pacific islands such as Papua New Guinea and Solomon Islands becoming occupied from approximately 50,000 years ago, and the Micronesian and Polynesian islands from c. 4000 years ago (Irwin 1998).

The people of the Pacific are thus rooted in the ocean. Their history and various lifestyles have adapted to, and been influenced by, the ocean. This is best appreciated standing on a low-level atoll, less than a metre above mean sea level, watching life around as the tide rises and falls. The atoll island may be only tens or hundreds of metres wide. It is often hard to define where the ocean ends and land begins with the rising and falling tides.

14.3   Protecting Marine Environments

Fundamental to the ambitions of **SDG 14** is protecting marine environments. For example, Target 14.1 focuses on reducing marine pollution, 14.2 on protecting marine and coastal ecosystems to achieve healthy and productive oceans, and 14.3 on addressing the impacts of ocean acidification. Coastal and marine environments are affected by a range of environmental stresses, and understanding these is critical to ensuring a sustainable future for oceans. This requires interdisciplinary marine science (Government Office for Science 2018), including those with geological expertise. In this section, we set out three broad environmental challenges facing the oceans, and the role of geoscientists in understanding and managing these: marine pollution (Sect. 14.3.1), sea-level rise (Sect. 14.3.2), and ocean acidification (Sect. 14.3.3).

14.3.1 Marine Pollution and Sustainable Development

Beiras (2018, p. 3) note marine pollution to be *'the introduction of substances or energy from humans into the marine environment, resulting in such deleterious effects as harm to living resources, hazards to human health, hindrance to marine activities, including fishing, impairment of quality for use of seawater, and reduction of amenities'*. Marine pollution includes a wide range of substances from human activities including toxic chemicals, pharmaceutical products, metals, gases, solid wastes, plastics, increased nutrients (e.g., from agricultural runoff into the ocean), sewage, ocean ship discharges, oil spills, and fishing nets. Of these, oil pollution is an example of a substance with an extractive industrial origin, and plastic is a tracer for how widespread and pervasive human pollution within the ocean environment has become.

Global oil production, in 2018, was around 2.2 million barrels per day, or 4700 million tonnes equivalent annually, up from 4000 million tonnes in 1993 (BP, 2019). Of this, around one-third is extracted from strata beneath the ocean and coastal waters. Just under two-thirds of all oil produced is transported by sea. Oil spills have numerous causes, from leaking ships, oil tanker accidents, to oil platform blowouts. One of the worst incidents of modern times was the BP-owned Deepwater Horizon blowout incident that occurred on 20 April 2010, in the Gulf of Mexico, with a total discharge of 4.9 million barrels of oil. The incident killed eleven oil platform workers and produced an oil spill that affected 180,000 km^2 of ocean, or a similar size to Oklahoma or Cambodia, as well as thousands of kilometres of coastline. Managing this environmental catastrophe involved dispersal, containment, and removal activities, involving 47,000 people and 7000 vessels (Liu et al. 2011). The impacts of oil pollution are numerous and particularly distressing to marine and coastal wildlife. For example, oil can destroy the insulating ability of fur-bearing mammals, the water repellence of birds' feathers, and be ingested by shellfish, fish, cetaceans, birds, and other marine wildlife (NOAA 2019b).

Considering the huge volumes of oil that are extracted from, or transported, through the oceans, there are relatively few large-scale oil-related environmental pollution incidents. Roser (2019) notes a decreasing number of oil spills from tankers between 1976 and 2016, averaging 1.7 large (>700 tonnes of oil) spills in the 2010s. There is no room for complacency, however, given the significant impact even small spills can have on ecosystems.

Another pollutant, plastics, are a key indicator of the *Anthropocene*, a proposed epoch where the activities of humans are the dominating influence on Earth's systems (see Waters et al. 2016). Plastics are particularly useful 'tracer' materials for the global impact and reach of human industrialisation. Geyer et al. (2017) estimate that, historically, 8300 million metric tonnes (Mt) of plastic were produced up to 2015, of which 79% have been accumulated in landfills or the natural environment, with every indication that these figures are set to increase. Plastic production has exponentially increased from the 1950s and shows few signs of slowing down. The sheer scale and extent of plastic pollution, from the deepest oceans, to the ocean surface, and many terrestrial environments is a particularly sobering and instructive story of the human impact on the planet, including oceans and marine life (Fig. 14.4).

Ocean gyres are large areas where ocean currents concentrate plastic, either at or close to the ocean surface. All oceans have examples of these plastic-concentration zones. The characteristic that makes plastics so useful (their durability and resistance to being chemically altered)

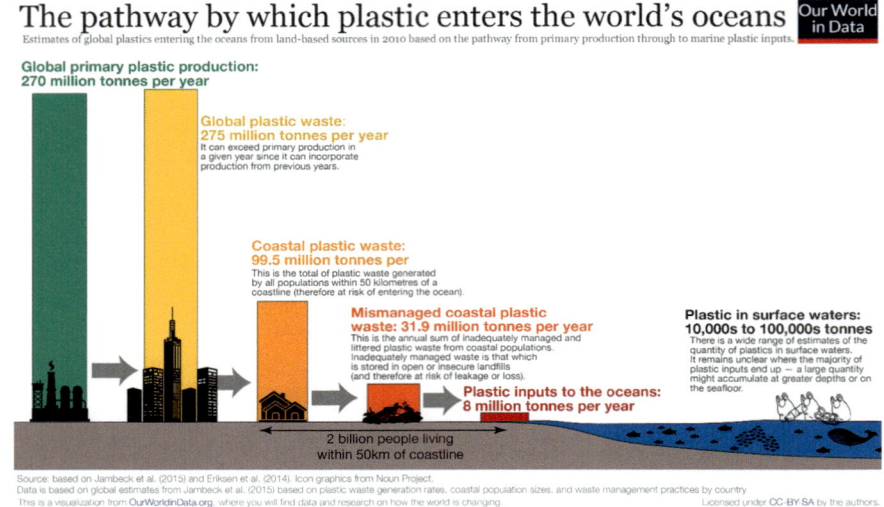

Fig. 14.4 **Cartoon illustration of plastic production and resultant waste streams, and potential paths of a range of plastic waste streams**. The oceans, being the largest basins on Earth and covering 71% of the Earth's surface, have a high probability of becoming the final repository of plastic waste. Image by Ritchie and Roser (2018), using data from Jambeck et al. (2015) and Eriksen et al. (2014). Reproduced under a CC-BY-SA Licence (https://creativecommons.org/licenses/by-sa/2.0/)

makes them particularly difficult within the ocean environment. Ocean processes, such as storms, wave action, hydration, and surface exposure to the atmosphere and ultraviolet radiation, tend to break plastic particles into ever-decreasing sizes, rather than organically digest plastic substances. Eriksen et al. (2014) estimated the total number and weight of plastic particles from 24 expeditions between 2007 and 2013, across all five subtropical ocean gyre plastic concentration zones. Their calculations estimated a minimum value of 5.25 trillion particles weighing almost 300,000 tonnes.

Eriksen et al. (2014) also hypothesised that ocean processes remove size fractions less than 4.75μm from the ocean surface. Microplastics on the ocean floor are forming modern geological sedimentary strata and will be subject to all the usual tectonics of the ocean: eventually being subducted into the mantle and released into the deep mantle earth system. Woodall et al. (2017) showed that microplastics were up to four orders of magnitude more abundant in deep-sea sediments from the Atlantic and Indian Oceans than in contaminated surface waters. They argued that this high quantity of plastic in deep-sea

sediments accounts for the volumes of 'missing plastic' identified from mass balance calculations concerning the known amount of plastic entering the ocean, compared with measured quantities of plastic at the ocean surface.

Through links between the geosphere, hydrosphere, biosphere, and anthroposphere, plastic is becoming part of a complex web of biogeochemical cycles, being ingested by living organisms from bacteria to cetaceans. Wilcox et al. (2015) performed a spatial risk analysis for 186 seabird species worldwide to model exposure to plastic debris. They estimated that up to 90% of seabirds currently digest plastic in some form, increasing to 99% by 2050. As humans, we are also likely consuming plastic, or pollutants from plastics, through digestion and breathing.

Plastics are one of our largest environmental existential challenges, with a solution to the management of waste plastic and its environmental impacts appearing to be remote at the time of writing. The ultimate solution to the plastic pollution challenge will be through reducing or eliminating its use, particularly single or limited use plastics, together with improved management of plastic waste and a surface ocean clean-up.

14.3.2   Changing Sea Levels and Sustainable Development

Global sea levels have varied significantly in geological time. Key controls with respect to global sea levels include the presence or absence of large ice sheets (linked to changing climates) and the volume of mid-ocean ridge activity which displaces large quantities of ocean. The geological record (e.g., ancient shoreline features, microfossils in sediment cores) can be used to plot how sea levels have changed in the past, demonstrating that sea level can change rapidly as ice melts or forms on the continents (USGS 1995). One of the most widely quoted diagrams that estimates sea-level variation over the last 540 million years comes from Hallam et al. (1992) (Fig. 14.5a). Sea levels can vary from around 400 m higher than the present day

(during the Ordovician) to around 120 m lower than the present day (at the peak of the last ice age—20,000 years ago). Sea-level rises since the Industrial Revolution are attributed to anthropogenically induced climate change (see **SDG 13**), with a rise in sea levels of over 80 mm between 1993 and 2019 (Fig. 14.5b). Global warming results in two processes contributing to sea-level rise: (i) added water from melting ice, and (ii) thermal expansion of seawater (NASA 2019). Sea-level rise can result in the inundation of land, increase the risk of flooding and coastal erosion, and contamination of groundwater supplies through the landward migration of the interface between saltwater and freshwater.

Variation in Sea-Level Rise

Current estimates suggest sea levels will rise between 0.25 and 1 m by 2100, depending on the extent of carbon emissions. Sea-level rises of

Fig. 14.5 (top, a) Sea-level changes during the Phanerozoic Period, showing a 500 m total variation in sea levels (from Robert, A. Rhode, Global Warming Art, reproduced under a CC-BY-SA 3.0 licence, https:// creativecommons.org/ licenses/by-sa/3.0/) and **(bottom, b) sea-level variations 1993–2019 (mm)** from NASA satellite observations, NASA Goddard Space Flight Center

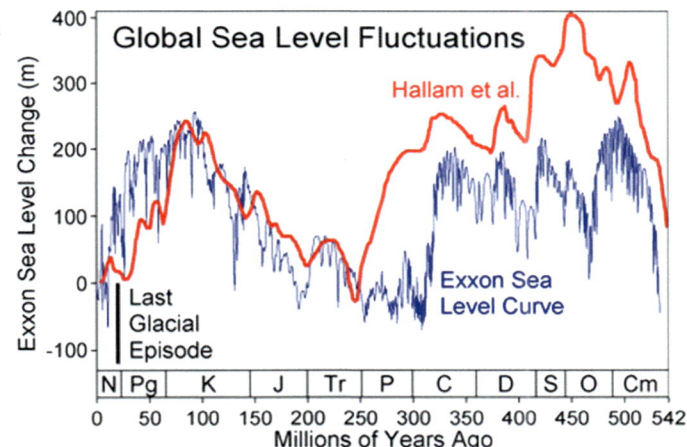

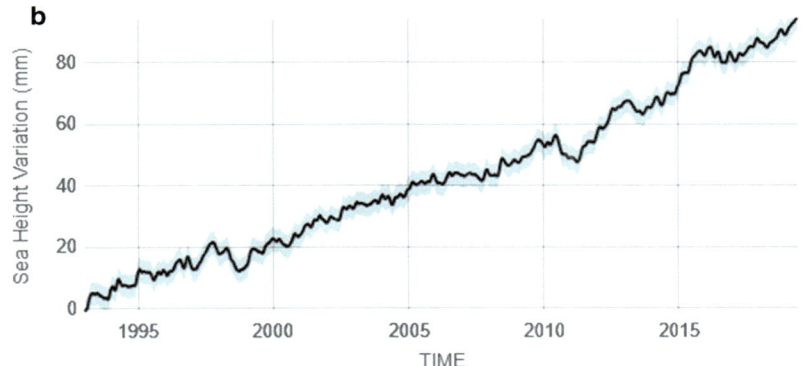

0.25 m, 0.5 m, or 1 m are, however, global averages, with much lower and higher values in some regions. Daigle and Gramling (2018) synthesise five reasons why this variation exists:

- *Expanding seawater.* Thermal expansion of seawater contributes to sea-level rise. Spatial variations in temperature changes will therefore affect thermal expansion and sea-level rise.
- *Glacial rebound.* Isostatic rebound following historical glaciations results in some land masses rising and some sinking, and therefore relative sea-level rise can differ from one place to another.
- *Sinking land.* Tectonic activity and subsidence can result in changes to land surface levels. The latter may be due to natural or anthropogenic processes. For example, 25 years of groundwater abstraction in the Mekong delta (Vietnam) has resulted in subsidence of ∼ 18 cm (Minderhoud et al. 2017).
- *Earth rotation.* The Coriolis effect can cause variation in water height, with higher water levels in some regions and lower water levels in others.
- *Melting ice sheets.* This results in a weakening of the gravitational pull from glaciers on nearby waters and therefore a drop in water levels near the glacier.

This combination of factors set out in Daigle and Gramling (2018) results in complex regional variations in sea-level rise, and a need for modelling to determine how any given place may be affected.

Sea-Level Rise and Atolls

Atolls are a key indicator of sea-level rise. Becker et al. (2012) provide a comprehensive analysis of observed sea-level change between 1950 and 2009 in the West Pacific region. Tide gauge measurements reveal an average sea-level rise of ∼ 1.7 mm/year and satellite altimetry data suggest ∼ 3.3 mm/year. In some regions, the sea-level rise was around three times more than the global average. Becker et al. (2012) also use GPS

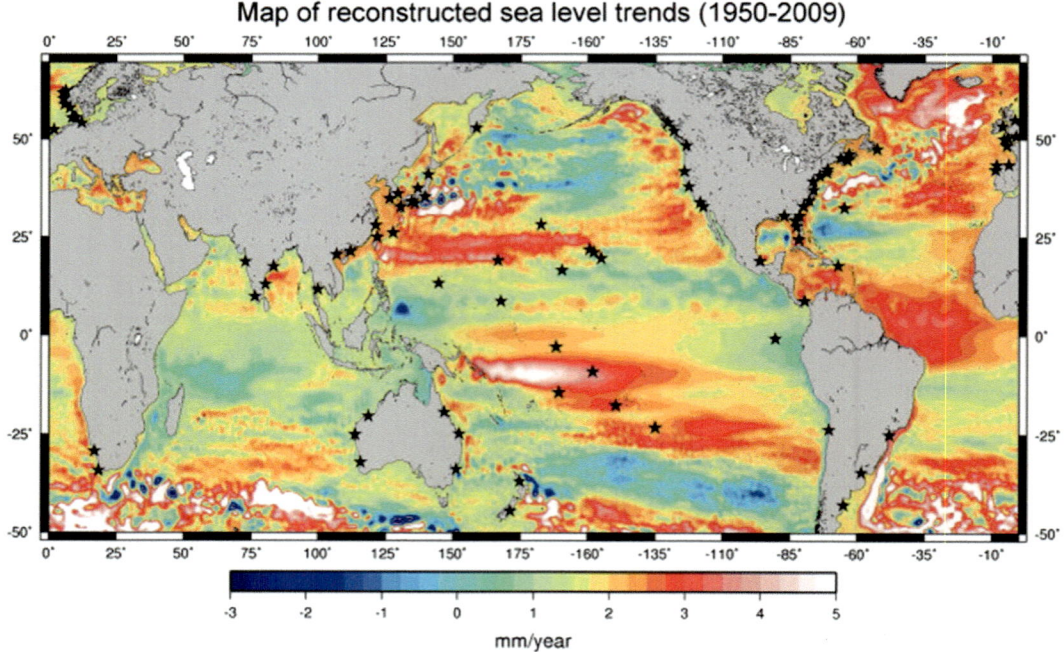

Fig. 14.6 Sea-level trends during 1950–2009 (mm/year) from DRAKKAR-based reconstruction of sea level (uniform trend of 1.8 mm/year included). Reprinted from Global and Planetary Change, 80, Becker et al, Sea-level variations at tropical Pacific islands since 1950, Copyright (2012), with permission from Elsevier

stations to measure variations in vertical uplift and subsidence, taking these into account when determining changing sea-levels. Figure 14.6 indicates that the reconstructed West Pacific sea-level rise can be as high as 3 times that of global mean sea-level rise, for the 1950-2009 period (indicated by regions of white, pink, and red to the east of Australia). Figure 14.6 shows that the areas of highest sea-level rise occur in two distinct regions: from Papua New Guinea in the West, eastwards through Honiara/Solomon Islands to Tuvalu and around the Federated States of Micronesia and the Marshall Islands in the North Pacific.

The Future of Atolls

Scientific discussion with respect to the future of atolls is divided into those that propose a bleak future for atolls as they become overwhelmed by rising seas, and those that contend that most atolls will still be present by the end of the century. For example, Storlazzi et al. (2015, 2018) claim that atolls will be uninhabitable within decades, arguing that as sea levels rise the marine platforms surrounding reefs are increasingly subject to deeper submarine conditions. This reduces their ability to mitigate the energy of incoming waves. Hence, the island itself bears the main force of incoming waves. More frequent inundations will allow saltwater to regularly ingress into freshwater aquifers as well as impacting on the island itself through erosion and general disturbance. Storlazzi et al. (2018) predict that Kwajalein atoll, Marshall Islands, and similar atolls that are dependent on groundwater will become uninhabitable by 2030–2040. The cumulative impacts of regular floods will make the groundwater unpotable as chloride levels rise above the maximum safe limit of drinking (250 mg/l).

In contrast, others suggest that atolls change shape and are dynamic over time, but with a general trend of stability or modest island growth (Kench et al. 2005, 2006, 2014, 2015, 2018; Webb and Kench 2010; Biribo and Woodruffe 2013; Mclean and Kench 2016). They argue that most atolls will still be present in 2100, and many islands will have grown in size, or stayed the same. A key factor in the persistence of atolls over time is sediment availability. Many atolls were largely built when sea level was higher (approximately 2000–5000 years ago). The fall in sea levels 2000–4000 years ago has depleted sediment supply. Atolls have been supplied mainly from sediment transport between islands, or more extreme events, such as cyclones or tsunamis, that not only erode parts of atolls but also tend to build up the height of atoll interiors (Kench et al. 2006). As sea levels rise again, Kench and his co-workers argue that sediment supply could increase and the higher frequency of extreme weather events could be a constructive island-building process. Each ocean-atoll environment must be considered in relation to the variables that affect atolls: ocean climate, wave type, wavelength, height and direction, tidal variations, ocean currents, wind direction/strength/variability, ENSO-related variations, the geometry and particle-size composition of atolls and individual islands, local sea-level rise rates, urbanisation/human interventions, and sediment supply. Detailed analysis of an atoll system within its individual oceanic environment is therefore key to determining a scientific prognosis of island shape/size and viability. One challenge is urbanised atolls, which can constrain natural processes and inhibit the natural changing shape of islands. Once urban infrastructures, such as airports, are constructed, they must be protected from environmental risks such as erosion.

Integrating Scientific Advice into Decision-Making

Differing perspectives on the future of atolls present a difficult dilemma for atoll leaders and decision makers. Either atolls will survive for only another 20–40 years (worst-case prediction) or well into the twenty-second century (best-case prediction). How do leaders usefully use such contradictory scientific advice to inform sustainable development decision-making? Policy responses to these scenarios are illustrated by two end members of a spectrum of possibilities

- ***Migration with Dignity***. Migration may take the form of internal (within country) and external (to another sovereign territory) migration. The migration of a whole culture/nation because of environmental changes caused by anthropogenic climate change is a radical and controversial issue. The Dhaka Principles for Migration with Dignity (United Nations 2012) set out good practice for the treatment of migrant workers. There are many challenges that migrants and host nations face following mass migration, including the creation of large camps and tent cities, hostility expressed by host country communities, and differential rights and treatments for migrants with respect to their host-country citizens. Migrants may have strong attachments to their original place of origin, and challenges adapting to new livelihood and subsistence options. Resettlement and migration programmes should be carefully planned, with detailed preparations for the early years of relocation (Hagen 2012; Edwards 2013, 2014). Education in adapting to a new physical and social environment and new ways of making a livelihood are particularly vital alongside approaches to retaining social and cultural cohesion and integration with the host community. Land security, new livelihoods, and the support of the local host community (to resettlement) are key to migration success (Edwards 2013).

- **Building to Defend**. This requires physical modification of the island environment through improved sea defences, an increased island elevation, and land reconstruction/reclamation. Green and 'soft' coastal defence options include reopening lagoon gaps in barrier reef islands to allow the natural flows of sediment, tides, and currents to redistribute material within the system (Fig. 14.7). Many engineers and scientists struggle to persuade the general public to opt for 'softer', rather than harder, sea defences, even if scientific evidence strongly supports the deployment of soft engineering options. Projects such as these are expensive and maintaining the structural integrity of reclaimed land may also be a challenge. Land reclamation can also pose significant threats to biodiversity and ecosystems.

Migration is the policy option that is most likely to be chosen if Pacific decision makers were to take the geoscience advice of Storlazzi et al. (2018) as the most important/realistic scientific advice. If Pacific leaders follow the advice of Mclean and Kench (2016), they are more likely to follow the latter, adopting green or 'soft engineering' approaches to allow natural processes to proceed relatively unimpeded by urban development. Intermediate planning options could include dispersing populations, creating a range of semi-urbanised island centres, and focused sea wall defence and island rampart building programmes. Inhabited floating artificial islands are also actively discussed at Pacific development meetings. It is likely that public pressure will not allow a *'do as little as possible'* response to rising sea levels in low-lying islands, and this may result in the decision being between hard engineering solutions and mass migration. To date, there is no rapid or sudden move for a mass migration option, however, these attitudes may change if there is a serious coastal inundation event resulting in significant casualties and/or infrastructure damage. A strong urge to remain within country by the majority of the population will probably push decision makers into negotiations with the international community with respect to exploring and developing major engineering island-elevation increase and land reclamation options. Geoscientists have a critical role to play in communicating a range of options derived from the latest geoscientific research (complemented by other disciplines) as described in Petterson (2019). This requires engagement with Pacific Leaders, communities, and decision makers to help inform decision-making.

Box 14.2. Responses to Sea-Level Rise in the Pacific Ocean

Kiribati has purchased land on Vanua Levu, Fiji, as a potential migration site for part of their population. They are also developing Kiritimati (or Christmas Island), the site of British and USA nuclear testing during the late 1950s/early 1960s. Kiritimati is the world's largest coral atoll with an area of 388 km^2 and is unusually high for an atoll, bring 13 m above sea

Fig. 14.7 Lagoon View, Tarawa, Kiribati. Note the intimate environmental links between ocean, lagoon, and island, and the shallow form of large areas of the lagoon that could be infilled to provide larger island living space (© Michael G Petterson)

level at its highest point. Banaba is a raised coral atoll, also with the potential for migration. Banaba has been extensively mined for phosphate and would have to be rehabilitated from its current post-mining situation if it were to become a suitable new home for migrants from other parts of Kiribati. Most of its original population was resettled to Rabi Island (Fiji). *Marshall Islanders* have a right to a green card and residency within the USA, and *Cook Islanders* can travel freely to New Zealand.

Land reclamation is common within Pacific islands and, in some areas has a long historical precedence. In Malaita (*Solomon Islands*) indigenous people have constructed artificial islands for centuries out of coralline materials. Reclaimed land engineering has been a key factor in development on the

main island of Viti Levu (*Fiji*), and South Tarawa (*Kiribati*). Jacobs, an international engineering group, is developing preliminary feasibility plans to reclaim the Temaiku area of Tarawa (*Kiribati*) which is currently uninhabited and poorly drained. The project would use sand and aggregates from the Tarawa lagoon to raise the height of the area, increase aquifer volumes, and provide living space for the country. More populated parts of Tarawa, Marshall Islands, and Tuvalu could, in theory, attract improved and raised sea-ramparts and defences and extend the period of atoll habitation significantly, despite rising sea levels and consequent wave impacts.

The Pacific Community (SPC) examined potential options for sea defences and groundwater protection on the island of

Lifuka Ha'apai Group, Tonga (SPC 2014). Lifuka had experienced 24 cm of vertical subsidence of the land relative to the sea, after a M_w 7.9 earthquake in 2006. Geological studies suggested that the overall longer term sustainability of the Lifuka coastline was better-served through the adoption of soft-engineering coastal management practices. Existing hard structures were starving parts of the coastline of sediment replenishment leading to coastal erosion.

14.3.3 Ocean Acidification

Approximately one-third of the carbon dioxide released into the atmosphere by human activity is dissolved into oceans, rivers, and lakes, resulting in acidification (a drop in its pH) (Sabine et al. 2004). CO_2 can react with water to produce carbonic acid, which, in turn, partially dissociates to produce a bicarbonate (HCO_3^-) and a hydrogen ion (H^+), thus increasing ocean acidity. Since the start of the industrial revolution, the acidity of surface ocean waters has increased by around 30% (NOAA 2019a). Future scenarios are currently pessimistic, predicting significant rises in ocean pH and acidification, alongside sea-level rise and ocean warming. Assuming business as usual in terms of carbon emissions, projections indicate that pH could drop by 0.2–0.5 units by 2100, an increase in acidity of 60–140% (Cummings et al. 2011). The oceans have not experienced this level of acidity for 14 million years, with the impacts being a significant disruption to marine ecosystems and marine biogeochemical cycles (Sosdian et al. 2018). Ocean acidification results in a net decrease in available carbonate ions making it harder for calcifying organisms (e.g., oysters, clams, corals, molluscs, and some plankton) to form biogenic calcium carbonate which forms the hard, protective shells (NOAA 2019a).

The whitening of coral (coral bleaching) and longer term degeneration of coral reef systems as a result of ocean acidification reduce habitats for perhaps one-quarter of ocean species, including fish, crustaceans, and marine plants. Reef systems are particularly common in tropical and subtropical waters, such as the Pacific islands, Indonesia, Philippines, SE Asia, the Caribbean, and the Maldives. The largest reef systems in the world are located close to Australia (the Great Barrier Reef), extending over 350,000 km^2, equivalent to the size of Germany. A key aim of **SDG 14** is the monitoring of the health of reef systems and related ocean acidification processes. Reefs, like atoll islands, are early warning indicators of oceanic change. Geoscientists, together with marine and bio-scientists play a significant part in monitoring through collecting chemical, physical, and oceanographic measurements on reef systems. The health of reef systems links closely to the health of atoll islands, as atolls depend on their sediment supply and fish resources (Sect. 14.3.2).

Ocean acidification is not uniform, with variation in sea-surface pH (Fig. 14.8). There is a general correlation of acidity with seawater temperatures and latitude/surface sea temperatures. Colder, high latitude waters can dissolve more carbon dioxide and will therefore suffer the greater falls in pH. High latitude waters also have lower saturation levels of carbonate (CO_3^{2-}). This combination of higher carbon dioxide solubility and lower bicarbonate saturation levels means that high latitude, southern and northern waters will experience the greatest rates of change, resulting in reduced calcification rates (e.g., of brachiopods, bivalves, and gastropods), and a decline in mollusc shell weights over time (Fabry et al. 2008; Cummings et al. 2011). Ocean currents and the dilution of seawater close to large continental rivers also have their impacts on ocean pH. Anthropogenic nitrogen and sulphur deposition to the ocean surface, as a result of fossil fuel burning and agriculture, can also result in increased acidity, particularly affecting coastal waters due to inputs from freshwater sources (Doney et al. 2007). Given the importance of coastal ecosystems to food security, resilience to natural hazards, and economic development, this is particularly concerning (Doney et al. 2007).

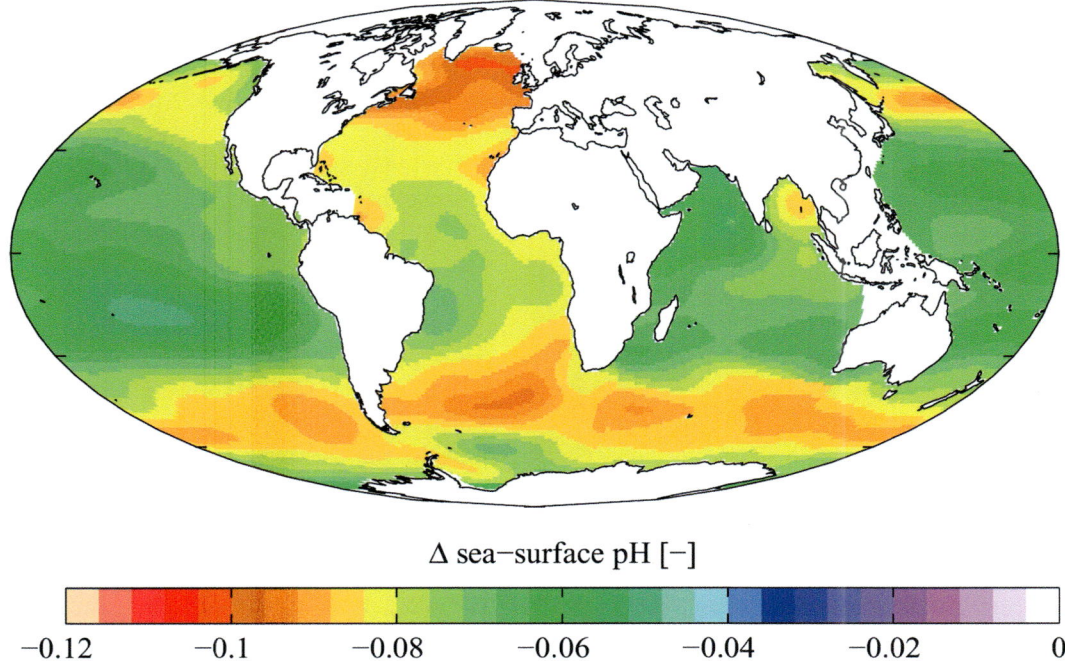

$$\Delta \text{ sea–surface pH } [-]$$

$$-0.12 \qquad -0.1 \qquad -0.08 \qquad -0.06 \qquad -0.04 \qquad -0.02 \qquad 0$$

Fig. 14.8 **Estimated change in seawater pH caused by anthropogenic carbon dioxide emissions between the 1700s and 1990s**. Note how colder high latitude waters have higher acidities, resulting from the greater ability of colder water to dissolve carbon dioxide. Global Ocean Data Analysis Project (GLODAP) and the World Ocean Atlas. *Credit* Plumbago (CC BY-SA 3.0, https://creativecommons.org/licenses/by-sa/3.0)

14.4   Ocean Resources for Sustainable Development

An ambition of SDG 14 is to '*increase the economic benefits to Small Island developing States and least developed countries from the sustainable use of marine resources, including through sustainable management of fisheries, aquaculture, and tourism*'. (Target 14.7). Much of the focus may be on living resources (e.g., fish stocks, and coral reefs), but there is a need to consider non-living resources both in the way that they interact with living resources and in their own right. The holistic management of marine space in 3-dimensions will become an increasing part of implementing **SDG 14**, as ocean activities become ever more prominent. Three important examples, discussed in this section, are marine parks conserving bio- and geodiversity (Sect. 14.4.1), the ocean as a source of minerals (Sect. 14.4.2), and renewable energy (Sect. 14.4.3).

14.4.1   Marine Parks

A common approach in marine conservation and managing a range of competing activities within marine space is the declaration of *marine conservation parks* with associated environmental and planning restrictions (e.g., no-mining zones). The Cook Islands, for example, is moving towards the declaration of a marine conservation zone that will have an area of approximately 1 million km², or approximately half their EEZ (Petterson and Tawake 2018).

Geoscientists can help shape the design of marine conservation parks, advising on the geodiversity that a park should include. For example, ocean landscapes such as the Ontong Java and Manihiki Ocean Plateaus, rift valleys within the plateau, abyssal plains and basins, and ocean seamounts form the backdrop and substrate to ocean ecosystems. These topographic-geological elements support unique ecosystems and environments with associated biota.

14.4.2 Ocean Seabed Mineral Resources

Mining controversies tend to focus on the balance between mining benefits and environmental degradation/damage caused by mining, and the contentious issue of benefit distribution. Who actually benefits from mining generated wealth? From a Pacific developmental standpoint, there is little to gain from mining if it does not deliver tangible and lasting benefits at local, regional, and national scales. This section discusses the issue of seabed minerals as a potential marine resource to support economic growth. Figure 14.9 presents the distribution of seabed mineral resources.

Three main types of seabed minerals are present in the Pacific islands region: polymetallic sulphides, cobalt-rich crusts, and manganese nodules (Petterson and Tawake 2016, 2018).

- *Polymetallic sulphide deposits* form through hydrothermal activity in active tectonic settings and are particularly rich in copper, lead, zinc, gold, and silver, occurring at depths of between approximately 1000 and 4000 m beneath sea level. The Exclusive Economic Zones (EEZs) of Papua New Guinea, Solomon Islands, Vanuatu, Fiji, Tonga, and New Zealand have a high potential for polymetallic sulphide mining, with a mine potentially beginning in 2019–20 (New Britain area of Papua New Guinea).
- *Cobalt-rich crusts* (CRC's) form on sediment-free rock surfaces within the ocean, forming layers up to 26 cm thick. These crusts and are generally found at water depths of between 600 and 7000 m deep, on the flanks of seamounts and undersea volcanoes, plateaus, and similar features. Crusts are rich in not only cobalt, but also nickel, copper, tellurium,

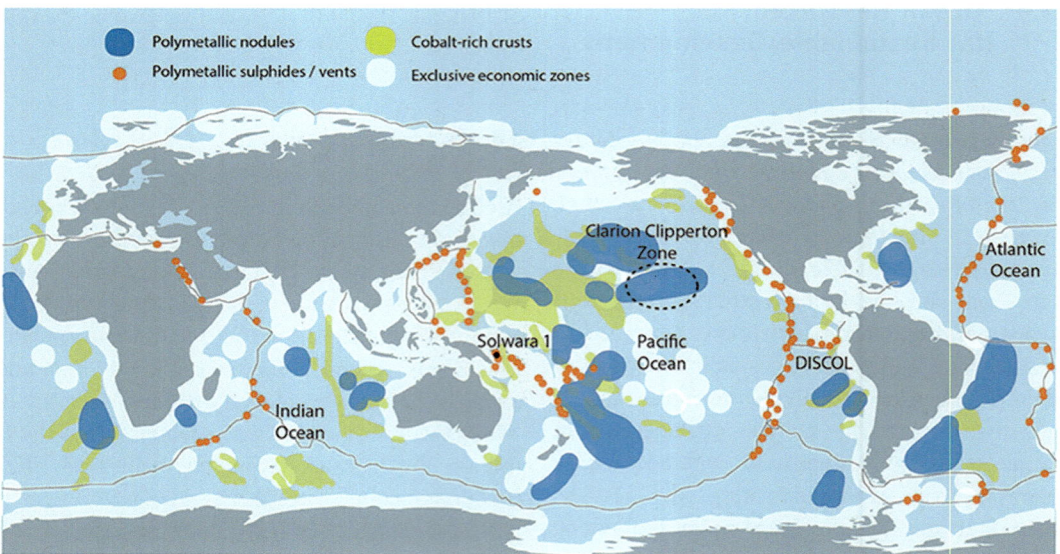

Fig. 14.9 Location of three primary marine mineral deposits: Polymetallic nodules (blue); polymetallic or seafloor massive sulphides (orange); and cobalt-rich ferromanganese crusts (yellow). From Miller et al. (2018), https://doi.org/10.3389/fmars.2017.00418, CC BY 4.0 (https://creativecommons.org/licenses/by/4.0/)

Fig. 14.10 Polymetallic manganese nodule from the Pacific Ocean . Nodules grow incrementally at rates of 1-10 mm per million years. Image width approximately 20 cm. *Credit* Koelle (CC BY-SA 3.0, https:// creativecommons.org/ licenses/by-sa/3.0)

platinum, zirconium, niobium, tungsten, and rare earths. They are particularly abundant close to the Federated States of Micronesia, Marshall Islands, Kiribati, Tuvalu, Cook Islands, and French Polynesia.

- *Polymetallic manganese nodules* (Fig. 14.10) form at depths more than 4000 m and up to approximately 6500 m, from cold seawater (hydrogenetic nodules) or from ocean floor sediment pore waters (diagenetic nodules). Most nodules are 4–14 cm in diameter, and vary in shape from spheroidal/sub-spheroidal to nodular/irregular. Polymetallic manganese nodules contain cobalt, copper, nickel, rare earths, molybdenum, lithium, and yttrium. Polymetallic manganese nodules are present within the EEZs of Kiribati, Cook Islands, and French Polynesia. The Clarion-Clipperton Zone (CCZ) in the Eastern-North Pacific is the region of highest abundance for polymetallic manganese nodules known on earth.

The Cook Islands is taking the prospect of seabed mining very seriously and has developed a range of governance tools, legislation, and expertise to prepare for a future possible industry (McCormack 2016; Petterson and Tawake 2018). The richest deposits of polymetallic manganese nodules in the Cook Islands are mainly within the Penrhyn Basin region of the EEZ, on the Eastern

side of the Manihiki Plateau. The Cook Islands Government is targeting this region for mining and beginning to plan how mines can operate within the basin in a way that (i) does not impact too much on the ocean floor and its biota; (ii) does not skew the economy of the Cook Islands too much; and (iii) provides a long-term sustainable source of funding for the Cook Islands, perhaps for a century or more.

This approach demonstrates the integrated approach needed for policymaking, integrating geoscience into broader socio-economic development discourses (e.g., Petterson 2019). Geoscience provides the data for the spatial extent and grade of the mineral deposit. It also provides data for the topographic and geological model of the ocean floor. Together with bioscience, geoscience assists with the understanding of ocean floor ecosystems. Oceanography provides data and an understanding of the ocean currents and biochemical variations within the water column. This informs our understanding of the origin of manganese nodules and how waste water is managed during the mining process, as well as in broader environmental assessment regarding the distribution of biota within the EEZ and the design of complementary marine parks (see Sect. 14.4.1). Geoscience and mining engineering knowledge are combined in developing a range of mining operational strategies recommended for minimising ocean floor disturbance, and management of ore and waste material transportation. Set-aside areas are being considered that are in adjacent sites, down-current of mining activities (McCormack 2016). These areas will be 'no-mining zones' and be used to compare and contrast ecosystems and environments with adjacent mined zones.

14.4.3   Renewable Energy Resources

As previously noted, many SIDS lack reliable, and sufficient power, with a dependence on oil or the use of diesel generators. Power provision often cannot rely on costly gridded electricity networks alone, but will most likely look towards local grids in numerous locations or networks that rely on no grids or limited village-level connections. Increasing use of renewable technologies, including new ocean energy technologies, could bring greatly improved, reliable and increased output to provide for future development. Examples of land-based renewable technologies include solar, hydropower, wind, biomass, and geothermal.

- *Solar energy technologies* have dropped exponentially in price over the past 10–20 years and battery life now extends to over eight hours or so (e.g., Asian Development Bank 2013). Solar energy is therefore becoming an increasingly attractive option in remote Pacific island locations, as well as supplementing power supply in urban regions.
- *Hydropower* will become increasingly important. Pacific countries such as Fiji, Papua New Guinea and Samoa all utilise hydropower to a significant extent already as part of their overall energy mix. For example, in Fiji hydropower already contributes 60% of the installed energy capacity.
- *Wind energy* has hardly been realised in the Pacific, although a few countries have invested in small wind farms (e.g., Fiji and Vanuatu). The use of wind energy will undoubtedly grow with time as it has in Europe over the past decade, for example.

- *Biomass-generated power plants* are another option, as tropical countries have rapid biomass growth rates. One biomass plant will shortly come on line in Fiji, developed by South Korea.
- *Geothermal energy* is possible in some countries, with Fiji, Papua New Guinea, Samoa, Solomon Islands, Tonga, and Vanuatu having moderate to high geothermal potential, for example.

Both solar and wind can be deployed in oceans, requiring expertise in offshore geotechnical engineering. Other sources of energy derived from ocean activity include *wave and tide energy*, and ocean thermal energy conversion (OTEC). The latter uses cold deep seawater as a heat sink, and uses warm surface seawater as a heat source to produce electricity. In closed OTEC systems, a low-temperature boiling working fluid (refrigerant), such as difluoromethane (R32) or ammonia, is vaporised at low temperature by the heat of warm ocean surface water, and condensed by cold deep ocean water. In open OTEC systems, seawater itself is the working fluid. This creates a continuous flow of working fluid, which rotates a turbine and produces electricity (Fig. 14.11). OTEC plants can be based on land with pipes extending to around 1 km below sea level, or fully marine platforms.

OTEC can generate surplus electricity when there is a temperature difference higher than 17 °C, between the surface and deeper waters. In the equatorial region, the temperature difference between ocean surface water and deep water (at about 1 km depth) is in the range of 20–25 °C throughout the year. Kiribati and Marshall Islands are examples of Pacific atoll countries with high ocean thermal conversion potential. Figure 14.12 shows that large areas of the ocean have the potential to generate electricity by OTEC, including Mexico, the Caribbean, the Philippines, Indonesia, Malaysia, coastal regions of east and west Africa, and small island states.

OTEC boasts little to no seasonal variation throughout the day and seasons. For remote islands and coastal villages that have no power grids, OTEC can provide clean, self-reliant,

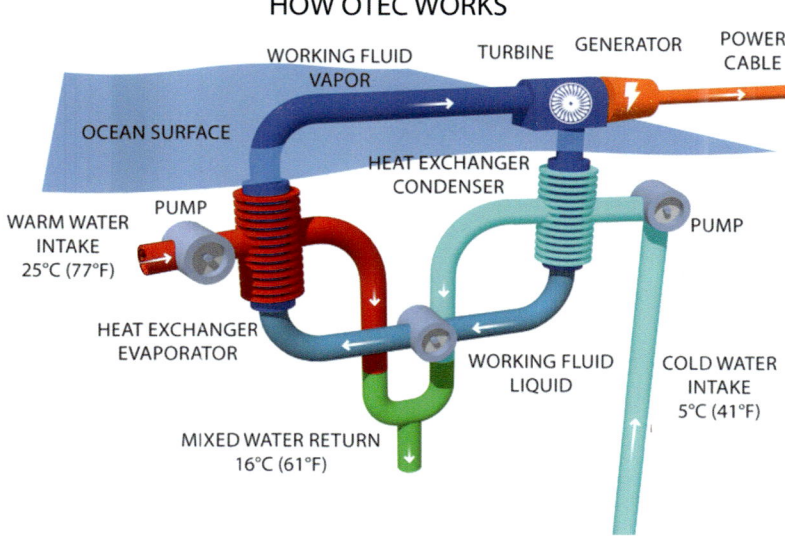

Fig. 14.11 Principles of OTEC generation. Vapours from warm seawater are used as the operational working fluid in an open-OTEC situation to drive an electricity generating turbine, then cooled and condensed, using cold seawater. The system can also generate freshwater and waters of differing salinity/temperature/water depth for aquaculture and agriculture purposes Adapted with permission from KRISO

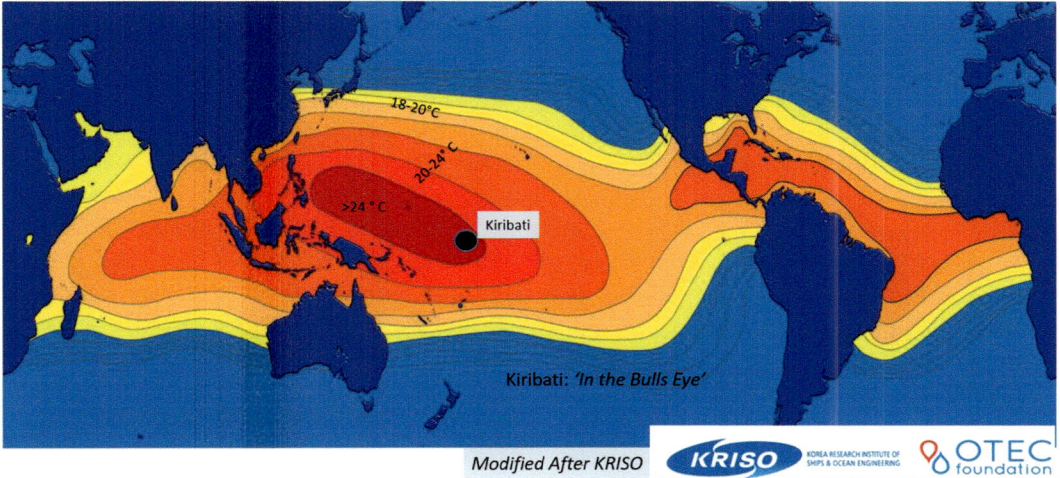

Fig. 14.12 Global map of temperature differences between surface waters and 1 km deep waters in the global ocean. KRISO plan to deploy a 1 MW land-based OTEC plant in Kiribati, in 2020. Adapted with permission from KRISO

sustainable energy. Deep seawater and surface seawater required for the operation of OTEC power can be used on land for multipurpose uses before returning to the sea (e.g., agriculture, refrigeration, and aquaculture). Deep seawater, still cold, can be used for cooling of surrounding

Fig. 14.13 1 MW Ocean Thermal Energy Conversion plant field experiment in Korea waters (© Hyeon-Ju Kim, Korea Research Institute of Ships and Ocean Engineering)

buildings and houses, and, because it is clean, it can be used for seawater desalination, hydroponics, and aquaculture.

OTEC, as a concept, is not new, but the realisation of scaled-up plants that produce significant amounts of electricity has been slow. Most OTEC plants are land-based. Experiments and demonstrations on full, ocean-sited plants (i.e., a power station set within a full ocean environmental setting) have been limited and have met with little success to date. One current developmental full, ocean-sited plant is the NEMO project, supported by the EU with an aim of installing a 16 MW OTEC plant on the Atlantic island of Martinique. Korea Research Institute of Ships and Ocean Engineering (KRISO) plans to deploy a 1 MW land-based OTEC plant for one year in South Tarawa, Kiribati. If tests are successful, this may lead to longer term projects and fully ocean deployed projects (Fig. 14.13).

14.5 Science Capacity for Ocean Management

Increasing scientific knowledge, developing research capacity, and transferring marine technologies to improve ocean health and support development of SIDS and the least developed countries are central to **SDG 14**. This requires enhanced research and development capacity in all countries (see **SDG 9**), effective scientific institutions at national and regional levels (see **SDG 16**), and effective partnerships for development (see **SDG 17**). These themes are explored in depth in these respective chapters, and we refer the reader to them. Below we set out some examples that integrate geoscience and ocean management.

- The **Pacific Community (SPC)**[1] is the principal scientific and technical agency

---

[1]https://gsd.spc.int/index.php.

supporting development in the Pacific, owned and governed by its 26 members. Its Geoscience Division aims to *'apply geoscience and technology to realise new opportunities for improving the livelihoods of Pacific communities'*. The Geoscience for Development Programme provides applied ocean, island and coastal geoscience services to member countries, providing expertise in oceanography, coastal processes and geomorphology, and hydrodynamic modelling (SPC 2019).

- The **Caribbean Community (CC or CARICOM)**[2] has its headquarters in Georgetown, Guyana. This grouping of fifteen member states, and five associate members, supports and engages in geoscience and environmental activities from research to practice, and policy advice. The CARICOM/Caribbean Call to Action for SDG 14 included commitments to addressing the plastic pollution issue, establishing marine conservation areas, and developing a range of marine and coastal governance policies.

- The **Coordinating Committee for Geoscience Programmes in East and Southeast Asia (CCOP)**[3] is based in Bangkok, Thailand. This intergovernmental organisation has 15 member countries from the East/SE Asia region. CCOP exists to promote the contribution of geoscience throughout the region and beyond for economic and social wellbeing, including links to **SDG 14**. CCOP develops conferences, and professional development and training activities aimed at sharing knowledge and expertise among its member countries.

- The **One Ocean Hub,**[4] funded by UK Research and Innovation (UKRI) through the Global Challenges Research Fund (GCRF), aims to transform our response to the urgent challenges facing our ocean, developing an integrated approach to managing how they are used. The Hub specifically addresses the challenges and opportunities of South Africa, Namibia, Ghana, Fiji and Solomon Islands, and will share knowledge at regional (South Pacific, Africa, and Caribbean) and international levels. The Hub is led by the University of Strathclyde (UK), with more than 50 partners around the world, including those with geological expertise.

- **Significance of Modern and Ancient Submarine Slope LandSLIDEs (S4LIDE)**[5] is an International Geoscience Programme (IGCP) project, supported by UNESCO and the International Union of Geological Sciences. S4LIDE brings together geoscientists from academia and industry to form an international and multidisciplinary platform to develop a more cohesive understanding of submarine landslides.

- The **Commonwealth Marine Economies Programme**[6] aims to support 17 Commonwealth SIDS to develop their marine (or 'blue') economies in a sustainable, resilient, and integrated way. The project is led by the UK Foreign and Commonwealth Office, with partners including the UK National Oceanography Centre, UK Hydrographic Office, and the UK Centre for Environment, Fisheries and Aquaculture Science. In Grenada, for example, this programme includes coastal vulnerability mapping, and an aim to integrate with regional and global natural hazard monitoring networks (FCO 2018).

- The **Joint Group of Experts on the Scientific Aspects of Marine Environmental Protection (GESAMP)**[7] provides advice to the UN system, working under the auspices of 10 key UN agencies (e.g., the International Maritime Organisation, Food and Agriculture Organisation, World Meteorological Organisation, and UN Environment). The group has

---

[2]https://caricom.org/.

[3]http://www.ccop.or.th/.

[4]https://www.strath.ac.uk/research/strathclydecentreenvironmentallawgovernance/oneoceanhub/.

[5]http://www.unesco.org/new/en/natural-sciences/environment/earth-sciences/international-geoscience-programme/igcp-projects/geohazards/project-640-new-2015/.

[6]https://www.gov.uk/guidance/commonwealth-marine-economies-programme.

[7]http://www.gesamp.org/.

working groups on themes such as marine geoengineering, trends in coastal pollution, and impacts of mining waste.

- The **International Seabed Authority (ISA)**[8] is a United Nations body established by the UN Convention on the Law of the Sea in 1994, and based in Kingston, Jamaica. The ISA has 167 members, and the European Union as a collective member. The ISA is responsible for the issuance of exploration and mining licences for seabed mineral-related activities within international waters, and the development of legal and regulatory systems for seabed mining.

These examples of institutions and initiatives demonstrate that geoscientists are connected to existing groups working to deliver the ambitions of **SDG 14**. The geological record provides a fundamental source of information on how our planet has previously responded to higher global temperatures, the melting of continental ice, and the acidification of oceans. Biogeochemists are pioneering approaches to understand the cycling of carbon (and other elements) through the natural environment (including interactions between the lithosphere, atmosphere, and hydrosphere). The development challenges of many SIDS have a strong connection to the geological environment—access to potable water, access to a reliable and sustainable energy supply, resilience to multi-hazard environments, and management of infrastructure and waste in confined spatial areas. While marine exclusive economic zones are demarcated by borders (as illustrated in Fig. 14.2), many of the challenges affecting our oceans are global and require collaborations across disciplines, sectors, and nations if we are to deliver appropriate solutions to address these.

14.6   Key Learning Concepts

- **SDG 14** focuses on the sustainable management and development of the oceans. Key focus areas include marine pollution, marine

environment, impacts of environmental and climate change, ocean health, and sustainable development of resources such as fisheries, minerals, and energy. This chapter highlights the close, symbiotic links between oceans and ocean/coastal communities, and how geoscience can help to address and deliver the ambitions of **SDG 14**.

- Small Island Developing States, or SIDS, exist in all the world's oceans, particularly the Caribbean and western Pacific regions. SIDS are characterised by relatively small land areas, large to very large ocean areas, geographical isolation, an archipelago geography, and limited opportunities for economic development. They also support unique cultures and ways of life, with a particularly close connection between humans and the open ocean.

- Oceans are the lowest basins on Earth and will receive much of the pollution produced by humans, as rivers and gravity tend to move material to the lowest point of gravitational potential energy (the oceans). Plastics are an excellent tracer for human pollution. Within the last 60 to 70 years, plastics have found their way into the deepest parts of the ocean and are a new anthropogenic ocean sediment. All oceans have plastic concentration zones at their surface. Plastics have serious health and mortality impacts on marine life. Human consciousness related to plastic pollution is growing rapidly, and many parts of the world are controlling/reducing plastic waste, perhaps for the first time in history.

- Sea-level rise occurs due to water thermal expansion and the melting of ice. Current sea-level rise is a consequence of human-induced climate change, with low-lying atoll islands particularly susceptible to inundation. There are a range of scientific opinions about the future of low-lying atolls. Some scientists predict that they will respond to changing ocean conditions and survive into the 2100 s and beyond, with others indicating that atolls may be uninhabitable by 2030–2040. Policies must be developed to optimise responses to sea-level rise within atoll countries.

[8]https://www.isa.org.jm/.

- The absorption of carbon dioxide into the oceans results in acidification. This process dissolves calcium carbonate, the key constituent of marine mollusc shells, making it difficult for marine molluscs to bio-generate new shells, and adversely impacts reef ecosystems.
- The holistic management of marine space in 3-dimensions will become an increasing part of implementing **SDG 14**, as ocean activities become ever more prominent. Geoscientists can help shape the design of marine conservation parks, advising on the geodiversity that a park should include.
- Three types of seabed minerals occur on the ocean floor: seabed sulphides produced by hydrothermal vent mineralisation, and cobalt-rich crusts and manganese (or polymetallic) nodules, which grow slowly on the ocean floor or seamount summits. These mineral deposits contain a wide range of metals, including those needed for new 'green' technologies. Mining of seabed minerals may occur within the next decade, needing consideration of how environmental challenges will be managed and avoided.
- The ocean also provides energy resources, with a lack of access to sustainable energy supplies being a particular challenge in SIDS. Solar and wind can be deployed in oceans, requiring expertise in offshore geotechnical engineering, and ocean activity can be converted to electricity using *wave energy, tide energy*, and ocean thermal energy conversion (OTEC). The latter uses cold deep seawater as a heat sink, and uses warm surface seawater as a heat source to produce electricity.
- SDG 14 emphasises the need for increasing scientific knowledge, research capacity, and marine technology transfer. Many initiatives are helping to connect the expertise and skills of geoscientists with ocean management. These approaches often bring together partners from across countries and disciplines. The concept of interconnected geoscience links the application of geoscience to contextual developmental and environmental situations.

14.7 Educational Ideas

In this section, we provide examples of educational activities that connect geoscience, the material discussed in this chapter, and scenarios that may arise when applying geoscience (e.g., in policy, government, private sector international organisations, and NGOs). Consider using these as the basis for presentations, group discussions, essays, or to encourage further reading.

- Examine bathymetric and oceanographic maps noting the sheer variety of the ocean floor seascapes (e.g., ocean ridges and mountains, seamounts and volcano chains, trenches, island arc systems, ocean islands, abyssal plains, and fracture zones), contributing to ecological diversity. Imagine you were designing a 'marine geopark' to profile geological diversity and its relationship with biodiversity. What features would you like this to include, and what impact may they have on the biodiversity that thrives in the region?
- Study the scattered distribution of ocean islands and small island states in the Pacific, Indian, and Atlantic Oceans, and their archipelago nature. In small groups, select an example and prepare a five-minute talk on their history of settlement, geology, and economic and developmental links to the ocean. Have a class discussion on similarities and differences between the examples presented, the challenges in governing and administering such island nations, and how geoscientists are actively contributing to sustainable development in SIDS.
- Study a global map of coral reefs, noting their distribution patterns and the rich biodiversity linked to reefs. Undertake a study identifying which reefs are affected and unaffected by the

effects of climate change (e.g., ocean acidification). What actions are needed to reduce ocean acidification? Explore the options and prepare a one-page summary of recommendations for policymakers.

- What overlap is there between the metals found in mobile phones and the metals found in seabed minerals? What challenges will seabed mining bring in environmental and technological terms?

- What linkages exist between SDG 14 and other SDGs (i.e., how can delivering SDG 14 help to achieve other SDGs, how can progress in other SDGs help achieve SDG 14)? Map out these relationships and identify opportunities for geoscientists in diverse sectors to help deliver the ambitions of SDG 14. What other sectors and disciplines would geoscientists need to partner with to deliver solutions to the challenges facing oceans around the world?

Further Reading and Resources

Beiras R (2018) Marine pollution: sources, fate and effects of pollutants in coastal ecosystems. Elsevier, 408 p. https://doi.org/10.1016/C2017-0-00260-4

Church JA, Clark PU, Cazenave A, Gregory JM, Jevrejeva S, Levermann A, Merrifield MA, Milne GA, Nerem RS, Nunn PD, Payne AJ, Pfeffer WT, Stammer D, Unnikrishnan AS (2013) Sea level change. Stocker TF, Qin D, Plattner G-K, Tignor M, Allen SK, Boschung J, Nauels A, Xia Y, Bex V, Midgley PM (eds) In: Climate change 2013: the physical science basis. Contribution of working group I to the fifth assessment report of the intergovernmental panel on climate change. Cambridge University Press, Cambridge, UK and New York, NY, USA

Heidkamp CP, Morrissey J (eds) (2019) Towards coastal resilience and sustainability. Routledge, Oxon, 360 p

Miller KA, Thompson KF, Johnston P, Santillo D (2018) An overview of seabed mining including the current state of development, environmental impacts, and knowledge gaps. Front Mar Sci 4:418

Petterson MG (2019) Interconnected geoscience for international development. Episodes 42(3):225–233. https://doi.org/10.18814/epiiugs/2019/019018

Petterson MG, Tawake AK (2016) Toward inclusive development of the Pacific region using geoscience. In: Wessel GR, Greenberg JK (eds) Geoscience for the public good and global development: toward a sustainable future. Geological Society of America Special Paper 520, pp 459–478

United Nations (2019) Small Island developing states. Available at: https://sustainabledevelopment.un.org/topics/sids. Accessed 2 Aug 2019

Stewart RH (2008) Introduction to physical oceanography, 358 p. Available at: https://open.umn.edu/opentextbooks/textbooks/introduction-to-physical-oceanography. Accessed 1 Oct 2019

References

Asian Development Bank (2013) Vanuatu National Energy Road Map. Government of Vanuatu, 99pp

Becker M, Meyssignac B, Letetral C, Llovel W, Cazenave, Delcroix T (2012) Sea level variations at tropical Pacific Islands since 1950. Glob Planet Chang 80–81:85–98

Beiras R (2018) Marine pollution: sources, fate and effects of pollutants in coastal ecosystems. Elsevier

Biribo N, Woodroffe CD (2013) Sustain Sci 8:345. https://doi.org/10.1007/s11625-013-0210-z

BP (2019) BP statistical review of world energy, 68th edn. BP PLC, 1 St James Square, London, UK

Cummings V, Hewitt J, Rooyen AV, Currie K, Beard S, Thrush S, Norkko J, Barr N, Heath P, Halliday J, Sedcole R, Gomez A, McGraw C, Metcalf S (2011) Ocean acidification at high latitudes: potential effects on functioning of the Antarctic Bivale, Laternula elliptica. PLoS One 6(1):e16069. https://doi.org/10.1371/journal.pone.0016069. Published online 5 Jan 2011

Daigle K, Gramling C (2018) Why sea level rise varies from place to place. Available at: https://www.sciencenews.org/article/why-sea-level-rise-varies-place-place. Accessed 15 Oct 2019

Dickinson WR (2009) Pacific atoll living, how long already and until when? GSA Today 19:4–9

Doney SC, Mahowald N, Lima I, Feely RA, Mackenzie FT, Lamarque JF, Rasch PJ (2007) Impact of anthropogenic atmospheric nitrogen and sulfur deposition on ocean acidification and the inorganic carbon system. Proc Natl Acad Sci 104(37):14580–14585

Edwards JB (2013) The logistics of climate-induced resettlement: lessons from the Carteret Islands, Papua New Guinea. Refug Surv Q 32(3):52–78. https://doi.org/10.1093/rsq/hdt011

Edwards JB (2014) Phosphate mining and the relocation of the Banabans to northern Fiji in 1945: lessons for climate change-forced displacement. Le Journal de la Société des Océanistes 138–139. https://doi.org/10.4000/jso.7100

Eriksen M, Lebreton LC, Carson HS, Thiel M, Moore CJ, Borerro JC, Galgani F, Ryan PG, Reisser J (2014) Plastic pollution in the world's oceans: more than 5 trillion plastic pieces weighing over 250,000 tons afloat at sea. PLoS ONE 9(12):e111913

Fabry VJ, Seibel BA, Feely RA, Orr JC (2008) Impacts of ocean acidification on marine fauna and ecosystem processes. ICES J Mar Sci 65(3):414–432

FAO (2018) The state of the world's fisheries and aquaculture. Available at: http://www.fao.org/state-of-fisheries-aquaculture/2018/en/. Accessed 15 Oct 2019

FCO (2018) Commonwealth Marine Economies Programme—Grenada Country Review. https://assets. publishing.service.gov.uk/government/uploads/ system/uploads/attachment_data/file/769184/ Commonwealth_Marine_Economies_Programme_-_ Grenada_Country_review.pdf. Accessed 21 Aug 2019

Geyer R, Jambeck JR, Law KL (2017) Production, use and fate of all plastics ever made. Sci Adv 3: e1700782. https://doi.org/10.1126/sciadv.1700782

Government Office for Science (2018) Foresight - Future of the Sea. Available at: https://assets.publishing. service.gov.uk/government/uploads/system/uploads/ attachment_data/file/706956/foresight-future-of-the-sea-report.pdf. Accessed 1 Oct 2019

Hagen K (2012) Resilience, response, recovery and ethnicity in post disaster processes. SPC SOPAC Technical report PR 182. Suva, Fiji

Hallam A (1992) Phanerozoic sea-level changes. Columbia University Press, USA

Halpern BS, Longo C, Hardy D, McLeod KL, Samhouri JF, Katona SK, Kleisner K, Lester SE, O'Leary J, Ranelletti M, Rosenberg AA, Scarborough C, Selig ER, Best BD, Brumbaugh DR, Chapin FS, Crowder LB, Daly KL, Doney SC, Elfes C, Fogarty MJ, Gaines SD, Jacobsen KI, Karrer LB, Leslie HM, Neeley E, Pauly D, Polasky S, Ris B, St Martin K, Stone GS, Sumaila UR, Zeller D (2012) An index to assess the health and benefits of the global ocean. Nature 488 (7413):615

Hoegh-Guldberg O (2015) Reviving the Ocean Economy: the case for action—2015. WWF International, Gland, Switzerland, Geneva, 60pp

House of Lords (2019) The future of seaside towns. https://publications.parliament.uk/pa/ld201719/ ldselect/ldseaside/320/320.pdf. Accessed 21 Aug 2019

ILO (2015) Cyclone PAM causes devastating impact on employment and livelihoods. https://www.ilo.org/ suva/public-information/WCMS_368560/lang–en/ index.htm. Accessed 15 Oct 2019

Irwin G (1998) The colonisation of the pacific plate: chronological, navigational and social issues. J Polyn Soc 107(2):111–144

Jambeck JR, Geyer R, Wilcox C, Siegler TR, Perryman M, Andrady A, Narayan R, Law KL (2015) Science 347(6223):768–771. https://doi.org/10.1126/ science.1260352

Kench PS, McLean RF, Nichol SL (2005) New model of reef-island evolution: Maldives, Indian Ocean. Geology 33(2):145–148. https://doi.org/10.1130/g21066.1

Kench PS, McClean RF, Brander RW, Nichol SL, Smithers SG, Ford MR, Aslam M (2006) The Maldives before and after the Sumatran tsunami. Geology 34 (3):177–180. https://doi.org/10.1130/g21907.1

Kench PS, Owen SD, Ford MR (2014) Evidence for Coral Island formation during rising sea level in the Central Pacific Ocean. Geophys Res Lett. https://doi.org/10. 1002/2013gl059000

Kench PS, Thompson D, Ford MR, Ogawa H, McLean RF (2015) Coral islands defy sea-level rise over the past century: records from a central Pacific atoll. Geology 43(6):515–518. https://doi.org/10. 1130/g36555.1

Kench PS, Ford MR, Owen SO (2018) Patterns of island change and persistence offer alternate adaptation pathways for atoll nations. Nat Commun 9:1–7. https://doi.org/10.1038/s41467-018-02954-1

Liu Y, MacFadyen A, Ji Z-G, Weisberg RH (2011) Monitoring and modelling the deepwater horizon oil spill: a record-breaking enterprise. Washington DC American Geophysical Union Geophysical Monograph Series. Geophysical Monograph Series, vol 195. https://doi.org/10.1029/gm195

McCormack G (2016) Cook Islands seabed minerals. A precautionary approach to mining. Rarotonga, Cook Islands: Cook Islands Natural Heritage Trust, 33pp. ISBN 978-982-98133-1-2

McLean R, Kench P (2016) Destruction or persistence of coral atoll islands in the face of 20th and 21st century sea-level rise? Wiley Interdiscip Rev-Clim Chang 6 (5):445–463. https://doi.org/10.1002/wcc.350

Minderhoud PSJ, Erkens G, Pham VH, Bui VT, Erban L, Kooi H, Stouthamer E (2017) Impacts of 25 years of groundwater extraction on subsidence in the Mekong delta, Vietnam. Environ Res Lett 12(6)

NOAA (2019a) What is Ocean Acidification? https:// www.pmel.noaa.gov/co2/story/What+is+Ocean +Acidification%3F. Accessed 20 Aug 2019

NOAA (2019b) How does oil impact marine life? https:// oceanservice.noaa.gov/facts/oilimpacts.html. Accessed 10 Oct 2019

Petterson MG, Tawake AK (2018) The cook islands (South Pacific) experience in governance of seabed manganese nodule mining. Ocean Coast Manag 167:271–287. https://doi.org/10.1016/j.ocecoaman. 2018.09.010

Roser M (2019) Oil Spills. https://ourworldindata.org/oil-spills'. Accessed 15 Oct 2019

Ritchie, H. and Roser, M. (2018) Plastic pollution. https:// ourworldindata.org/plastic-pollution. Accessed 15 Oct 2019

Sabine CL, Feely RA, Gruber N, Key RM, Lee K, Bullister JL, Wanninkhof R, Wong CSL, Wallace DW, Tilbrook B, Millero FJ (2004) The oceanic sink for anthropogenic CO2. Science 305(5682):367–371

Sosdian SM, Greenop R, Hain MP, Foster GL, Pearson PN, Lear CH (2018) Constraining the evolution of

Neogene ocean carbonate chemistry using the boron isotope pH proxy. Earth Planet Sci Lett 498:362–376

SPC (2014) Assessing vulnerability and adaptation to sea-level rise: Lifuka Island Ha'apai, Tonga. Pacific Community Geoscience Division

SPC (2019) Pacific community geoscience division. https://gsd.spc.int/index.php. Accessed 21 Aug 2019

Storlazzi CD, Elias EPL, Berkowitz P (2015) Many atolls will become uninhabitable within decades due to climate change. Sci Rep 5:14546. https://doi.org/10.1038/srep14546

Storlazzi CD, Gingerich SB, van Dongeren A, Cheriton OM, Swarzenski PW, Quataert E, Voss CI, Field DW, Annamalai H, Piniak GA, McCall R (2018) Most atolls will be uninhabitable by the mid-21st century because of sea-level rise exacerbating wave-driven flooding. Sci Adv 4(4):p.eaap9741

United Nations (2012). Sustainable supply chains. The dhaka principles for migration with dignity. http://supply-chain.unglobalcompact.org/site/article/141

United Nations (2015) The millennium development goals report. www.un.org/millenniumgoals/2015_MDG_Report/pdf/MDG%202015%20rev%20(July%201).pdf. Accessed 13 Aug 2019

United Nations (2016) Progress towards the sustainable development goals: report of the secretary-general. UN Economic and Social Council. E/2016/75

United Nations (2019) SDG 14—life below water. https://www.un.org/sustainabledevelopment/oceans/. Accessed 21 Aug 2019

USGS (1995) Sea level change: lessons from the geologic record. U.S. Geological Survey, U.S. Department of the Interior. FS-117–95

Wahl T (2017) Sea-level rise and storm surges, relationship status: complicated. Environ Res Lett 12 (11):111001

Waters CN, Zalasiewicz J, Summerhayes C, Barnosky AD, Poirier C, Gałuszka A, Cearreta A, Edgeworth M, Ellis EC, Ellis M, Jeandel C (2016) The anthropocene is functionally and stratigraphically distinct from the Holocene. Science 351(6269): aad2622

Webb AP, Kench PS (2010) The dynamic response of reef islands to sea-level rise: Evidence from multi-decadal analysis of island change in the Central Pacific. Glob Planet Chang 72(3), 234–246. https://doi.org/10.1016/j.gloplacha.20https://doi.org/10.05.003

Wilcox C, Sebille EV, Hardesty BD (2015) Threat of plastic pollution to seabirds is global, pervasive, and increasing. Proc Natl Acad Sci 112(38):11899–11904. https://doi.org/10.1073/pnas.150218112

Woodall LC, Sachez-Vidal A, Canals M, Paterson GL, Coppock VS, Calafat A, Rogers AD, Narayanaswamy BE, Thomson RC (2017) The deep sea is a major sink for microplastic debris. R Soc Open Sci 1:140317. https://doi.org/10.1098/rsos.140317

World Bank (2019a) Mortality rate attributed to unsafe water, unsafe sanitation and lack of hygiene (per 100,000 population). https://data.worldbank.org/indicator/SH.STA.WASH.P5. Accessed 1 Oct 2019

World Bank (2019b) Prevalence of undernourishment (% of population). https://data.worldbank.org/indicator/sn.itk.defc.zs. Accessed 1 Oct 2019

Michael G. Petterson is Professor of Geology at the Auckland University of Technology. He has worked in the field of geoscience and international development for three decades, with residential/long term spells in Afghanistan, Pakistan, Solomon Islands, Fiji, Guyana, UK, New Zealand, and the Pacific islands region. He has worked in areas such as applying geoscience to disaster and risk, economic and sustainable development, infrastructure, governance and policy, and geoscience and education. He has mentored hundreds of people across the world and been active in many fields of geoscience and the SDGs.

Hyeon-Ju Kim has led the Seawater Utilization Plant Research Center at the Korea Research Institute of Ships and Ocean Engineering (KRISO) since 2005, and is managing the Ocean Thermal Energy Conversion (OTEC) project funded by the Korean Ministry of Oceans and Fisheries since 2010. He has a Ph.D. in Ocean Engineering from Pukyoung National University, and is an ocean engineering specialist, licensed in Korea. His professional experience ranges from theoretical analysis to experimental evaluation of seawater utilization systems for food, energy and water.

Joel C. Gill is International Development Geoscientist at the British Geological Survey, and Founder/Executive Director of the not-for-profit organisation *Geology for Global Development*. Joel has a degree in Natural Sciences (Cambridge, UK), a Masters degree in Engineering Geology (Leeds, UK), and a Ph.D. focused on multi-hazards and disaster risk reduction (King's College London, UK). For the last decade, Joel has worked at the interface of Earth science and international development, and plays a leading role internationally in championing the role of geoscience in delivering the UN Sustainable Development Goals. He has coordinated research, conferences, and workshops on geoscience and sustainable development in the UK, India, Tanzania, Kenya, South Africa, Zambia, and Guatemala. Joel regularly engages in international forums for science and sustainable development, leading an international delegation of Earth scientists to the United Nations in 2019. Joel has prizes from the London School of Economics and Political Science for his teaching related to disaster risk reduction, and Associate Fellowship of the Royal Commonwealth Society for his international development engagement. Joel is a Fellow of the Geological Society of London, and was elected to Council in 2019 and to the position of Secretary (Foreign and External Affairs) in 2020.

Life on Land

15

Eric O. Odada, Samuel O. Ochola, and
Martin Smith

E. O. Odada (✉)
African Collaborative Centre for Earth System
Science (ACCESS), College of Biological and
Physical Sciences, University of Nairobi, Chiromo
Campus, Riverside Drive, P.O. Box 30197-00100
Nairobi, Kenya
e-mail: eodada@uonbi.ac.ke; access@uonbi.ac.ke

S. O. Ochola
Department of Environmental Studies and
Community Development, Kenyatta University,
P. O. Box 43844-00100, Nairobi, Kenya

M. Smith
British Geological Survey, The Lyell Centre,
Research Avenue South, Edinburgh EH14 4AP,
Scotland, UK

© Springer Nature Switzerland AG 2021
J. C. Gill and M. Smith (eds.), *Geosciences and the Sustainable Development Goals*,
Sustainable Development Goals Series, https://doi.org/10.1007/978-3-030-38815-7_15

Abstract

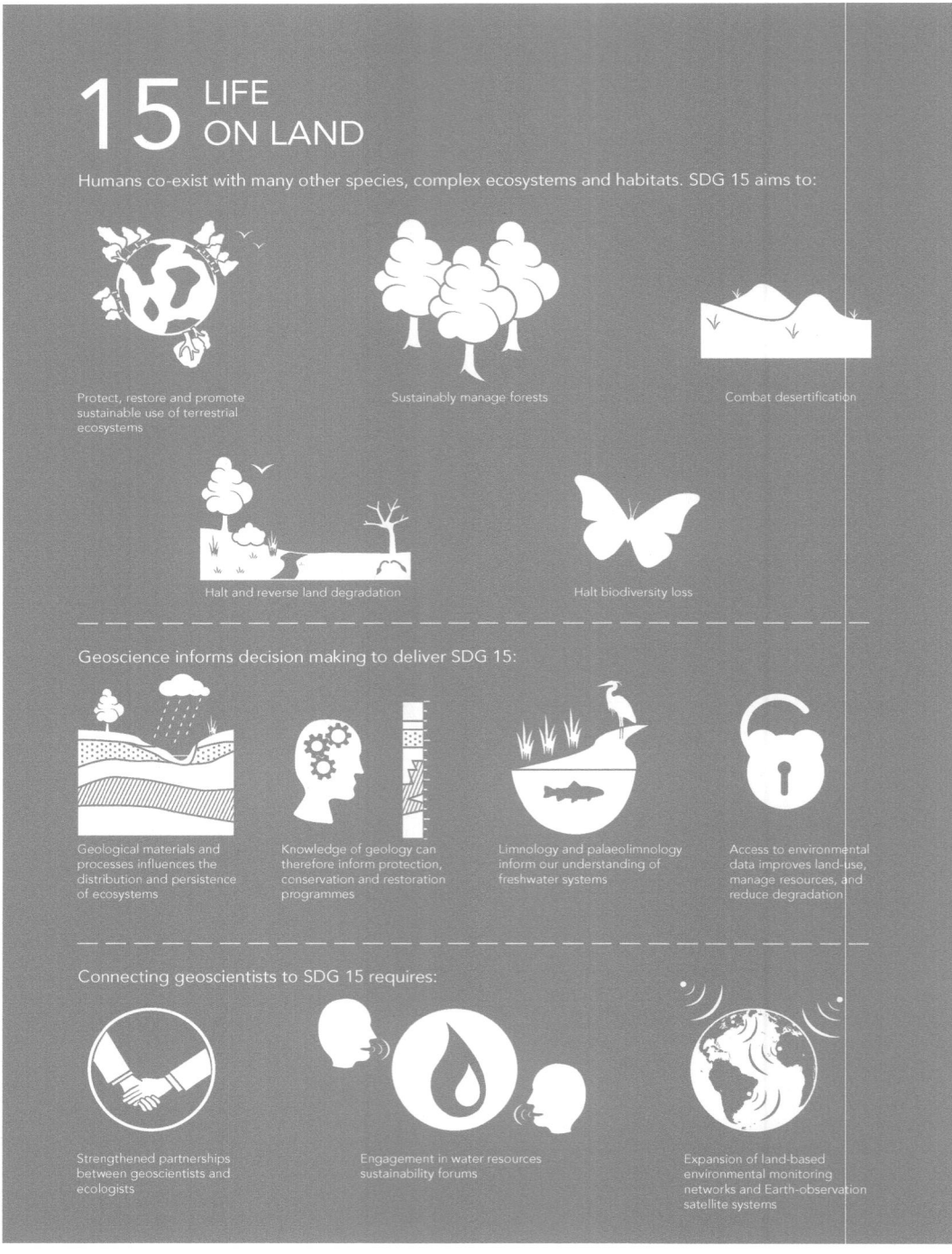

15.1 Introduction

A flourishing life on land is the foundation for sustenance and livelihood for all humankind. Humans do not live in isolation from the millions of other species that call Earth their home, the complex ecosystems they form, and a myriad of habitats (from forests to rock crevices, lakes to dune systems, savannahs to caves). Rather they coexist with complex interactions and dependencies. The United Nations (2015a, b) note: '*We are all part of the planet's ecosystem and we have caused severe damage to it through deforestation, loss of natural habitats and land degradation. Promoting a sustainable use of our ecosystems and preserving biodiversity is not a cause. It is the key to our own survival*'. Examples of the importance of diverse life on land to sustainable development objectives include

- Plants provide 80% of the human diet (FAO 2017), and approximately 2.6 billion people depend on agriculture for their livelihoods (UNDP 2019).
- Forests account for 30% of the Earth's surface, providing habitats to more than 80% of all terrestrial species of animals, plants, and insects (UNDP 2019). Forests contribute to clean air and water, and help to mitigate against climate change by sequestering carbon dioxide. The International Union for the Conservation of Nature (IUCN) suggest that restoring 3.5 million km^2 of deforested and degraded land could sequester up to 1.7 gigatonnes of carbon dioxide equivalent annually (IUCN 2019a). Nature-based climate solutions can contribute to a third of the CO_2 reductions needed by 2030 (UNDP 2019).
- Lewis et al. (2019) note that locking carbon in ecosystems through more forests has broader benefits, including helping to manage water resources and create jobs. The value of ecosystems to human livelihoods and well-being is valued at US$ 125 trillion per year (UNDP 2019). Around 1.6 billion people depend on forests for their livelihoods (UNDP 2019).

- Mountain regions are a critical source of freshwater, with all of the world's major rivers having their headwaters in mountainous regions (FAO 2003). Mountain regions are delicate due to extreme environments (e.g., altitude and temperature), and therefore ecosystems associated with these regions (and the services they provide) are particularly vulnerable. Mountains are hubs of biological and cultural diversity, influencing the climate at many scales, and are home to approximately 12% of the global population (IUCN 2019b).

The ecosystems and services hosted and provided by forests, mountains, and freshwater resources are therefore critical to supporting and sustaining the long-term economic and social development of communities around the world. It is therefore counterproductive to pursue measures that degrade these in the pursuit of a short-term gain. Human activities, however, continue to erode the health of ecosystems on which all species (our own included) depend. Land-use changes, including deforestation, result in a loss of valuable habitats, a decrease in clean water, land degradation, soil erosion, and the release of carbon into the atmosphere. They contribute to the loss of valuable economic assets and livelihood opportunities. As set out by the Food and Agriculture Organization (2019), natural resources are deteriorating, ecosystems are stressed, and biological diversity is being lost across the globe. **SDG 15** therefore sets the goal to '*protect, restore and promote sustainable use of terrestrial ecosystems, sustainably manage forests, combat desertification, and halt and reverse land degradation and halt biodiversity loss*' (United Nations 2015a). This is essential to providing environmental services for ensuring safe and sustainable water supplies, supporting sustainable food systems, and mitigating climate change.

Table 15.1 shows 12 targets and three means of implementation relating to the **SDG 15**. These include ambitions to protect freshwater ecosystems and their services, improve forest management, combat desertification, and conserve mountain ecosystems. Its targets and 12 indicators draw heavily from and build on the 2011–

Table 15.1 SDG 15 Targets and Means of Implementation

Target	Description of Target (15.1 to 15.9) or Means of Implementation (15.A to 15.C)
15.1	By 2020, ensure the conservation, restoration and sustainable use of terrestrial and inland freshwater ecosystems and their services, in particular forests, wetlands, mountains and drylands, in line with obligations under international agreements
15.2	By 2020, promote the implementation of sustainable management of all types of forests, halt deforestation, restore degraded forests and substantially increase afforestation and reforestation globally
15.3	By 2030, combat desertification, restore degraded land and soil, including land affected by desertification, drought and floods, and strive to achieve a land degradation-neutral world
15.4	By 2030, ensure the conservation of mountain ecosystems, including their biodiversity, in order to enhance their capacity to provide benefits that are essential for sustainable development
15.5	Take urgent and significant action to reduce the degradation of natural habitats, halt the loss of biodiversity and, by 2020, protect and prevent the extinction of threatened species
15.6	Promote fair and equitable sharing of the benefits arising from the utilization of genetic resources and promote appropriate access to such resources, as internationally agreed
15.7	Take urgent action to end poaching and trafficking of protected species of flora and fauna and address both demand and supply of illegal wildlife products
15.8	By 2020, introduce measures to prevent the introduction and significantly reduce the impact of invasive alien species on land and water ecosystems and control or eradicate the priority species
15.9	By 2020, integrate ecosystem and biodiversity values into national and local planning, development processes, poverty reduction strategies and accounts
15.A	Mobilize and significantly increase financial resources from all sources to conserve and sustainably use biodiversity and ecosystems
15.B	Mobilize significant resources from all sources and at all levels to finance sustainable forest management and provide adequate incentives to developing countries to advance such management, including for conservation and reforestation
15.C	Enhance global support for efforts to combat poaching and trafficking of protected species, including by increasing the capacity of local communities to pursue sustainable livelihood opportunities

2020 Strategic Plan for Biodiversity and its Aichi Targets, to be reached by 2020 (see Box 15.1).

Central to the SDGs is the notion that development and environmental protection are not contradictory, but should move forward in parallel. For example, innovation in agriculture (**SDG 2**), pursuit of 'green' economic growth (**SDG 8**), and resource efficiency in industrialisation and infrastructure development (**SDG 9**) can not only reduce their impacts on the natural environment, but also take positive steps to reinforce environmental standards. Given this focus, the progress being made towards **SDG 15** is a critical measure of overall progress on the UN 2030 Agenda, and a key enabler for many other SDGs and targets, including supporting food security (**SDG 2**) and climate action (**SDG 13**). Actions to implement **SDG 15** directly affect the lives and well-being of many indigenous communities, pastoralists, and others traditionally viewed

as excluded, marginalised, or at risk of being left behind (United Nations 2018a).

This chapter characterises how geoscientists, in partnership with others, can help to deliver the targets in Table 15.1, and ensure the thriving of diverse ecosystems. Unsurprisingly, given the title 'life on land' the **SDG 15** targets strongly link to the biotic aspects of the natural environment, but it is impossible to understand terrestrial and freshwater ecosystems without also recognising interactions between biotic and abiotic features. The underlying geology of a region shapes the landscapes that form, including the topography, vegetation, and diversity of natural habitats (Fig. 15.1). Understanding the distribution of ecosystems (from wetlands to forests) and taking measures to restore them requires decision-making to consider geological materials and processes, as outlined through this chapter.

Fig. 15.1 Biological environment associated with a limestone outcrop. Photo from Singing Sands, Bruce Peninsula National Park, Canada (*Credit* Arbitrarily0, CC-BY-SA 3.0, https://creativecommons.org/licenses/by-sa/3.0/)

Many of the large geoscience unions, such as the European Geosciences Union (EGU), include divisions focused on biogeosciences, hydrology, and soil system sciences, all at the interface between geodiversity and biodiversity. Contributions from others, including hydrogeologists, geomorphologists, Earth observation specialists, and contaminated land scientists, also contribute to our understanding of landscapes, land degradation, and the preservation and restoration of natural habitats. The EGU General Assembly in 2019 included 32 sessions with 'biodiversity' in the title or abstract, including diverse themes such as '*mountain building, volcanism, climate and biodiversity in the Andes*', '*coastal wetlands: their processes, interactions and future*', and '*biogeochemical cycles and ecohydrology in changing tropical systems*'.[1] While engagement by ecologists, zoologists, and botanists is

imperative to delivering **SDG 15,** the need for geoscientists should not be overlooked.

In Sect. 15.2, we set out context to SDG 15, examining progress and key challenges guiding actions towards the targets. In Sect. 15.3, we explore the role of geoscientists in informing actions towards Targets 15.1 to 15.5. This section therefore includes a discussion of geoscience and freshwater ecosystems, forests, land degradation and desertification, and mountain ecosystems. The results broadly relate to the protection and restoration of diverse natural habitats. Section 15.4 reflects on the examples within this chapter to make recommendations to strengthen the links between ecology and geology communities to support SDG 15.

15.2 Progress Towards SDG 15 and Remaining Challenges

The Millennium Development Goals (2000–15) included a broad goal focused on ensuring environmental sustainability, including by reversing

---

[1]https://meetingorganizer.copernicus.org/EGU2019/sessionprogramme.

the loss of environmental resources (e.g., forests) and reducing biodiversity loss. Achievements during this period included a substantial increase in the extent of terrestrial protected areas, for example, from 8.8 to 23.4% between 1990 and 2014 in Latin America and the Caribbean (United Nations 2015b). Actions catalysed by this goal were supported by the *Strategic Plan for Biodiversity 2011–2020*, including the Aichi Biodiversity Targets (Box 15.1).

Box 15.1. Strategic Plan for Biodiversity 2011–2020

Agreed by the 10th Conference of the Parties to the Convention on Biological Diversity (a multilateral treaty with 195 states and the European Union), the Strategic Plan for Biodiversity included a set of targets to implement between 2011 and 2020, known as the Aichi Biodiversity Targets (CBD 2010). This plan was seen as a step towards a longer term vision of living in harmony with nature by 2050, with biodiversity *'valued, conserved, restored and wisely used, maintaining ecosystem services, sustaining a healthy planet and delivering benefits essential for all people'* (CBD 2010).

The Aichi Targets include 5 strategic goals (summarised below), with 20 targets underneath these:

A. Address the underlying causes of biodiversity loss by mainstreaming biodiversity across government and society
B. Reduce the direct pressures on biodiversity and promote sustainable use
C. Improve the status of biodiversity by safeguarding ecosystems, species, and genetic diversity
D. Enhance the benefits to all from biodiversity and ecosystem services
E. Enhance implementation through participatory planning, knowledge management, and capacity building

Action on these goals is primarily through local to national activities, with support from regional and global organisations and partnerships. In 2020, the Convention on Biological Diversity will adopt a new framework to support efforts to advance progress towards the 2050 Vision.

Despite some achievements, and an increase in global awareness of the value of protecting biodiversity, significant challenges remain. The most recent Global Biodiversity Outlook report, published in 2014, highlighted progress is not enough to achieve the overall ambitions set in the Strategic Plan (CBD 2014). The concluding Millennium Development Goals report (United Nations 2015b) noted that species are declining overall in numbers and distribution. It is estimated that between 15% and 37% of terrestrial species may be lost by 2050 (Bradford and Warren 2014), and the remaining species likely will migrate to higher latitudes and altitudes to create novel animal assemblages. Reliable prediction of which species will go extinct, where they will relocate to, and subsequent associations can only be achieved through significant advances in our understanding of the physical and biological world (Bradford and Warren 2014).

The United Nations (2015b) also estimates that 5.2 million hectares of forest were lost each year from 2000 to 2010 (the same regional extent as Costa Rica). While this marks a decrease in the rate of forest loss (down from 8.3 million hectares in the 1990s), it is still significant with particularly high losses in South America and sub-Saharan Africa (United Nations 2015b). Figure 15.2 shows the changing extent of forest as a share of the land area in 1990 and 2015, with both evidence of reforestation (e.g., in Europe) and deforestation (e.g., in South America). The Food and Agriculture Organization's latest reporting on 'The State of the World's Forests' (FAO 2018a, b) also notes increasing forest cover in parts of Asia.

Forest area as share of land area, 1990

Forest area is land under natural or planted stands of trees of at least 5 meters in situ, whether productive or not, and excludes tree stands in agricultural production systems (for example, in fruit plantations and agroforestry systems) and trees in urban parks and gardens.

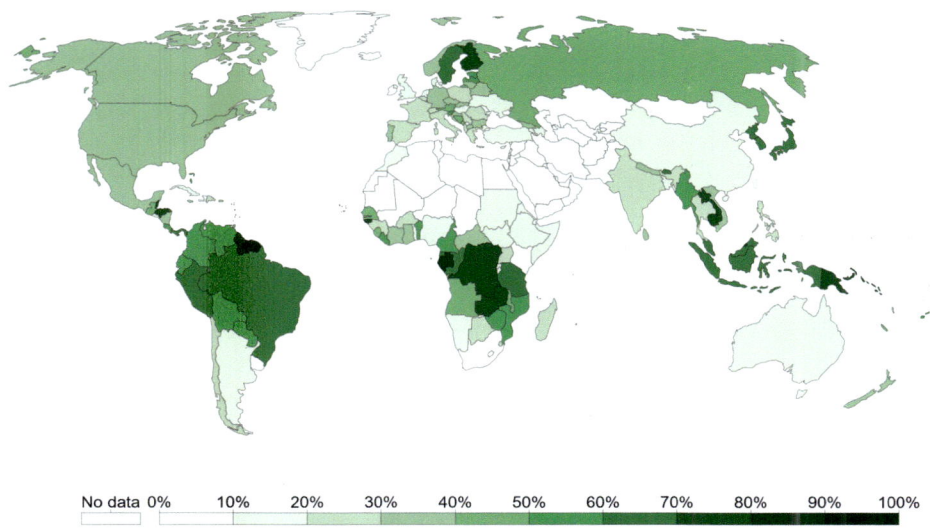

No data 0%   10%   20%   30%   40%   50%   60%   70%   80%   90%   100%

Source: UN Food and Agriculture Organization (FAO)   OurWorldInData.org/forests/ • CC BY

Forest area as share of land area, 2015

Forest area is land under natural or planted stands of trees of at least 5 meters in situ, whether productive or not, and excludes tree stands in agricultural production systems (for example, in fruit plantations and agroforestry systems) and trees in urban parks and gardens.

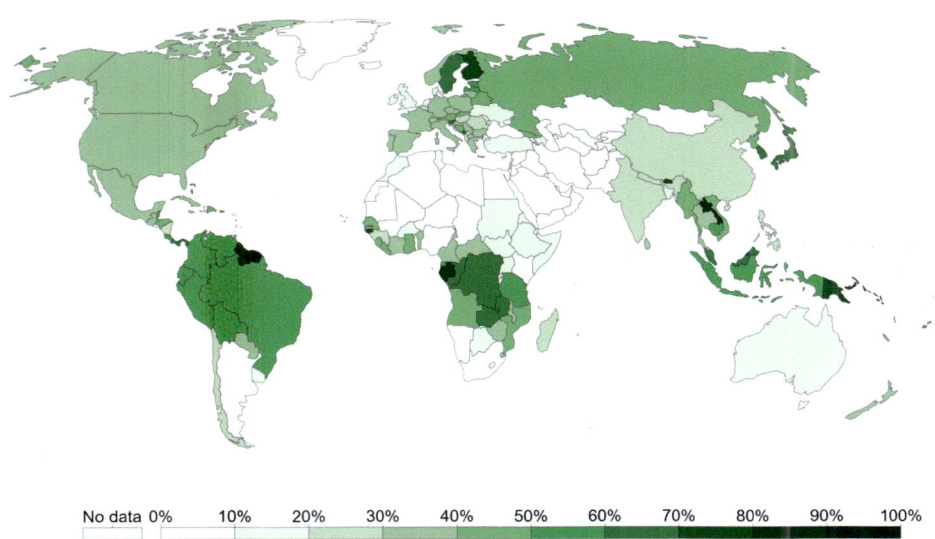

No data 0%   10%   20%   30%   40%   50%   60%   70%   80%   90%   100%

Source: UN Food and Agriculture Organization (FAO)   OurWorldInData.org/forests/ • CC BY

Fig. 15.2 Forest area as a share of Land Area in (top) 1990 and (bottom) 2015. From Roser (2019). Reproduced under a CC-BY license (https://creativecommons.org/licenses/by/4.0/)

Challenges extend beyond the loss of biodiversity and forest cover, to include also drought and desertification, affecting migration and the livelihoods of the global poor (Tittensor et al. 2011; UNECA 2018). About one-fifth of the Earth's land surface covered by vegetation shows persistent and declining trends in productivity in the decades between 1999 to 2013, threatening the livelihoods of over one billion people (United Nations 2018b). Land is one of the few productive assets owned by the rural poor, and almost all such households engage in some form of agriculture (Barbier and Hochard 2016). Between 2000 and 2010, the numbers of rural poor living on degrading agricultural land increased in low-income countries, and in sub-Saharan Africa and South Asia (Barbier and Hochard 2016). There was a substantial loss of arable land, with sub-Saharan Africa showing some of the most severe land reductions (Nkonya et al. 2016). Deforestation and loss of biodiversity have become risks reported in many countries across Africa (Nkonya et al. 2016) and South America (Fig. 15.3). Deforestation and

loss of biodiversity and food cause diverse health effects such as undernourishment and skin disease and loss of medicinal plants. In Gabon, for instance, the decline of agricultural production has led to a change in dietary habits with increased consumption of bushmeat. Deforestation also favours freshwater snails carrying schistosomiasis and mosquitoes carrying malaria.

It is set against these challenges and trends that the SDGs were agreed, and the ambitions of **SDG 15** were shaped. Since 2015, progress towards the stated targets in Table 15.1 for protecting and preserving critical and fragile ecosystems for human survival reveals mixed trends (United Nations 2019a). Land degradation continues, with biodiversity loss at an alarming rate. Poaching and wildlife trafficking are hindering efforts to protect ecosystems (United Nations 2019a). Expansion of major agribusiness frontiers (e.g., for cattle, soy, palm oil, and cocoa) into tropical forest regions is another cause for concern (Weisse and Goldman 2018). The links between deforestation, biodiversity loss and land degradation with climate change

Fig. 15.3 Deforestation in the Maranhão state of Brazil, 2016. *Credit* Felipe Werneck - Ascom/Ibama (CC-BY 2.0, https://creativecommons.org/licenses/by/2.0/)

(**SDG 13**), tenure insecurity (**SDG 10**), food insecurity (**SDG 2**), gender inequality (**SDG 5**), and poverty (**SDG 1**) are well-documented (Van Haren and van Boxtel 2017). Taking action to implement **SDG 15**, therefore, remains crucially important to the attainment of other SDGs.

In setting the scene for SDG 15, it is also important to note that current discourses around tackling land degradation and ecosystem and biodiversity loss focus on *local-* or *national*-scale actions. The local/national perspective is obviously very important, particularly to ensure we leave no-one behind, and the SDGs are delivered in an inclusive manner. Responsibility to address challenges, however, falls on the governments within whose borders these issues persist, rather than them being seen as global challenges, linked to our shared global economy, and requiring shared responsibility (and leadership) in their resolution. The conversion of forest to agricultural land may be to meet the demands of export markets in the wealthier countries of Europe and North America, and emerging middle classes in India and China. This increased agricultural production does not make any significant contribution to local livelihoods or food security.

Van Haren and van Boxtel (2017) also note that the emphasis on 'local' within the SDG framework allows states to set their own definition of *'sustainable development'* and to pursue it according to national priorities and capacities. This results in trade-offs between social, economic, and environmental pillars of sustainable development, and a lack of integrated thinking regarding the environmental interactions with social and economic dimensions (i.e., environmental protection as an enabler of human and economic development). Addressing **SDG 15** by 2030 requires recognition of a shared responsibility for ecosystem damage, and actions at both local and global levels to reverse this.

15.3 Role of Geoscience in Ecosystems and Planning

Geoscience is intimately linked with the biological, chemical, and physical sciences and investigates the past, measures the present, and models the future behaviour of our planet. The link between geoscientists and sustainability has been highlighted previously (Gill 2017; Rogers et al. 2018), together with recognition of the relevance of geodiversity and geoconservation (Gordon et al. 2017; Schrodt et al. 2019). We build on these analyses in this section to illustrate how diverse geoscience activities can help to address the challenges described in Sect. 15.2, and meet the targets expressed within **SDG 15**. Through the monitoring of diverse environmental processes and materials (e.g., lake sediments, the rock record, variations in depth to groundwater), geoscientists develop a holistic, multiscale, multidisciplinary understanding of Earth systems that can help underpin environmental policy making and assess the impact of diverse policies on the natural environment. Geoscience input is particularly relevant for Targets 15.1 (conserve and restore terrestrial and freshwater ecosystems), 15.2 (end deforestation and restore degraded forests), 15.3 (end desertification and restore degraded land), 15.3 (ensure conservation of mountain ecosystems), and 15.5 (protect biodiversity and natural habitats).

15.3.1 Biodiversity: Freshwater Ecosystems

In both a biological and a societal sense, water is the basis for life on land as we know it. It is and will remain a crucial factor in the many challenges that our world faces, and its protection and management cut across many of the SDGs (Bhaduri 2016), such as health (**SDG 3**), gender equality (**SDG 5**), and sustainable cities (**SDG 11**). The IUCN (2019c) note that over 140,000 described species (including 55% of all fish species) rely on freshwater habitats, with these species going extinct more rapidly than terrestrial and marine species due to habitat loss, introduction of alien species, pollution, and overharvesting. Achieving targets associated with the conservation, restoration, and sustainable use of freshwater ecosystems and their services therefore requires (i) expansion of officially protected areas, particularly key biodiversity areas (KBAs),

(ii) reduction in pollution, and (iii) increases in environmental data to inform decisions about use and restoration of ecosystems.

The protection and restoration of inland freshwater ecosystems can be guided by understanding how these ecosystems are shaped by geological environments and processes. For example, Arbuckle and Downing (2002) demonstrate relationships between freshwater mussels (abundance and species richness) and geology. They found that alluvial deposits improve groundwater flux to streams, contributing to relatively stable stream flows in alluvial watersheds, and helping to improve mussel persistence. An implication of this particular study is the knowledge that characterising geoscience can inform conservation efforts. Arbuckle and Downing (2002) found that mussel conservation efforts are most critical in highly sloping landscapes with less permeable soils, where low groundwater flows might lead to unfavourable conditions. Systematic studies of how geological materials (e.g., rock types) and processes favour or hinder the success of particular species could guide broader conservation efforts, the selection of sites for restoration (where chances of success are greater), and the identification of similar sites around the world where rare biodiversity may be found (and should be protected).

Official freshwater protected areas are recognised to achieve the long-term conservation of nature. UNEP-WCMC and IUCN (2016) estimate that just under 15% of the world's terrestrial and inland waters are Protected Areas. A fundamental measure of their efficacy is the extent to which protected areas overlap with places that contribute significantly to the maintenance of global biodiversity, including KBAs. Safeguarding KBAs around the globe in inland freshwater ecosystems (as well as terrestrial and mountain ecosystems) is critically important for maintaining genetic resources, species, and ecosystem diversity, and protecting the benefits they provide to people.

One factor hampering the understanding of hydrological systems and sustainable management of freshwater ecosystems in many countries in the Global South, including many across

Africa, is a decline in the networks of hydrological observing stations and water quality measurements. The World Bank (2018) estimate that 68% of hydrometeorological observation networks in so-called developing countries are in a poor or declining state, with a further 14% inadequate to meet all user needs. Stations may be neglected and abandoned, with reductions in budgets for field maintenance and inspection, and insufficient discharge measurements being made to understand the system. Consequences include a lack of real-time data for monitoring the progress of droughts and floods, and insufficient long-term data for the design of water-related schemes and for the integrated management of large multinational river basins, planning and implementing projects. Furthermore, without effective environmental monitoring, it is difficult to assess the sources, migration, and impact of pollution on freshwater ecosystems. For example, Förstner and Prosi (1979) set out diverse origins of heavy metal pollution in freshwater ecosystems, including weathering of bedrock.

Systematic environmental data collection, management, integration, and access are a widely recognised development priority, informing policy coherence and resource management (Gill et al. 2019). This includes monitoring of groundwater resources (see **SDG 6**). Using groundwater resources in an effective and environmentally sensitive manner requires long-term monitoring programmes, good understanding of the geological environment, and effective partnerships and donor practices (Contestabile 2012; Langenberg 2012).

Box 15.2. Limnology—Lakes and Sustainable Development

Case study contributed by: Keely Mills (British Geological Survey) and Laura Hunt (British Geological Survey/University of Nottingham)

Although <0.01% of the Earth's freshwater occurs as surface water (i.e., in lakes, swamps, and rivers), these systems are critically important to the environment, including the ecosystems and human populations that they support. Freshwater lakes (lentic systems) have a high Natural Capital, and provide an array of ecosystem services including clean water for drinking, water for domestic use and sanitation, water for industry and agriculture (e.g., irrigation), supporting fish stocks (aquaculture), ecotourism, and recreation. Lake systems are vulnerable to Earth's changing climate, and are also under increasing stress from anthropogenic impacts, both directly (e.g., through the abstraction of water, or the introduction of fish for aquaculture) and indirectly (e.g., as a result of anthropogenic modification of its catchment, including urbanisation) (Fig. 15.4). Geoscientists working in the disciplines of limnology and palaeolimnology aim to understand how freshwater

systems change through space and time which is crucial to ensuring global-scale resource availability at a time when humans are increasingly driving environmental change (Mills et al. 2017; Dubois et al. 2018).

In its broadest definition, limnology is the study of the functional relationships and productivity of freshwater communities, and how they are affected by their physical, chemical, and biotic environment. Geoscientists, who study palaeolimnology, seek to apply limnological knowledge to understand how lake systems have functioned in the past (Wetzel 1983). Lakes are of particular interest to geoscientists as they accumulate sediments over time, recording changes that occur in the lake and its catchment at the time they were deposited. These sediments can be

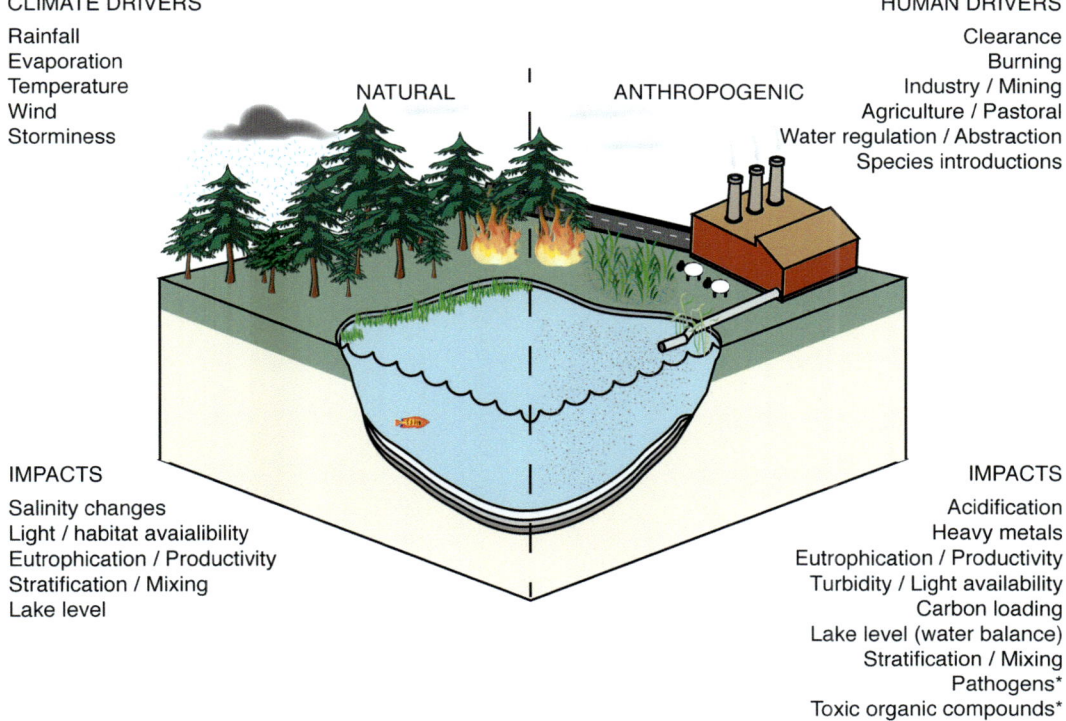

CLIMATE DRIVERS

Rainfall
Evaporation
Temperature
Wind
Storminess

NATURAL | ANTHROPOGENIC

HUMAN DRIVERS

Clearance
Burning
Industry / Mining
Agriculture / Pastoral
Water regulation / Abstraction
Species introductions

IMPACTS

Salinity changes
Light / habitat avaialibility
Eutrophication / Productivity
Stratification / Mixing
Lake level

IMPACTS

Acidification
Heavy metals
Eutrophication / Productivity
Turbidity / Light availability
Carbon loading
Lake level (water balance)
Stratification / Mixing
Pathogens*
Toxic organic compounds*

Fig. 15.4 Climatic and anthropogenic drivers, and the impacts they have on lake systems. From Mills et al. 2017. © BGS/NERC (used with permission)

extracted (as a sediment core) and analysed for their chemical, physical, and biotic parameters, allowing past environmental conditions to be inferred.

For example, current limnological and palaeolimnological research in western Uganda is investigating how small crater lakes respond to past climate change, and recent human impacts (Fig. 15.5). Despite the provision of groundwater pumps in western Uganda, surface waters still provide drinking water to many remote communities, and the crater lakes play a crucial role in the livelihoods of many people. Exponential population growth in the region is placing unprecedented pressures on all water resources, which is coupled with uncertainty relating to the impact of future climate change scenarios on water balance in tropical Africa. The intensity and magnitude of land-use changes in the region are already impacting the water balance and water quality of the crater lakes. Future changes in climate will only serve to exacerbate negative feedback processes, which has implications for achieving a number of the SDGs: including zero hunger (**SDG 2**), clean water and sanitation (**SDG 6**), and life on land (**SDG 15**).

The sensitivity of the lake systems in western Uganda can be understood through contemporary monitoring (limnology), but records in the region are sparse, and often incomplete. In the absence of monitoring records, analysis of the crater lake sediments (palaeolimnology) provides information on past changes in hydrology and water quality (on the order of tens to thousands of years).

Fig. 15.5 Lake Nyungu, Uganda. The catchment of this crater lake has experienced natural vegetation clearance to create space for small-scale banana plantations. © Keely Mills (used with permission)

Research in the region has shown that the lakes are sensitive to past changes in hydroclimate, and many phases of drought in the region have been linked to political unrest and the abandonment of settlements. Conversely, some lakes are known to have persisted during long-lived periods of rainfall deficit. Quantifying the nature of past changes has major implications that can inform the Earth system, and models that allow the identification of areas and communities which will be particularly vulnerable to future hydrological stress (Mills et al. 2018).

For geoscientists to support efforts to protect freshwater biodiversity, there is a primary need to provide adequate and accurate data and information for water resources management and development activities at national levels (river and basin). This includes undertaking, integrating, and expanding geoscientific research data collection and knowledge management capacity on water resources to support improved policymaking, public awareness, and multi-stakeholder mobilisation. Geoscientists can help to establish effective and sustainable environmental monitoring networks across the Global South. This will help improve the capacity for effective water resources management at all levels and inclusion systematically in integrated water resource management plans. Geoscientists can also advance and communicate research into the geological links to freshwater species persistence and diversity, helping to integrate this knowledge into the design of conservation and restoration programmes, or selection of sites to protect.

15.3.2 Biodiversity: Forests

Forests act as a source of food, medicine, and fuel for more than a billion people. In addition to helping to respond to climate change and protect soils and water, forests hold more than three-quarters of the world's terrestrial biodiversity, provide many products and services that contribute to socio-economic development, and are particularly important for hundreds of millions of people in rural areas, including many of the world's poorest. The increasing demand for food and natural resources will place enormous pressure on the way we use productive land, particularly in the Global South where the overwhelming majority of the world's 800 million poor and hungry people are concentrated. Deforestation, chiefly caused by the conversion of forest land to agriculture and livestock areas, damages biodiversity, increases land degradation, removes natural protection against hazards such as landslides, and hinders efforts to tackle climate change. It also threatens the livelihoods of those who depend on the forest, including many indigenous communities. In 2019, there was a surge in the number of fires in the Brazilian Amazon rainforest, with this increase consisting of large, intense, and persistent fires associated with land clearing (NASA 2019). Figure 15.6 shows the extent of these, using an image from the MODIS instrument on NASA's Terra satellite. Meeting the food and resource needs of future generations, without reducing forest area, is one of the great challenges of our times.

Forests are shaped by the underlying geology of the region. Landscapes are determined by geological processes—from ancient volcanic activity and faulting to the large glaciations that have carved valleys and transported large volumes of material from one place to another. The chemical and physical nature of soils is also determined by the underlying geology. Soils evolve as biological material combines with the products of rock weathering and erosion. Different underlying rocks result in different soils with different structures and chemistry. For example, Fayolle et al. (2012) note that species distribution in a study area in central African moist forests is strongly driven by geological substrate, climate, and recent history of human disturbance. They note, for example, a large sandstone plateau ($\sim 25,000$ km^2) having a dominating role in shaping tree distribution patterns. Higgins et al. (2011) suggest that

Fig. 15.6 Fire detections from the MODIS instrument on NASA's Terra satellite (15 to 22 August, 2019). *Credit* NASA Earth Observatory (public domain)

Amazonian forests are shaped by geological formations and their influence on the soil.

Given the relationship between geology and soils, and soils and the resulting forest development, understanding the subsurface is an important part of forest management. The Land Management Plan for the Forest of Dean (Forestry England 2019) reflects on the complex geology beneath the forest, and its impact on the vegetation covering the ground and the trees growing. It notes that (i) limestones generally lead to alkaline, well-drained, and often quite shallow soils; (ii) sandstones, sands, and gravels lead to more acidic, well-drained brown, 'podzolic' soils (characterised by a high sand content, low nutrients and moisture, and therefore poorer agriculture); and (iii) fine-grained rocks, such as clays,

mudstones, and shales, lead to poorly drained soils (Forestry England 2019). Understanding the geological history of the region, and access to appropriate geological maps, provides insights into soil types and properties, and the likely impact on vegetation. This can guide decision-making in forest conservation and restoration, ensuring strategies to reverse the decline in forests articulated in Sect. 15.2 that are informed by our best understanding of the natural environment. While the underlying geology of forest regions may be understood at a high resolution in many parts of the world, this is not the case in the Global South, indicating the importance of improved geological mapping to help deliver **SDG 15**.

In addition to informing the restoration of forests, geoscientists also have a critical role to play in managing erosion and slope stability, and potential adverse effects of deforestation. Vegetation can help increase the stability of a slope, by removing water from the system and increasing the resistive forces in the slope (providing an anchor). Removing vegetation therefore reduces the shear strength of slopes and promotes slope failure. Deforestation can also result in increased flooding, due to soil erosion. Geoscientists can help to identify areas that are particularly prone to landslides, interpreting the landscapes to assess for past mass movements, and characterising the geology to determine failure susceptibility. This information can inform disaster risk reduction and management approaches, with 'understanding all components of risk' being a key facet of the Sendai Framework for Disaster Risk Reduction (see **SDGs 1** and **13**). It can also help to embed disaster risk reduction into the actions supporting **SDG 15**, informing sites for forest restoration that also increase community resilience to natural hazards.

15.3.3 Land Degradation and Desertification

Target 15.3 aims to '*combat desertification, restore degraded land and soil, including land affected by desertification, drought and floods, and strive to achieve a land degradation-neutral world*'. **Land degradation** is defined as the '*temporary or permanent decline in the productive capacity of the* **land**, *and the diminution of the productive potential, including its major* **land** *uses (e.g., rain-fed arable, irrigation, forests) and farming systems (e.g., smallholder subsistence), and its value as an economic resource*' (Stocking 2001). Land degradation can occur due to natural processes, inappropriate land use, or other human activities and habitation patterns that contribute to contamination, soil erosion, and the destruction of vegetation through deforestation or overgrazing (United Nations 1997). For example, soil degradation may occur due to the use of fertilisers, the dumping of industrial waste, leaching of chemicals from mining activities (both artisanal and larger scale). Land

degradation is affecting one-fifth of the Earth's land area and the lives of one billion people (United Nations 2019a). This represents a significant loss of services essential to human wellbeing.

Desertification is a specific type of land degradation, where fertile land transitions to desert as a result of '*the persistent degradation of dryland ecosystems by human activities—including unsustainable farming, mining, overgrazing, and clear-cutting of land—and by climate change*' (United Nations 2019b). Such activities contribute to a process where vegetation binding soils are removed, soil nutrients are depleted by farming, and/or the original nutrient-rich topsoil is eroded away by animals, human activities, wind, and water. What is left is a highly infertile mix of dust and sand (United Nations 2019b), resulting in eventual 'desertification'. The impacts are widespread, including damage to ecosystems, food security, livelihoods, and the potential displacement of communities, creating environmental refugees (Fig. 15.7).

Despite the crucial role that land plays in human welfare and development, investments in sustainable land management are low, especially in the Global South (FAO 2012). The cost of land degradation due to land use and cover change accounts for 78% of the US$ 300 billion total global cost of land degradation (Nkonya et al. 2016), highlighting the need to prioritise sustainable land management through effective policies and planning. These should protect high-value biomes and ecosystem services that benefit both local and global communities. Policies and planning should be informed by environmental data and landscape characterisation (see Box 15.3, for example) to ensure resources are used wisely and actions do not exacerbate existing environmental challenges.

Box 15.3 Desertification and Groundwater

Geomorphological and hydrogeological investigations can enhance our understanding of the processes triggering desertification and the actions needed to reverse this.

Fig. 15.7 Telly, Mali. Desertification is a major environmental issue in Mali, with drought, soil erosion, and deforestation all contributing. *Credit* Ferdinand Reus (CC-BY-SA 2.0, https://creativecommons.org/licenses/by-sa/2.0/)

The Hunshandake Sandy Lands (Inner Mongolia, China) are one of the largest areas of sand in China, undergoing rapid desertification. A study by Yang et al. (2015) in the *Proceedings of the National Academy of Sciences* traces the origins of deserts in northern China. They describe desertification triggered by climate change-induced changes in the hydrological and geomorphological systems approximately 4000 years ago. One aspect of the hydrological change described by Yang et al. (2015) is the sapping of groundwater away from the region, diverting water resources into the Xilamulun River, and resulting in a drop in the water table of approximately 30 m. Rapid desertification ensued, having a devastating impact on the Hongshan Culture. This example illustrates the potential for climate change to result in rapid and intense desertification. It also highlights the need to understand the environmental factors contributing to desertification. Yang et al. (2015) note that the irreversible regional geomorphic and hydrological change will hinder any rehabilitation efforts in this region, with resources better used to support land restoration efforts elsewhere.

More generally, improved monitoring and improved management of groundwater resources play an important role in understanding and tackling desertification. Excessive abstraction can exacerbate desertification, while increased access and careful management can improve irrigation and support reforestation programmes. Chebaane et al. (2004) illustrate this through a study in the *Hydrogeology Journal* that outlines groundwater management options in Jordan. Groundwater over-exploitation is causing environmental damage, soil salinisation and could contribute to desertification. Management strategies include limiting the amount of groundwater that can be used for irrigation. While this would tackle the problems of over-exploitation, it is possible that a reduction in irrigation will also contribute to desertification. In this case, water management practices need to be accompanied by land restoration projects, with land that is not going to be irrigated restored to its original condition using land management practices.

15.3.4 Mountain Ecosystems

Mountains cover 25% of the world's land surface, and directly support 12% of the world's population (IUCN 2019b), and as articulated in Sect. 15.1 and Target 15.4, conservation of mountain ecosystems is integral to sustainable development. Mountains have unique biodiversity, contribute to essential freshwater and clean air, and are homes to rich sources of cultural diversity, leisure activities, and geological landscapes. They provide important income sources for communities through agriculture, tourism, and use of natural resources. The *Mountain Green Cover Index* measures changes in the area of green vegetation (forest, shrubs, pastures, and cropland) in mountain areas, and is used to monitor progress on Target 15.4. This index is an indicator of the extent to which mountains are efficiently managed, taking into consideration the inherent trade-offs and delicate balance between conservation and sustainable use of mountain resources. The Earth observation community will contribute to the data needed to monitor and report on this indicator (Fig. 15.8).

Fig. 15.8 Mountain Ecosystems in the Sahale Arm in the North Cascades, Washington, United States. *Credit* Jeff Pang (CC-BY 2.0, https://creativecommons.org/licenses/by/2.0/)

As with other ecosystems described in Sects. 15.3.1 and 15.3.2, there are links between species distribution and diversity and the underlying geological materials and processes in mountain regions. Significant biodiversity in mountain regions is strongly associated with the bedrock geology. Mountain substrates are often very different to those in lowland basins, and more heterogeneous, contributing to a greater range of species (Rahbek et al. 2019). Badgley et al. (2017) note that topographically complex regions feature hotspots of biodiversity that reflect geological influences on ecological and evolutionary processes. Riehl et al. (2018) and Antonelli et al. (2018) also highlight how species richness correlates with erosion rates and heterogeneity of soil types, with geomorphologists and geologists contributing to the understanding of both in diverse regions around the world. Biodiversity is also influenced by the location and orientation of mountain ranges in relation to air circulation patterns, and how species diversification, dispersal, and refugia (a location supporting an isolated population of a once more widespread species) respond to climate change. In summary, understanding biosphere–lithosphere interactions, and their responses to climate change, enriches our understanding of the patterns and evolution of mountain ecosystems and biodiversity (Badgley et al. 2017).

15.4 Strengthening Links Between Ecological and Geological Sciences

Freshwater ecosystems, forests, and mountain ecosystems all provide critical services to support human development and flourishing. Their protection and restoration underpin progress on many of the SDGs, and cannot be separated from the economic and social development that many choose to prioritise. Throughout this chapter, we have illustrated how these diverse 'land-based' ecosystems and their biodiversity are shaped (in part) by the underlying geology and geological

processes, and how land degradation (including through desertification) can be exacerbated by activities associated with the geosciences (e.g., mining, groundwater exploitation), and therefore also addressed through improved management of geological resources. This chapter, therefore, advocates for strengthened links between ecological and geological sciences. This is important for three reasons:

1. The *systematic collection, management, integration, and availability of (geo)environmental data can inform decision*-making on issues such as land use, groundwater abstraction, and pollution management. In many regions, environmental monitoring networks lack capacity, or data is not stored and made available in a way that others can access it. Geoinformatics experts can help to improve decision-making through developing and implementing appropriate data standards.

2. Understanding of the distribution and properties of geological materials can *inform protection, conservation, and restoration programmes*, informing decisions about site selection and regions where comparable biodiversity may be located.

3. Research into geological processes (e.g., erosion and, slope stability) can also *inform protection, conservation, and restoration programmes*, informing decisions about site selection and how actions to deliver **SDG 15** can align with initiatives to reduce disaster risk.

Reviews of progress towards **SDG 15** reveal mixed trends for life on land, with key challenges remaining unaddressed, and undesirable trade-off of SDG-targets and business as usual approaches not being challenged (United Nations 2018a). Meeting the ambitious targets of **SDG 15** will require integrated approaches, and engagement by geoscientists. Progress depends on access to high-quality environmental data, knowledge sharing, and new interdisciplinary and multisectoral partnerships (see **SDG 17**). In reflecting on the examples set out in this chapter, we make the following recommendations.

- **Ground geosciences in the sustainability of life on land**. Links between SDG 15 and geoscience should be made clearer in the training and organisational structures in the geoscience community, by emphasising the value of partnerships with ecologists and introducing more teaching on geological influences on ecosystems and biodiversity. Furthermore, increasing public awareness of the role of geoscience in SDG 15 is needed, with research results being transparent and comprehensible.
- **Transition to water** sustainability. Geoscientists will have to address the water-related science, policy and societal questions regarding global environmental change and the pathways towards sustainable futures (see **SDG 6** for further discussion). Going forward, geosciences should strive to support a transition to water resources sustainability and enhance understanding of how the global water system may change in the future, helping to improve protections for freshwater ecosystems. Studies of freshwater systems can inform risk assessments and be used to develop strategies to better promote the protection of water systems.
- **Expand monitoring**. Both traditional land-based environmental observation networks and state-of-the-art Earth observation satellite systems can provide detailed observations to inform policy and planning around **SDG 15** (e.g., for groundwater, erosion, and land degradation). Data should be collected in a systematic way, managed carefully, and made available with appropriate metadata to support integration and long-term data sustainability.

These contributions will support efforts to deliver **SDG 15**, halt land degradation, and protect and restore valuable ecosystems, with wide-ranging positive consequences on human security for generations to come.

15.5   Key Learning Concepts

- **SDG 15** aims to '*protect, restore, and promote sustainable use of terrestrial ecosystems, sustainably manage forests, combat desertification, and halt and reverse land degradation and halt biodiversity loss*', recognising that the environmental services they provide facilitate sustainable and resilient communities. Humans do not live in isolation from the millions of species that call Earth their home, the complex ecosystems they form, and a myriad of habitats. Humans coexist with complex interactions and dependencies.
- Addressing **SDG 15** by 2030 requires recognition of a shared responsibility for ecosystem damage, and actions at both local and global levels to reverse trends such as continued deforestation, biodiversity loss, and land degradation.
- Ecosystems and their biodiversity are shaped (in part) by the underlying geology and geological processes. Understanding of the distribution and properties of geological materials can inform protection, conservation, and restoration programmes, informing decisions about site selection and regions where comparable biodiversity may be located.
- Land degradation (including through desertification) can be exacerbated by activities associated with the geosciences (e.g., mining, and groundwater exploitation), and therefore also addressed through improved management of geological resources. For example, improved monitoring and management of groundwater resources play an important role in tackling desertification.
- Many actions to support **SDG 15** would benefit from access to environmental data and strengthened environmental monitoring networks. The systematic collection, management, integration, and availability of (geo) environmental data can inform decision-making on issues such as land use, groundwater abstraction, and pollution management.

15.6 Educational Ideas

In this section, we provide examples of educational activities that connect geoscience, the material discussed in this chapter, and scenarios that may arise when applying geoscience (e.g., in policy, government, private sector international organisations, and NGOs). Consider using these as the basis for presentations, group discussions, essays, or to encourage further reading.

- Select a forest near you and investigate (a) the geology beneath it, and (b) the primary tree and plant species found in the forest. Prepare a public information leaflet for visitors to the forest that outlines the relationship between geoscience and the forest ecosystems.
- Explore the impacts of desertification on other SDGs. For example, how may desertification affect **SDG 2** (food security), **SDG 8** (decent work and economic growth), and **SDG 11** (sustainable communities)? Is desertification being exacerbated by attempts to deliver any of the other SDGs?
- This chapter highlights challenges in maintaining environmental monitoring networks to inform decision-making. Research the existence of groundwater monitoring networks in the least developed country, a small island developing state, an upper-middle-income country, and a high-income country. Characterise their spatial distribution, the time range of the information available, how information is presented, and what parameters are included.
- How could Earth observation support ground-based data collection networks? Review the availability of open-access Earth observation data and outline how these could inform decision-making linked to **SDG 15**. What are the primary gaps in terms of the availability and resolution of open-access Earth observation data?

Further Reading and Resources

FAO (2018) The state of the world's forests. http://www.fao.org/publications/sofo/en/. Accessed 21 Oct 2019

FAO (2019) Mountain partnerships. http://www.fao.org/mountain-partnership/en/. Accessed 21 Oct 2019

UNEP-WCMC, IUCN and NGS (2018). Protected Planet Report 2018. UNEP-WCMC, IUCN and NGS: Cambridge UK; Gland, Switzerland; and Washington, DC, USA

Zech W (2016) Geology and soils. In Pancel, L., and Kohl, M (Eds). Tropical forestry handbook. Springer. 1–191

References

Antonelli A, Kissling WD, Flantua SGA, Bermúdez MA, Mulch A, Muellner-Riehl AN, Kreft H, Linder HP, Badgley C, Fjeldså J, Fritz SA, Rahbek C, Herman F, Hooghiemstra H, Hoorn C (2018) Geological and climatic influences on mountain biodiversity. Nat Geosci 11:718–725. https://doi.org/10.1038/s41561-018-0236-z

Arbuckle KE, Downing JA (2002) Freshwater mussel abundance and species richness: GIS relationships with watershed land use and geology. Can J Fish Aquat Sci 59(2):310–316

Badgley C, Smiley TM, Terry R, Davis EB, DeSantis LR, Fox DL, Hopkins SS, Jezkova T, Matocq MD, Matzke N, McGuire JL (2017) Biodiversity and topographic complexity: modern and geohistorical perspectives. Trends Ecol Evol 32(3):211–226

Barbier EB, Hochard JP (2016) Does land degradation increase poverty in developing countries? PLoS ONE 11(5):e0152973

Bhaduri A, Bogardi J, Siddiqi A, Voigt H, Vörösmarty C, Pahl-Wostl C, Bunn SE, Shrivastava P, Lawford R, Foster S, Kremer H, Renaud FG, Bruns A, Osuna VR (2016) Achieving sustainable development goals from a water perspective. Front Environ Sci 4:64. https://doi.org/10.3389/fenvs.2016.00064

Bradford MA, Warren RJ (2014). Terrestrial Biodiversity and Climate Change. In: Freedman B (eds) Global environmental change. Handbook of global environmental pollution, vol 1. Springer, Dordrecht

CBD (2010) The strategic plan for biodiversity 2011–2020 and the aichi biodiversity targets. Conference of the parties to the convention on biological diversity: Tenth meeting. UNEP/CBD/COP/DEC/X/2. www.cbd.int/doc/decisions/cop-10/cop-10-dec-02-en.pdf. Accessed 21 Oct 2019

CBD (2014) Global Biodiversity Outlook ¬4: A mid-term assessment of progress towards the implementation of the Strategic Plan for Biodiversity 2011–2020. https://www.cbd.int/gbo4/. Accessed 21 Oct 2019

Chebaane M, El-Naser H, Fitch J, Hijazi A, Jabbarin A (2004) Participatory groundwater management in Jordan: development and analysis of options. Hydrogeol J 12(1):14–32

Contestabile M (2012) 'Water at a crossroads', Nature climate change. Nature Publishing Group, a division of Macmillan Publishers Limited. All Rights Reserved., 3, p 11. https://doi.org/10.1038/nclimate1780

Dubois N, Saulnier-Talbot É, Mills K, Gell P, Battarbee R, Bennion H, Chawchai S, Dong X, Francus P, Flower R, Gomes DF, Gregory-Eaves I, Humane S, Kattel G, Jenny J-P, Langdon P, Massaferro J, McGowan S, Mikomägi A, Ngoc NTM, Ratnayake AS, Reid M, Rose N, Saros J, Schillereff D, Tolotti M, Valero-Garcés B (2018) First human impacts and responses of aquatic systems: A review of palaeolimnological records from around the world. The Anthropocene Rev 5:28–68

FAO (2003) Mountains and Freshwater. http://www.fao.org/english/newsroom/focus/2003/dec_idm_3.htm. Accessed 22 Oct 2019

FAO (2012) Voluntary Guidelines on the responsible governance of tenure of land, forests and fisheries. http://www.fao.org/docrep/016/i2801e/i2801e.pdf

FAO (2017) Plant health and food security. http://www.fao.org/3/a-i7829e.pdf. Accessed 1 Oct 2019

FAO (2018) The state of the world's forests. http://www.fao.org/state-of-forests/en/. Accessed 22 Oct 2019

Fayolle A, Engelbrecht B, Freycon V, Mortier F, Swaine M, Réjou-Méchain M, Doucet JL, Fauvet N, Cornu G, Gourlet-Fleury S (2012) Geological substrates shape tree species and trait distributions in African moist forests. PLoS ONE 7(8):e42381

Forestry England (2019) Our shared forest – Forest of dean land management plan. https://www.forestryengland.uk/sites/default/files/documents/Our%20Shared%20Forest%20-%20Forest%20of%20Dean%20Land%20Management%20Plan%20published%20June%202019.pdf. Accessed 22 Oct 2019

Förstner U, Prosi F (1979) Heavy metal pollution in freshwater ecosystems. In Biological aspects of freshwater pollution (pp. 129–161). Pergamon

Gill JC (2017) Geology and the Sustainable Development Goals. Episodes: (40):70–76. https://doi.org/10.18814/epiiugs/2017/v40i1/017010

Gill JC, Mankelow J, Mills K (2019) The role of earth and environmental science in addressing sustainable development priorities in Eastern Africa. Environ Develop 30:3–20

Gordon J, Crofts R, Díaz-Martínez E, Sik Woo K (2017) Enhancing the role of geoconservation in protected area management and nature conservation. Geoheritage. https://doi.org/10.1007/s12371-017-0240-5

Higgins MA, Ruokolainen K, Tuomisto H, Llerena N, Cardenas G, Phillips LO, Vasquez R, Rasanen M (2011) Geological control of floristic composition in Amazonian forests. J Biogeogr. https://doi.org/10.1111/j.1365-2699.2011.02585.x

IUCN (2019a) Forests and climate change. https://www.iucn.org/resources/issues-briefs/forests-and-climate-change. Accessed 22 Oct 2019

IUCN (2019b) Mountain ecosystems. https://www.iucn.org/commissions/commission-ecosystem-management/our-work/cems-specialist-groups/mountain-ecosystems. Accessed 22 Oct 2019

IUCN (2019c) Freshwater biodiversity. https://www.iucn.org/theme/species/our-work/freshwater-biodiversity. Accessed 22 Oct 2019

Langenberg H (2012) 'Hydrology: Complex water future', nature geoscience. Nature Publishing Group, a division of Macmillan Publishers Limited. All Rights Reserved., 5, p 849. https://doi.org/10.1038/ngeo1658

Lewis SL, Wheeler CE, Mitchard ET, Koch A (2019) Restoring natural forests is the best way to remove atmospheric carbon. Nature 568:25–28

Mills K, Schillereff D, Saulnier-Talbot É, Gell P, Anderson NJ, Arnaud F, Dong X, Jones M, McGowan S, Massaferro J, Moorhouse H, Perez L, Ryves DB (2017) Deciphering long-term records of natural variability and human impact as recorded in lake sediments: a palaeolimnological puzzle. WIREs Water 4:e1195

Mills K, Vane C, Lopes Dos Santos RA, Ssemmanda I, Leng MJ, Ryves DB (2018) Linking land and lake: using novel geochemical techniques to understand biological response to environmental change. Quatern Sci Rev 202:122–138

NASA (2019) fire information for resource management system (FIRMS). Accessed October 22, 2020

Nkonya E, Johnson T, Kwon HY, Kato E (2016) Economics of Land Degradation in Sub-Saharan Africa. In: Nkonya E, Mirzabaev A, von Braun J (eds) Economics of land degradation and improvement – A global assessment for sustainable development. Springer, Cham

Rahbek C, Borregaard MK, Antonelli A, Colwell RK, Holt BG, Nogues-Bravo D, Rasmussen CM, Richardson K, Rosing MT, Whittaker RJ, Fjeldså J (2019) Building mountain biodiversity: geological and evolutionary processes. Science 365(6458):1114–1119

Riehl AN, Kreft H, Linder HP, Badgley C, Fjeldså J, Fritz SA, Rahbek C, Herman F, Hooghiemstra H, Hoorn C (2018) Geological and climatic influences on mountain biodiversity. Nat Geosci 11(10):718–725. https://doi.org/10.1038/s41561-018-0236-z

Rogers SL, Egan SS, Stimpson IG (2018) Tracking Sustainability concepts in geology and earth science teaching and learning, Keele University, UK. J Acad Develop Educ (10). http://dx.doi.org/10.21252/KEELE-0000028

Roser (2019) Forests. https://ourworldindata.org/forests. Accessed 21 Oct 2019

Schrodt F, Bailey JJ, Kissling WD, Rijsdijk KF, Seijmonsbergen AC, Van Ree D, Hjort J, Lawley RS, Williams CN, Anderson MG, Beier P, Van Beukering P, Boyd DS, Brilha J, Carcavilla L, Dahlin KM, Gill JC, Gordon JE, Gray M, Grundy M, Hunter ML, Lawler JJ, Mongeganuzas M, Royse KR, Stewart I,

Record S, Turner W, Zarnetske PL, Field R (2019) To Advance Sustainable Stewardship, We Must Document Not Only Biodiversity But Geodiversity. Proceedings Of The National Academy Of Sciences Of The United States Of America, 116, 16155–16158.

Stocking M (2001) Land degradation. Int Encycl Soc Behav Sci 8242–8247

Tittensor DP, Rex MA, Stuart CT, McClain CR, Smith CR (2011) Species-energy relationships in deep-sea molluscs. Biol Lett 7:718–722

UNDP (2019) Goal 15: Life on land. https://www.undp.org/content/undp/en/home/sustainable-development-goals/goal-15-life-on-land.html. Accessed 22 Oct 2019

UNEP-WCMC and IUCN (2016) Protected Planet Report 2016. Cambridge UK and Gland, Switzerland, UNEP-WCMC and IUCN

United Nations (1997). Glossary of environment statistics, studies in methods, Series F, No. 67, New York

United Nations (2015a): Transforming our world: the 2030 agenda for sustainable development, Goal 15, A/RES/70/1. http://www.un.org/ga/search/view_doc.asp?symbol=A/RES/70/1&Lang=E. Accessed 22 Oct 2019

United Nations (2015b) Millennium development goals report 2015. https://www.un.org/en/development/desa/publications/mdg-report-2015.html. Accessed 22 Oct 2019

United Nations (2018a) HLPF Review of SDG 15. https://sustainabledevelopment.un.org/content/documents/196552018backgroundnotesSDG15.pdf. Accessed 30 Oct 2019

United Nations (2018b) The sustainable development goals report 2018. United Nations, New York

United Nations (2019a) The sustainable development goals report 2019. United Nations, New York. ISBN 978-92-1-101403-7

United Nations (2019b) Desertification. https://www.un.org/en/events/desertificationday/desertification.shtml. Accessed 22 Oct 2019

van Haren N, van Boxtel K (2017) Grounding sustainability: land, soils and the sustainable development goals. iSQAPER/Both ENDS

Weisse M, Goldman ED (2018) 2017 Was the second-worst year on record for tropical tree cover loss. World Resources Institute. https://www.wri.org/blog/2018/06/2017-was-second-worst-year-record-tropical-tree-cover-loss.Accessed 30 Oct 2019

Wetzel RG (1983) Limnology. Saunders College Publishing, USA

World Bank (2018) Assessment of the state of hydrological services in developing countries. https://www.gfdrr.org/sites/default/files/publication/state-of-hydrological-services_web.pdf. Accessed 19 Oct 2019

Yang X, Scuderi LA, Wang X, Scuderi LJ, Zhang D, Li H, Forman S, Xu Q, Wang R, Huang W, Yang S (2015) Groundwater sapping as the cause of irreversible desertification of Hunshandake Sandy Lands, Inner Mongolia, northern China. Proc Natl Acad Sci 112(3):702–706

Eric O. Odada is a full Professor of Geology at the University of Nairobi. He obtained his Ph.D. in Marine Geochemistry from Imperial College London (1986) and has been with the University of Nairobi since 1989. Prior to this he served as Chief Research Officer at the Kenya Marine and Freshwater Research Institute and Principal Geologist at the Mines and Geology Department, Ministry of Environment and Natural Resources, Kenya. Professor Odada is a Fellow of the Geological Society of Kenya, an Honorary Fellow of the Geological Society of London, and a Fellow of the World Academy of Arts and Sciences. He has served as Director of the UN University Regional Centre for Water Education in Nairobi, and has been a member of several international committees, including the UN Secretary General's Advisory Board on water and sanitation.

Samuel O. Ochola is a lecturer at Kenyatta University. He holds a Ph.D. in Geosciences (University of Heidelberg), an M.Sc in Geological Risks and Management (University of Geneva), an M.Sc in Environmental Geology and Management (University of Nairobi) and a B.Sc in Geology (University of Nairobi). Samuel is an expert on climate change and variability and disaster resilience, with research interests on the nexus between Climate-Disasters-Livelihoods and Food-Energy-Water. Samuel was the Coordinating Lead Author on Sustainable Natural Resources Management for Kenya's Ministry of Agriculture's 2019–2029 strategy. He has been a consultant for USAID, UNEP, IUCN, DFID, the African Union and others.

Martin Smith is a Science Director with the British Geological Survey and Principle Investigator for the BGS ODA Programme Geoscience for Sustainable Futures (2017–2021). He has a first degree in Geology (Aberdeen) and a Ph.D. on tectonics (Aberystwyth, UK). A survey geologist by training Martin has spent a career studying geology both in the UK and across Africa and India. As Chief Geologist for Scotland and then for the UK he has worked closely with government and industry on numerous applied projects including in the UK on national crises, major infrastructure problems, decarbonisation research and urban geology and overseas for DFID-funded development projects in Kenya, Egypt and Central Asia. Martin is a Chartered Geologist and fellow of the Geological Society of London. He was awarded an MBE for services to geology in 2016.

Peace, Justice, and Strong Institutions

16

Joel C. Gill, Amel Barich, Nic Bilham,
Sarah Caven, Amy Donovan, Marleen de Ruiter,
and Martin Smith

J. C. Gill (✉)
British Geological Survey, Environmental Science
Centre, Nicker Hill, Keyworth, Nottingham NG12
5GG, UK
e-mail: joell@bgs.ac.uk; joel@gfgd.org

J. C. Gill · N. Bilham
Geology for Global Development, Loughborough,
UK

A. Barich
Geothermal Research Cluster (GEORG),
Grensásvegur 9, 108 Reykjavík, Iceland

N. Bilham
University of Exeter Business School, SERSF
Building, Penryn Campus, Penryn TR10 9FE,
Cornwall, UK

N. Bilham
Camborne School of Mines, University of Exeter,
Penryn Campus, Penryn TR10 9FE, Cornwall, UK

S. Caven
Independent minerals and sustainability consultant,
Vancouver, British Columbia, Canada

A. Donovan
Department of Geography, University of Cambridge,
Downing Place, Cambridge CB2 3EN, UK

M. de Ruiter
Institute for Environmental Studies, Vrije
Universiteit Amsterdam, De Boelelaan, 1111
Amsterdam, The Netherlands

M. Smith
British Geological Survey, The Lyell Centre,
Research Avenue South, Edinburgh EH14 4AP, UK

© Springer Nature Switzerland AG 2021
J. C. Gill and M. Smith (eds.), *Geosciences and the Sustainable Development Goals*,
Sustainable Development Goals Series, https://doi.org/10.1007/978-3-030-38815-7_16

Abstract

16 PEACE, JUSTICE AND STRONG INSTITUTIONS

Geoscience diplomacy

Science can facilitate international diplomacy and inform foreign policy objectives

Achieving the SDGs will involve significant international collaboration, including some diplomacy

Geoscience collaborations help to understand or manage transboundary geological features (e.g., volcanoes) or resources (e.g., aquifers)

Geoscience diplomacy takes time and commitment

Corruption and geoscience

Corruption undermines the rule of law and takes vital resources from development

Corruption increases vulnerability to natural hazards and reduces the effectiveness of risk reduction

The Extractives Industry Transparency Initiative promotes accountable management of natural resources

Effective geoscience instituitions

Science-based institutions can help to support sustainable development. For example:

Improving environmental data collection

Advance public understanding of Earth Science

Improve training and professional developments

Science-based institutions should shape and model accepted behaviors

Financial transparency

Inclusive decision making

Safe environments free from harassment and discrimination

 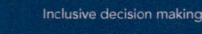 

16.1 Introduction

Strong institutions, access to justice, and peaceful communities are key for sustainable development progress. Conflict, corruption, and weak institutions provide significant barriers to the delivery of national development strategies and the SDGs. Ensuring access to health services (**SDG 3**) or education (**SDG 4**), building resilient infrastructure (**SDG 9**) and sustainable cities (**SDG 11**), and promoting prosperity (**SDG 8**) all depend on security of contractors, access to and good management of financial resources, and transparent, evidence-based decision-making. **SDG 16** aims to enable this safe and secure environment, at all scales, supporting the delivery of public services and sustained economic development. **SDG 16** includes 10 targets and 2 means of implementation (Table 16.1).

The promotion of peaceful and inclusive societies, ensuring access to justice for all, and the building of effective, accountable, and inclusive institutions may seem distantly related to the everyday work of Earth and environmental scientists. Many links exist, however, including with sectors employing geoscientists and the activities that geoscientists are regularly engaged in. Some examples include

- Many geoscientists investigate natural hazards, risk, and the impacts of disasters. Disasters disproportionately affect the poor, and those living in states affected by violence and active conflict (Harris et al. 2013; Peters 2017). Conflict may result in heightened vulnerability to hazardous events, or the displacement of people into areas more exposed to natural hazards. For example, the Rohingya refugee camps in Bangladesh are home to more than 700,000 people who have fled persecution in Myanmar. The refugee camps are affected by heavy monsoon rains, triggering flooding and landslides, which killed 170 people in 2017 (BBC 2019). Geoscientists may, therefore, come into direct contact with the impacts of conflict, and their work may be shaped by the likelihood of conflict changing exposure or vulnerability.

- Geoscientists working across multiple sectors may be exposed to corruption and bribery. Corruption undermines the rule of law and prevents vital resources from being spent on development projects (Fig. 16.1). In just one

Table 16.1 SDG 16 Targets and Means of Implementation

Target	Description of Target (16.1 to 16.10) or Means of Implementation (16.A to 16.B)
16.1	Significantly reduce all forms of violence and related death rates everywhere
16.2	End abuse, exploitation, trafficking and all forms of violence against and torture of children
16.3	Promote the rule of law at the national and international levels and ensure equal access to justice for all
16.4	By 2030, significantly reduce illicit financial and arms flows, strengthen the recovery and return of stolen assets and combat all forms of organised crime
16.5	Substantially reduce corruption and bribery in all their forms
16.6	Develop effective, accountable, and transparent institutions at all levels
16.7	Ensure responsive, inclusive, participatory, and representative decision-making at all levels
16.8	Broaden and strengthen the participation of developing countries in the institutions of global governance
16.9	By 2030, provide legal identity for all, including birth registration
16.10	Ensure public access to information and protect fundamental freedoms, in accordance with national legislation and international agreements
16.A	Strengthen relevant national institutions, including through international cooperation, for building capacity at all levels, in particular in developing countries, to prevent violence and combat terrorism and crime
16.B	Promote and enforce non-discriminatory laws and policies for sustainable development

Fig. 16.1 Exhorting the public to say no to corruption in Zambia. *Credit* Lars Plougmann, CC BY-SA 2.0 (https://creativecommons.org/licenses/by-sa/2.0/), available at: https://commons.wikimedia.org/w/index.php?curid=3215552

decade (2005–2014), USD$36–69 billion was lost from Africa in illegal financial flows (Global Financial Integrity 2017). To put this into context, the cost of ensuring safely managed water and sanitation across sub-Saharan Africa is estimated to be approximately US$24.7 billion per year between 2015 and 2030 (Hutton and Varughese 2016). Corruption can occur in all parts of mineral and hydrocarbon value chains, at all stages of the extraction life cycle, and in both big and small operations. In a study of 496 mines in Africa, with more than 92,000 survey respondents, Knutsen et al. (2017) found that the opening of new mines was systematically related to increase in bribery and corruption perceptions and the hampering of local-level institutional quality.

- Many geoscientists work in science-based institutions, which have an important role to play in supporting sustainable development. Examples include national geological surveys, ministries of mining and the environment, and academic or research institutes. Institutions take diverse forms and exist for a myriad of

purposes. Irrespective of their function, if they are not accountable and inclusive they can propagate injustices, making it impossible to achieve sustainable and resilient societies. Institutions help to shape and model accepted behaviours, such as financial transparency, ensuring inclusive decision-making, and creating safe environments free from harassment and discrimination. During a survey of academic fieldwork experiences (from the life, physical, and social sciences disciplines), 22% of respondents reported being the victim of sexual assault (Clancy et al. 2014; St John et al. 2016). Organisational cultures can help to stop such behaviour, ensuring safe, open and respectful environments, and a safe means of reporting bad practice with the victim having confidence that appropriate action will be taken.

These examples demonstrate why **SDG 16** matters to the professional work of geoscientists, and suggest that achieving **SDG 16** will require engagement by geoscientists and those sectors employing geoscientists. In this chapter, we, therefore, focus on three broad themes within **SDG 16**, and their relationship to geoscience. We begin in Sect. 16.2 by profiling the role of geoscience in facilitating international diplomacy, and peacefully settling disputes in accordance with the Charter of the United Nations (Chapter VI). In Sect. 16.3, we explore the importance of tackling corruption if we are to reduce disaster risk and ensure the benefits of the extractives sector are felt by all. In Sect. 16.4, we move to the theme of effective, accountable, and transparent institutions. We also point the reader to other relevant chapters, including **SDGs 1**, **10**, and **17**.

16.2 Geoscience Diplomacy and the SDGs

16.2.1 From Science Collaboration to Science Diplomacy

A group of national academies across Europe declared in 2016 that '*science is global*', emphasising the importance of international science and research collaborations in tackling global challenges. Positive, equitable partnerships play a critical role in scientific capacity building, opening up access to science, strengthening national institutions, and fostering dialogue between groups. Scientists around the world share sets of professional values that help to foster trust and benefit sharing. The Royal Society (2010) report 'New Frontiers in Science Diplomacy' notes that

> `Scientific values of rationality, transparency and universality are the same the world over. They can help to underpin good governance and build trust between nations. Science provides a non-ideological environment for the participation and free exchange of ideas between people, regardless of cultural, national or religious backgrounds.`

In addition to science being an enterprise that different groups can unite around, science is critical to informing a nation-state's response to international humanitarian and development challenges, and enhancing relationships across a politically sensitive divide. The Royal Society

(2010) report, therefore, defines three types of science diplomacy:

- *Science for diplomacy*, which involves scientific collaboration specifically for the purpose of improving diplomatic relations. For example, in 1961 the USA-Japan Committee on Science Cooperation was announced, aiming to strengthen dialogue between the two countries (Turekian and Neureiter 2012). This programme has now existed for more than 50 years, bringing together scientists, senior ministers and policy advisors from the US and Japanese administrations.

- *Science in diplomacy*, in which scientific advice is required to inform foreign policy objectives, including active crisis situations. For example, the United Kingdom has a Scientific Advisory Group for Emergencies (SAGE), which brings together leading experts to ensure timely and coordinated scientific advice to support UK cross-government decisions (Cabinet Office 2012). Recent activations of SAGE have been triggered by the 2015 Nepal earthquake, the 2011 Fukushima nuclear emergency, and the 2010 eruption of the Eyjafjallajökull volcano in Iceland (Fig. 16.2). In all of these contexts (and others), scientific advice helped to inform the UK's response to these international situations. A link to minutes of a SAGE meeting relating to volcanic ash disruptions caused by the eruption of Eyjafjallajökull is included in the further reading.

- *Diplomacy for science*, in which considerable political and cross-cultural groundworks have to be undertaken in order to facilitate scientific endeavour. For example, Mount Paektu stands on the border between the Democratic People's Republic of Korea (DPRK, also known as North Korea) and China, with the international border running through the summit crater. The Mount Paektu Geoscientific Group (MPGG) is a UK scientific collaboration with scientists from DPRK, which requires considerable diplomacy and negotiation of international sanctions. This example is discussed in detail in Sect. 16.2.2.

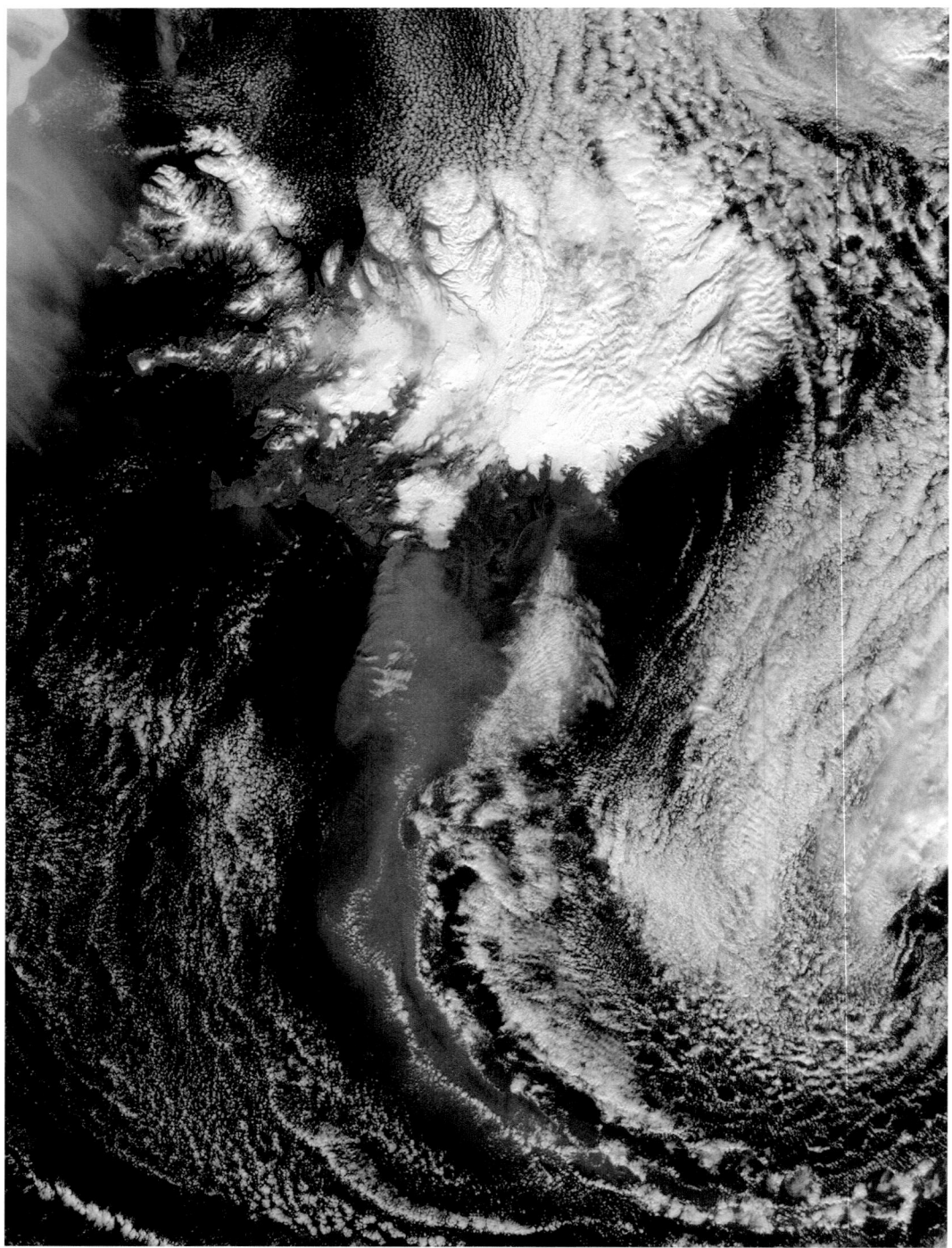

Fig. 16.2 Ash from the eruption of Iceland's Eyjafjallajökull Volcano in 2010. This image was taken on 19 April 2010 by the Moderate Resolution Imaging Spectroradiometer (MODIS) on NASA's Terra satellite. The main ash plume is seen as a brown streak south of Iceland. NASA image courtesy Jeff Schmaltz, MODIS Rapid Response Team at NASA GSFC

These three categories of science diplomacy are not mutually exclusive, with some diplomatic engagement likely to combine two or more of the above. Political groundwork may be required in sensitive areas to start new scientific projects or host scientific conferences, with improved diplomatic relations being among the longer term benefits of that scientific collaboration. The Synchrotron-Light for Experimental Science and Applications in the Middle East (SESAME) independent laboratory, based in Jordan, is an example of a complex science diplomacy project. This facility is unique in the region, offering scientists the opportunity to access a centre of excellence to advance research in diverse fields (e.g., biology, chemistry, medicine, the environment, archaeology). The project, created under the auspices of UNESCO and officially opened in 2017, required significant political groundwork to identify a location and secure the necessary support. This facility now brings together representatives of Cyprus, Egypt, Iran, Israel, Jordan, Pakistan, the Palestinian Authority, and Turkey, aiming to promote peace through scientific collaboration.

Gluckman et al. (2017) suggest a reframing of the three categories above, focusing instead on the purpose of the scientific engagement—actions that seek to directly advance the *needs of a nation-state*; actions seeking to deal with *cross-border challenges*; and actions that seek to *meet global needs and challenges*. The latter two categories particularly open up opportunities for geoscientists to engage in science diplomacy. Collaborations may

- Help to understand resources and manage risks associated with cross-border rivers, tectonic faults, or volcanoes that are close to a border (Donovan and Oppenheimer 2019).
- Work to characterise or extract natural resources (e.g., minerals, groundwater) from a site that transgresses a border.
- Bring together complementary expertise and resources to develop a combined satellite monitoring programme.

- Enhance the scientific capacity of multiple nation states, through shared learning, research, and innovation.

These examples demonstrate how scientific engagement or partnerships on issues linking to hydrogeology, climate change, engineering geology, minerals, hydrocarbons, and geological hazards can all be related to diplomacy and the strengthening (or establishing) of diplomatic relations between two or more countries.

16.2.2 Case Study: The Mount Paektu Geoscientific Group

An example of a long-running scientific project that requires considerable diplomacy and negotiation of international sanctions, among other factors, is the Mount Paektu Geoscientific Group (MPGG), which began in 2011 (Hammond 2016). Mount Paektu stands on the border between DPRK and China (Fig. 16.3), with the international border running through the summit crater. The MPGG is a UK collaboration with scientists from the DPRK. It was initiated following a request from the DPRK to the American Association for the Advancement of Science (AAAS). The AAAS reached out to a British scientist, and this led to a meeting in Pyongyang in 2011 that generated a research proposal. However, as DPRK is under international sanctions, it took 2 years to get the relevant permissions from the UK and US governments. Relations with DPRK were volatile at this time, and this was the first time that institutions like the UK Foreign Office had had to deal with such requests for exemption. Hammond (2016) notes that two elements were central in finally succeeding with the project: a strong scientific focus, and enthusiastic scientists on all sides.

The MPGG does not work directly with the partners in DPRK. It works through two NGOs —one in Beijing and one in Pyongyang. Communication can be time-consuming, particularly with language differences, but over the years a

Fig. 16.3 Lake Chon (summit caldera lake on Mount Paektu) from the Korean side of the volcano. © Amy Donovan (University of Cambridge)

very good relationship has built up. In addition, a Memorandum of Understanding (MoU) was signed between relevant Korean institutions, the AAAS and the Royal Society—this was critical in the success of the project. The role of the Foreign Secretary of the Royal Society was very important here—the UK universities involved were nervous about signing an MoU with DPRK institutions, but the Royal Society brought considerable international experience and expertise. This demonstrates the important role of effective scientific institutions in delivering the Sustainable Development Goals (see Sect. 16.4).

Once work was underway and scientific data was being gathered and processed, the partners had to be flexible. In different geographical contexts, science can differ in terms of access to resources, approaches used, and priorities. Scientists in DPRK had not visited active volcanoes elsewhere in the world and had limited access to

scientific literature (Donovan 2019). Expectations also differ, and the restrictions resulting from international sanctions affected the kinds of work that could be done. However, several scientific papers were produced from this first phase of the project, and another is in review—this time co-authored by Chinese scientists as well as UK and Korean partners.

Volcano monitoring is best achieved by a network that covers the entire feature; however, China and DPRK currently monitor Paektu separately and do not share data. The primary control on data sharing is restriction by the Chinese government; however, after several years of discussions, it has become possible to share some partially processed data. The next stages of the MPGG will seek to expand on this cooperation, focussing on a cross-border seismic experiment and the reconciliation of the stratigraphy, which is interpreted differently by different groups of authors (Donovan 2019). Ultimately, this project hopes to reduce disaster risk around the volcano through engagement and training in understanding and reducing social vulnerability as well as hazard modelling and scenario planning. As Hammond (2016) notes, the driving force behind this project is to understand the volcano and the risks it poses —it is focussed on scientific discovery, and is facilitated by diplomacy, open communication, and flexibility. These ensure that goals are achievable and that good relationships are maintained, and also require some patience and multiple funding proposals—a single project, even a five-year one, would not be enough to build the necessary relationships for this kind of work.

16.2.3 Integrating Science Diplomacy into a Career as a Geoscientist

There are many opportunities for international, transboundary geoscience projects in challenging political contexts. Many international boundaries are defined or characterised by geological and geomorphological features, including rivers (DPRK and China), mountain ranges (Chile and Argentina), and lake boundaries (Tanzania and the Democratic Republic of Congo) as well as volcanoes (DPRK and China) and oceans (Box 16.1). The International Groundwater Resources Assessment Centre[1] estimates that there are 366 identified transboundary aquifers. Regions where multiple countries have a vested interest for political or scientific reasons also provide rich opportunities for science diplomacy. The Arctic Science Agreement[2], came into force in 2018, and helps to facilitate access to research areas for data collection, supports open access to scientific data, and promotes opportunities for students and early-career scientists. The agreement was signed by all eight member states of the Arctic Council: Canada, Denmark, Finland,

Iceland, Norway, Russian Federation, Sweden, and the United States of America.

> **Box 16.1 International Ocean Discovery Programme (IODP)**
>
> This international marine research collaboration brings together scientists from 23 eligible countries to collect seafloor sediments and rocks, and monitor environments below the sea floor, to explore Earth's history and dynamics. Key themes are climate change, deep life, planetary dynamics, and geohazards. Scientists from eligible countries (identified below) are selected to staff IODP research expeditions, conducted in diverse ocean settings (Fig. 16.4).
>
> *Platform Providers:* National Science Foundation (United States of America, Canada); Ministry of Education, Culture, Sports, Science and Technology (Japan); and European Consortium for Ocean Research Drilling (Austria, Denmark, Finland, France, Germany, Ireland, Italy, Netherlands, Norway, Portugal, Spain, Sweden, Switzerland, United Kingdom).
>
> *Additional Funding Partners:* Ministry of Science and Technology (China); Institute of Geoscience and Mineral Resources (Korea); Australia-New Zealand IODP Consortium; Ministry of Earth Science (India); and Coordination for Improvement of Higher Education Personnel (Brazil).

For geoscientists wanting to work at this interface between science and international relations, it is important to recognise that this requires investment in a range of skills and competencies beyond those in technical geoscience (Gill 2016). Geoscientists require cultural understanding, patience, and the ability to communicate across cultures and disciplines. It is likely that geoscientists will need to understand more about public policy than is currently included in traditional geoscience courses (see SDG 1), recognising that science

---

[1]https://www.un-igrac.org/areas-expertise/transboundary-groundwaters

[2]https://arctic-council.org/index.php/en/our-work2/8-news-and-events/488-science-agreement-entry-into-force

Fig. 16.4 IODP Arctic expedition September 2004. © NERC (used with permission)

diplomacy may bring them into contact with policy makers, politicians, and diplomats. These skills could be integrated into the education and continued professional development of geoscientists (formal education), but are more likely to be developed informally, through extracurricular reading and activities. The gap between university geoscience courses and the skills and understanding required to contribute to sustainable development objectives, including peace and justice, requires new learning approaches and resources.

The complexity of doing science and building science partnerships in politically sensitive areas can make it challenging to secure funding. Those providing finance may not be persuaded that a project is achievable, that the time frame is realistic, or that institutions (in all countries involved) will persevere to bring the project to fruition. The SESAME facility (described in Sect. 16.2.1) was opened in 2017, but there was significant work in the two decades before that to manage

the process, identify a physical location for the facility, and bring people together. This is also the case for a lot of IODP projects (and the International Continental Scientific Drilling Programme) which can take decades to put together, involving negotiations across several countries. Building partnerships of any type takes time and commitment (see **SDG 17**), particularly when there are complex political, cultural, or religious sensitivities to navigate. The United Kingdom, through the Newton Fund, provides *Researcher Links* and *Institutional Links* grants for the development of collaborations between the UK and partner countries (e.g., Brazil, China, India, Jordan, Philippines) that tackle local development challenges. This type of funding can help to start conversations, build trust, and explore shared perspectives on societal challenges and science priorities. The evidence generated at this preliminary stage can then inform longer term engagement and funding applications. In many contexts, however, more time and preliminary funding will be needed to establish the relationships necessary for larger projects. International scientific conferences and workshops also provide an informal opportunity to contribute to science diplomacy, building networks and relationships that span different geographies and sectors.

Science diplomacy, broadly conceived, can enhance international relationships and can strengthen capacity in the Global South. An important aspect of capacity strengthening is through the sharing of knowledge, expertise, and scientific resources—providing training, lending or giving equipment, and providing opportunities for wider interaction through conferences and workshops. Indeed, achieving many of the SDGs will involve significant international scientific collaboration, much of which is likely to require some level of science diplomacy. While SDGs 16 and 17 are explicitly linked to peace and partnerships, engagement with challenging political contexts and across ideological divides will be necessary for success in achieving all the SDGs.

16.3 Corruption and Geoscience

Corruption involves the abuse of power in return for wealth, status, or access. It hinders a nation or region from reaching its economic, social, and environmental ambitions (Murshed and Mredula 2018). For example, comparing the Corruption Perception Index (a measure of how corrupt experts perceive a country to be) with the Human Development Index (a measure looking at health, education, and living standards) suggests that countries perceived to be less corrupt generally have higher measures in the Human Development Index (Ortiz-Ospina and Roser 2019) (Fig. 16.5).

Two examples of corruption and their effects that we explore in this chapter are

- Bribes paid to government officials by construction workers, in return for ignoring violations of building codes. *This can result in buildings being vulnerable to natural hazards,*

loss of life, and a higher reconstruction bill for the city.
- Bribes paid to government officials by industry, in exchange for privileged access to natural resources. *This can result in poor deals and lost revenue for the taxpayer, with a risk that firms with poor performance and procedures for health and safety, environmental sustainability or human rights are given lucrative contracts.*

Other examples include bribing an official in return for paperwork approving the export of research samples, exploiting power to give a job to a personal contact (i.e., a family member or friend) even if they are not qualified, and the collusion of parties in the public or private sector to fix prices and increase consumer costs.

16.3.1 Tackling Corruption to Improve Disaster Risk Reduction

SDG 1 introduced the concept of disaster risk, highlighting components of hazard, exposure, and vulnerability. Changing one risk component causes a change in the overall risk. For example, decreasing vulnerability, while hazard and exposure remain the same, decreases the overall risk (Adger 2006). Vulnerability can be understood as '*the conditions determined by physical, social, economic and environmental factors or processes which increase the susceptibility of an individual, a community, assets or systems to the impacts of hazards*' (UNDRR 2017). Vulnerability consists of four main components: physical (e.g., the built environment), socio-economic (e.g., household income), environmental (e.g., soil quality), and institutional (e.g., levels of corruption) (de Ruiter et al. 2017). Corruption *increases* vulnerability and *decreases* the positive effects of disaster risk reduction measures, which together *increases* risk (Kenny 2012; Lewis 2011). Tackling corruption can, therefore, help to decrease risk, and perpetuating corruption can increase risk (Leeson and Sobel 2008).

Human Development Index vs. Corruption Perception Index, 2017

The vertical axis shows scores in the UN Human Development Index (lower values reflect lower development). The horizontal axis shows scores in Transparency International's Corruption Perception Index (lower values reflect higher perceived corruption).

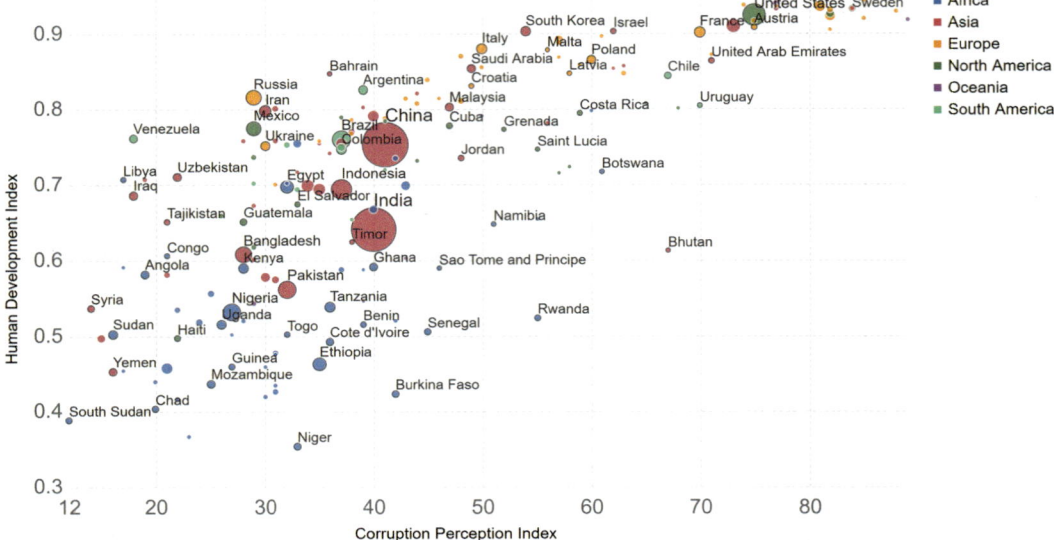

Source: UNDP (2018), Transparency International (2018), Population by country, 1800 to 2100 (Gapminder & UN)
OurWorldInData.org/corruption/ • CC BY

Fig. 16.5 Human Development Index versus Corruption Perception Index, 2017. Images created by Ortiz-Ospina and Roser (2019), using data from UNDP (2018), Transparency International (2018), Gapminder and UN (2019). Reproduced under a CC BY license (https://creativecommons.org/licenses/by/4.0/)

The institutional dimension of vulnerability refers to the functioning of governments and the effectiveness of policies such as risk mitigation strategies, and regulation control and law enforcement (Ciurean et al. 2013). Governments play an important role in managing a community's risk level and can decrease risk by implementing disaster risk reduction measures or developing policies such as mandatory building standards (Ahrens and Rudolph 2006; Green 2005). Trust in official information sources is often cited as being of high importance to the level of institutional vulnerability that exists (Werg et al. 2013). For example, governments may issue evacuation notices if they believe that a volcanic eruption is imminent. If people do not have trust in the integrity and knowledgeability of government institutions, these warnings may not be heeded. A key factor affecting trust in official institutions is

the extent to which those institutions are corrupted or perceived to be corrupted.

Corruption is difficult to quantify (Ambraseys and Bilham 2011), and it is, therefore, hard to prove a causal relation between corruption and disaster risk. There does seem to be a strong correlation between corruption and the extent of impacts from a disaster, such as an earthquake (Alexander 2016; Ambraseys and Bilham 2011; Escaleras et al. 2007). Over the past 30 years, 83% of the deaths attributed to collapsed buildings due to earthquakes occurred in anomalously corrupt countries (Ambraseys and Bilham 2011).

The construction industry has been shown to be the most corrupt industry of the global economy (Betts and Farrell 2009). Corruption in this sector increases disaster risk by decreasing the ability of the built environment to withstand negative disaster impacts. In other words, corruption (institutional vulnerability) decreases the quality of the built environment (increasing physical vulnerability). Physical vulnerability is shaped by how well a building is designed and constructed, and how well building codes are implemented. Corruption can compromise these factors, resulting in buildings that are poorly constructed and in breach of building codes, increasing a built environment's vulnerability (Bosher and Dainty 2011).

Corruption in the construction industry mainly arises from two aspects:

1. *Regulation Difficulties.* This can be due to the large number of companies involved in one construction project, the often-changing composition of partners per project and the lack of repetition in projects.
2. *Informal Labour.* The relatively low barriers in entering the industry's labour market make it prone to become an informal labour market.

These factors aggravate the industry's corruption-proneness, which increases the built environment's vulnerability to extreme events (Lewis 2003). Corruption not only directly influences the quality of the construction of buildings and the enforcement of building codes; it can also negatively influence policy agendas and foster inequality (Lewis 2008). Although still debated, studies have shown that corruption has stronger negative impacts on sustainable development in developing countries compared to developed countries (Freckleton et al. 2012; Lewis 2011). Corruption is not just a causal factor in disasters occurring, but can also be a consequence of disasters and their damage. Disasters can increase corruption in the public sector as governments struggle to maintain oversight during the often chaotic post-disaster phase.

Box 16.2 Building Codes and Earthquake Damage in Turkey

In Turkey, corruption has resulted in a mismatch between building codes that exist on paper and the actual state of the built environment, revealed by the number of earthquake-damaged buildings (Akarca and Tansel 2012; Alexander 2008, 2016). In 1999, the northwestern part of Turkey was successively hit by two major earthquakes, both causing significant damage and many deaths. At the time, Turkey already had strict building codes and regulations in place. The extensive building damage was inconsistent with these strict regulations. While many older buildings survived the earthquakes, many recently constructed buildings were damaged. During the post-disaster recovery phase, a study found that local politicians had been taking bribes to authorise the construction of sub-standard buildings. This explained why many of the old buildings remained standing while the newer buildings were severely damaged: the newer buildings were inappropriately designed, poorly constructed, and located in high-risk zones that were deemed not safe for construction. During the subsequent elections of 2002, nearly half of the electorate voted for new parties that had not been in power during the 1999 earthquakes (Akarca and Tansel 2012) (Fig. 16.6).

There are different ways of addressing corruption and thereby improving Disaster Risk Reduction (DRR). The most important components of a governance structure to best support risk reduction are *awareness, accountability, participation, predictability,* and *transparency* (Ahrens and Rudolph 2006). By strengthening each of these governance components, we can help to address the challenge of corruption (Kaufmann et al. 2003). In the context of the construction sector, this means

- Simplifying regulations so that complying with them is easier for engineers and construction employees during construction, as well as for governmental assessors to evaluate compliance with building codes after construction (Kenny 2012).

Fig. 16.6 Istanbul. This city of approximately 15 million people in Turkey is close to the North Anatolian Fault, responsible for several earthquakes throughout the history of Istanbul. Image by Şinasi Müldür from Pixabay

- Engaging local communities in DRR to create shared information, ownership, and capacity building (Bosher and Dainty 2011; Petal et al. 2008).
- Designing building standards and relevant policies in agreement with people working in the industry, and providing easily accessible information about these standards (Alexander 2008).
- Reforming the professionalisation of the local construction industry (Alexander 2008), and transitioning from a fragmented approach to a holistic approach where the different components of construction are better integrated (Bosher and Dainty 2011).

For geoscientists, understanding the relationship between corruption and risk, the ways in which corruption manifests itself in the built environment and the actions that can be taken to strengthen governance is critical. Geoscience informs the development of appropriate building codes, recognising local ground conditions that can affect the characteristics of a given hazard. Geoscientists may engage directly with the public through education, outreach, and research activities. This provides an opportunity to share knowledge regarding how hazards affect the built environment and the consequences of corruption on their safety. Finally, strong professional geoscience organisations and learned societies—while not exempt from corruption—can provide independent knowledge and information that helps to hold governments (with responsibility for approving changes in land use, building inspections and other relevant factors) to account, while also ensuring that geoscientists are held to high professional standards (see Sect. 16.4.3).

16.3.2 Tackling Corruption to Improve the Extractives Sector

Corruption can occur in all parts of mineral and hydrocarbon value chains, and at all stages of the extraction life cycle. These include the awarding of mining, oil, and gas rights, formulation and implementation of contracts, operations phases (including with regard to regulation), and commodities trading. Corruption in the extractives sector often involves bribery (indeed, the OECD reports that one in five cases of transnational bribery occurs in this sector), but it can take other forms, including embezzlement, extortion, misappropriation and diversion of public funds, abuse of office and trading in influence. It occurs and has impacts across the public, private, and civil society sectors, and is recognised as a major impediment to development. Governments, companies in all parts of the value chain and civil society all have a role in preventing corruption (OECD 2016).

Responsible mining and hydrocarbon extraction (and responsible sourcing of other raw materials) means minimising the negative social, environmental, and economic impacts of resource extraction and maximising the positive impacts (Wall et al. 2017). Corruption constitutes just one set of potential negative impacts, and it can exacerbate and be closely interlinked with a wide range of other impacts (Goodland 2012; Church of England 2017; Ayuk et al. 2019). A particular concern is the link between corruption, illegal mining and trading of commodities, and conflict and human rights abuses such as forced and child labour (Bleischwitz et al. 2012). Corruption around the awarding of access rights and the formulation and implementation of contracts can also lead to significant harm to environments and communities.

There are many initiatives, schemes, and standards relating to aspects of responsible mining and hydrocarbon extraction. The most significant one directly addressing corruption and financial transparency is the Extractive Industries Transparency Initiative (EITI) (see Box 16.3). These challenges are also being actively addressed by the Organisation for Economic Co-operation and Development (OECD), the International Monetary Fund, and the World Bank, among others. Corruption and financial transparency are also addressed alongside other social and environmental impacts of mining, as part of many schemes and initiatives. These include the International Council on Mining and Metals (ICMM) (a coalition of major mining companies), the Responsible Mining Index (RMI) and the Initiative for Responsible Mining Assurance (IRMA) (both independent non-profit multi-stakeholder organisations), whole-chain single-commodity schemes such as the Aluminium Stewardship Initiative (ASI), and those relating to particular geographical areas (particularly the Democratic Republic of Congo and surrounding countries) such as the Regional Initiative against the Illegal Exploitation of Natural Resources (RINR) under the auspices of the International Conference on the Great Lakes Region (ICGLR).

Box 16.3 Extractive Industries Transparency Initiative (EITI)

What is it? EITI sets and administers the global standard to promote open and accountable management of oil, gas, and mineral resources, recognising that these resources belong to a nation's citizens. It initially focused on revenue transparency but, from 2013 onwards, the EITI Standard has addressed a wider range of financial and legal transparency and good governance objectives, across government and industry. The EITI Standard requires timely and accurate reporting by countries on matters such as how licenses are allocated, the tax, royalties and in-kind social donations made by corporations involved in extractives, and how revenues make their way through national and local governments (EITI 2019a).

Purpose: The EITI seeks to strengthen public and corporate governance, promote understanding of natural resource management, and provide the data to inform reforms for greater transparency and accountability in the extractives sector, including curbing corruption.

Who is involved? There are currently 52 implementing countries, including many nations with significant extractive industries. Implementation of the EITI standard in each implementing country is led by a multi-stakeholder group of representatives from government, companies, and civil society, working together towards a shared vision of transparency and accountability. Multi-stakeholder national groups set country-specific objectives for EITI implementation and oversee the reporting process to the international EITI board. The EITI is jointly funded by industry and governments.

Assessment and Evaluation: The EITI standard is applied in the same way to all implementing countries. Validation is the process by which disclosed data is reviewed and assessed alongside a broader consultation process. Countries may be deemed to be making inadequate, meaningful or satisfactory progress, and the next validation timetabled for 3 months to 2.5 years later. If progress is inadequate, a country is temporarily suspended from the EITI. If no progress has been made, a country is delisted and must reapply. Since 2016, seven members have been deemed to be making inadequate progress and temporarily suspended (Afghanistan, Azerbaijan, Iraq, Kyrgyz Republic, Niger, Solomon Islands, and Tajikistan). Seven members have been assessed as making satisfactory progress (Colombia, Mongolia, Nigeria, Norway, Philippines, Senegal, Timor-Leste) (EITI 2019b).

EITI Impact: Assessments of the impact of EITI have been generally (though not universally) positive. A report from the Chr.

Michelsen Institute reviewing 50 evaluations of EITI's work recognised the scale of the challenge, the relatively small scale of the organisation, the long-term nature of its developmental objectives, and the difficulty of attributing change to a single intervention. It concluded that good progress has been made in many areas, including operational and promotional objectives, and that there is some evidence of success regarding development goals, though this is context-dependent. The lack of participation of a number of very significant resource-rich countries, including some regarded as highly corruption-prone, is a notable weakness (Lujala et al. 2017). Nonetheless, the EITI has already made a significant contribution to tackling corruption and associated problems in the extractives sector, and has scope to develop further. Among other positive impacts, data collected through the EITI process has been used across the public and private sectors to (i) inform legal and fiscal reforms, (ii) strengthen tax collection, (iii) create financial models to deal with high and low commodity prices, (iv) monitor whether contracts are being adhered to, and (v) clarify the investment environment for companies (EITI 2019b) (Fig. 16.7).

Geoscientists can play a vital role in reporting and addressing corruption. Some, particularly those who take on senior management roles in companies in the extractives sector, have significant influence over company policies and governance, establishment of joint ventures and selection of partner companies, procurement rules for goods and services, negotiation of access rights and operational compliance with regulatory requirements—all vital considerations in seeking to identify and control corruption (OECD 2016). More generally, geoscientists often find themselves at the front end of exploration and operational phases on the ground, and may come into

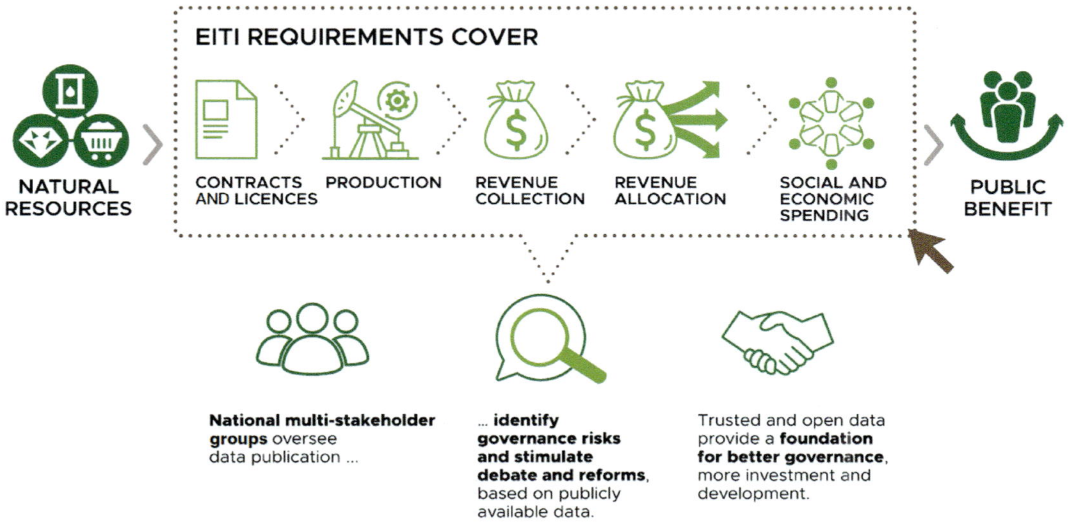

Fig. 16.7 EITI Process and Impact. *Credit* EITI (2019b), EITI Progress Report 2019, prepared by the EITI International Secretariat (eiti.org)

contact with improper practices including actual or potential corruption. It is incumbent on them to report such practices, and to conduct their own professional activities according to high ethical standards (see discussion of professional bodies and codes of conduct in Sect. 16.4.3). It is also essential that companies encourage their employees (geoscientists and others) to behave in this way, and for geoscientists in leadership positions to promote supportive and inclusive professional environments in which such matters can be discussed openly (Peppoloni et al. 2019).

Minerals pass through long, global, complex and entangled supply chains and value chains on their way from mines to manufacturers, end-products, and consumers. This constitutes a major challenge to those further along these chains who want to know where their raw materials have come from and what the impacts of their extraction are. Many manufacturing companies are increasingly paying attention to these questions, but have no way of accessing the information which would allow them to be answered. Certification of the standards under which minerals have been extracted and processed and their traceability through supply chains is thus a shared challenge for mining companies, investors, manufacturing companies, and other stakeholders. A number of the schemes and standards discussed above are seeking to address this challenge, but with the exception of a few commodities such as gemstones and certain 'conflict minerals', limited progress has been made so far. This challenge is the subject of ongoing interdisciplinary research. Alongside societal, institutional, and managerial approaches, new technologies (including blockchain and related mechanisms) may have a role to play. Geoscientific innovation may also contribute to monitoring mining activities and their impacts (including corruption, whether directly or indirectly), and the subsequent tracking of materials, for instance, through 'big data' approaches to in situ sensing data made publicly available online in real time, or by developing novel remote sensing techniques and applications.

16.4 Effective Geoscience Institutions

16.4.1 Characteristics and Types of Geoscience Institution

SDG 16 sets out key characteristics of effective institutions, noting them to be accountable and transparent, having responsive, inclusive, participatory, and representative decision-making, and (for relevant global organisations) having meaningful participation of developing countries. Geoscientists are involved in a range of local, national, and international institutions with a responsibility to embed these characteristics. Some of these (such as national geological surveys, government departments, and agencies) are public sector bodies and have a crucial role in improving environmental data collection, management, integration and access to support sustainable development objectives. Others (including most learned societies and professional bodies) exist as not-for-profit organisations or charitable bodies, working to advance the study of Earth science, strengthen the profession and improve societal access to Earth science. There are several main categories of geological or geoscientific institution:

- **Global and continental unions and societies** (for example, the International Union of Geological Sciences (see Box 16.4), the Geological Society of Africa and the European Geosciences Union). We discuss the contribution of this type of institution to sustainable development objectives in Sect. 16.4.2.
- **National academies of science**, whose scope goes beyond geoscience but which may have geoscience sections (examples include the US National Academy of Sciences and the Tanzania Academy of Sciences—see the Interacademy Partnership website[3] for a partial list).
- **National geological societies**, sometimes referred to as learned societies (examples include the Geological Society of India and the Geological

Society of Australia). We discuss the contribution of this type of institution to sustainable development objectives in Sect. 16.4.3.
- **National professional geoscience bodies**, which are usually separate from national geological societies (for example, the Ilustre Colegio Oficial de Geólogos in Spain and the Institute of Geologists of Ireland), although both functions may be fulfilled by the same body (as by the Geological Society of London in the UK). Such institutions are also discussed in Sect. 16.4.3.
- **Specialist geoscientific societies, associations and groups**, whether at international or national level, including specialist subgroups of broader geoscience organisations (such as the International Association of Hydrogeologists, the Seismological Society of America, and the Forensic Geoscience Group of the Geological Society of London).
- **National geological surveys** (such as the Geological Survey of Botswana and the Colombian Geological Survey).
- **Government departments, agencies, and directorates** (examples in Kenya include the Ministries of Environment and Forestry (including the National Environment Management Authority), Mining and Petroleum (including the Directorate of Geological Surveys), Water and Sanitation, and Energy).
- **Voluntary groups** focused on the interface between geoscience and topics such as development and ethics (such as Geology for Global Development and the International Association for Promoting Geoethics, together with its national chapters).
- **Regional, local and student societies, groups, chapters, and clubs**.

Many of these institutions have formal or informal links and working relationships with one another, and shared membership and objectives. They all have great potential to bring together, represent and build connections between local, national, and international communities of geoscientists, and to stimulate and empower these communities to play a significant part in delivering the SDGs.

---

[3]http://www.interacademies.org/31841/Members

16.4.2 International Scientific Institutions and Sustainable Development

International scientific institutions, such as unions operating at a continental or global level, contribute to sustainable development progress through promoting research, training, and education. Institutions help to

- *Advance scientific understanding*, supporting scientific discovery, innovation, and dissemination;
- *Strengthen public understanding of science* and trust in scientific advice;
- *Build links with other institutions* (e.g., other scientific disciplines, think tanks, intergovernmental agencies, national governments), helping to develop interdisciplinary solutions to complex challenges, provide a united voice (e.g., on climate change), and embed science into all forms of decision-making;
- *Facilitate communication and collaboration* among geoscientists.

Many of these activities link to issues of global sustainability, helping to inform and shape societal transformations towards sustainability at individual to global scales. For example, global geoscience institutions can help to connect geology with other natural or social sciences to generate solutions to lack of resource access, natural hazards, rapid urbanisation, and geodiversity and biodiversity conservation.

Transparency and good governance of scientific institutions are inherently connected. The more the organisation's leadership is transparent, the greater the opportunity for a culture of trust to be established, an organisation to thrive, its mission to be better delivered, and its societal benefit to be maximised. In practice this requires the following:

- *Clear, accessible, and adhered to 'rules'.* Rules may govern how an organisation operates and how members can get involved (e.g., through standing for elected positions). Such

rules should tackle injustices (e.g., encouraging representation from marginalised groups) and not perpetuate injustices. There should be clear processes in place to respond when rules are violated, and mechanisms to ensure action can be and is taken.
- *Management of Competing Interests.* Those with decision-making responsibilities should declare any competing interests (financial or otherwise), so as to ensure decision-making focuses solely on the best interests of the institution and society at large.
- *Financial and Project Transparency.* Organisations should publish an annual report that describes and characterises key financial information (e.g., income and expenditure, assets) and how resources have been used to advance the aims and objectives of the organisation. This information should be easily accessible, presented in a form that stakeholders can understand, and invite and encourage dialogue. Reporting of progress against objectives throughout the year helps to build interest in an organisation, and supports accountability.
- *Inclusive Decision-Making.* Diversity and inclusiveness are key to achieving the SDGs (see **SDG 5, SDG 10**). Scientific institutions should, therefore, actively model these values. Inclusiveness in institutional representations, especially in science, takes different forms and meanings. It can be correlated with regional, gender, ethnicity, or youth representation in the decision-making system of a given scientific institution, among many other factors and characteristics. Representation should be meaningful (not tokenistic) and participatory.
- *Open Data.* Open data plays an important role in improving governance and achieving scientific objectives relevant to sustainable development. Institutions may support open access to their own data and reports (e.g., the International Union of Geological Sciences, Box 16.4, publishes a fully open-access scientific journal), or actively partner in broader open science initiatives. The *Digital Earth Africa* programme, for example, processes freely available Earth Observation (EO) data to produce decision-ready products to

improve lives across Africa. This is funded by the Leona M. and Harry B. Helmsley Charitable Trust and the Australian Government, and includes partners such as the Committee on Earth Observation Satellites and the Group on Earth Observations.

These characteristics should not be unique to large global or regional scientific institutions. Such organisations, however, are often viewed as global leaders and, therefore, have a responsibility to develop, model, and catalyse good practice. Global geoscience organisations can develop resources or provide advice on governance that national organisations can adopt to improve their effectiveness and accountability. There are likely to be many contexts, however, in which national institutions are more advanced in their pursuit of openness and inclusivity, and can help drive improvements at a global scale.

Box 16.4. International Union of Geological Sciences (IUGS)

The International Union of Geological Sciences (IUGS), established in 1961, plays a unique leadership role in the global geological community. IUGS *'promotes and encourages the study of geological problems, especially those of world-wide significance, and supports and facilitates international and interdisciplinary cooperation in the Earth sciences'*. The IUGS is a member of the International Science Council (ISC, formerly ICSU), which plays a leading role in coordinating the UN Major Group on Science and Technology (giving scientists a voice in UN processes). Membership of the ISC, if active and sustained, therefore, helps to give geoscientists a voice in sustainable development planning and decision-making.

Website: www.iugs.org/.

Membership: IUGS membership is at a national level, through a so-called

'Adhering Organisation', such as a Geological Survey, Geological Society or National Academy. IUGS also has 'Affiliated Organisations', which are typically international societies (often with a regional or thematic focus) representing sections of the geoscience community. At the time of writing, IUGS had 57 active adhering national members (with 8 active pending members), and 56 affiliated organisations (e.g., American Geophysical Union, Geological Society of Africa, Geology for Global Development, International Association for Promoting Geoethics, International Consortium on Landslides, Society of Economic Geologists).

What does IUGS do? IUGS fosters dialogue and communication among the global Earth science community by organising international projects and meetings, sponsoring symposia and scientific field trips, and producing publications. The topics addressed through these activities relate to fundamental research, economic and industrial applications, environmental and societal challenges, and educational and developmental problems.

Examples of Activities:

- *Commissions, Task Groups, and Initiatives.* These are concerned with a wide range of geological research and practice, often of interest to governments, industry, and academia.
- *Scientific Publishing.* The journal *Episodes* covers developments of regional and global importance in the Earth sciences and is distributed worldwide to scientists in more than 150 countries. Research outputs from IUGS initiatives and conferences are also published from time to time as standalone Special Publication volumes, usually by the Geological Society of London on behalf of IUGS.

- *International Geoscience Programme (IGCP)*. Since 1972, IUGS has co-sponsored, with UNESCO, the IGCP, helping to build geoscience capacity around the World, especially in developing countries. The programme aims at fostering North–South and, especially, South–South cooperation between geoscientists on key thematic areas including geohazards, use of natural resources and climate change.
- *International Geological Congress*. This is a major gathering of the Earth science community, taking place every four years and supported by the IUGS. The IGC includes the 'Geohost' programme providing financial support to early-career scientists and those from low-income countries, enabling them to participate fully in the congress.
- *Resourcing Future Generations*. This flagship initiative seeks to engage the international geological community in a global effort to meet the world's future resource needs sustainably. It has stimulated novel research agendas, interdisciplinary links, and a major IUGS conference on the topic in 2018. It arose in part from the success of the *International Year of Planet Earth* in 2008 (with activities running from 2006 to 2009), which was organised by IUGS in partnership with UNESCO and other UN agencies.

16.4.3 National Geological Societies and Professional Bodies

National Geological Societies and Professional Bodies (NGSPBs) have an especially important role to play, given their focal position in national geoscience communities and the scope they have to reach out beyond these communities. This is a particular responsibility for those that are larger and relatively well resourced. This is recognised,

for instance, by the Geological Society of London, whose strategy for 2017–2027 includes the aim of *'promoting the role of geoscience in sustainable global development'* (Geological Society of London 2017). NGSPBs can and should perform a number of functions that can contribute to the delivery of the SDGs.

Within their own memberships, NGSPBs can highlight the many ways in which geological research, education, and professional practice can help to deliver the SDGs, as described throughout this volume. They can also play a wider leadership role within their national geological communities, and can communicate to those studying geoscience (whether at school or university level), public audiences (including via social media and traditional media), and policy-makers the vital role of geoscience in sustainable global development. By working with their own members with relevant expertise, NGSPBs are ideally placed to raise awareness among these audiences of our dependence on the Earth for a vast range of resources, including energy, minerals, and water, and to make the case for ensuring that they are sourced, managed, and used in an environmentally, socially, and economically responsible way. They can highlight the detrimental social and environmental impacts that extraction and use of natural resources can have, but also the potential for these activities to support sustainable livelihoods and economic growth (**SDG 8**), to provide energy and materials for sustainable patterns of consumption and infrastructure development (**SDGs 6, 7, 9, 11,** and **12**), and to facilitate the necessary transition to low carbon energy systems (**SDG 13**). Similarly, they can communicate the need for high-quality geoscience to underpin environmental management and protection of ecosystems (**SDGs 14** and **15**), sustainable agriculture (**SDG 2**), protection against natural hazards (**SDGs 1** and **11**), and engineering activities to develop infrastructure (**SDGs 9** and **11**).

A key role for NGSPBs is to engage with national and regional governments and regulators, urging them to develop and put into effect policies and regulatory frameworks across the domains outlined above which are informed by the best available geoscientific evidence, and helping them to access the requisite expertise and information via their members and the wider geoscience community. This may take the form of responding to external policy initiatives such as departmental consultations and parliamentary inquiries, or more proactive efforts to highlight matters of concern identified within the geoscience community. Effective development and implementation of policy is essential for reducing bribery and corruption (SDG 16.5) and ensuring accountability and transparency (SDG 16.6) in the natural resources sectors, and NGSPBs can help governments and other decision-makers connect to those with expertise and experience relevant to addressing these challenges.

Professional bodies have as their core purpose the role of setting, certifying, and policing high professional standards in geoscience. This is done by a variety of means, often including the award of professional titles such as *Chartered Geologist* (Geological Society of London), *Certified Professional Geologist* (American Institute of Professional Geologists), *Professional Geoscientist* (Geoscientists Canada), *Professional Geologist* (Institute of Geologists of Ireland), *Perito Geólogo* (Ilustre Colegio Oficial de Geólogos), and *European Geologist* (European Federation of Geologists, working with its national member professional bodies), which are typically available to suitably qualified individuals from outside as well as within the country of the awarding body. Such bodies (as well as many national geological societies) require title holders and other members to abide by professional ethics codes or codes of conduct, and promulgate these codes as standards to which all professional geoscientists should adhere. A partial list of these codes is available on the website of the International Association for Promoting Geoethics[4].

Significant efforts have been made to promote common professional standards in geoscience internationally, including through mutual recognition agreements between professional bodies, the work of the European Federation of Geologists (which acts as an umbrella body for national professional bodies across Europe), and the work of a Task Group on Global Geoscience Professionalism (TG-GGP) within the International Union of Geological Sciences (Peppoloni et al. 2019). The awarding of professional titles and associated activities of professional bodies are intended primarily to ensure that high professional standards are met in the practice of geoscience, for the public good, rather than to benefit individual geoscientists. Setting and seeking to enforce high standards in the extraction and management of natural resources, infrastructure development and protection against natural hazards are all essential to sustainable and equitable global development. Furthermore, efforts to extend the availability of professional titles and the impact of associated standards internationally support the education and professional development of a global geoscience workforce equipped to meet these global needs, and help to build capacity in countries beyond those currently able to support their own national professional titles and bodies.

Professional bodies also play a vital role in helping to develop and ensure adoption of external standards, including mineral reporting codes such as Pan-European Reserves & Resources Reporting Committee (PERC), Australasian Code for Reporting of Exploration Results, Mineral Resources and Ore Reserves (JORC), and those overseen by other member bodies of the Committee for Mineral Reserves International Reporting Standards (CRIRSCO). Adherence to such standards and their adoption by stock exchanges and other relevant bodies are essential for ensuring accountability and transparency in the mining sector and in downstream minerals supply and value chains (see Sect. 16.3.2).

NGSPBs can contribute directly to the delivery of **SDG 16** through the good governance of their own affairs (see the principles outlined in Sect. 16.4.2) and their interactions with others. By maintaining and applying rigorous governance documents and procedures, and keeping these under review to make sure they are fair and fit for purpose, they can help to engender a

---

[4]http://www.geoethics.org/codes

culture of rule of law and equal access to justice for all (SDG 16.3) at a sub-governmental level, within and beyond geoscience communities, and ensure accountability and transparency (SDG 16.6). They should be responsive, inclusive, and representative in their decision-making (SDG 16.7), and actively encourage participation of all members in this, having regard to potential barriers to participation including implicit discrimination and unconscious bias. They can work with others to reduce corruption and bribery (SDG 16.5), by directly addressing such ethical issues in their professional codes, through collaboration with bodies such as the International Association for Promoting Geoethics and fellow NGSPBs, and by demanding high ethical standards from sponsors, stakeholders and partner organisations across the public, private and voluntary sectors.

Many NGSPBs and other geoscience institutions have recently recognised the need to pay far greater attention to issues of discrimination and harassment (SDG 16.B), and have implemented or initiated measures to create an environment in which such behaviours are unacceptable and those experiencing them feel confident in reporting them. These measures include development of new codes of conduct (for example, regarding attendance at conferences) and amendment of existing ones. There is also a renewed focus on diversity, equality, and inclusion in many geoscience institutions, although much remains to be done to ensure a truly level playing field.

Finally, NGSPBs can do a great deal to promote international cooperation and collaboration, and to support capacity building, including through helping national geoscience bodies in developing countries to develop and grow their activities (SDG 16.A). Through initiatives such as the Associated Societies schemes of the Geological Society of America and the Geological Society of London, as well as through bilateral relationships and informal cooperation, better resourced national bodies can share experience regarding governance and organisation, provide access to members with relevant expertise and experience, share materials, support participation in national and international initiatives (such as Earth Science Week) and offer moral support and friendship.

16.4.4 National Geological Surveys

A third group of geoscience institutions which contribute to sustainable development is national geological surveys. These are typically government sector, not-for-profit research organisations, tasked with

- Observing, characterising, and understanding the Earth's natural systems.
- The provision of reliable, impartial scientific data and advice.
- Delivering projects that focus on public-good science, to support national planning, economic growth, and social well-being.

The longevity of many geological surveys has enabled them to play an internationally important role in delivering geoscience for, the responsible use of natural resources, building resilience to natural hazards and environmental change and fundamentally to act as the national body for the collection, management and long-term storage of geodata (Fig. 16.8).

Some of the oldest surveys have played a key influencing role in establishing and building national survey activities around the globe. For example, the British Geological Survey (BGS) founded in 1835, the US Geological Survey founded in 1879, the Argentinian Geological and Mining service created in 1904, and in 1912 the South African Council for Geoscience (formerly the Geological Survey of South Africa). In contrast, the China Geological Survey was only re-founded in 1999 but today has many links with geological surveys across Africa and Asia.

The mission statements of national geological surveys highlight their roles in contributing to economic growth. Most have yet to formally adopt the SDGs within their programmes. One exception is the British Geological Survey which in its new research strategy 2019–23 (BGS 2019) seeks to align its global activities and research challenges to the SDGs. Coordination of research

Fig. 16.8 The Core Store at the British Geological Survey. Part of the UK National Geological Repository. © NERC (used with permission)

by geological surveys is also enhanced by global partnership initiatives including OneGeology (2019), the Global Earthquake Model (2019), the Global Volcano Model (2019), the Commission for the Geological Map of the World (2019) (websites for which are all listed in the references), all of which directly or indirectly underpin **SDG 16** and its targets.

16.5 Key Learning Concepts

- Conflict may result in heightened vulnerability to hazardous events, or the displacement of people into areas more exposed to natural hazards.
- Science can help to facilitate international diplomacy and inform foreign policy objectives, to support the peaceful resolution of disputes and conflict. Geoscientists have a particular role in collaborating to understand and/or help manage transboundary geological features (e.g., volcanoes) or resources (e.g., aquifers). The sharing of resources (including

expertise) both North-South and South-South is important in reducing risk.
- Corruption undermines the rule of law, and prevents vital resources from being spent on development projects and the effective regulation of infrastructure development. Corruption, therefore, increases vulnerability to natural hazards, and reduces the effectiveness of disaster risk reduction interventions. Hazardous events can also lead to (increased) corruption as a result of the post-disaster chaos.
- The Extractives Industry Transparency Initiative (EITI) is an example of a global initiative that promotes open and accountable management of hydrocarbon and mineral resources. National-level multi-stakeholder groups are responsible for setting objectives and implementing these to improve transparency and accountability.
- Science-based institutions have an important role to play in supporting sustainable development. Institutions help to shape and model

accepted behaviours, such as good governance and financial transparency, ensuring inclusive decision-making, and creating safe environments free from harassment and discrimination. They also work to ensure that high professional standards for geoscientists are set and implemented, and can be a vital source of independent and authoritative geological information and expertise.

16.6 Educational Ideas

In this section, we provide examples of educational activities that connect geoscience, the material discussed in this chapter, and scenarios that may arise when applying geoscience (e.g., in policy, government, private sector international organisations, NGOs). Consider using these as the basis for presentations, group discussions or essays, or to encourage further reading.

- Select a transboundary aquifer using the IGRAC map of transboundary aquifers (https://apps.geodan.nl/igrac/ggis-viewer/viewer/tbamap/public/default), and research how this shared resource is currently managed. If you wanted to conduct research on this aquifer, what political and cultural factors may you need to consider prior to starting your work?
- Conduct a volcanic crisis simulation to explore the challenges of decision-making in a volcanic crisis. There are various options available on the Internet, which can be tailored to your particular group. Another option is Hazagora, a board game that has been developed to engage with risk-sensitive development decisions (https://serc.carleton.edu/introgeo/roleplaying/examples/125523.html, https://portal.opendiscoveryspace.eu/et/osos-project/eruption-web-based-simulation-managing-volcanic-crisis-850927, https://games4sustainability.org/gamepedia/hazagora/).
- Haiti and the Dominican Republic share the same island. Nonetheless, the impacts of disasters on the population and the built environment of these two countries are vastly different. In 2004, Hurricane Jeanne made landfall at the east side of the Dominican Republic and flooded both countries. In the Dominican Republic, this took the life of 19 people while in Haiti approximately 3,000 people were killed. How can these differences in vulnerability between the two countries be explained and can you think of any other places where similar events had such different impacts?
- Review what geoscience institutions exist in your country, and the opportunities for students to get involved in their activities. Write a short media article about how one of these institutions is contributing to the SDGs, thinking carefully about the audience, the language you use, and your sources of information.

Further Reading and Resources

Bobbette A, Donovan A (eds) (2019) Political geology: active stratigraphies and the making of life. Palgrave Macmillan, 379 p

Center for Science Diplomacy. Available at:https://www.aaas.org/programs/center-science-diplomacy

Extractives Industry Transparency Initiative. Available at: https://eiti.org/

The Royal Society (2010a) New frontiers in science diplomacy. The Royal Society, London, p 44

UK Parliamentary Office of Science and Technology (2018) Science Diplomacy. https://researchbriefings.files.parliament.uk/documents/POST-PN-0568/POST-PN-0568.pdf

UK Scientific Advisory Group for Emergencies (2010) Volcanic Ash Disruptions – Meeting Minutes. https://webarchive.nationalarchives.gov.uk/20130705051929, https://www.bis.gov.uk/assets/goscience/docs/s/10-1371-sage-volcanic-ash-minutes-21-april-2010.pdf

References

Adger WN (2006) Vulnerability. Glob Environ Change 16(3):268–281. https://doi.org/10.1016/j.gloenvcha.2006.02.006

Ahrens J, Rudolph PM (2006) The importance of governance in risk reduction and disaster management. J Contingencies Crisis Manag 14 (4):207–220. https://doi.org/10.1111/j.1468-5973. 2006.00497.x

Akarca A, Tansel A (2012) Turkish voter response to government incompetence and corruption related to the 1999 earthquakes. SSRN Electronic Journal. https://doi.org/10.2139/ssrn.1934339

Alexander DE (2008) Mainstreaming disaster risk management, 38–54. https://doi.org/10.4324/9780203938 720-11

Alexander DE (2016) The game changes: "Disaster Prevention and Management" after a quarter of a century. Disaster Prevent Manag: Int J 25(1):2–10. https://doi.org/10.1108/DPM-11-2015-0262

Ambraseys N, Bilham R (2011) Corruption kills. Nature 469(7329):153–155. https://doi.org/10.1038/469153a

Ayuk E, Pedro A, Ekins P (2019) Mineral resource Governance in the 21st century: gearing extractive industries towards sustainable development. Summary for policymakers and business leaders. https://www. resourcepanel.org/reports/mineral-resource-governance-21st-century. Accessed 22 Aug 2019

BBC (2019) Rohingya refugees in Bangladesh battle monsoon landslides and floods. https://www.bbc.co. uk/news/world-asia-48905031. Accessed 10 July 2019

Betts M, Farrell S (2009) Global construction 2020: a global forecast for the construction industry over the next decade. London : Global Construction Perspectives and Oxford Economics. https://vu.on.worldcat. org/search?queryString=no%3A+489637787#/oclc/ 489637787

BGS (2019) Science strategy. www.bgs.ac.uk/about/ strategy.html. Accessed 11 Oct 2019

Bleischwitz R, Dittrich M, Pierdicca C (2012) Coltan from Central Africa, international trade and implications for any certification. Resour Pol 37(1):19–29. https://doi.org/10.1016/j.resourpol.2011.12.008

Bosher L, Dainty A (2011) Disaster risk reduction and 'built-in' resilience: towards overarching principles for construction practice. Disasters 35(1):1–18. https:// doi.org/10.1111/j.1467-7717.2010.01189.x

Cabinet Office (2012) Scientific advisory group for emergencies (SAGE). https://www.gov.uk/ government/publications/scientific-advisory-group-for-emergencies-sage. Accessed 17 July 2019

Commission for the Geological Map of the World (2019). https://ccgm.org/en/. Accessed 11 Oct 2019

Church of England (2017). Extractive industries: the policy of the national investing bodies of the church of england and the ethical investment advisory group's advisory and theological papers. https://www. churchofengland.org/more/media-centre/news/church-englands-national-investing-bodies-launch-policy-investing-extractive. Accessed 22 Aug 2019

Ciurean R, Schröter D, Glade T (2013) Conceptual frameworks of vulnerability assessments for natural disasters reduction. approaches to disaster management - examining the implications of hazards, emergencies and disasters, 3–32. https://doi.org/10.5772/55538

Clancy KB, Nelson RG, Rutherford JN, Hinde K (2014) Survey of academic field experiences (SAFE): Trainees report harassment and assault. PLoS ONE 9(7): e102172

de Ruiter MC, Ward PJ, Daniell JE, Aerts JCJH (2017) Review article: a comparison of flood and earthquake vulnerability assessment indicators. Nat Hazards Earth Syst Sci Discuss 1(2011):1–34. https://doi.org/10. 5194/nhess-2017-45

Donovan A, Oppenheimer C (2019) Volcanoes on borders: a scientific and (geo)political management challenge. Bull Volc . https://doi.org/10.1007/s00445-019-1291-z

Donovan A (2019). Politics of the lively geos: volcanism and geomancy in Korea. In Political geology. Palgrave Macmillan, Cham, pp 293–343

EITI (2019a). The EITI standard 2019. https://eiti.org/ sites/default/files/documents/eiti_standard2019_a4_ en.pdf. Accessed 22 Aug 2019

EITI (2019b) EITI progress report 2019. https://eiti.org/ sites/default/files/documents/eiti_progress_report_ 2019_en.pdf. Accessed 22 July 2019

Escaleras M, Anbarci N, Register CA (2007) Public sector corruption and major earthquakes: a potentially deadly interaction. Public Choice 132(1–2):209–230. https:// doi.org/10.1007/s11127-007-9148-y

Freckleton M, Wright A, Craigwell R (2012) Economic growth, foreign direct investment and corruption in developed and developing countries. J Economic Stud 39(6):639–652. https://doi.org/10.1108/014435812 11274593

Geological Society of London (2017) A Strategy for the Geological Society, 2017–2027. https://www.geolsoc. org.uk/strategy. Accessed 13 Aug 2019

Gill JC (2016) Building good foundations: skills for effective engagement in international development. Geol Soc Am Spec Pap 520:1–8

Global Financial Integrity (2017) Illicit financial flows to and from developing countries: 2005–2014. https:// secureservercdn.net/45.40.149.159/34n.8bd. myftpupload.com/wp-content/uploads/2017/04/GFI-IFF-Report-2017_final.pdf?time=1562760146. Accessed 10 July 2019

Gluckman PD, Turekian VC, Grimes RW, Kishi T (2017) Science diplomacy: a pragmatic perspective from the inside. Sci Dipl 6(4)

Goodland R (2012) Responsible mining: the key to profitable resource development. Sustainability 4 (9):2099–2126. https://doi.org/10.3390/su4092099

Green P (2005) Disaster by design: corruption, construction and catastrophe. Br J Criminol 45(4):528–546. https://doi.org/10.1093/bjc/azi036

Global Earthquake Model (2019) Available at: https:// www.globalquakemodel.org/. Accessed 11 Oct 2019

Global Volcano Model (2019) https:// globalvolcanomodel.org/. Accessed 11 Oct 2019

Gapminder and UN (2019) Population information. http:// www.gapminder.org/. Accessed 9 Oct 2019

Hammond JO (2016). Understanding volcanoes in isolated locations: engaging diplomacy for science. Sci Dipl 5(1)

Hutton G, Varughese M (2016) The costs of meeting the 2030 sustainable development goal targets on drinking water, sanitation, and hygiene. Water and sanitation program: technical paper. World Bank Group. http://documents.worldbank.org/curated/en/415441467988938343/pdf/103171-PUB-Box394556B-PUBLIC-EPI-K8543-ADD-SERIES.pdf. Accessed 10 July 2019

Kaufmann D, Kraay A, Mastruzzi M (2003) Governance matters III: governance indicators for 1996–2002. The World Bank. https://doi.org/10.1596/1813-9450-3106

Kenny C (2012) Disaster risk reduction in developing countries: costs, benefits and institutions. Disasters 36 (4):559–588. https://doi.org/10.1111/j.1467-7717.2012.01275.x

Knutsen CH, Kotsadam A, Olsen EH, Wig T (2017) Mining and local corruption in Africa. Am J Polit Sci 61(2):320–334

Leeson PT, Sobel RS (2008) Weathering Corruption. J Law Econom 51(4):667–681. https://doi.org/10.1086/590129

Lewis J (2003) Housing construction in earthquake-prone places: Perspectives, priorities and projections for development. Aust J Emer Manag 18(2)

Lewis J (2008) The worm in the bud: corruption, construction and catastrophe. In: L Bosher (ed) Hazards and the built environment: attaining built-in resilience. Taylor and Francis, Abingdon, pp 238–263. https://doi.org/10.4324/9781315689081-13

Lewis J (2011). Corruption: the hidden perpetrator of under-development and vulnerability to natural hazards and disasters. www.datum-international.eu

Lujala P, Rustad SA, Billon PL (2017) Has the EITI been successful? Reviewing evaluations of the extractive industries transparency initiative (U4 Brief 2017:5). https://www.cmi.no/publications/file/6300-has-the-eiti-been-successful.pdf. Accessed 22 Aug 2019

Murshed M, Mredula F (2018) Impacts of corruption on sustainable development: a simultaneous equations model estimation approach. J Account, Finan Econom 8(1):109–133

OECD (2016) Corruption in the extractive value chain: typology of risks, mitigation measures and incentives. https://read.oecd-ilibrary.org/development/corruption-in-the-extractive-value-chain_9789264256569-en. Accessed 22 Aug 2019

Ortiz-Ospina E, Roser M (2019) Our world in data - corruption. https://ourworldindata.org/corruption. Accessed 10 July 2019

OneGeology (2019) http://www.onegeology.org/. Accessed 11 Oct 2019

Peppoloni S, Bilham N, Di Capua G (2019) Contemporary Geoethics Within the Geosciences. In: Bohle M (ed) Exploring geoethics: ethical implications, societal contexts, and professional obligations of the geosciences. Springer Nature, Cham, pp 25–70

Petal M, Green R, Kelman I, Shaw R, Dixit A (2008) Community-based construction for disaster risk reduction. In Hazards and the built environment: attaining built-in resilience. Routledge, pp 191–217. https://doi.org/10.4324/9780203938720

Peters K (2017) The next frontier for disaster risk reduction: tackling disasters in fragile and conflict-affected contexts. Overseas Development Institute Report, London

St John K, Riggs E, Mogk D (2016) Sexual harassment in the sciences: a call to geoscience faculty and researchers to respond. J Geosci Educ 64(4):255–257

The Royal Society (2010b) New frontiers in science diplomacy. R Soc, London, p 44

Transparency International (2018) Corruption perception index. https://www.transparency.org/cpi2018. Accessed 9 Oct 2019

Turekian VC, Neureiter NP (2012) Science and diplomacy: The past as prologue. Chemistry in Australia, p 26

UNDRR (2017) Terminology. Available at: https://www.undrr.org/terminology. Accessed 29 October 2020

UNDP (2018) Human development index. http://hdr.undp.org/en/indicators/137506#. Accessed 9 Oct 2019

Wall F, Rollat A, Pell R (2017) Responsible sourcing of critical metals. Elements 13(5):313–318. https://doi.org/10.2138/gselements.13.5.313

Werg J, Grothmann T, Schmidt P (2013) Assessing social capacity and vulnerability of private households to natural hazards - integrating psychological and governance factors. Nat Hazards Earth Syst Sci 13(6):1613–1628. https://doi.org/10.5194/nhess-13-1613-2013

Harris K, Keen D, and Mitchell T (2013) When disasters and conflicts collide. Improving links between disaster resilience and conflict prevention. Available at: https://www.odi.org/sites/odi.org.uk/files/odi-assets/publications-opinion-files/8228.pdf (accessed on 28 October 2020)

Joel C. Gill is International Development Geoscientist at the British Geological Survey, and Founder/Executive Director of the not-for-profit organisation Geology for Global Development. Joel has a degree in Natural Sciences (Cambridge, UK), a Masters degree in Engineering Geology (Leeds, UK), and a PhD focused on multi-hazards and disaster risk reduction (King's College London, UK). For the last decade, Joel has worked at the interface of Earth science and international development, and plays a leading role internationally in championing the role of geoscience in delivering the UN Sustainable Development Goals. He has coordinated research, conferences, and workshops on geoscience and sustainable development in the UK, India, Tanzania, Kenya, South Africa, Zambia, and Guatemala. Joel regularly engages in international forums for science and sustainable development, leading an international delegation of Earth scientists to the United Nations in 2019. Joel has prizes from the London School of Economics and Political Science for his teaching related to disaster risk reduction, and Associate Fellowship of the Royal Commonwealth Society for his international development engagement. Joel is a Fellow of the Geological Society of London, and was elected to Council in 2019 and to the position of Secretary (Foreign and External Affairs) in 2020.

Amel Barich holds a PhD in petrology and geochemistry from the University of Granada (2015). During her graduate studies, she developed a passion for science diplomacy. Since 2010, Amel has served the board of several international geoscience organisations, including as councillor for the International Union of Geological Sciences (2014–2018). Amel was the first early career geoscientist to join the IUGS executive board. She was secretary general of the International Association of Geoethics (2016–2018), and Vice President of the Young Earth Scientists Network (2010–2016). Amel is currently holding a project manager position at the Geothermal Research Cluster (GEORG) in Iceland and manages various international geothermal projects. She is interested in strategic growth and branding/marketing of international organisations focused on serving society through geoscience.

Nic Bilham researches responsible sourcing of minerals, the relationship between mining and the circular economy, and the challenge of assuring environmental and social impact standards across complex value chains and production-consumption networks. Until 2018, he worked at the Geological Society of London for over 20 years, most recently as Director of Policy and Communications. Nic is an Executive Council member of the International Association for Promoting Geoethics (IAPG), and chair of trustees of Geology for Global Development (GfGD). He holds degrees in History and Philosophy of Science (BA, University of Cambridge) and Science and Technology Policy (MSc, University of Sussex).

Sarah Caven Having started out in mineral exploration, Sarah now draws upon a diversity of global experience spanning private sector, government, social enterprise, and international development. A natural translator across scales and sectors, she contributes to artisanal and small scale mining programs through to regional prospectivity projects. In response to the resourcing future generations challenge, Sarah is passionate about enhancing collaboration, equity and business innovation in mining to unlock development opportunities. Sarah holds a master's in geology (University of Leicester), a Master of Business Administration, (University of British Columbia), and participated in Columbia University's executive education program, Extractive Industries and Sustainable Development.

Marleen de Ruiter an Assistant Professor (Multi-Risk) at the Institute for Environmental Studies (IVM), Vrije Universiteit Amsterdam. She has a background in Environmental and Resource Management from the University of British Columbia (Canada) specialising in disaster vulnerability and post-disaster recovery. Her current research concentrates on improving our scientific understanding of multi-hazard risk and assessing the impacts of adaptation measures on disaster risk reduction. She particularly focuses on the dynamics of vulnerability in the face of consecutive disasters. She is a co-science officer of the European Geosciences Union's multi-hazard sub-division and coordinates the VU University master programme Global Environmental Change and Policy.

Amy Donovan is an interdisciplinary geographer working at the interface of human and physical geography. She is particularly interested in the nature and (geo)politics of scientific knowledge in the context of disaster risk. Much of her work has been on active volcanoes, but she also works on related climate hazards and earthquakes.

Martin Smith is a Science Director with the British Geological Survey and Principle Investigator for the BGS ODA Programme Geoscience for Sustainable Futures (2017-2021). He has a first degree in Geology (Aberdeen) and a PhD on tectonics (Aberystwyth, UK). A survey geologist by training Martin has spent a career studying geology both in the UK and across Africa and India. As Chief Geologist for Scotland and then for the UK he has worked closely with government and industry on numerous applied projects including in the UK on national crises, major infrastructure problems, decarbonisation research and urban geology and overseas for DFID-funded development projects in Kenya, Egypt and Central Asia. Martin is a Chartered Geologist and fellow of the Geological Society of London. He was awarded an MBE for services to geology in 2016.

Partnerships for the Goals

17

Strengthen the Means of Implementation, Revitalise the Global Partnership for Sustainable Development

Susanne Sargeant, Joel C. Gill, Michael Watts, Kirsty Upton, and Richard Ellison

S. Sargeant (✉) · K. Upton
British Geological Survey, The Lyell Centre,
Research Avenue South, Edinburgh EH14 4AP, UK
e-mail: slsa@bgs.ac.uk

J. C. Gill · M. Watts · R. Ellison
British Geological Survey, Environmental Science
Centre, Nicker Hill, Keyworth, Nottingham NG12
5GG, UK

J. C. Gill
Geology for Global Development, Loughborough,
UK

© Springer Nature Switzerland AG 2021
J. C. Gill and M. Smith (eds.), *Geosciences and the Sustainable Development Goals*,
Sustainable Development Goals Series, https://doi.org/10.1007/978-3-030-38815-7_17

Abstract

17 PARTNERSHIPS FOR THE GOALS

Overview

Partnerships between diverse groups are critical to achieving the SDGs

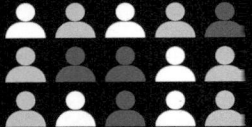

Effective partnerships take time, effort, commitment, and ongoing reflection and learning

SDG 17 aims to create an environment in which partnerships can flourish

Building Partnerships:

Three partnership types have a vital role to play, each with benefits and challenges:

North-South: Cooperation between richer industrialised and poorer developing countries

Capacity building is embedded into many partnerships, requiring clear objectives and accountability

South-South: Cooperation between two or more developing countries

Partnerships may involve multiple sectors and disciplines

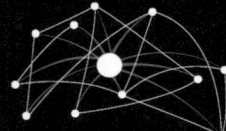

Triangular: Driven by Southern partners, with external support (e.g., from World Bank)

Effective partnerships integrate perspectives from communities and indigenous knowledge

Engaging with national and international organisations and frameworks:

Requires awareness of political, economic, social, technological, legal and environmental context

Helps to complement existing approaches and priorities, rather than duplicate or undermine these

Engaging in the UN Technology Facilitation Mechanism helps to improve use of geoscience in decision making Examples include:

- Applying for UNESCO International Geoscience Programme funding
- Engaging with UN Major Groups (e.g., science and technology)
- Proactive dissemination of geoscience research and its impacts
- Engagement in the UN forum on science, technology and innovation for the SDGs

17.1 Introduction

Real, meaningful partnership and cooperation between diverse groups is critical to achieving all of the SDGs and is the focus of **SDG 17**. Partnership is one '*way of organising*' social interactions (Harriss 2000, p 225) where the people or organisations involved agree to cooperate in order to further their mutual interests. In practice, effective partnership takes time, effort, deliberate reflection and learning, and sustained commitment to achieve. For geoscientists, global challenges such as clean water and sanitation (**SDG 6**), affordable and clean energy (**SDG 7**), making cities and human settlements inclusive, safe, resilient and sustainable (**SDG 11**), and climate action (**SDG 13**) require us to be part of collaborations and partnerships with people from all sections of society to undertake research and exchange knowledge. This includes working in partnership with researchers from other disciplines, people from other sectors such as government or the private sector, or with the public. Add to this the often frequent need for these partnerships to extend across national boundaries, and to bridge cultural and language differences, sometimes in fragile settings, and the potential rewards and challenges that **SDG 17** presents start to emerge.

SDG 17 aims to create an '*enabling environment*' in which cooperative relationships can flourish at all levels through coordinated policies and frameworks, and by revitalising existing partnerships (United Nations 2015a). What does this mean in practice, and what does it mean for the geoscience community and how we work? We address these questions in this chapter, and introduce the targets of **SDG 17**.

The importance of partnerships in facilitating development has long been recognised. Positive, transformative change that leads to improvements in people's lives, livelihoods, health, education, and well-being requires different groups to work together. In 2000, UN Member States identified the need for a '*global partnership for development*' in order to create a '*conducive environment for development at the national and local levels*'[1]. This led to Millennium Development Goal (MDG) 8: Develop a Global Partnership for Development. MDG 8 focused strongly on creating an environment in which trading between countries was open, predictable, and rules-based, and where the specific needs of developing countries were addressed, debt problems were tackled and where people living in developing countries would have access to affordable essential drugs and be able to benefit from technological advances (e.g., mobile cellular signal) (United Nations 2013). The authors of United Nations (2015b) reflect on the progress made during the MDGs and note that Official Development Assistance (ODA) from developed countries increased by 66%, 95% of the world's population was covered by a mobile-cellular signal, and internet penetration grew from approximately 6% of the world's population in 2000 to 43% in 2015 (United Nations 2015b).

The ambitions of MDG 8 run through **SDG 17** and are reflected in its 19 targets (Table 17.1). These are grouped into five themes: finance, technology, capacity strengthening, trade, and 'systemic issues' (policy coherence, for example). Targets are geared towards creating an environment that encourages partnership and equality, meaning **SDG 17** has a wide range of themes.

- *Finance* **targets (17.1–17.5)** aim to lead to strengthened mobilisation of resources within a country (e.g., capacity to collect tax), between countries (e.g., developed countries fulfilling their official development assistance —or ODA—commitments), address issues relating to debt such as debt financing, relief and restructuring policies, and promoting investment.
- *Technology* **targets (17.6–17.8)** bring the role of science, technology and innovation in achieving the SDGs to the fore. It also highlights the role of various UN-level mechanism to help support coordination and knowledge sharing. We will consider the different forms

---

[1]http://www.un.org/en/development/desa/policy/untaskteam_undf/faqs.pdf.

Table 17.1 SDG 17 Targets (United Nations 2015a). Targets are grouped into five themes: finance, technology, capacity building, trade, and systemic issues

Target		Description
Finance	17.1	Strengthen domestic resource mobilisation, including through international support to developing countries, to improve domestic capacity for tax and other revenue collection
	17.2	Developed countries to implement fully their official development assistance commitments, including the commitment by many developed countries to achieve the target of 0.7 per cent of ODA/GNI to developing countries and 0.15–0.20 per cent of ODA/GNI to least developed countries; ODA providers are encouraged to consider setting a target to provide at least 0.20 per cent of ODA/GNI to least developed countries
	17.3	Mobilize additional financial resources for developing countries from multiple sources
	17.4	Assist developing countries in attaining long-term debt sustainability through coordinated policies aimed at fostering debt financing, debt relief, and debt restructuring, as appropriate, and address the external debt of highly indebted poor countries to reduce debt distress
	17.5	Adopt and implement investment promotion regimes for least developed countries
Technology	17.6	Enhance **North–South, South–South, and triangular cooperation on and access to science, technology and innovation** and enhance knowledge sharing on mutually agreed terms, including through improved coordination among existing mechanisms, in particular at the UN level, and through a **global technology facilitation mechanism**
	17.7	Promote the development, transfer, dissemination, and diffusion of environmentally sound technologies to developing countries on favourable terms, including on concessional and preferential terms, as mutually agreed
	17.8	Fully operationalise **the technology bank and science, technology and innovation capacity-building mechanism** for least developed countries by 2017 and enhance the use of enabling technology, in particular information and communications technology
Capacity Building	17.9	Enhance international support for implementing effective and targeted **capacity building** in developing countries to support national plans to implement all the sustainable development goals, including through **North–South, South–South, and triangular cooperation**
Trade	17.10	Promote a universal, rules-based, open, non-discriminatory and equitable multilateral trading system under the World Trade Organization, including through the conclusion of negotiations under its Doha Development Agenda
	17.11	Significantly increase the exports of developing countries, in particular with a view to doubling the least developed countries' share of global exports by 2020
	17.12	Realise timely implementation of duty-free and quota-free market access on a lasting basis for all least developed countries, consistent with World Trade Organization decisions, including by ensuring that preferential rules of origin applicable to imports from least developed countries are transparent and simple, and contribute to facilitating market access
Systemic Issues	17.13	Enhance global macroeconomic stability including through policy coordination and policy coherence
	17.14	Enhance **policy coherence** for sustainable development
	17.15	Respect each country's policy space and leadership to establish and implement policies for poverty eradication and sustainable development.
	17.16	Enhance the global partnership for sustainable development, complemented by **multi-stakeholder partnerships** that mobilise and share knowledge, expertise, technology, and

(continued)

Table 17.1 (continued)

Target		Description
		financial resources, to support the achievement of the sustainable development goals in all countries, in particular developing countries
	17.17	Encourage and promote effective public, public–private and civil society partnerships, building on the experience and resourcing strategies of partnerships
	17.18	By 2020, enhance capacity-building support to developing countries, including for least developed countries and Small Island Developing States, to increase significantly the availability of high-quality, timely and reliable data disaggregated by income, gender, age, race, ethnicity, migratory status, disability, geographic location, and other characteristics relevant in national contexts
	17.19	By 2030, build on existing initiatives to develop measurements of progress on sustainable development that complement gross domestic product, and support statistical capacity building in developing countries

that cooperation is envisaged to take in Sect. 17.2 and UN mechanisms in Sect. 17.5.

- The *Capacity Building* **target (17.9)** focuses on enhancing international support for capacity strengthening in developing countries through the various forms of cooperation. The UNDP views capacity strengthening as '*the process through which individuals, organisations and societies obtain, strengthen and maintain the capabilities to set and achieve their own development objectives over time*' (UNDP 2009). Strengthening geoscience capacity is an important part of this and is discussed more in Sect. 17.3.
- *Trade* **targets (17.10–17.12)** aim to address the institutions that support trade, exports from developing countries, and access to markets. These are not discussed further in this chapter, but increasing exports from developing countries links to issues like the sustainable use of natural resources where geoscientists do have a role to play (see **SDGs 8** and **12**).
- *Systemic Issues* **targets (17.13–17.19)** aim to tackle issues relating to 'policy and institutional coherence' (17.13–17.15), 'multi-stakeholder partnerships' (17.16–17.17), and 'data, monitoring and accountability' (17.18–17.19). Building effective multi-stakeholder partnerships and some of the related issues that a geoscientist might encounter, for example, around working with difference, integrating scientific knowledge and local

experience, working with researchers from other disciplines, and ethics are tackled in Sect. 17.4

Good partnerships have the '*ability to bring together diverse resources in ways that can achieve more: more impact, greater sustainability, increased value to all*' (Stibbe et al. 2018, p7). They also bring together unique individuals, each with their own experiences, skills, and perspectives on the characteristics of an effective partnership. One of the fundamental challenges when it comes to achieving **SDG 17** is building relationships between people with different backgrounds, from different cultures, with different values and worldviews, who may face different day-to-day challenges, are often separated by large distances and may not always agree.

The authors recognise in developing this chapter and reflecting on their own experiences of partnerships that it is not possible to separate what we write from our own particular worldview and perspectives, shaped by (among other things) where we live and work and the rules, norms and values that shape the society that we live in. This chapter should be read with this in mind, and we encourage readers to critically reflect on and evaluate what we have written. Our perspectives on partnership may differ to your own. That is not to say that either group is wrong but it might take time to understand each perspective and find a shared definition that has meaning for both.

Stibbe et al. (2018) argue that partnerships are not currently creating enough of an impact for sustainable development. They suggest that there are two reasons for this. First, there are not enough of them and the systems needed to develop them at the scale required to deliver the SDGs are not in place. Second, many of the partnerships that do exist are not fulfilling their potential because they might not be running efficiently or be fit for purpose or the context. This chapter aims to respond to that second point by considering what a geoscientist might need to think about when participating in a partnership for development (or indeed any partnership). To do this, we unpack some of the main concepts in **SDG 17** and explore what they mean for geoscientists and how we work, using examples to illustrate them along the way.

17.2 Partnership Types and Characteristics

The UN System Task Force on the post-2015 UN Development Agenda recognised the importance of other partnership types besides the 'global partnership', acknowledging that cooperative arrangements such as North–South, South–South,

and triangular partnerships (see Targets 17.6 and 17.9) all have a vital role to play. They each provide a way to contribute to global development and each bring particular benefits and challenges.

17.2.1 North–South Cooperation

North–South partnerships refer to cooperation between richer industrialised countries in the 'Global North' (i.e., North America, Western Europe, Australia and New Zealand, and parts of Asia) and poorer developing countries in the 'Global South' (i.e., developing Asia, Africa, the Middle East, and Latin America). Table 17.2 (taken from Dodson 2017) sets out the benefits and challenges that may arise from North–South partnerships. Dodson (2017) argues that there are benefits to partners from both the North and South, such as access to resources and expertise, the potential for learning and exchanging knowledge, capacity strengthening, and increasing profile and esteem.

Managing these partnerships can be complex, with potential challenges including power imbalances, cultural differences, or different ways of working. For example, some scientists in the Global South may find it difficult to challenge

Table 17.2 Benefits and challenges of North–South Partnerships (from Dodson 2017, integrating perspectives from Academy of Medical Sciences (2012) and Horton et al. (2009)). Reproduced under the Open Government Licence v.3. http://www.nationalarchives.gov.uk/doc/open-government-licence/version/3/

Benefits	Challenges
• Better access to scientific resources (laboratories, equipment, expertise) and talent, expertise and ideas, including access to increasingly complex (and often large-scale) instrumentation • Mutual learning and knowledge exchange between partners that may lead to broadened perspectives and new solutions to key challenges • Greater access to financial resources • Enhanced research impact • Capacity strengthening for individuals, institutions, and national research systems • Improved quality, cost efficiency, and productivity of research programmes • Improved institutional and individual profile and esteem • Long-term relationship and continuity that is not dependent on individuals	• More complex management and decision-making processes • Additional workload required to maintain the partnership over and above existing responsibilities • Higher financial costs and difficulty in overhead recovery • Power imbalance and research agenda dominated by the Northern institution • Side-lining of local and long-term research agendas • Diversion of staff and resources away from parts of the Southern institution not involved in the partnership • Logistical challenges (visas, international travel, and difficulty transporting samples between countries) • Tensions due to cultural differences • The wider political and social context

the wishes of more powerful partners (especially if they are providing finance), for fear of harming future access to resources (Nordling 2019). Maintaining partnerships requires significant effort (see **SDG 16**), which may bring higher financial costs and affect the timescale on which a partnership's goals can be achieved. For a North–South research partnership, the potential for these challenges to emerge should be anticipated at the start and factored into any resourcing plans or work timetables.

Example: The Global Challenges Research Fund (UK)

The UK's Global Challenges Research Fund (GCRF) is a £1.5 billion fund to support research directed at addressing socio-economic challenges that recipients of *Official Development Assistance*[2] face (GCRF 2017). The research funded by the GCRF is intended to be challenge-led, potentially interdisciplinary and strengthen capacity for research, innovation, and knowledge exchange in the UK and developing countries through partnership (GCRF 2017). It has funded many different research programmes and other activities since it was announced in 2015, including the *'Building Resilience to Environmental Hazards'* programme (Box 17.1).

- *Resilience or resistance? Negotiated mitigation of landslide risks in informal settlements in Medellin,* a project that piloted community-based monitoring and mitigation of landslide risk in Colombia (a collaboration between Heriot-Watt University, the University of Edinburgh and Universidad Nacional de Colombia).
- *Resilience in Groundwater Supply Systems (RIGSS): integrating resource-based approaches with agency, behaviour and choice in West Africa* (a collaboration between Cardiff University, the University of Ibadan and the University of Maiduguri in Nigeria, the Skat Foundation in Switzerland and the British Geological Survey).
- *Socioecological resilience to soil erosion driven by extreme climatic events: past, present, and future challenges in East Africa* (a collaboration between the University of Plymouth, UK, and the Nelson Mandela African Institution of Science and Technology, Arusha, Tanzania) (Fig. 17.1).

Box 17.1. Building Resilience to Environmental Hazards

This GCRF-funded programme supported short-duration research projects that took an interdisciplinary approach to understanding resilience to natural and human-made environmental hazards in a range of Global South contexts (Sargeant et al. 2018). Many geoscientists were involved and most of the projects involved North–South partnerships. Examples of these projects include

A workshop was held in London in 2018 when the Building Resilience to Environmental Hazards programme (see Box 17.2) ended. Here, a group of the researchers who had been involved, including geoscientists (and mostly from the UK), reflected on their experiences and generated a set of points that they thought should be considered good practice for future interdisciplinary challenge-led research projects. When it came to fostering equitable North–South partnerships in which everyone is treated fairly, these were some of the points that emerged from their discussion (see Sargeant et al. (2018) for full details):

---

[2]http://www.oecd.org/dac/financing-sustainable-development/development-finance-standards/daclist.htm.

Fig. 17.1 Disciplines involved in the Socioecological resilience to soil erosion driven by extreme climatic events: past, present, and future challenges in East Africa study. This figure illustrates the interconnections between disciplines and their position in the soil erosion-land degradation-community resilience challenge. From Blake et al. (2018), used under a CC BY 3.0 license (https://creativecommons.org/licenses/by/3.0/)

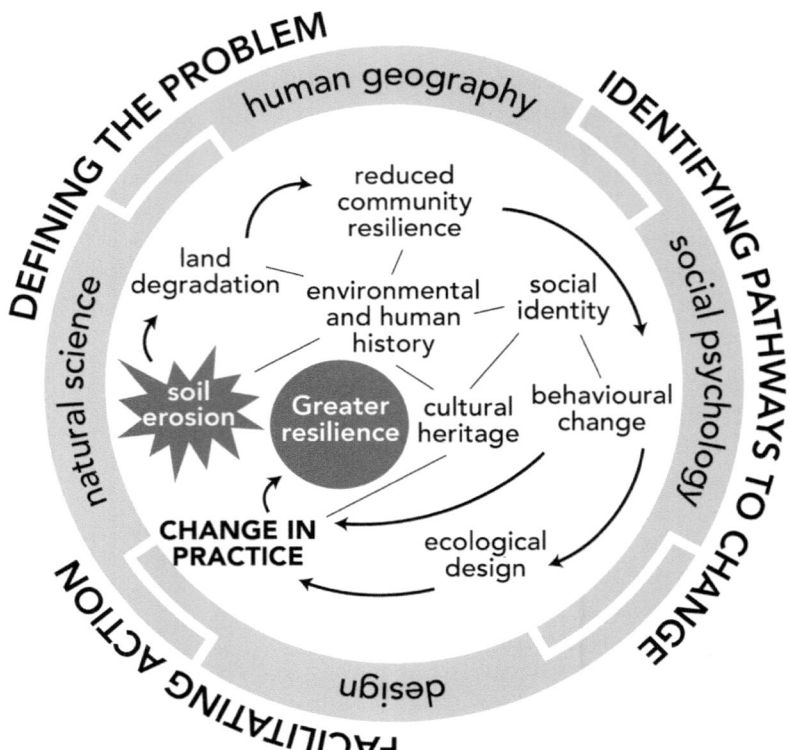

- Ensure that the role and responsibilities of each partner is clearly defined and manage the expectations of all those involved.
- Include activities throughout the project that build trust between partners.
- Allow for learning throughout projects and ensure that there is space for people's perceptions of each other to evolve, and for assumptions and misconceptions to be challenged.

These points represent the perspectives of those from the Global North (one side of the partnership), and, therefore, must be examined alongside perspectives from the Global South to achieve a holistic view of what makes North–South partnerships equitable. For example, Gill et al. (2017) discussed partnerships with Earth and environmental scientists in Kenya and characteristics of greatest importance to them were (i) sharing of project outputs, (ii) sharing of data, (iii) being treated as equals by other members of the partnership, and (iv) access to training and capacity strengthening.

17.2.2   South–South Cooperation

South–South cooperation refers to collaboration between two or more developing countries to '... *pursue their individual and/or shared national capacity development objectives through exchanges of knowledge, skills, resources and technical know-how and through regional and interregional collective actions, including partnerships involving Governments, regional organizations, civil society, academia and the private sector, for their individual and/or mutual benefit within and across regions. South-South cooperation is not a substitute for, but rather a complement to, North-South cooperation*' United Nations (2012).

South–South cooperation occurs in all sectors and encourages regional integration, sharing of knowledge, learning, expertise and other resources, and technology transfer (see United Nations 2012). In South–South cooperative relationships, the people and countries involved are assumed to have more experiences in common such as historical events, development pathways, and shared challenges (United Nations 2012) than might be the case in a North–South partnership (UNDP 2009). South–South partnerships are, therefore, considered to be potentially more 'horizontal' than North–South arrangements where the project might be driven from the North (United Nations 2012). It is reasonable to expect South–South cooperation might be characterised by greater equity and mutuality between partners, but (as in any partnership) differences and inequalities between the partners may exist and need to be managed.

Example: Sustainable access to water and sanitation in Africa

Sustainable management and provision of access to water and sanitation are regarded as an important pathway to social and economic development and poverty eradication in Africa (see **SDG 6**). As with many aspects of the physical environment, surface and groundwater resources do not follow international boundaries (see Fig. 17.2) and so cooperation between the countries in which they occur is vital for these resources to be managed sustainably.

Such efforts take place around the world, and there are various initiatives in this space that might be considered to be South–South collaborations. Examples include

- *Lake and River Basin Organisations (L/RBOs).* For example, the Orange-Senqu River Commission[3] brings together expertise from Botswana, Lesotho, Namibia, and South Africa to promote the equitable and sustainable development of the resources of the Orange-Senqu River (Fig. 17.3).
- *Cooperation agreements for transboundary aquifer management.* For example, the North-Western Sahara Aquifer System, which straddles Algeria, Libya, and Tunisia, and the Nubian Sandstone Aquifer System, which cuts across Chad, Egypt, Libya, and Sudan (Nijsten et al. 2018).

At the continental scale, the African Ministers' Council on Water (AMCOW)[4] was set up in 2002 to '*promote cooperation, security, social and economic development and poverty eradication among* [over 50] *member states through the effective management of the continent's water resources and provision of water supply services*'. AMCOW supports transboundary water management on the continent, along with many other initiatives that strive to support the Sharm El-Sheikh commitments to accelerate the achievement of water and sanitation goals in Africa (see African Union 2008). One of AMCOW's main initiatives is Africa Water Week, held every two years. This brings together those with an interest in the issue (including government, civil society, the private sector, international partners, and the scientific community) to discuss how to meet Africa's water and sanitation challenges. In '2018 Libreville Multistakeholders' Declaration on Achieving Water Security and Safely Managed Sanitation for Africa' help to direct action. Of interest to geoscientists here is the Declaration's call for greater knowledge sharing and focus on increasing understanding of surface and groundwater resources, addressing capacity issues and improving data monitoring networks. Another South–South partnership in this area includes the Africa Groundwater Network[5]. This is mainly a capacity strengthening organisation for the groundwater sector in Africa. An example of one of their initiatives (although this also involved partners from the North) is a training manual on management of transboundary groundwater resources (AGW-Net 2014).

17.2.3 Triangular Cooperation

Triangular cooperation is the third configuration referred to in **SDG 17**. These are '*southern-driven partnerships that are supported by one or more developed countries or multilateral organisations (e.g., the World Bank, the United*

---

[3]http://orasecom.org/.

[4]https://www.amcow-online.org/.

[5]http://www.agw-net.org.

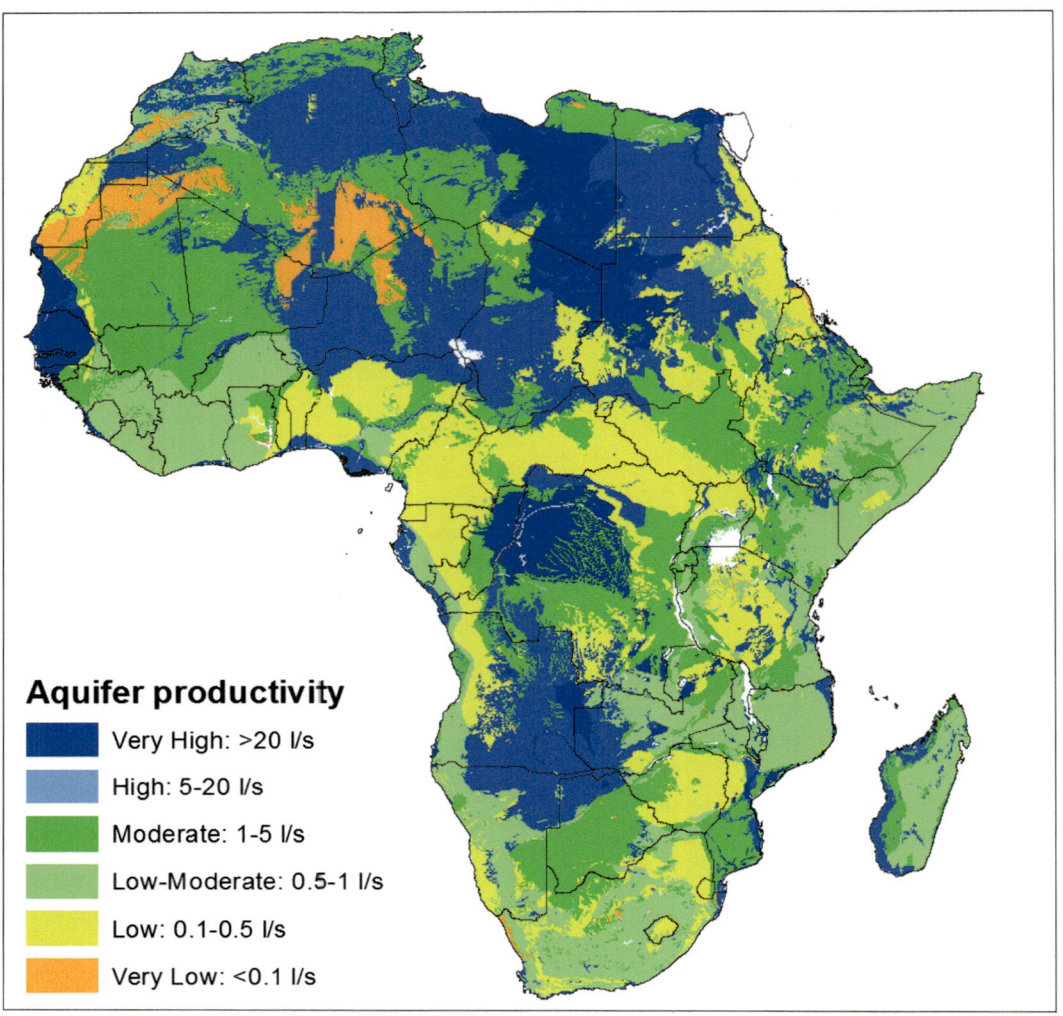

Fig. 17.2 Map of aquifer productivity for Africa. This map shows the likely interquartile range in sustainable yields for boreholes drilled and sited using appropriate hydrological techniques and expertise (MacDonald et al. 2012). Country boundaries from GADM, Version 3.6 (https://gadm.org/). © BGS/NERC 2012 (used with permission)

Nations), to implement development cooperation projects and programmes'.

Example: Strengthening African capacity in soil geochemistry

Soil geochemistry has a profound effect on agriculture (**SDG 2**) and human health (**SDG 3**) so understanding the processes involved is vital for supporting agricultural policies and identifying regions at risk of micronutrient deficiencies and toxicities. In this example, an initial partnership linking research institutes and universities in Africa and the UK in order to develop practical scientific outputs and strengthen capacity (see Sect. 17.3) has evolved into a number of networks to broaden African scientific influence and avenues for sustainable development beyond the initial 5-year project.

Fig. 17.3 Orange River (Southern Namibia). *Credit* Hp.Baumeler CC BY-SA 4.0 (https://creativecommons.org/licenses/by-sa/4.0/)

The starting point for this is a project, which runs from 2015 to 2020, that is funded by the Royal Society and the UK Department for International Development (DFID). It is a collaboration between researchers from Zimbabwe, Zambia, Malawi, and the UK (Fig. 17.4). By working together, they aim to increase knowledge and strengthen African capacity in soil geochemistry so that African research leaders are collaborators at national, regional, and international levels (through improved resourcing and increased scientific visibility). The project centres on three core, complementary PhD projects, one located in each of the three African partner institutions with secondments to the UK for technical training and laboratory analyses. To these, two aligned UK and one Malawian PhD projects were established to work in tandem and promote two-way learning between the PhD cohort.

Cross-country knowledge exchange activities and annual network/training events rotating through the African countries were crucial in the first three years of the project to build close partnerships, from which spin-out competitive grant-funded projects have been won. These projects are enabling interactions to be sustained beyond the life of the Royal Society-DFID project. Project meetings evolved into training camps from year 3 to 5 for permanent laboratory staff at the partner research institutes and universities to ensure sustainable capacity strengthening beyond the project and to reinforce the capital investment made in laboratory equipment. Additional trainees were drawn in from the BGS '*Geoscience for Sustainable Futures*' programme activities in Kenya to broaden the laboratory network. Each of the institutes has had one of their laboratory staff visit BGS for training fellowships, funded by the Commonwealth Scholarship Council, reinforcing the training received

Fig. 17.4 Delegates at a Royal Society-Department For International Development (RS-DFID) soil geochemistry for agriculture and health network/training event in Harare © UKRI

in-country. The members of the network have developed their own form of communication through WhatsApp groups to develop laboratory documentation (standard operating procedures, health & safety) and ask for technical help with instruments. This is resulting in greater self-support for Africa-Africa interactions, hopefully reducing reliance on help from the UK.

This laboratory network has developed further through involvement in an UN-FAO Global Soil Laboratory Network (GLOSOLAN[6]), which is committed to harmonising soil data production around the world through free-to-access standard operating procedures. The laboratory network, developed initially from the Royal Society-DFID project, has now joined a cohort of more than 230 laboratories around the world and specifically the regional network-AFRILAB[7]. This so far comprises over 20 African countries and more than 30 institutes. Alongside harmonisation of methods, training events within each region and

directed capacity strengthening will be supported through to 2027, providing support for sustainable development of technical capacity in partner countries.

Partner scientists and laboratory personnel have also been involved in international conferences, notably the 2018 International Conference for the Society of Environmental Geochemistry and Health (SEGH[8]) in Livingstone, Zambia. African partners were integral in forming an African section for SEGH, from which they have extended their scientific network internationally, across disciplines and across Africa. Since the conference, a number of competitive grant proposals have been developed as a result of the connections that were made between people at the meeting. Early Career Researchers (ECRs) from Africa are now also connected through a mentoring programme with international ECRs. Watts et al. (2019) gives an account of current SEGH initiatives and future aspirations.

---

[6]http://www.fao.org/global-soil-partnership/pillars-action/5-harmonization/glosolan/en/.

[7]https://www.afrilabs.com/.

[8]www.segh.net.

Table 17.3 Issues affecting capacity strengthening, and the relationship to geoscience activities. Modified from UNDP (2009)

Core issue	What does this mean?	Geoscience examples
Institutional Arrangements	The rules, norms, and values that govern and shape how a group operates	This might be the ethical policies that set out how research should be conducted (the rules), the strategy that defines how a research department operates, or scientific values such as scientific integrity, honesty, and curiosity
Leadership	'The ability to influence, inspire and motivate others to achieve or go beyond their goals'. This can either be formal or informal	This could be the Principal Investigator on a research project, the leader of a work package or even a thought leader, who is recognised as an authority in their field
Knowledge	At an individual level, this includes education, on-the-job training and life experience. However, it could also be the knowledge held within an organisation or by a group of people	The knowledge and experience gained through academic training, undertaking fieldwork, contained within geoscience journals or within organisations such as geological surveys or mining bureaus
Accountability	An individual or organisation being held responsible for their actions	Project progress reports for science funders, research conduct

17.3 Capacity Building

Capacity building (see Targets 17.8, 17.9, and 17.18), often referred to as 'capacity strengthening' in recognition of the fact that significant 'capacity' will in fact already exist, is critical to human development and *the process through which individuals, organisations and societies obtain, strengthen and maintain the capabilities to set and achieve their own development objectives over time*' (UNDP 2009, p5). In some senses, effective capacity strengthening echoes the goals of partnerships for development in that they both aim to bring about some kind of change, tackle problems and produce outcomes that are locally led and sustainable (UNDP 2009). Capacity strengthening can take on a variety of forms. It might be at the level of state governance, or the way an organisation operates and helping to create an environment that enables people to act or for something to happen. Or it might be at a smaller scale—the level of an individual, in which case, capacity strengthening might refer to efforts to increase a person's skills or knowledge in order for them to operate effectively in some way, through training, for example (UNDP 2009).

As with other aspects of **SDG 17,** a lot of complexity lies behind the words and significant effort will be required to ensure positive outcomes. UNDP (2009) identifies four 'core issues' that have the greatest bearing on capacity strengthening at any of these levels. These are summarised in Table 17.3.

The UNDP approach to capacity development identifies the following elements that should shape any capacity strengthening project: (i) clear understanding of the purpose of the project and the development objective/s that the project is responding to; (ii) who the focus of the project is; and (iii) what capacities need to be developed to achieve the development objective, and what the current capacities are (UNDP 2009). With this understanding, a capacity strengthening project that has the potential to be effective and impactful can then be designed and implemented. Evaluating and monitoring the project and its outcomes is also an important part of the process. While funders will often focus on the deliverables from the project, there is an argument for greater emphasis on understanding how the partnerships that underpin the project's activities are working.

Example: Institutional Strengthening of the Afghanistan Geological Survey (2004–2008)

Like other government institutions in Afghanistan, the geological survey was severely weakened by the impact of decades of conflict. Following the fall of the Taliban regime in 2001, efforts soon began to restore the Afghanistan Geological Survey (AGS) to a working survey. The main purpose of the project was to build up knowledge of the mineral resources in the country so that these could be promoted and exploited. One project, commissioned by the UK's DFID, was undertaken by the British Geological Survey (BGS) between 2004 and 2008. The objective of the project was to develop new or strengthen existing facilities, processes, and practices at the AGS (institutional strengthening). Significant effort at the start of the project went into understanding the wider context and the needs that the project had to address. BGS worked closely with the Afghanistan Minister of Mines and his staff to develop an initial project proposal, which was then further developed to take account of views of stakeholders. Objectives were refined as the team developed their on-the-ground understanding of what was necessary. In the first six months, activities included refurbishment of part of the AGS building, and sourcing project vehicles and field equipment. A 'business training needs assessment' for professional staff at AGS was also completed, which could be used as the basis for a training plan. Eventually, the project consisted of a comprehensive programme of capacity strengthening, geological mapping, evaluation of mineral resources, reinstating of working laboratories (Fig. 17.5) and archives, and development of databases and geographic information systems. 54 training courses were given. BGS staff were based at AGS throughout the project (as the security situation allowed), which meant that AGS staff could benefit from being able to have ad hoc interactions with their BGS counterparts, and vice versa. Having a continuous presence in the country (rather than people flying in and out) was considered to have worked well. Bringing in a broad range of geoscience expertise, providing English language training and training for women (aided by the establishment of a crèche) were also highlights. The final stage of the project was to prepare tenders for international companies to bid into to develop key deposits such as the Aynak copper deposit in eastern Afghanistan. Long-term sustainability of any capacity strengthening project is a challenge. In this example, there was a need for more graduate geologists to sustain what the project had achieved. However, it is also important to note that this project is one among many geoscience capacity strengthening efforts that have been undertaken in Afghanistan since 2001. Further aspects of this collaboration are discussed in the chapter exploring **SDG 7** (energy).

> ## Box 17.2. Reflections on achieving effective capacity strengthening
>
> The geologist who led the project in Afghanistan has these suggestions based on his experiences:
>
> - Network with anyone you can find because having a wide range of contacts is vital. Connect with other groups (e.g., NGOs) who are working in the same place—they may well be working with similar people (e.g., government ministers).
> - Carry out an in-depth risk assessment and work with security professionals.
> - Choose your project team carefully—they should be willing volunteers who are good team players.
> - Have clear lines of communication.
> - Have a very clear project plan and try to stick to it (making small changes if necessary).
> - Keep revisiting your objectives to make sure that the project does not 'drift'.
> - Give thought to what happens after the project ends (e.g., would a Memorandum of Understanding that will outlive the project be useful?).

Fig. 17.5 Training activities carried out by the Afghanistan project. Processing samples in the reinstated laboratories. © UKRI

17.4 Systemic Issues

The last group of targets cover the 'systemic issues' that contribute towards making the 'system' (taken here to mean the environment in which development happens) work more effectively (see Targets 17.14 and 17.18). Policy coherence is considered to be key to achieving the SDGs (OECD 2019). However,

> *"This is no easy feat: it requires meaningful collaboration and coordinated action across policy sectors (horizontal coherence), as well as between different levels of government (vertical coherence). It also requires balancing short-term priorities with long-term sustainability objectives and taking into account the impact of domestic policies on global well-being outcomes."*
> (OECD 2019)

For a geoscientist involved in designing, implementing, or contributing to development-related activities, this requires us to have an awareness of the wider context in which the project is being implemented so that it can be tailored to the specific situation. It is important to recognise that there will be probably be a multitude of short- and long-term processes or projects that will relate in some way to what you are doing, some of which might have very similar objectives to you and/or involve the same partners. Having an awareness of these activities allows you to design your approach in a way that complements rather than duplicates or undermines other activities. There are various methods that can be used to analyse a particular context depending on what is required. PEST (LE) analysis is one relatively simple way to investigate the **P**olitical, **E**conomic, **S**ocial, **T**echnological, **L**egal, and **E**nvironmental aspects of a context. More complicated frameworks exist which focus on the knowledge-policy-practice space, and which are useful if your aim is to improve the impact of knowledge on policy (see Jones et al. 2012).

The following example illustrates the potential impact of limited coordination between projects trying to do a very similar thing with the same partner (as part of bigger research projects). In this case, there had been efforts by several different projects driven from the Global North to strengthen the seismic monitoring capacity of a national agency in a developing country. Each project brought different types of equipment (seismometers and acquisition systems, etc.), which led to a situation where the agency was left with an incredibly complex seismic monitoring network comprising different types of equipment that did not easily 'fit together'. This made it very difficult for the agency staff to operate and maintain the network and meant that each capacity strengthening project's effort did not perhaps achieve the full impact that had been hoped for. It also put a lot of pressure on the national agency staff by significantly increasing their workload. One could argue that greater coordination between the capacity strengthening projects might have helped to avoid these issues and delivered a greater, more sustainable benefit.

Other aspects of the systemic issues are discussed below.

17.4.1 Multi-stakeholder Partnerships

Stibbe and Prescott (2016) define *Multi-Stake-holder Partnerships* (MSPs) as 'voluntary undertakings operating under their own principles'. Great emphasis is put on the value of MSPs for tackling development challenges because these tend to be complex, involve multifaceted issues and each stakeholder will hold a different piece of the puzzle. Stibbe and Prescott (2016) go on to say that MSPs are '*highly context-specific, building on the interests, capacities, resources and leadership of the partners involved*' (Stibbe and Prescott 2016, p2). This is highlighted in Table 17.4, which shows what a selection of stakeholders, including geoscientists, might bring to a development challenge.

In theory, by working together to tackle an issue of shared interest, an MSP is likely to achieve more than each stakeholder could achieve by working alone. Recommendations on how to make such partnerships effective are given in Box 17.3.

> **Box 17.3: How to make multi-stakeholder partnerships effective (from Stibbe and Prescott 2016)**
>
> 1. Question and move past assumptions and preconceptions about each other
> 2. Recognise and value diversity as an asset rather than a problem
> 3. Value the different contributions that a partner brings
> 4. Develop skills in partnership-building, brokering collaborations, and leadership
> 5. Understand the systems and contexts in which the partnerships operate
> 6. Apply the highest standards, rigour, and accountability to all partnering endeavours
> 7. Invest in the partnering process to create the conditions for effective relationships to be developed.

There is growing recognition that it is necessary for researchers from different disciplines to come together to investigate development-related issues because of their multifaceted nature. As with MSPs, it is recognised that bringing a range of disciplinary knowledge and expertise together allows for a potentially more complete characterisation of a complex problem. More than ever, geoscientists are finding themselves working on collaborative projects with colleagues from the arts and humanities, economics and the social sciences, and engineering, for example. The forms that research collaborations might take (if not single disciplinary) are summarised below. There are many ways to define these types of research, and here, we use the definitions of Toomey et al. (2015) in a briefing note that is available on the SDG Knowledge Platform.

- *Multidisciplinary*: This type of research '*draws on knowledge from different disciplines but stays within their boundaries*' (Choi and Pak 2006).

Table 17.4 What different sectors can bring to address development-related challenges (after Stibbe and Prescott 2016) with some of what geoscience and geoscientists can provide added

NGOs and Civil Society	Business	Government/Parliamentarians
• Technical knowledge/capacity • Access to and deep knowledge of communities • Legitimacy/social capital • Passion and people-focus	• A market-based/commercial/value creation approach • A market-based/commercial/value creation approach • Power of the brand and access to customer base • Technical and process innovation	• Regulatory framework (e.g., licenses for water, etc.) • Integration with public systems/long-term planning • Taxation policy • Capacity strengthening (e.g., agricultural extension services) • Provision of land and supporting infrastructure • Democratic legitimacy
International Agencies/UK	Donors and Foundations	Geoscientists
• Technical support, knowledge, and experience • Legitimacy and impartiality • Access to a global network • Political access	• Funding and support • In many cases foundations can be less risk averse and support more experimental and innovative approaches, providing proof of concept that can be expanded by more traditional donors	• Scientific method • Knowledge of surface and subsurface systems • Understanding of the nature and potential impact of geohazards and other processes • Expertise in using spatial information (e.g., maps, satellite imagery)

- *Interdisciplinary*: This type of research '*analyses, synthesises and harmonises links between disciplines into a coordinated and coherent whole*' (Choi and Pak 2006). Toomey et al. (2015) point out that interdisciplinarity takes an 'integrated approach to answering a question' (Toomey et al. 2015, p1) that allows a wider view than a more 'compartmentalised system' of research permits (Toomey et al. 2015, p1). The blending of the various disciplines from the very start of the research is crucial to this.
- *Transdisciplinary*: This kind of research project goes '*beyond the bridging of divides within academia to engaging directly with the production and use of knowledge outside of the academy*'. This could then mean that the various stakeholders in Table 17.4 might be researchers in a transdisciplinary project (which is, therefore, also an MSP).

For any partnership that crosses disciplinary divides, the recommendations in Box 17.3 apply.

Example: Supporting self-recovery and promoting safer building after disasters

After a disaster, the majority of families will rebuild their homes with little or no support from the international humanitarian community to encourage safer building practices (Parrack et al. 2014). In the humanitarian shelter sector, this is called 'self-recovery' but the process of self-recovery and how it can be supported is relatively poorly understood. In Twigg et al. (2017), a group of researchers from different disciplines (social scientists, geoscientists, and engineers) and organisations (universities, a development think tank and a geological survey) worked with humanitarian shelter practitioners from an international NGO to investigate self-recovery, focussing on what happened after two typhoons in the Philippines and the 2015 Gorkha earthquake in Nepal. The group developed their research questions together at the start of the project so that the questions reflected what each discipline and sector wanted to find out or viewed as important. Carrying out fieldwork together (e.g., interviews and focus groups with community members, geomorphological assessment of the landscape, engineering surveys) allowed the people in the team to get to know each other and learn more about the other perspectives, knowledge, and expertise within the team.

From a geoscience point-of-view, Twigg et al. (2017) found that despite the difference in the contexts in which recovery takes place, there are some common barriers to self-recovery in a substantially changed and dynamic multi-hazard environment (e.g., disruption to water supply, impacts of relatively small-scale geohazard events and the availability of technical advice). This would not have been possible without the range of perspectives embodied in the team. Further research is necessary to understand how the humanitarian sector and the geoscience community could assist in tackling these barriers. However, through the relationships between geoscientists, social scientists and humanitarian practitioners developed in the project and a greater understanding of each group's ways of working and priorities, it might be possible to find a way forward together in order to find ways to better support people who self-recover after disasters.

17.4.2 Working with Communities

Table 17.4 highlights the need to bring together NGOs and civil society, business, government, international agencies, donors and foundations, and geoscientists to address development-related challenges. One important stakeholder not included in Table 17.4 are members of the public who are directly impacted by the development project. Their engagement can help to identify local priorities, and bring critical local knowledge into the design and implementation of

development programmes such as understanding of culture and environmental dynamics, past histories of development engagement (including successes and failures), and perspectives of appropriate technologies and interventions.

Community engagement is fundamental to effective and sustainable development activities, with user participation essential if a development project is to be deemed acceptable by a community (Gill 2016). For example, the failure of many projects aiming to provide access to clean drinking water has long been attributed to (amongst other things) a lack of community engagement. Elmendorf and Isely (1981) note that projects that do not capture the interest of communities, provide the community with maintenance training, or establish the necessary community water management groups are likely to result in failure. Failure to listen to a community's perspectives on, for example, the siting of a water project, can result in a project being poorly used by a community and, therefore, failing. In contrast, a key feature of successful water projects is user (i.e., the community) participation at every stage of project implementation (Narayan 1995; WaterAid 2011).

Communities can also play an important role in research partnerships, including those with an applied focus aiming to inform sustainable development interventions. Silliman et al. (2009) set out an excellent example of this scientist-community partnership, with local community members in the village of Adourékoman (Bénin) being trained to collect and test groundwater samples on a weekly basis, over a period of more than three years. It was not feasible for academic or government researchers in Bénin to complete weekly water testing from wells within Adourékoman, but community members could do so. It took time to build mutual trust and respect, but this resulted in an excellent water quality data set proving to be reliable. Silliman et al. (2009) demonstrate the efficacy of relying on a local community to collect reliable environmental data.

Engagement with communities—whether as research partners or research subjects—requires careful consideration of research ethics and safeguarding, and adherence to relevant standards. Research ethics governs the way any research involving interaction between a researcher and other humans (or their data) is designed, managed, and conducted. The prime purpose of considering research ethics is to minimise risk to participants, researchers, and third parties and to ensure research respects the dignity, rights, and welfare of all those involved. In some contexts (e.g., academic departments), a research ethics application may be required, with this requiring the review and approval of a research ethics committee before work can begin. Research ethics is not typically integrated into the training of geoscientists (compared to geographers, or zoologists, for example), but should be emphasised given the many interactions between geoscience and people (as illustrated by this book). Resources to support geoscientists to understand research ethics and embed these principles into their work will ensure better partnerships between researchers and communities.

17.4.3 Bringing Indigenous and Scientific Knowledge Together

The UN estimate that globally there are approximately 370 million Indigenous Peoples (5% of the world's population) belonging to 5000 groups in 90 countries (United Nations 2019a). A key strength of multi-stakeholder partnerships is that they create the opportunity to bring together diverse types of knowledge from across disciplines and sectors. The Indigenous Peoples Major Group advocates for recognition of the value of indigenous and traditional knowledge, with this treated equally to science and other knowledge systems (IPMG n/d). At the UN Forum on Science, Technology and Innovation for the SDGs (see Sect. 17.5), a session typically focuses on the importance of indigenous knowledge in helping to achieve the SDGs.

Indigenous and traditional knowledge is essential to fully understand environmental dynamics and change, extending beyond available instrumental data both spatially and temporally. Environmental history, passed between generations through storytelling, for example, can be an important source of information that enriches our understanding of environmental processes and impacts. Indigenous peoples are, therefore, 'custodians of knowledge systems that can, alongside formal science, offer solutions to intractable development challenges' (UNDESA 2019). The value of indigenous and traditional knowledge, and approaches to integrate this with other knowledge systems, however, are not typically included in the curricula of geoscientists. This may hinder the extent to which geoscientists accept the validity of indigenous knowledge, proactively engage with this as a source of evidence, and integrate it into their work.

The importance of indigenous knowledge to understanding environmental hazards has been well documented in many regions, with examples including the Solomon Islands (Cronin et al. 2004), Bangladesh (Howell 2003), and Pakistan (Dekens 2007). Mercer et al. (2010) set out a framework to integrate indigenous and scientific knowledge for disaster risk reduction. This framework sets out an approach that enables communities to establish potential solutions to their vulnerability to environmental hazards. It consists of four steps:

1. Community engagement to determine if and how a community consider their vulnerability to environmental hazards, and their desire to identify an integrated strategy to reduce this.
2. Identification of extrinsic and intrinsic vulnerability factors by the community.
3. Identification of indigenous and scientific strategies, used both in the past and in the present to cope with intrinsic factors affecting vulnerability. Past strategies may emerge as relevant and beneficial.
4. Participatory development of an integrated strategy, by analysing the data from Steps 2 and 3 to negotiate and develop an integrated strategy to reduce vulnerability to hazards.

We refer the reader to Mercer et al. (2010) for detail on each step, including potential problems and limitations. What this framework demonstrates, however, is that approaches exist that enable geoscientists to bring together their understanding with indigenous understanding and unite this to address a challenge, such as exposure and vulnerability to geological hazards. Any framework to bring indigenous and scientific knowledge together—whether for disaster risk reduction, climate change adaptation, or natural resource management—requires trust, communication, and genuine acceptance of the importance of both knowledge systems to environmental strategies (Mercer et al. 2010). The principles of equitable and respectful partnerships set out throughout this chapter, therefore, apply whether those partners are professional scientists in different countries, or scientists and indigenous peoples from different places.

17.5   UN Mechanisms to Build Science Partnerships for Development

The Global Technology Facilitation Mechanism[9] (TFM, see Target 17.6) has been set up to support the implementation of the SDGs and to '*facilitate multi-stakeholder collaboration and partnerships through the sharing of information, experiences, best practices and policy advice among Member States, civil society, the private sector, the scientific community, United Nations entities and other stakeholders*' (United Nations 2019b). The TFM has three components, each of which provides opportunities for geoscientists to support the implementation of the SDGs:

1. A **UN Interagency Task Team (ITT) on Science, Technology and Innovation for the SDGs**, with geoscience represented through the involvement of UNESCO. The ITT is supported by a 10-member group of representatives from civil society, the private sector, and the scientific community.

2. An annual **Multi-Stakeholder Forum on Science, Technology and Innovation (STI) for the SDGs**. This is expressed in an annual gathering at the UN headquarters in New York, discussing cooperation around thematic areas for the implementation of the SDGs. The STI Forum aims to facilitate interactions, networks, and partnerships to identify and examine needs and gaps in technologies, scientific cooperation, innovation, and capacity strengthening to support the SDGs. Member states (official national representatives), civil society, the private sector, the scientific community, and United Nations entities (e.g., UNESCO, UN Water) attend this forum, with geoscientists welcome. To date, each STI Forum has focused on a subsection of the SDGs, grouped around a theme. In 2018, the Forum discussed the science required for '*transformation towards sustainable and resilient societies*', including SDGs 6 (water and sanitation), 7 (energy), 11 (sustainable cities), 12 (responsible consumption and production), and 15 (life on land). The 2019 Forum theme was '*science, technology and innovation for ensuring inclusiveness and equality*', exploring SDGs 4 (quality education), 8 (decent work and economic growth), 10 (reduced inequalities), 13 (tackling climate change), and 16 (peace, justice, and strong institutions).

3. An **online platform** providing information on existing science, technology and innovation initiatives, mechanisms and programmes. This platform aims to (a) act as a gateway for information on existing STI initiatives within and beyond the UN, (b) facilitate access to relevant information and learning, and (c) facilitate dissemination of relevant open access scientific publications generated worldwide.

Engagement in the TFM by geoscientists can help to improve access to, and use of, geological science in decision-making. For example, increasing recognition of the mineral requirements needed for the scaling up of green technologies (**SDG 7**), or emphasising the importance of characterising the subsurface if we

---

[9]https://sustainabledevelopment.un.org/TFM.

are to have sustainable and resilient cities (**SDG 11**). Gill et al. (2019) suggested that many development strategies relating to contexts in eastern Africa would benefit from greater engagement with geoscientists. Coherent, environmental policies underpin good environmental management, critical to the implementation of many of the SDGs. Greater engagement with and by the Earth science community in the TFM could help to address this. Active engagement in spaces beyond the traditional scientific conference, such as the STI Forum, can help to build bridges between geoscience and policy communities.

How can geoscientists engage in the Technology Facilitation Mechanism?

There are many opportunities for geoscientists to engage in the TFM through UNESCO's involvement. Geoscientists from around the world are involved in activities coordinated by UNESCO, thus creating a knowledge base that can inform contributions to the UN Interagency Task Team. For example, the International Geoscience Programme (IGCP) promotes collaborative geoscience projects with a special emphasis on creating benefit to society, building capacity, and sharing knowledge between scientists through effective North–South and South–South cooperation. Each year, UNESCO encourages new applications to the IGCP providing participants with the opportunity to work with and share learning (through reporting) with UNESCO and the wider UN community.

Being part of active coalitions can help to facilitate engagement in the STI Forum and other UN processes. This amplifies voices and increases the likelihood of key messages being conveyed and captured. Existing coalitions that geoscientists may already be members of include national geological societies and international geoscience unions. UN Stakeholder or Major Groups are another form of coalition, helping to encourage active participation and coordinate the contributions of different stakeholders in UN activities, including the STI Forum. Stakeholder or Major Groups include ageing, business and industry, children and youth, education and academia, farmers, indigenous peoples, local authorities, NGOs, persons with disabilities, scientific and technological community, volunteers, women, and workers and trade unions. Two particular groups of interest to the readers of this chapter are

- The **UN Major Group for Children and Youth (UN MGCY)**[10] is the UN General Assembly-mandated, official, formal, and self-organised space for children and youth (considered to be those under 30, and, therefore, including many early-career scientists). UN MGCY acts as a bridge between young people and the UN system in order to ensure their right to meaningful participation is realised. They have a cross-cutting science-policy interface platform and diverse working groups, including on disaster risk reduction, habitats, and sustainable consumption and production. For many Earth science students and early-career professionals, this is an excellent space to learn more about the UN and to contribute geoscience understanding to statements made on behalf of UN MGCY.

- The **Scientific and Technological Major Group**[11] is coordinated by the International Science Council and the World Federation of Engineering Organisations, thus bringing together a diverse community of natural scientists, social scientists, and engineers. This group integrates scientific and technological information to indicate what is scientifically and technologically feasible with respect to solutions for sustainable development.

Individuals and organisations can get involved in the work of Stakeholder and Major Groups, including writing and commenting on reports, standing for leadership positions, and being part of delegations at UN events. The specific opportunities and governance processes differ from one group to another.

---

[10]https://www.unmgcy.org/.

[11]https://sustainabledevelopment.un.org/majorgroups/scitechcommunity.

An additional way that geoscientists can contribute to the TFM is through the proactive dissemination of relevant scientific publications and reports. For example, PreventionWeb[12] is a knowledge platform for work relating to disaster risk reduction, including geological hazards. It allows individuals to share a range of content, including publications, policies and statements, educational materials, and maps (PreventionWeb 2019). Geoscientists can also contribute by sharing case studies that demonstrate the positive impacts of geoscience on society. Throughout the SDGs process, national governments are preparing Voluntary National Reviews on sustainable development objectives to feed into the UN High-Level Political Forum on Sustainable Development. In preparation for completing their 2019 review, the United Kingdom opened a call-for-evidence in 2018, requesting case studies of how groups were contributing to the SDGs and the impact of these activities. This call-for-evidence was open to all, including geoscientists, to contribute, and the final published review highlighted activities of the British Geological Survey (UK Government 2019).

Example: Early career geoscientists championing geoscience in UN-level discussions
Earth scientists can contribute to the STI Forum through formal interventions made during plenary sessions, the organisation of or contribution to side events, or the informal dissemination of resources and information. Each opportunity provides geoscientists with an opportunity to integrate their understanding of Earth dynamics, natural resources, and environmental change into discussion of the SDGs.

To date, the not-for-profit organisation *Geology for Global Development* (GfGD[13]) has played a leading role in representing the global geoscience community at the STI Forum. In 2019, they led an international delegation of early-career geoscientists to the Forum, funded by the *International Union of Geological Sciences* and *International Geoscience Programme* Project 685 (Fig. 17.6). This delegation aimed to increase the visibility of the Earth science community in sustainable development discussions, championing the importance of understanding the natural environment, enhancing public understanding of Earth systems and resources, and building strong professional communities of Earth and environmental scientists.

During the Forum GfGD emphasised key themes, including.

Environmental Education for Sustainable Development. In a formal intervention, GfGD emphasised the need for an understanding of the natural environment to be at the heart of a reshaped education to support sustainable development, noting that increased public understanding of the dynamics of environmental systems can help to encourage actions to secure a resilient and sustainable future for all.

Environmental Implications of Technologies. Through formal and informal activities, GfGD highlighted the need to consider the natural resource (e.g., minerals, water) requirements to scale up green technologies, and both social and environmental challenges associated with this. This includes the building of strong scientific communities, environmental institutions, and Earth science networks, particularly in the Global South.

GfGD's efforts led to their points being included in some of the outputs from the forum (e.g., UN ECOSOC 2019) that will guide decision-making globally. This example illustrates how important it is for geoscientists to participate in these high-level discussions.

17.6 Concluding Thoughts

SDG 17 brings to the fore the relationships between people that will be key to delivering the SDGs. In some ways, partnerships are the 'glue' of the SDGs and as we have seen in the chapter, partnerships can take many forms. North–South partnerships bring people and organisations from the Global North and the Global South together.

---

[12]https://www.preventionweb.net/.

[13]www.gfgd.org/.

Fig. 17.6 GfGD-led delegation at the 2019 UN STI Forum. © Geology for Global Development 2019 (used with permission)

There are benefits to both sets of partners but these types of partnerships can bring issues that must be acknowledged and managed carefully such as potential power imbalances and cultural differences. South–South partnerships offer potential opportunities for regional integration, knowledge sharing, and technology transfer with the expectation that these cooperative relationships may be characterised by greater equity and mutuality than North–South arrangements. Triangular partnerships are somewhat of a hybrid, potentially bringing the benefits of North–South and South–South collaborations into the picture.

SDG 17 also recognises the vital importance of capacity strengthening so that development stakeholders, be they individuals, organisations, or societies, can build their capability to achieve their development objectives. Focused projects responding to clear needs have the potential to be effective and leave a lasting legacy. **SDG 17** also sets out some of the systemic issues that must be addressed to create an environment that enables development. For example, through enhanced policy coherence and coordination, multi-stakeholder partnerships which bring different groups together and are potentially able to become more than the sum of their parts. UN initiatives such as the Technology Facilitation Mechanism provide other opportunities for geoscientists to engage and influence international development processes

As we have journeyed through what **SDG 17** might mean for geoscientists and how we work, there are recurring themes that emerge. These include the importance of trying to understand the context in which you are working, networking, recognising and valuing diversity, building trust, and creating spaces for reflection and learning. Being mindful of the ethical dimension of the work you are doing is also important but is something that traditionally geoscientists have not engaged with a lot as a whole. In fact, reflecting on partnership is not something that is traditionally done by geoscientists either. What we have presented here is just our perspective on what **SDG 17** might mean for geoscientists and how we work and we hope that we have provided you with a starting point for your own exploration of the topic. To help with this, we have provided suggestions for further reading and activities to think further about partnership.

Building relationships and working in partnership can be daunting and sometimes difficult. It can often require you to move out of your comfort zone and challenge you in ways that feel uncomfortable or scary. Therefore, it is extremely helpful to have people around you who can act as sounding boards, who can share their experiences with you and help you navigate your way through this complex space. All that said, working in the kinds of partnerships that we have covered in this chapter can be enormously rewarding and a lot of fun with the opportunity to develop long-lasting collaborations (and friendships) with people from around the world.

17.7 Key Learning Concepts

- **SDG 17** aims to strengthen the means of implementation to deliver SDGs 1 to 16, supporting a broad range of partnership types and revitalising the global partnership for sustainable development. Real, meaningful partnership and cooperation between diverse groups (e.g., countries, sectors, and disciplines) is critical to achieving all of the SDGs.
- Building partnerships can take time and resources, as it involves bringing people together from different backgrounds, from different cultures, with different values and worldviews, who may face different day-to-day challenges, are often separated by large distances and may not always agree.

- Partnership types include North–South, South–South, and triangular partnerships, each contributing to global development. Potential challenges when bringing partners together from across countries include power imbalances, cultural differences, or different ways of working. Maintaining partnerships requires significant effort but there are enormous benefits to all from constructive and positive partnerships, including access to resources and expertise, the potential for learning and exchanging knowledge, capacity strengthening, and increasing profile and esteem.
- Capacity strengthening involves individuals and organisations developing the skills that they need to set and achieve their own development objectives. Capacity strengthening is often done through North–South or South–South partnerships, and requires clear understanding of roles and responsibilities, effective leadership, access to knowledge and accountability.
- Tackling development challenges requires partnerships that bring together disciplines and different groups in society (e.g., the public and indigenous voices). Community engagement is fundamental to effective and sustainable development activities, with user participation essential if a development project is to be deemed acceptable by a community. Bringing a range of disciplinary knowledge and expertise together allows for a potentially more complete characterisation of a complex problem and the development of appropriate solutions.
- There are multiple opportunities for geoscientists at all stages of their career to engage in United Nations mechanisms to build and support science partnerships, and to improve knowledge exchange to inform implementation of the SDGs. Examples include the Multi-Stakeholder Forum on Science, Technology and Innovation (STI) for the SDGs, engagement with UN Major Groups (e.g., the scientific and technological major group), and contribution to Voluntary National Reviews submitted to the UN.

17.8 Educational Ideas

In this section, we provide examples of educational activities that connect geoscience, the material discussed in this chapter, and scenarios that may arise when applying geoscience (e.g., in policy, government, private sector international organisations, NGOs). Consider using these as the basis for presentations, group discussions, essays, or to encourage further reading.

- Conduct mini-interviews with five people (friends, family members, or colleagues) about partnership and ask them these questions:
    - What does 'partnership' mean to you?
    - What do you think is necessary to make it successful?
    - What are the advantages of working in partnership?
    - What do you think the difficulties of working in this way might be?

  Discuss the main themes that arise in the interviewees' responses. What do they tell us about what is necessary for SDG 17 to be achieved and what are the implications of this for geoscientists?
- Investigate a partnership or collaboration of your choosing (it could be a research collaboration, a sports team, band, etc.). How did it begin and what made it work (or not)? Consider how learning from this example could inform your own science-for-development partnerships.
- In small groups, research and discuss the *indicators* for **one** of the targets (https://sustainabledevelopment.un.org/sdg17). Would you change them? Why? What would you use to measure success instead?

Acknowledgements We are very grateful for the support of Bob Macintosh and Brighid Ó Dochartaigh whose experience and insight were vital in the development of this chapter.

Further Resources

Hartung FE (1951) Science as an institution, Philosophy of Science, vol 18, no 1, 35–54 – a reminder that science is just one way of viewing the world and that other people will have different worldviews – it's about finding a way to bring them all together

Heffernan M (2014) A bigger prize: why competition isn't everything and how we do better. Simon and Schuster, ISBN-10 1471100758; ISBN-13 978-1471100758 – an engaging look at the power – and necessity – of collaboration for operating in a complex world

The Partnering Initiative. https://thepartneringinitiative.org/

References

Academy of Medical Sciences, the Royal College of Physicians, the Wellcome Trust, the Bill and Melinda Gates Foundation and Universities UK (2012). Building institutions through equitable partnerships in global health: Conference Report

African Union (2008) Sharm el-Sheik Commitments for accelerating the achievement of water and sanitation goals in Africa, Assembly/AU/Decl. 1 (XI), https://www.susana.org/_resources/documents/default/2-2004-sharmel-sheikhdecisionsenglish1.pdf. Accessed 27 Aug 2019

AGW-Net (2014) Integration of groundwater management into transboundary basin organizations in Africa - a Training Manual by AGW-Net, BGR, IWMI, CapNet, ANBO, & IGRAC. https://www.un-igrac.org/sites/default/files/resources/files/Training%20Manual%20Integration%20of%20GW%20Management.pdf. Accessed 27 Aug 2019

Blake WH, Rabinovich A, Wynants M, Kelly C, Nasseri M, Ngondya I, Patrick A, Mtei K, Munishi L, Boeckx P, Navas A (2018) Soil erosion in East Africa: an interdisciplinary approach to realising pastoral land management change. Environ Res Let 13(12):124014

Choi CK, Pak AWP (2006) Multidisciplinarity, interdisciplinarity and transdisciplinarity in health research, services, education and policy: 1. Definitions, objectives, and evidence of effectiveness. Clin Invest Med 29(6):351–364

Cronin SJ, Petterson MJ, Taylor MW, Biliki R (2004) Maximising multi-stakeholder participation in government and community volcanic hazard management programs; a case study from Savo, Solomon Islands. Nat Hazards 33(1):105–136

Dekens J (2007) The lost messengers? Local knowledge on disaster preparedness in Chitral district. Pakistan,

International Centre for Integrated Mountain Development, Kathmandu

Dodson J (2017) Building partnerships of equals – the role of funders in equitable and effective international development collaborations. https://www.ukcdr.org.uk/wp-content/uploads/2017/1./Building-Partnerships-of-Equals_-REPORT-2.pdf. Accessed 11 July 2019

Elmendorf M, Isely R (1981) The role of women as participants and beneficiaries in water supply and sanitation programs: water and sanitation for health project (USAID) Report 11, pp 1–28

GCRF (2017) UK strategy for the global challenges research fund (GCRF). https://www.ukri.org/files/legacy/research/gcrf-strategy-june-2017/. Accessed 11 July 2019

Gill JC (2016) Building good foundations: skills for effective engagement in international development. Geolo Soc Am Spec Pap 520:1–8

Gill JC, Mankelow J, Mills K (2019) the role of earth and environmental science in addressing sustainable development priorities in Eastern Africa. Environ Develop 30:3–20

Gill JC, Mills K, Mankelow J (2017) Workshop report: earth and environmental science for sustainable development (Nairobi, March 2017). Nottingham, UK, *British Geological Survey*, 28 pp. (OR/17/039) (Unpublished)

Harriss J (2000) Working together: the principles and practice of co-operation and partnership. In: Robinson D, Hewitt T, Harriss J (eds) Managing development: understanding inter-organizational relationships. Sage Publications in association with The Open University, London, pp 225–242

Horton D, Prain G and Thiele G (2009). Perspectives on partnership: A literature review. International Potato Center (CIP), Lima

Howell P (2003) Indigenous early warning indicators of cyclones: potential application in coastal Bangladesh. Benfield Hazard Research Centre, London

IPMG (n/d) Indigenous peoples major group policy brief on sustainable development goals and Post-2015 development agenda: a working draft. https://sustainabledevelopment.un.org/content/documents/7036IPMG%20Policy%20Brief%20Working%20Draft%202015.pdf. Accessed 23 Aug 2019

Jones H, Jones NA, Shaxson L, Walker D (2012) 'Knowledge, Policy and Power in International Development' Policy Press, ISBN 978-1-44730-095-3.

MacDonald AM, Bonsor HC, Dochartaigh BE, Taylor RG (2012) Quantitative maps of groundwater resources in Africa. Environ Res Let 7:024009. https://doi.org/10.1088/1748-9326/7/2/024009

Mercer J, Kelman I, Taranis L, Suchet-Pearson S (2010) Framework for integrating indigenous and scientific knowledge for disaster risk reduction. Disasters 34 (1):214–239

Narayan D (1995) Contribution of People's participation: evidence from 121 rural water supply projects. World Bank Environmentally Sustainable Development Occasional Paper 1, 122 p

Nijsten G-J, Christelis G, Villholth KG, Braune E, Bécaye Gaye C (2018) Transboundary aquifers of Africa: Review of the current state of knowledge and progress towards sustainable development and management. J Hydrol: Region Stud 20:21–34

Nordling L (2019) Africa's science academy leads push for ethical data use. Nature 570(7761):284

OECD (2019) Policy coherence for sustainable development: Empowering People and Ensuring Inclusiveness and Equality. OECD Publishing, Paris, https://doi.org/10.1787/a90f851f-en

Parrack C, Flinn B, Passey M (2014) Getting the message across for safer self-recovery in post-disaster shelter. Open House Int 39(3)

PreventionWeb (2019) About PreventionWeb - Share your content. https://www.preventionweb.net/about/contribute. Accessed 1 July 2019

Sargeant S, Hart A, Hart K, Hughes R (2018) GCRF building resilience event: summary report, https://nerc.ukri.org/research/funded/programmes/building-resilience/workshop-report/. Accessed 2 Aug 2019

Silliman S, Crane P, Boukari M, Yalo N, Azonsi F, Glidja F (2009). Groundwater quality monitoring in collaboration with rural communities in Bénin. In Groundwater and climate in Africa. Proceedings of the kampala conference, Uganda, 24–28 June 2008 (pp 27–35). IAHS Press

Stibbe D, Prescott D (2016) An introduction to multi-stakeholder partnerships. The Partnering Initiative, Oxford

Stibbe DT, Reid S, Gilbert J (2018) Maximising the impact of partnerships for the SDGs, the partnering initiative and UN DESA. https://sustainabledevelopment.un.org/content/documents/2564Maximising_the_impact_of_partnerships_for_the_SDGs.pdf. Accessed 11 July 2019

Toomey AH, Markusson N, Adams E, Brockett B (2015) Inter- and trans-disciplinary research: a critical perspective. GSDR 2015 Brief

Twigg J, Lovell E, Schofield H, Miranda Morel L, Flinn B, Sargeant S, Finlayson A, Dijkstra T, Stephenson V, Albuerne A, Rossetto T, D'Ayala D (2017) Self-recovery from disasters: an interdisciplinary perspective. Overseas Development Institute Working Paper 523

UK Government (2019) Voluntary national review of progress towards the sustainable development goals (United Kingdom of Great Britain and Northern

Ireland, June 2019). https://assets.publishing.service.gov.uk/government/uploads/system/uploads/attachment_data/file/813501/UKVNR-web-accessible.pdf. Accessed 1 July 2019

UN ECOSOC (2019) STI Forum 2019. https://sustainabledevelopment.un.org/content/documents/231772019_Forum_summary_advanced_unedited_version.pdf. Accessed 1 July 2019

UNDESA (2019) Session background notes the 4th annual multi-stakeholder forum on science, technology and innovation for the SDGs. https://sustainabledevelopment.un.org/content/documents/22703Background_Notes_for_STI_Forum_2019.pdf. Accessed 23 Aug 2019

UNDP (2009) Capacity development: a UNDP primer. www.undp.org/content/dam/aplaws/publication/en/publications/capacity-development/capacity-development-a-undp-primer/CDG_PrimerReport_final_web.pdf. Accessed 11 July 2019

United Nations (2012) Framework of operational guidelines on United Nations support to South-South and triangular cooperation: note/by the Secretary-General. SSC/17/3 (2012), New York, 31 p

United Nations (2013) A renewed global partnership for development. https://sustainabledevelopment.un.org/content/documents/833glob_dev_rep_2013.pdf. Accessed 28 Aug 2019

United Nations (2015a) Transforming our world: the 2030 agenda for sustainable development, Goal 15, A/RES/70/1. http://www.un.org/ga/search/view_doc.asp?symbol=A/RES/70/1&Lang=E. Accessed 22 Oct 2019

United Nations (2015b) Millennium development goals report 2015. https://www.un.org/en/development/desa/publications/mdg-report-2015.html. Accessed 22 Oct 2019

United Nations (2019a) Indigenous peoples major group. https://sustainabledevelopment.un.org/majorgroups/indigenouspeoples. Accessed 23 Aug 2019

United Nations (2019b) Technology facilitation mechanism. https://sustainabledevelopment.un.org/tfm. Accessed 1 July 2019

WaterAid (2011) Sustainability framework. www.wateraid.org/what-we-do/our-approach/research-and-publications/view-publication?id=0b45ec09-e7d2-43e1-9423-c00f5ff4e733. Accessed 28 Aug 2019

Watts MJ, An T, Argyraki A, Arhin E, Brown A, Button M, Entwistle JA, Finkelman R, Gibson G, Humphrey OS, Huo X, Hursthouse AS, Marinho-Reis AP, Maseka K, Middleton DRS, Morton-Bermea O, Nazarpour A, Olatunji AS, Osano O, Potgieter-Vermaak S, Saini S, Stewart A, Tarek M, Torrance K, Wong MH, Yamaguchi KE, Zhang C, Zia M (2019) The society for environmental geochemistry and health (SEGH): building for the future. Environ Geochem Health. https://doi.org/10.1007/s10653-019-00381-9

Susanne Sargeant Susanne is a seismologist at the British Geological Survey. She has a background in seismic hazard assessment and has worked on many commercial seismic hazard projects for engineering applications in the UK and internationally, and has co-authored the national seismic hazard maps for the UK and the United Arab Emirates. Susanne has always been passionate about the human aspects of science and disasters and since 2012, her work has focused on finding ways to build bridges between geoscience (and geoscientists) and decision-making for reducing disaster risk and supporting recovery after disasters, mostly in developing countries.

Joel C. Gill is International Development Geoscientist at the British Geological Survey, and Founder/Executive Director of the not-for-profit organisation Geology for Global Development. Joel has a degree in Natural Sciences (Cambridge, UK), a Masters degree in Engineering Geology (Leeds, UK), and a PhD focused on multi-hazards and disaster risk reduction (King's College London, UK). For the last decade, Joel has worked at the interface of Earth science and international development, and plays a leading role internationally in championing the role of geoscience in delivering the UN Sustainable Development Goals. He has coordinated research, conferences, and workshops on geoscience and sustainable development in the UK, India, Tanzania, Kenya, South Africa, Zambia, and Guatemala. Joel regularly engages in international forums for science and sustainable development, leading an international delegation of Earth scientists to the United Nations in 2019. Joel has prizes from the London School of Economics and Political Science for his teaching related to disaster risk reduction, and Associate Fellowship of the Royal Commonwealth Society for his international development engagement. Joel is a Fellow of the Geological Society of London, and was elected to Council in 2019 and to the position of Secretary (Foreign and External Affairs) in 2020.

Michael Watts is Head of Inorganic Geochemistry at the British Geological Survey and is an Associate Professor with the University of Nottingham through the joint Centre for Environmental Geochemistry. His research interests on geochemistry and 'health' interactions employs analytical chemistry for research on pollution pathways via 'natural' or anthropogenic geochemical sources and mineral nutrient dynamics in soil-crop-human/animal systems. Increasingly the research is multidisciplinary with greater emphasis towards challenges and partnerships in developing countries. Michael is currently the President for the Society for Environmental Geochemistry and Health. Michael has extensive experience in capacity strengthening projects from Afghanistan to Africa, in particular in the design of laboratory systems and training programmes for technical and laboratory activities.

Kirsty Upton is a hydrogeologist with the British Geological Survey, focussing on applied and interdisciplinary groundwater research in the UK and sub-Saharan Africa. Her research focuses on methods for groundwater resource assessment and understanding groundwater availability during droughts. She is a lead author of the Africa Groundwater Atlas and has worked with a variety of stakeholders, both in the UK and Africa, to understand ways in which groundwater research can be translated to inform policy and practice.

Richard Ellison has more than 40 years of experience as a geologist with the British Geological Survey where he is currently an Honorary Research Associate, having retired in 2014. Richard held numerous positions, most recently Head of BGS Global Geoscience (2013–2014), Regional Manager for the Middle East, Asia and the Far East (2006–2012), and Project Director for the United Arab Emirates Project (2006–2012). Areas of expertise include geological mapping, stratigraphical analysis, and work on sedimentary sequences and Quaternary processes. Richard worked widely in the UK, and spent extended periods in Peru, Hong Kong, Afghanistan, Tajikistan, Oman, and the United Arab Emirates.

Reshaping Geoscience to Help Deliver the Sustainable Development Goals

18

Joel C. Gill

18.1 Our Shared Future

The SDGs (and frameworks embedded into the SDGs[1]) have the potential to transform society, giving human beings everywhere dignity and equality, meeting the needs of present and future generations in a responsible manner, and ensuring a healthy planet where environmental protection is valued and prioritised. This is an exciting vision, and one that communities across the globe would recognise as being positive and fundamental to their successful future. This vision is also highly ambitious and enormously complex and challenging to deliver by 2030, in all contexts, leaving no one behind.

Chapters relating to **SDGs 1–17** have highlighted that while the world has made steps towards ending poverty, and improving health, gender equality, and access to drinking water, significant hurdles remain. 'Business as usual'

will not realise the vision expressed through the SDGs (Spangenberg 2017). The Overseas Development Institute (ODI) SDG Scorecard 2030 forecasted that unless there are significant changes, we would not achieve any of the SDGs (Nicolai et al. 2015). Data and analyses published since then have examined progress towards the goals as a whole (e.g., an SDG Tracker[2] by Ritchie et al. 2018), in specific national contexts (e.g., through voluntary national reviews[3]), and in sector-specific reports (e.g., FAO reports on The State of Food Security and Nutrition in the World[4]). In the case of some regions, we are decades (if not centuries) away from realising specific ambitions of the SDGs. For example, all thing being equal, the World Economic Forum (2018) indicates that it will take more than 160 years to achieve gender parity in East Asia and the Pacific, and North America (see **SDG 5**). Achieving universal access to even a basic sanitation service (**SDG 6**) by 2030 will require the current annual rate of progress to be doubled (United Nations 2019).

Decisive actions, new approaches, and a willingness to change can all support progress. This responsibility extends beyond governments, to also require engagement (in terms of active participation in the design, promotion, implementation, monitoring, and evaluation of

---

[1]Examples include the Sendai Framework for Disaster Risk Reduction, and the Paris Climate Change Agreement.

J. C. Gill (✉)
British Geological Survey, Environmental Science Centre, Nicker Hill, Keyworth, Nottingham NG12 5GG, England
e-mail: joel@gfgd.org; joell@bgs.ac.uk

J. C. Gill
Geology for Global Development, Loughborough, UK

[2]https://sdg-tracker.org/.

[3]https://sustainabledevelopment.un.org/vnrs/.

[4]https://www.fao.org/state-of-food-security-nutrition/en/.

© Springer Nature Switzerland AG 2021
J. C. Gill and M. Smith (eds.), *Geosciences and the Sustainable Development Goals*,
Sustainable Development Goals Series, https://doi.org/10.1007/978-3-030-38815-7_18

activities) by individuals, businesses, and special interest groups, including the science and technology community. As has been demonstrated throughout this book, geoscientists possess skills and understanding to advance progress and support the transition to sustainability. The relevance of geoscience to the SDGs is not limited to a few explicitly environmental goals (e.g., **SDGs 13, 14**, and **15**), or those focused on increasing access to water (**SDG 6**) or energy (**SDG 7**). Geoscientists understanding of Earth resources, dynamics, and systems can help (in partnership with others) to tackle major social challenges, the provision of essential services, the growth of green and diverse economies, the development of sustainable and resilient cities and infrastructure, and effective protection of local, national, and global environmental systems.

While our contributions and potential contributions to sustainability are significant, the geoscience community currently has a low profile in the sustainable development arena, and is less represented in sustainability discourses (Mora 2013; Stewart and Gill 2017), particularly when comparing with scientists focused on biotic

aspects of the planet. The 2019 Global Sustainable Development Report (titled *The Future is Now: Science for Achieving Sustainable Development*) does not mention geology, geoscience, or Earth science (IGS 2019) in the main text. This is likely a reflection of only a few geoscientists engaging and contributing to the call-for-evidence for this report. While increased awareness of climate change, biodiversity loss, and other environmental challenges has increased, awareness of the role of geoscientists in addressing these and other social and economic challenges remains low. There is still a long way to go to persuade people that future planning requires an understanding of geological processes, systems, and resources.

The urban, coastal community represented in the cartoon illustration in Fig. 18.1 is typical of many contexts around the world. There are flows of resources and waste products in and out of the city, and competing demands for energy, aggregates and minerals, water, and land. There are complex interactions between Earth processes, surface activities, and the subsurface, between the natural and the built environments, and

Fig. 18.1 SDGs and Urban, Coastal Community. This cartoon image demonstrates how the SDGs sit together in a single region—a coastal city and its wider catchment—with interdependencies between goals. Education (SDG 4), research and innovation (SDG 9), equality (SDGs and 10), strong institutions (SDG 16), and effective partnerships (SDG 17) all support management and restoration of this environment to support sustainability

between urbanised zones and the wider catchment in which the city sits. Changes in land use, anthropogenic activities, or natural hazards can have cascading consequences. Progress in one SDG can drive progress in another SDG, or result in emerging challenges that require mitigation. Delivering sustainable development in this context provides geoscientists from across all sectors and sub-disciplines with an opportunity to engage and contribute to sustainability.

This book not only demonstrates *why* geoscientists should be engaged in sustainable development dialogues, but also seeks to equip them to do this more effectively by providing socio-economic context, introductions to key international mechanisms and processes, and a diverse set of case studies (many drawn from the Global South). Given that 'business as usual' is not enough to realise the ambitions of the SDGs, this book also sets out how the geoscience community could evolve and adapt to enhance the relevance and impact of our contribution. Chapters in this book make several recommendations of changes to education programmes, ongoing professional development and training, data collection, research agendas, industry practice, and engagement with non-governmental, governmental, and intergovernmental organisations. We synthesise and reflect on these in Sect. 18.3, commenting on some emerging themes relating to the role of geoscience institutions, the availability of data, the training of geoscientists, and the communication of geoscience to the public and others engaged in sustainable development initiatives.

18.2 Beyond 2030: Delivering and Maintaining Sustanability

Our hope is that the descriptions of 'current progress' articulated in this volume rapidly become outdated as immediate and effective actions are taken to help deliver the SDGs. We believe, however, that the central messages of this book and recommendations for geoscientists, geoscience-based sectors, and geoscience institutions will not diminish in their relevance and importance. Beyond 2030, communities will develop new infrastructure, there will be new demands on natural resources for emerging technologies, and we will likely identify new environmental links to health that inform the policy responses to promote well-being.

The ambitions of the SDGs not only require concerted action today, and in the months and years to 2030, but an ongoing commitment to pursue knowledge and adhere to frameworks that enable humankind to live sustainably beyond 2030. Increasing and sustaining Earth and environmental science education can strengthen understanding of how the natural environment is responding to anthropogenic activities (**SDG 4**). Improving and sustaining environmental monitoring (highlighted in **SDG 15**, for example) and connecting the analysis of environmental data to those groups developing and shaping local, national, and regional policies will be as important in 2050 as it is in 2020. Safe and secure work environments (**SDG 8**), equality of opportunity for all (**SDGs 5** and **10**), and responsible consumption and production (**SDG 12**) are ongoing commitments embedded in the SDGs, and not just short-term goals.

Communities living sustainably in 2030 is not a guarantee of sustainability in perpetuity. The world is changing rapidly, with new technologies and insights to support sustainable development. Yang et al. (2019), for example, sets out an approach to enable rapid charging of electric vehicles, giving them a 200-mile range in 10 minutes of charging. Such research, when commercialised, could facilitate the mainstream adoption of battery electric vehicles and help to decarbonise transport (**SDGs 7, 12,** and **13**). At the same time, there are also emerging threats and challenges. New conflicts could hinder access to or the flow of raw materials, or human error could result in technological accidents that contaminate critical water supplies. Complacency, or political, social, and environmental changes may, therefore, undo or reverse progress made to 2030 and result in cascading impacts through our heavily interconnected societies, as exemplified through the Covid-19 pandemic.

This book focuses on 17 interdependent sustainability goals, that may or may not be recognised as the 'Sustainable Development Goals' in the years following 2030, but their themes will still be pertinent to society as we commit to ongoing interventions to support sustainable development.

18.3 Integrating Learning to Inform Recommendations

There is great value in what each of the chapters exploring **SDG 1–17** have set out individually, and the global suite of case studies presented. These highlight the range of geological studies that can support the SDG targets, and initiatives where geoscientists could get involved. Integrating reflections from the 17 chapters also enables us to consider how to catalyse greater geoscience engagement in sustainable development, and to set out specific actions to foster equity, improve knowledge exchange, and encourage interdisciplinarity. These three topics are emphasised in the 2030 Agenda, and highlighted in the introduction to this book. We have grouped our reflections into six themes (Sects. 18.3.1–18.3.6), and outline recommendations (Sect. 18.3.7) where we believe concerted action could help strengthen the contribution of geoscientists to the SDGs.

18.3.1 Global Challenges Require Integrated Solutions

While expressed in 17 individual goals, the overall ambitions of the 2030 agenda are an integrated, 'indivisible' whole (Nilsson 2016). Firmly embedded into the 2030 Agenda, and reinforced throughout this book, is the narrative that achieving the SDGs requires an integrated approach. Each chapter has set out how geoscience can help to address different sustainability challenges, but there are dependencies between them that can result in trade-offs and reinforcements.

- Access to safe water and sanitation (**SDG 6**) is essential for social well-being, supporting outcomes in health (**SDG 3**), education (**SDG 4**), livelihoods (**SDG 8**), and gender equality (**SDG 5**).
- Progress in many goals (from education to health, clean water to infrastructure) can support **SDG 8** (decent work and economic growth). Addressing this goal generates investment for basic services (e.g., improved sanitation facilities), improvements in social development, and an enhanced natural environment (all of which reinforce progress in **SDG 8**).
- Protecting and restoring terrestrial ecosystems (**SDG 15**) can support food security (**SDG 2**) and climate action (**SDG 13**), as well as improve the lives of communities that are excluded, marginalised or at risk of being left behind (**SDG 10**).

Similar relationships exist for the other SDGs (Pradhan et al. 2017). The International Science Council have published a comprehensive review of interactions within the SDGs, determining to what extent they reinforce or conflict with each other (ISC 2017). Understanding these interactions helps to guide decision-making and ensure that policies are coherent so as not to undermine progress.

The example of land use in Box 18.1 highlights how significant integration when addressing the SDGs can help ensure wise investments and the mitigation of unintended consequences. Geoscientists can support this approach, drawing on their existing thematic knowledge and cognitive skills (outlined in **SDG 4**) to inform decision-making from the perspective of the natural environment (Fig. 18.2).

Box 18.1. Geoscience and Land-Use Planning for Sustainability

Actions to deliver many of the SDGs will require land use or cover changes. This includes increasing urbanisation (**SDG 11**), the development of infrastructure (**SDG 9**),

Fig. 18.2 Implementing the SDGs will increase demand for land. Geoscience can help to inform decision-making to support economic growth, social development, and environmental protection. Image by Hans Braxmeier from Pixabay.

or increased demand for food (**SDG 2**) and mineral resources for renewable energy technologies (**SDG 7**).

The use of land that is currently forests and woodlands, supporting diverse ecosystems, to meet these demands may hinder efforts to protect terrestrial and inland freshwater ecosystems (**SDG 15**), and risk the degradation of essential services they provide, from tackling climate change (**SDG 13**) to improving health (**SDG 3**).

Integration across SDGs (and disciplines) can, therefore, help to ensure land is stewarded wisely, with the right function allocated to the right land. Environmental datasets (including the underlying geology, active geological processes, geochemical characteristics of soils, hydrogeochemistry)

and analyses can support planners, policymakers and politicians in this decision-making process, and inform choices regarding:

- *Parcels of land to protect or restore because of their capacity to host unique biodiversity or their contributions to essential ecosystem services.* For example, limestone pavements have distinct surface features (e.g., fissures) that often support rare plant species (Cottle 2004). Certain glacier-fed freshwater ecosystems are significant annual CO_2 sinks, due to chemical weathering processes (St. Pierre et al. 2019).
- *Parcels of land for subsurface infrastructure development.* The use of the

subsurface environment for transport networks, car parking, shopping centres, or waste management systems is guided by having a three-dimensional understanding of the subsurface. This includes characterising the geological materials to determine their suitability for excavation, tunnelling, and hosting different types of infrastructure.

- *Parcels of land hosting critical natural resources.* Developing a new urban environment over a major ore deposit is a costly mistake, preventing access to critical materials that support economic development. Geoscientists can help to integrate understanding of resources (e.g., minerals, groundwater, and industrial aggregates) into land-use planning to maximise efficiencies and minimise conflicts of use.
- *Parcels of land for productive agriculture.* Geochemical mapping can inform decisions about the siting of new agriculture, identifying soils rich or depleted in nutrients, and soils where previous land use has resulted in contaminants that affect plant, animal, and human health. Contaminated sites need remediation and are often better suited to the development of infrastructure or industrial facilities.

Understanding of geological hazards (e.g., landslides, earthquakes, groundwater flooding) also informs land-use decisions, how hazards affect different activities on different geological substrates, and how human activities affect these hazards. For example, converting a forested hillside into agricultural land has the potential to increase the magnitude or frequency of landslides, with serious consequences on those using the land at the foot of the slopes.

It is also critical to integrate environmental data with an understanding of social and economic factors to inform decision-making. For example, some land has cultural and spiritual significance to certain groups, and any change to its use will be deeply problematic. Other land formations generate significant revenue to a local region through ecosystem services, geotourism or leisure activities—with changes in land use being counterproductive and economically costly.

Land is just one resource, with **SDG 12** characterising a nexus that also involves food, water, energy, and material resources (e.g., minerals, aggregates), with many interdependencies between these, as illustrated by the discussion of land in Box 18.1. Delivering many of the SDGs has a resource implication, potentially resulting in conflicting demands for water or energy, industrial aggregates or food. Resolving these conflicts needs strong institutions and effective resource governance (**SDG 16**), underpinned by access to reliable environmental data and analyses. Integrated resource management has a cascading effect on educational demands (**SDG 4**), work opportunities and challenges (**SDG 8**), and the development of interdisciplinary communities of practice that promote open exchange of science.

18.3.2 Integrated Solutions Require Interdisciplinary and Multisectoral Partnerships

Integrated solutions to multifaceted problems require interdisciplinary partnerships. The contribution of geoscientists is inseparable from political, economic, social, technological, legal, cultural, and (other) environmental context. Contributions from different sectors should complement and build of one another, rather than undermine and contradict one another. For example, if we consider tackling energy poverty (**SDG 7**) in the Global South and envisage a global transition to a net-zero economy by 2050 to tackle climate change (**SDG 13**). The solutions will require behavioural change (*psychology*), new technologies (*design and engineering*), and economic shifts (*economics*), but also access to and improved management of natural resources, ground characterisation, and environmental modelling (*geoscience, ecology, forestry*). Bringing disciplines together to co-design strategic and coherent plans and policies, informed by diverse evidence, is more likely to result in integrated solutions to sustainability challenges.

Embracing interdisciplinarity requires geoscientists to engage proactively in new forums and settings, building partnerships and enhancing communication with a wider range of institutions and disciplines. We anticipate that skills such as feeding into policy making processes (Sect. 18.3.5) will become increasingly important for geoscientists, and, therefore, need to be better reflected in geoscience education and training opportunities (Sect. 18.3.6), national survey programmes and learned societies. Others contributing to sustainable development will benefit from the skills and knowledge that geoscientists can bring. To aid this process, geoscientists should take steps to reach out and help raise awareness by improving the messaging associated with geoscience events, initiatives and outputs. For example, badging of more geoscience meetings with information about relevant SDGs could help to demonstrate societal links between themes such as geochemistry and health, minerals and renewable energy technologies, or geohazards and sustainable cities in a clearer way, appealing to both geoscientists and other disciplines engaged in work on these SDGs.

In the chapter exploring **SDG 17**, the authors describe the time and resources needed to build effective interdisciplinary and multisectoral partnerships. While different disciplines need to come together to innovate, so do voices from diverse sectors and groups. There can be challenges and conflicting priorities and perspectives to navigate, but the benefits in terms of potential impact vastly outweigh the effort required to overcome these difficulties. All partnership development takes time and resources. The emphasis on building interdisciplinary research

partnerships for development by the UK Government (e.g., through the Global Challenges Research Fund[5]) is not common, but provides a model by which such partnerships could be encouraged.

18.3.3   Improve the Collection, Management, Integration, and Accessibility of Data

Many of the chapters in this book have highlighted how improved data collection, management, integration, and accessibility provides an opportunity to support development. In some contexts (e.g., **SDGs 2** and **15**), this focused on environmental data, with weaknesses in monitoring networks and data management capacity in the Global South identifed as hindering efforts to collect and synthesise environmental data to inform decision-making. The East Africa Community Vision 2050 notes improvements in environmental data collection as one of their priorities (EAC 2016). Workshops in Kenya, Zambia, and Tanzania, conducted in 2017 with representatives of 48 organisations, all converged on a common problem of lack of access to data, or different agencies holding different datasets that would be particularly useful when integrated (Gill et al. 2019). Participants highlighted problems including data being in analogue rather than digital form, and digital data not being backed up or placed on a secure server (Fig. 18.3).

Improving the management, integration, and access could help to (i) identify data gaps so data collection can be targeted, (ii) support analyses that rely on access to multiple datasets (e.g., spatial relationships between climate change, soil geochemistry, and disease prevalence), (iii) inform research questions to address societal priorities, and (iv) leverage additional funding to support greater data collection and strengthen monitoring networks. The Southern African Development Community (SADC) have also

---

[5]https://www.ukri.org/research/global-challenges-research-fund/.

Fig. 18.3 Discussing Environmental Data for Sustainable Development in Tanzania. Emphasised in many chapters of this book, and supported by dialogue with sustainability stakeholders in Eastern Africa, the strengthening of environmental data collection, management, integration, and access can support many SDGs. © UKRI (used with permission)

emphasised the need to '*build capacity for collection, management and exchange of information/data for the sustainable management of environment and natural resources*' (SADC 2005, p. 62).

In other contexts (e.g., **SDG 7**), the chapter author highlighted discrepancies in data characterising current levels of development between the Global North and Global South. For example, it is difficult to identify appropriate data on energy use in many countries in the Global South or the prevalence of postgraduate geoscience courses in sub-Saharan Africa. As voluntary national reviews on progress towards the SDGs are completed, data collection and availability may improve (e.g., for SDG indicators[6]). Portals to deposit data sets held by businesses or civil society groups could also help to improve access to data to characterise progress towards the SDGs. International geoscience unions could invest in targeted data collection on themes such as availability of geoscience education courses (**SDG 4**), or progress in diversity, equality and inclusion (**SDGs 5** and **10**), to support the community to implement actions for sustainable development.

Finally, while the aggregation and integration of information to understand systems and inform decision-making is clear, disaggregation of information (by region, age, sex, ethnicity, income, and other characteristics) can help to understand the impact of interventions aiming to deliver sustainability, assessed through change in the 232 individual indicators[7] agreed by the UN General Assembly. This will be increasingly important if we are to monitor our commitment to leave no one behind and improve inclusivity.

---

[6]https://unstats.un.org/sdgs/indicators/indicators-list/.

[7]https://unstats.un.org/sdgs/indicators/indicators-list.

18.3.4 Strengthen International and National Science Institutions to Catalyse and Resource Action

Many of the actions suggested through the chapters of this book require both global leadership (e.g., by continental and global geoscience organisations, large businesses, and key academic networks), as well as local implementation. Improving gender equality, public understanding of Earth science, equitable access to scientific knowledge and training, and equitable research partnerships requires strong commitment to these themes by diverse geoscience organisations (e.g., international unions, professional and learned societies, see **SDG 16**), with a concerted effort to catalyse and resource change in the broader community.

There are positive developments in recent years linked to ethics, tackling harassment, and improving diversity that indicate a growing acceptance that the geoscience community must take responsibility for securing change within our sphere of influence. Examples include

- *The AGU Ethics and Equity Center*[8] provides resources to support responsible scientific conduct and establish tools, practices, and data for organisations to foster a positive work climate in science.
- *The International Association for Promoting Geoethics*[9] has grown into a multidisciplinary, global platform. They published the Cape Town Statement on Geoethics in 2017, which is supported by at least 22 global organisations and translated into 35 languages.
- *Meeting Codes of Conduct.* Many large geoscience meetings and conference coordinators (e.g., the Geological Society of London[10]) now have codes of conduct that aim to provide a safe, open, and respectful environment for participants.

- *AGI Statement on Harassment in the Geosciences.* This statement, agreed by the American Geosciences Institute in 2018, makes recommendations to member societies regarding tackling harassment through intervention, enforcement and reporting.[11]
- *Girls into Geoscience.* This outreach initiative[12] promotes diversity and equality, and provides role models for aspiring female scientists across the geosciences.
- *International Association for Geoscience Diversity.* This non-profit is dedicated to creating access and inclusion for persons with disabilities in the Geosciences.[13]

These examples suggest positive action towards creating geoscience institutions, meetings, training, and places of employment that are inclusive, diverse, and safe, but there is of course scope for further work and engagement by a broader group of people. The actions of members should reflect the ambitions stated in institutional codes of conduct and live up to their commendable language around tackling inequalities (including gender, race, and income). Leaders should demonstrate, resource, and enforce these commitments. Delaying any wholehearted embrace of diversity, equality, and inclusion comes with far greater costs than actively championing and implementing this agenda.

Strong leadership and institutions help to tackle inequalities (**SDGs 5** and **10**), but also support safe and secure work environments (**SDG 8**), improve public understanding of science (**SDG 4**), strengthen research and development (**SDG 9**), and facilitate science partnerships for development (**SDG 17**). This is particular the case when international institutions have an inclusive leadership, with full, meaningful participation of scientists from all regions, including the Global South. Commitment of resources to support Global South scientists, and embracing

---

[8]https://ethicsandequitycenter.org/.

[9]https://www.geoethics.org/.

[10]https://www.geolsoc.org.uk/code-of-conduct-events.

[11]https://www.americangeosciences.org/content/agi-statement-harassment-geosciences.

[12]https://www.plymouth.ac.uk/research/earth-sciences/girls-into-geoscience.

[13]https://theiagd.org/.

technologies and virtual meeting spaces, can ensure Global South scientists are included. Where scientific meetings and events consider pricing, diversity of speakers, and the services provided to participants, they could help to strengthen scientific communities that may otherwise be disadvantaged. Breaking the isolation of scientists in the Global South was an 'imperative' set out by Berger (1991), and remains so nearly 30 years later.

Many of the institutions highlighted in this book are membership organisations. A proactive membership, contributing to the life of the organisation, will *strengthen* these institutions and empower them to effect positive change. Volunteering for committees, nominating people for awards and recognition (particularly underrepresented groups), and contributing to engagement activities can seem small contributions when set against the towering injustices described throughout this book. For institutional change to happen, however, all geoscientists need to embrace their individual responsibilities as citizens of a professional community, with the collective capacity to enact positive and lasting change.

18.3.5 Strengthen the Links Between Geoscientists and Decision-Making

International and national science institutions also help to bridge the gap between scientists and decision-makers, relevant to every SDG. The production of knowledge is not sufficient to drive change, but should be complemented with proactive and effective communication of results and their implications to decision-makers and those with the ability to influence decision-makers. Whether considering micronutrient deficiencies (**SDGs 2** and **3**), increasing geotourism potential (**SDG 8**), urbanisation (**SDG 11**), management of natural resources (**SDG 12**), or protecting the oceans (**SDG 14**), clear lines of communication between geoscientists and decision-makers can help 'knowledge' to result in a tangible and sustainable impact. In this

context, we define decision-makers broadly, and recognise that these will vary. It could be individuals in a community with the ability to deliver change at a household level, local or national politicians and diplomats, heads of department, industry leaders, or cultural and religious leaders.

Strengthened links between geoscientists and decision-makers (including those responsible for the development of local, national, or regional policies) will benefit from enhanced sociopolitical understanding (e.g., how government works), recognition of the complexity of policymaking (and engaging with those shaping policy), and accepting that geoscience is one form of evidence in the decision-making process (Boyd 2016; Gluckman 2016). Embedding this understanding and the skills to engage in the policy arena into the training of geoscientists could improve ongoing engagement in the science-policy-practice interface. This is also supported through understanding what information would help stakeholders (from community groups to policy makers), how they will use this information, and how best to present it to support these uses (Gill and Bullough 2017). Dialogue, investment in building relationships, and interdisciplinary partnerships can help to translate geoscience knowledge into tools to inform decision-making (Lubchenco et al. 2015).

Bridges between geoscientists and decision-makers are beneficial at local, national, regional, and global scales. National geoscience institutions (such as learned societies and professional institutions) are particularly well placed to help bridge the gap between geoscientists and decision-makers in many countries, with the ability to convene diverse expertise from across the membership and coordinate interdisciplinary responses with other scientific organisations. This does not mean they have the sole responsibility to engage with decision-makers. In many local contexts, individual geoscientists will work with decision-makers to inform research questions or tool development. Engaging stakeholders early in the research process will ultimately produce knowledge of greater use (Weichselgartner and Kasperson 2010).

At a global scale, the InterAcademy Partnership have described how scientists can input into policymaking. Their 2019 report '*Improving Scientific Input to Global Policymaking, with a Focus on the UN Sustainable Development Goals*' sets out routes by which the science community can support the SDGs, particularly focusing on UN processes, reports, and mechanisms (IAP 2019). **SDG 17** outlines examples of ways geoscientists can engage. Increasing engagement at this level will make an important contribution to increasing recognition of and demand for geoscientific data to inform decision-making.

18.3.6 Reshape Geoscience Education to Meet Future Demand

The topics and skills included within geoscience training should reflect the best available science and areas of debate, equip students to join the geoscience profession (both research and industry), and be transferable enough to allow geoscientists to effectively contribute to other professions. They should also align with societal demand in terms of the ability to provide advice and services for public benefit. To evaluate whether existing courses currently meet these criteria, we must look not only at employment opportunities and societal demands today, but also in the years ahead (10, 20, 30 years from now). In **SDG 8**, we highlighted some existing analyses of the future of work. These have uncertainties, but it is clear that as the emphasis society places on sustainable development increases, the opportunities available for geoscientists will change. Global commitments to decarbonisation, the circular economy, and restoring the natural environment will likely result in new expectations placed on professional geoscientists and new career paths available to them. Increased demands on land, water, and rocks and minerals will require geoscientists with both the technical knowledge to characterise these resources, but also the sustainability insights to inform their exploitation.

Analysing societal demand and changes is essential to understanding if geoscience education is shaped appropriately to help deliver the SDGs.

- *Availability of Courses.* Specialised postgraduate training programmes can help to develop the knowledge and capacity to tackle development challenges. For example, postgraduate courses in themes such as subsurface energy systems, renewable energy systems, sustainable energy, renewable energy and resource management, and applied geoscience (geoenergy) exist in the UK, and will prepare students to contribute to efforts to deliver many SDGs. There are many other regions, particularly in the Global South, where similar postgraduate training could boost regional skills capacity and support governments, industries, and civil society to deliver the SDGs. A systematic and comprehensive assessment of current geoscience postgraduate training programmes (by region and subject) is necessary to understand where to invest resources in new training initiatives to meet societal demands.
- *Modules and Skills for Sustainability.* The chapter exploring **SDG 4** makes a convincing case for geoscience to reform to support sustainable development. This is supported by all of the chapters in this book, which highlight concepts and skills that would enrich the education of geoscientists and their ability to apply their knowledge to address global challenges. Undergraduate programmes are often pressed for time with limited scope to add new content without removing existing content. Institutions should review whether all existing course content adequately prepares geoscientists for the jobs they are likely to be involved in, or if changes are helpful. Additional optional modules could be developed to focus on social geology or sustainable geoscience (Stewart and Gill 2017), providing an opportunity for geoscientists to enrich their understanding of the interdisciplinary approaches required so support sustainability. Professional skills modules should evolve

Fig. 18.4 Geology for Global Development Annual Conference 2015. Geology for Global Development is a registered charity providing opportunities for geoscientists to strengthen their understanding of how to deliver the SDGs and develop the skills to contribute effectively. Conferences and workshops complement technical training provided to students through university courses. © Geology for Global Development (used with permission)

from report writing and risk assessment, to also cover themes such as ethics, partnership building, and public policy engagement. Engagement with organisations such as *Geology for Global Development*[14] (Fig. 18.4) may also provide opportunities to strengthen skills for development.

Addressing both the themes above (availability of courses, and content of courses) would strengthen the geoscience profession and improve the way we serve society.

18.3.7 Next Steps and Recommendations

Bringing together the reflections in Sects. 18.3.1–18.3.6, we offer some suggested next steps and recommendations to support the global geoscience community to evolve and positively affect the delivery of the ambitions expressed in the 17 SDGs. We suggest that these steps will (i) increase the relevance and resilience of the geosciences as a discipline, (ii) prepare geoscientists for major societal transitions, and (iii) equip geoscientists to advocate for evidence-informed changes and influence decision-makers. There are many other recommendations set out in the individual chapters of this book which relate to specific geoscience research and activities that can help to deliver specific SDGs (with all the cascading effects, highlighted previously). Here, we focus on recommendations to help facilitate the impact of these (and other activities), and themes with major implications for the geoscience as a whole, cutting across national and sub-disciplinary boundaries.

1. *Proactively engage in new forums and settings.* While tens of thousands of geoscientists gather each year at major geoscience conventions (i.e., traditional scientific meetings and dissemination forums), they are often underrepresented at meetings focused on development themes. This results in

---

[14]https://www.gfgd.org/. See https://www.gfgd.org/education for a set of open-access Higher Education Learning Resources.

missed opportunities to influence agendas by sharing a geological perspective and build partnerships across disciplines. This engagement does not always require a physical presence. There are opportunities to contribute through joining networks, submitting contributions to major reports (e.g., the *Global Sustainable Development Report*,[15] commissioned by the United Nations), and applying for positions on development committees and advisory boards.

2. *Create opportunities for exchanges, visiting fellowships, and permanent positions for geoscientists in development institutes (and* vice versa). Encouragement and resources for geoscientists wishing to build links with organisations associated with development objectives will help to foster the integrated solutions supporting sustainability. This includes links with NGOs, government departments, think tanks and academic institutes, and intergovernmental agencies. There are multiple models for this, including exchange schemes, visiting fellowships, and honorary research associates. Building these opportunities may result in greater recruitment of those with a geoscience background directly into sustainability and development organisations.

3. *Improve impact messaging associated with geoscience events and outputs.* Hundreds of conferences and workshops are organised by the geoscience community every year, around the world. Very few of these explicitly note how these scientific meetings link to the SDGs (and to the Sendai Framework or the Paris Agreement), yet this could be done with some simple badging. This could also extend to academic papers, reports, and other tools and technologies produced by geoscientists. Not all science is focused on development impact, but a significant amount of geoscience has an applied focus. By improving how we communicate the relevance of our science we will help to raise awareness of the public, other development professionals, and policymakers of how geoscience connects with the SDGs.

4. *Improve the resourcing of environmental data collection, monitoring networks and centres for analysis and data interpretation, particularly in the Global South.* Effective decision-making linked to many of the SDGs depends on access to reliable environmental data. Geoscience organisations across the globe should support efforts to strengthen data collection (e.g., by integrating scientists from the Global South into initiatives and programmes on data collection and standards), improve data management, and ensure accessibility to those using the information to improve decision-making. This includes strengthening the means to conduct analyses and data interpretation in the Global South, creating and supporting world-class laboratories adhering to international quality standards, requiring larger scale investments. Geoscientists should build links with development banks and other funders to set out how such facilities can support development strategies.

5. *Bring together and resource innovative organisations leading on diversity, equality, and inclusion.* We highlight a number of examples of initiatives helping to improve diversity, equality, and inclusion in the geosciences. These organisations should not act in isolation, but know they have the full support of the wider geoscience community as they continue and expand their activities. Furthermore, by bringing representatives of these groups together, the global geoscience community can (i) capture learning, (ii) seek advice as to how to replicate activities in other contexts or scale activities up, and (iii) be held to account for progress across the community as a whole.

6. *Improve evaluation of the impact of past geoscience partnerships for development to inform future collaborations.* The global, field based, and resource nature of the geosciences means that geoscientists have long participated in capacity building, institutional strengthening, and the spectrum of

---

[15]https://sustainabledevelopment.un.org/globalsdreport/.

partnerships (e.g., North–South, South–South, Triangular) set out in **SDG 17**. This is sometimes through research partnerships, sometimes through government-to-government collaborations, and sometimes through the initiatives of scientific organisations and not-for-profits. Greater evaluation of the work that geoscientists have done to support development efforts is beneficial to helping shape future activities and collaborations. This evaluation could use the Organisation for Economic Co-operation and Development (OECD) Criteria for Evaluating Development Assistance reflecting on measures of relevance, effectiveness, efficiency, impact, and sustainability. An emphasis on equity, and how partnerships have reduced or exacerbated diverse inequalities, would also be beneficial.

7. *Commission a 'portal' to gather data and analyse the future of work for geoscientists.* Recognising the skills and understanding that geoscientists have (or could have, if educational reforms take place) a data portal such as this could collate sources of data, identify potential data gaps, and help assess the threats to and opportunities for the employment of geoscientists, disaggregated by region and theme. Capturing different perspectives from the Global North and Global South would be particularly helpful and ensure specific local contexts are considered. A data portal, together with analysis, focused reports and a taskforce with participation by educators, geological surveys, industry groups, and others would help to explore questions such as (i) what new roles will emerge for geoscientists, (ii) what existing roles (not currently filled by geoscientists as the normal) could be suitable for those with an understanding of geoscience (and some reskilling), (iii) what is the global distribution of specialised geoscience postgraduate training courses, and (iv) what are the significant thematic and geographic gaps for training and development?

8. *Reform geoscience education to enable the sustainability transition.* With a growing understanding of the skills and knowledge required of geoscience graduates in a world committed to decarbonisation, the circular economy, and environmental integrity, it is necessary to reform geoscience education. Themes such as engaging in public affairs, building partnerships (across countries, sectors, disciplines), and ethics should be included in diverse professional development courses and programmes. Approaches that embed the social and economic dimensions of sustainability into geoscience education and training, providing the context to engage effectively in international sustainability initiatives, should be encouraged. Geoscientists' training should introduce and celebrate interdisciplinarity at an early stage, with postgraduate opportunities to enhance these skills and explore research opportunities. Professional qualifications (including chartership) should embed values of sustainability and social responsibility.

18.4 An Opportunity to Reposition Geoscience

Geoscience is foundational to sustainability, and an enabler of inclusive economic growth, human development, and environmental protection. The SDGs provide geoscientists with an important opportunity and responsibility to increase access to and understanding of our science, and refresh our engagement with business leaders, politicians, scientists, development practitioners, and civil society to support sustainable development (Geological Society 2018). We need to shape a stronger public and political awareness of the benefits of geoscientists to sustainable development. Our message, however, must be matched by our actions—on ethics and integrity, diversity, equality, inclusion, and partnerships.

Whilst the geoscience community has a significant role to play in transforming our shared world, what society manages to achieve by 2030 will depend on factors far beyond the reach and influence of most geoscientists. We encourage readers around the world to reflect on the themes set out in volume, explore the opportunities presented by the SDGs, and work within their own communities to help catalyse geosciences' contribution to the SDGs. By embracing values and implementing ideas set out in this book, we can together help achieve sustainability, meet the needs of present and future generations, and secure environmental protection and restoration.

References

Berger AR (1991) Three imperatives for global geoscience. In: Stow DA, Laming DJC (Eds) Geosciences in development. AGID Report Series No. 14, pp 295–302

Boyd IL (2016) Take the long view. Nature 540 (7634):520–521

Cottle RA (2004) Linking geology and biodiversity. English Nature

EAC (2016) Vision 2050, East African Community, 57p

Geological Society (2018) Geology and the UN sustainable development goals. www.geolsoc.org.uk/~/media/shared/documents/policy/SDGs%20Note_FINAL.pdf?la=en. Accessed 4 Oct 2019

Gill JC, Mankelow J, Mills K (2019) The role of Earth and environmental science in addressing sustainable development priorities in Eastern Africa. Environ Dev 30:3–20

Gill JC, Bullough F (2017) Geoscience engagement in global development frameworks. Ann Geophys 60

Gluckman P (2016) The science–policy in-terface. Science 353(6303):969–969

IAP (2019) Improving scientific input to global policy-making with a focus on the UN sustainable development goals. https://www.interacademies.org/50429/SDGs_Report. Accessed 23 October 2020

IGS (Independent Group of Scientists appointed by the Secretary-General) (2019) Global sustainable development report 2019: the future is now—science for achieving sustainable development. United Nations, New York, p 252

ISC, International Science Council (2017) A guide to SDG interactions: from science to implementation. In:

Griggs DJ, Nilsson M, Stevance A, McCollum D (eds) International Council for Science, Paris. 239 p

Lubchenco J, Barner AK, Cerny-Chipman EB, Reimer JN (2015) Sustainability rooted in science. Nat Geosci 8 (7):741–745

Mora G (2013) The need for geologists in sustainable development. GSA Today 23(12):36–37

Nicolai S, Hoy C, Berliner T, Aedy T (2015) Projecting progress: reaching the SDGs by 2030. Overseas Development Institute, London, p 48

Nilsson M (2016) Understanding and mapping important interactions among SDGs. Background paper for Expert meeting in preparation for HLPF 2017. https://workspace.unpan.org/sites/Internet/Documents/UNPAN96735.pdf. Accessed 26 Oct 2019

Pradhan P, Costa L, Rybski D, Lucht W, Kropp JP (2017) A systematic study of Sustainable Development Goal (SDG) interactions. Earth's Future 5(11):1169–1179

Ritchie H, Roser M, Mispy J, Ortiz-Ospina E (2018) Measuring progress towards the sustainable development goals. https://sdg-tracker.org/. Accessed 31 Oct 2019

SADC (2005). Regional indicative strategic development plan, Southern African Development Community, 163p

Spangenberg JH (2017) Hot air or comprehensive progress? A critical assessment of the SDGs. Sustain Dev 25(4):311–321

St. Pierre KA, St Louis VL, Schiff, SL, Lehnherr I, Dainard PG, Gardner AS, Aukes PJK, Sharp MJ (2019) Proglacial freshwaters are significant and previously unrecognized sinks of atmospheric CO_2. Proc Natl Acad Sci USA 116(36):17690–17695

Stewart IS, Gill JC (2017) Social geology—integrating sustainability concepts into Earth sciences. Proc Geol Assoc 128(2):165–172

United Nations (2019) Progress of Goal 6 in 2019. https://sustainabledevelopment.un.org/sdg6. Accessed 1 Oct 2019

Weichselgartner J, Kasperson R (2010) Barriers in the science-policy-practice interface: toward a knowledge-action-system in global environmental change research. Global Environ Change 20(2):266–277

World Economic Forum (2018) The Global Gender Gap Report 2018. World Economic Forum

Yang XG, Liu T, Gao Y, Ge S, Leng Y, Wang D, Wang CY (2019) Asymmetric temperature modulation for extreme fast charging of lithium-ion batteries. Joule 3, 1–18, December 18, 2019 [a] 2019 Elsevier Inc. https://doi.org/10.1016/j.joule.2019.09.021

Joel C. Gill is International Development Geoscientist at the *British Geological Survey*, and Founder/Executive Director of the not-for-profit organisation *Geology for Global Development*. Joel has a degree in Natural Sciences (Cambridge, UK), a Masters degree in Engineering Geology (Leeds, UK), and a Ph.D. focused on multi-hazards and disaster risk reduction (King's College London, UK). For the last decade, Joel has worked at the interface of Earth science and international development, and plays a leading role internationally in championing the role of geoscience in delivering the UN Sustainable Development Goals. He has coordinated research, conferences, and workshops on geoscience and sustainable development in the UK, India, Tanzania, Kenya, South Africa, Zambia, and Guatemala. Joel regularly engages in international forums for science and sustainable development, leading an international delegation of Earth scientists to the United Nations in 2019. Joel has prizes from the London School of Economics and Political Science for his teaching related to disaster risk reduction, and Associate Fellowship of the Royal Commonwealth Society for his international development engagement. Joel is a Fellow of the Geological Society of London, and was elected to Council in 2019 and to the position of Secretary (Foreign and External Affairs) in 2020.

Index

Printed by Printforce, the Netherlands